PRINCIPLES OF TIMBER
DESIGN FOR ARCHITECTS
AND BUILDERS

PRINCIPLES OF TIMBER DESIGN FOR ARCHITECTS AND BUILDERS

DON A. HALPERIN

G. THOMAS BIBLE

JOHN WILEY & SONS, INC.

New York • Chichester • Brisbane • Toronto • Singapore

The drawings, tables, and descriptions in this book are presented in good faith, but the author, illustrator, and publisher, while they have made every reasonable effort to make this book accurate and authoritative, do not warrant, and assume no liability for, its accuracy or completeness or its fitness for any particular purpose. It is the responsibility of users to apply their professional knowledge to the use of information contained in this book, to consult original sources for additional information when appropriate, and to seek expert advice when appropriate.

This text is printed on acid-free paper.

Copyright © 1994 by John Wiley & Sons, Inc.

All rights reserved. Published simultaneously in Canada.

Reproduction or translation of any part of this work beyond that permitted by Section 107 or 108 of the 1976 United States Copyright Act without the permission of the copyright owner is unlawful. Requests for permission or further information should be addressed to the Permissions Department, John Wiley & Sons, Inc., 605 Third Avenue, New York, NY 10158-0012.

This publication is designed to provide accurate and authoritative information in regard to the subject matter covered. It is sold with the understanding that the publisher is not engaged in rendering legal, accounting, or other professional services. If legal advice or other expert assistance is required, the services of a competent professional person should be sought.

Library of Congress Cataloging in Publication Data:
Halperin, Don A.
    Principles of timber design for architects and builders / Don A. Halperin, G. Thomas Bible.
       p. cm.
    Includes index.
    ISBN 0-471-55768-4 (alk. paper)
    1. Building, Wooden. I. Bible, G. Thomas. II. Title.
TA666.H34 1994
721'.0448—dc20   93-40910

Printed in the United States of America

10 9 8 7 6 5 4 3 2 1

I wish to dedicate this work to Miriam, Elizabeth, Leo and Madeline whose patience and care has seen me through many long days.

G. Thomas Bible

CONTENTS

	Preface	ix
1	Material Properties	1
2	Beam Design	17
3	Axial Loads	55
4	Connections	78
5	Trusses	164
6	Glued—Laminated Arches	195
7	Plywood	235
Appendix A:	Lumber Properties	307
Appendix B:	Beam Design Tables	319
Appendix C:	Column Design Tables	344

Appendix D:	**Moment Calculations**	351
Appendix E:	**Wood Species and Specific Gravities**	359
Appendix F:	**Required Connectors for Uplift Overturning and Sliding**	361
Appendix G:	**Horizontal Diaphragms and Vertical Shearwall Tables**	375
Index		395

PREFACE

This text is intended for use by architects, builders, and engineers. It has a pragmatic approach, utilizing logic and reasoning to obtain first approximations to sizes of structural members rather than providing some mysterious trial size tendered by an author. Because standard timber sizes have large variations from one size to the next (a 2×6 is about 60% larger than a 2×4, for example), problem solutions are slightly simplified, since extreme accuracy is pointless. They are still correct, and yield very good economical results. In addition, tables are utilized to get answers much faster. Several of these tables have never before appeared in print.

The book strives to cover completely the most common types of timber construction while not drifting into engineering esoterica. In addition to the usual work on beams, diaphragms, columns, connections, and heavy trusses, it contains studies of joists, special beams, residential trusses, and arches. Special sections on the capacity of recommended fasteners and an interpretation of the recent *Uniform Building Code* seismic provisions have been developed in response to current concerns. All of these important elements might not be included in a course of design, but they are here for reference if needed.

The language is simple, the approach is straightforward, and many example problems are included. Since wood is our only renewable building resource, it should achieve more and more use as the material of choice in the primary structural building frame. Hence its study is essential today, to prepare for tomorrow.

We want to thank the following for their assistance in reviewing this work and for permission to reproduce certain tables and figures: American Forest and Paper Association, American Institute of Timber Construction, American Plywood Association, American Society for Testing and

Materials, American Lumber Standards Committee, Council of American Building Officials, Forest Products Laboratory, International Conference of Building Officials, SILVER TECO, Simpson Strong-Tie Company, Inc., Structural Insulated Panel Association, Truss Plate Institute, and Western Wood Products Association. Thanks to Dr. Robert Brungruber, Benson Woodworking, Inc. for reviewing the text. Thanks also to Michael Davin and Kevin Hoffman–Dobo who prepared the illustrations.

DON A. HALPERIN

Gainesville, Florida
February, 1994

G. THOMAS BIBLE

Cincinnati, Ohio
April, 1994

PRINCIPLES OF TIMBER DESIGN FOR ARCHITECTS AND BUILDERS

1

MATERIAL PROPERTIES

Wood is our only renewable building resources. In 1914 there were 42 billion board feet of lumber cut in the United States. By 1950 the U.S. consumption of lumber products had reached 57 billion board feet, and around 1970 the U.S. consumption peaked at over 70 billion board feet of lumber. During that year close to half of all wood harvested was used in construction. We are actually growing more wood as time goes on, due to better fertilization, scientific cutting practices, firefighting, and all the other aspects of forest management. New residential construction is historically the largest market for timber construction. About 38% of timber in construction is used for residential construction; 14% is employed in nonresidential, with about the same amount used in remodeling and repair; and the balance is used for other items, such as poles, boxes, and crates.[1]

WOOD SPECIES

There are two categories of trees, hardwood and softwood. The former are deciduous, or broad-leaved, and drop their leaves in cold weather. They constitute about 25% of all wood that is harvested, but are not used much in construction, being restricted to trim, flooring, finished

[1] Robert N. Stone and Thomas C. Marcin, *Research on U.S. Timber and Wood Products Requirements*, U.S. Department of Agriculture, Forest Service, Forest Products Laboratory, Washington, DC, 1984.

faces of doors, and so on. Examples of hardwood are birch, maple, oak, and walnut. As a group, they are denser than the other woods.

The softwoods are coniferous and are evergreen. They constitute about 75% of all lumber, and most of the construction lumber. About 70% of the softwoods are western woods—Douglas Fir, Hemlock, Larch, Redwood, Spruce, and others. The balance of the evergreens are eastern woods, such as Eastern White Pine, Cypress, and Southern Pine, both longleaf and shortleaf. The two species most commonly used in construction are Douglas Fir–Larch (DF-L) and Southern Pine (SP), sometimes referred to as Southern Yellow Pine (SYP). These two species have about the highest strengths available and therefore will usually afford the longest spans. We use these species in the examples in this book. Mixed species, where two or more species are grown and harvested together, may prove more economical. These include Spruce–Pine–Fir (SPF) grown in the northern United States and Canada. Mixed Southern Pine (MSP) and Western Woods (WW) from the west coast.

Because the strength of wood depends in part on its density, hardwoods tend to be stronger than softwoods. Hardwoods are used infrequently in building structures, primarily because of the cost. Hardwood is harder to harvest, takes longer to grow, and is more susceptible to disease. Historically, until recent, hardwoods were preferred over softwoods. Many historical wood structures, such as barns, houses, and wood bridges from a hundred years ago, were build of hardwoods. Our interest in preserving these structures, along with the resurgence of timber-frame construction, has created a market for hardwood construction. Starting in 1990 the Northeastern Lumber Manufacturers Association, one of the grading agencies in North America, began listing wood strengths for six species of hardwoods.

MANUFACTURE

The pieces of structural lumber are cut into even-foot lengths: 6 ft long, 8 ft long, 10 ft long, and so on. Widths and thickness are stated in even inches, such as 2 in., 4 in., 6 in., and so on, except that 1-in. and sometimes 3-in. thicknesses are also standard. These dimensions (e.g., 1 in., 2 in., 4 in.) are the width and thickness as the pieces come from the sawmill. They will have burrs and splinters, and perhaps some unevenness. This "rough" wood is sometimes planed smooth on two sides (S2S), or, more usually, on all four sides (S4S). This results in the 1-in. thickness becoming $\frac{3}{4}$ in., the 2 in. becoming $1\frac{1}{2}$ in., the 4 in. reducing to $3\frac{1}{2}$ in., and the 6 in. become $5\frac{1}{2}$ in. Thereafter the width or thickness is reduced by $\frac{3}{4}$ in. for those pieces with one of the dimensions less than 5 in., so that a 4 × 8 would actually measure $3\frac{1}{2}$ in. × $7\frac{1}{4}$ in. If both width and thickness are greater than 5 in., the reduction is $\frac{1}{2}$ in. for each side, so that a 6 × 10 would measure $5\frac{1}{2}$ in. × $9\frac{1}{2}$ in. *Throughout this book all wood will be considered to be S4S unless otherwise noted.* The various standard sizes are shown in Table A.1 in the Appendix. This table also shows the geometric properties of the cross section, the section modulus S, the area A, and the moment of inertia I.

The last column is the table in the *board measure*, which is a measure of relative volume used in pricing the wood. The board measure is given in units of *board feet*. One board foot is a piece of wood that is nominally 1 ft (12 in.) wide, nominally 1 in. thick, and actually 1 ft long. Thus a 2 × 4, 24 ft long, would contain 16 board feet $[2 \times (4/12) \times 24] = 16$. The actual piece is, of course, only $1\frac{1}{2}$ in. × $3\frac{1}{2}$ in. × 24′-0″.

Wood pieces are classified based on size; the classifications are used in grading and buying material. They are defined as shown in Table 1.1.

TABLE 1.1 SIZE CLASSIFICATION AND GRADES

Category	Thickness (in.)	Width (in.)	Grades
Dimensional lumber	2–4	Any width	
Structural light framing	2–4	2–4	Select Structural Nos. 1, 2, 3
Light framing	2–4	2–4	Construction, Standard, Utility
Stud	2–4	2–6	Stud
Joists and planks	2–4	5 and wider	Select Structural Nos. 1, 2, 3[a]
Beams and stringers[b]	5 and thicker	Widths more than 2 in. greater than thickness	Select Structural Nos. 1, 2[a]
Posts and timbers[c]	5 and thicker	5 and wider	Select Structural Nos. 1, 2[a]

[a] Other strengths are available for some species, such as "Dense No. 1."
[b] Beams and stringers are intended to be used in bending.
[c] Posts and timbers are intended to be used in axial compression.

WOOD STRENGTHS

Steel can be thought of as the "engineer's material." It is manufactured and therefore can be designed and formulated to have a specific strength; it is uniform and isotropic. Wood, on the other hand, is none of these. Because it is a natural material, it is subject to a number of imperfections or limitations that affect its strength and final use. Thus one of the roles of the wood designer is to learn the natural characteristics of wood and to understand not only how to overcome its defects, but also how to capitalize on its strengths.

In the United States the American Forest and Paper Association gathers and disseminates most of the latest technical information concerning the use of wood in construction. One of their major activities is writing a recommendation for design titled *National Design Specification for Wood Construction* (NDS), 1991 Edition.[2]

[2] American Forest and Paper Association, 1111 19th St., N.W., Washington, DC 20036.

This design specification contains information on the strength of commonly used species and grades of wood, strengths of connectors, and standard engineering principles. The NDS is adopted by all the major code agencies and by most state and local building codes. Anyone involved in timber engineering will need to refer to the NDS and become familiar with it. Throughout this book we have reproduced the applicable portions of tables or charts from the NDS as an aid to the reader. In certain areas, most notably Chapter 4, the NDS tables on connections are so extensive that it would be impractical to reproduce them here. Instead, we refer the reader to the appropriate NDS tables. To that end the reader will need access to the 1991 edition of the NDS to follow the examples in Chapter 4.

GRADING

One major step in turning the natural material into an engineered product is the grading of the lumber. The basis of grading is to establish a relationship between the defects in wood and the strength of the wood. Up until about 1980 there was only one means of grading lumber: *visual grading*. The procedure, which is still used for the large majority of graded lumber, involves a carefully trained inspector assigning a grade to each piece of lumber based on a number of visually identifiable defects. Visually graded lumber is marked with a stamp that shows its species and grade as well as the mill identification number and the agency that oversees the grading at that mill (see Figure 1.1).

Accredited Agencies and Typical Grade Stamps as approved by the Board of Review of the American Lumber Standards Committee.

Interpreting Grade Stamps

Most grade stamps, except those for rough lumber or heavy timbers, contain five basic elements:

a. The trademark indicates agency quality supervision.
b. Mill Identification - firm name, brand or assigned mill number.
c. Grade Designation - grade name, number or abbreviation.
d. Species Identification - indicates species individually or in combination.
e. Condition of Seasoning at time of surfacing:
   S-Dry - 19% max. moisture content
   MC 15 - 15% max. moisture content
   S-GRN- over 19% moisture content (unseasoned)

TYPICAL GRADE STAMP

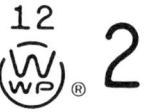

FIGURE 1.1 Typical grade stamps.

In the mid-1980s another procedure was deployed in some lumber mills, resulting in *machine stress rated* (MSR) lumber. Each piece of MSR is load-tested and an allowable bending stress, F_b, is determined, together with the modulus of elasticity, E. In the early 1990s a form of graded material referred to as MEL, *machine evaluated lumber*, became available in which the allowable bending stress is determined by testing, and all other properties are determined through the use of visual grading.

Wood properties vary even within a grade. If 100 pieces of No. 2 Southern pine 2 × 6's were tested, they would not all necessarily fail at the same load. If the design strength were based on the average strength of the 100 pieces, half of them would probably fail under the maximum load—such a design would be inherently unsafe. To establish a statistical measure of safety, design strengths are based on the 5% exclusion principle: Of 100 pieces, the strength is taken such that 95 pieces will not fail. In addition to this, a factor of safety in the range of 2.0 is applied to establish the design strength. For material that is machine stress rated (MSR) and for material used in glued-laminated beams, the average strengths are grouped much closer together, so there is a higher statistical confidence. The reduction in strength, represented by the 5% exclusion limit, does not have to be as great to afford the same factor of safety. Visually graded and MEL material is grouped together and referred to at times as $COV_E > 0.11$. MSR and material used in glued-laminated timbers have a COV_E value of 0.11 or less. COV_E designates the coefficient of variation for elasticity. Wood with a lower COV_E value has a greater certainty in the predicted strengths and elasticity. This distinction is important in applying certain design formulas for beams and columns.

The strength of wood depends on a number of factors, the most important being (1) species, (2) orientation of the forces, (3) natural (and human-made) defects, (4) moisture content, and (5) duration of load. Grading procedures that are used today address the first four of these characteristics.

ORIENTATION OF THE FORCES

All wood grows upward and from its core outward. Every year another layer is added to its outer shell, thereby creating another ring. The protective outer surface is the *bark* of the tree. Immediately beneath that is the *cambium*. The nourishing liquid that flows from roots to leaves, the *sap*, travels in small continuous tubes within the cambium.

Wood fiber is composed of wood cells. Each cell on a microscopic level looks like a thin soda straw, a hollow tube that is many times longer than it is wide. Cell walls are composed of cellulose, which are like fibers, and lignin, which glues the cellulose fibers together. The orientation of the cells is generally such that the axis of the cell aligns with the axis of the tree. Because the cells are strongest in their long direction, wood exhibits a high strength in compression when loaded along the axis of the tree. This strength is referred to as *compression parallel to grain* (Figure 1.2b). Because the tube shape of the cells is much weaker in the direction perpendicular to its axis, wood exhibits much less strength when compressed across the grain. You can experience this by compressing a bundle of soda straws: first attempting to crush them by pushing on their ends, then by pushing across their axis. Wood is weaker in *compression perpendicular to grain* (Fig 1.2a). Wood is loaded in compression perpendicular to grain when it is used in bearing, such as at the end of a beam.

Lignin bonds the cellulose together and thus bonds cell to cell. Because of its

6 MATERIAL PROPERTIES

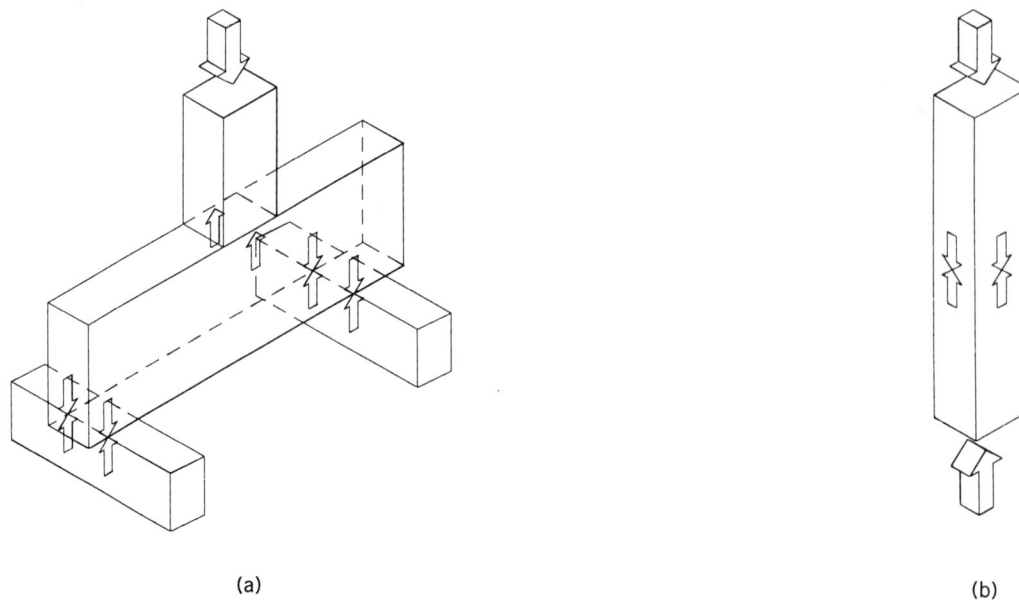

FIGURE 1.2 Compression: (a) perpendicular to grain, $F_{c\perp}$, and (b) parallel to grain, F_c.

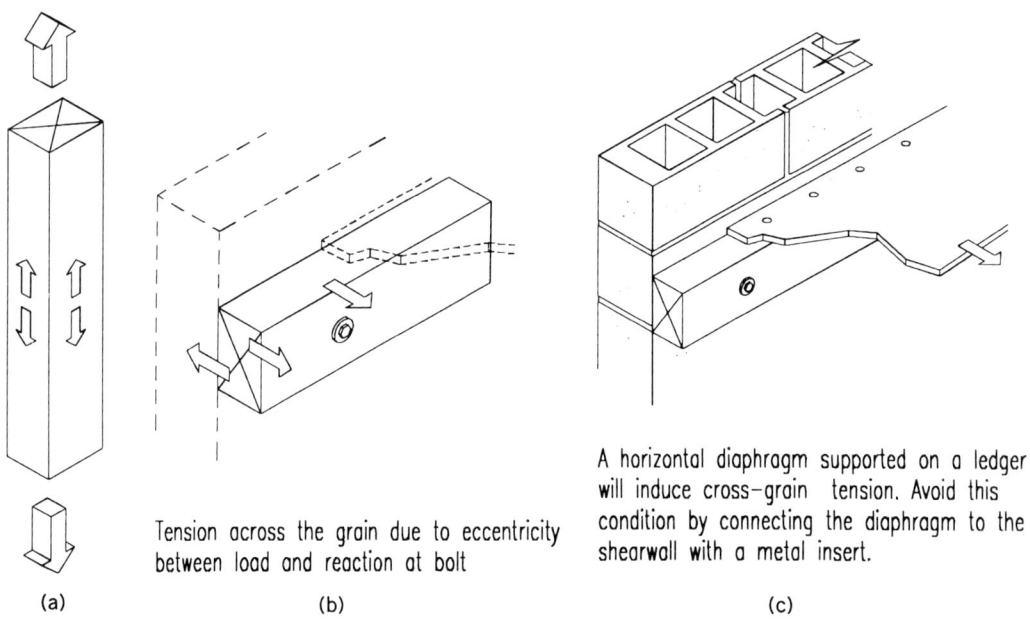

Tension across the grain due to eccentricity between load and reaction at bolt

A horizontal diaphragm supported on a ledger will induce cross-grain tension. Avoid this condition by connecting the diaphragm to the shearwall with a metal insert.

FIGURE 1.3 Tension: (a) parallel to grain, F_t; (b) across the grain, and (c) cross-grain tension due to eccentric load.

strength, wood can perform well under tension loads when applied parallel to the grain of the wood (along the axis of the tree). This strength is referred to as *tension parallel to grain* (Figure 1.3*a*). The strength of wood in cross-grain tension is so small that for design purposes we assume that wood has no strength in tension perpendicular to grain. All wood design codes forbid using wood in such a way that it will be loaded in tension perpendicular to grain (Figure 1.3*b*). This provision prohibits both *cross-grain tension* and *cross-grain bending*.

A member that is subject to bending forces must be able to resist both tension and compression with relatively equal ease. Because wood has good strength characteristics in both tension and compression it is an ideal bending member. The *bending strength* is usually controlled by the ability to carry tension.

Any beam that is loaded perpendicularly to its long axis will develop both bending and shear stress. Shear is composed of a vertical and a horizontal component. Wood has far less capacity to resist horizontal shear than it does vertical shear. *Horizontal shear* (Figure 1.4) causes the wood to split along the grain. It is common in some wood members, particularly short beams, that shear will govern the design.

DEFECTS IN LUMBER

A piece of wood with a straight grain without knots, holes, or cracks has the highest strength possible for its species. Such a piece of wood is available only in very limited dimensions. The reality of wood as a natural material is that it is full of a number of *growth characteristics* that affect the appearance and strength of the material. Visual stress grading correlates the size, quantity, and location of a variety of defects with their strength-reducing effects. Each piece of lumber that is visually graded is examined for all of these defects and assigned a grade and a set of strengths based on the most limiting of the defects. The characteristics that most affect the strength of wood are listed here. Table 1.2 lists some of the limiting characteristics for two grades of 2×10 joists. The definitions are taken directly from the *National Grading Rule for Dimension Lumber*.[3]

When a log is cut into pieces of lumber, each piece exhibits certain lines along its length called *grain*. The grain is never in perfectly straight lines because trees do not grow perfectly straight. *Slope of grain* is the deviation of the line of fibers from a straight line parallel to the sides of the piece (see Figure 1.5). Slope of grain is noted in the same manner as roof slopes. In the figure the line of the fibers deviates from the side of the piece by an inch over a length of the piece of 12 in. This slope is referred to as 1 in 12. As discussed above, the strength of wood in tension and compression is greater in the direction of the grain. If the grain aligns exactly with the sides of the piece, a piece of wood will

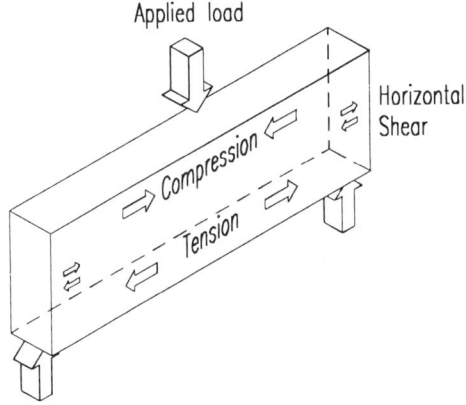

FIGURE 1.4 Bending and shear.

[3] American Lumber Standards Association, U.S. Department of Commerce, Washington, DC.

8 MATERIAL PROPERTIES

TABLE 1.2 GRADES AND LIMITING CHARACTERISTICS

Characteristic	Maximum Permissible	
	Select Structural	No. 2
Strength (as a percentage of clear wood strength)	65%	45%
Slope of grain	1 in 12	1 in 8
Knots	Sound, firm, encased, and pith knots if tight and well spaced, are permitted in sizes not to exceed the following, or equivalent displacement	Well-spaced knots of any quality are permitted in sizes not to exceed the following or equivalent displacement
Knot size		
Edge of wide face	$1\frac{7}{8}$ in.	$3\frac{1}{4}$ in.
Centerline wide face	$2\frac{5}{8}$ in.	$4\frac{1}{4}$ in.
Unsound knot or hole	$1\frac{1}{4}$ in.	$2\frac{1}{2}$ in.
Splits	Equal in length to the width of the piece (10 in.)	Equal in length to $1\frac{1}{2}$ times the width (15 in.)
Shake	If through at ends, limited as splits. Surface shakes up to 2 ft long.	If through at ends, limited as splits. Away from ends through shakes up to 2 ft long, well separated. If not through, single shakes shall not exceed 3 ft long or $\frac{1}{4}$ the length whichever is greater
Checks	Surface seasoning checks not limited; through checks at end are limited as splits	Seasoning checks not limited; through checks at end are limited as splits
Wane	$\frac{1}{4}$ the thickness, $\frac{1}{4}$ the width, full length; provided that wane not exceed $\frac{1}{2}$ the thickness or $\frac{1}{3}$ the width for $\frac{1}{4}$ the length	$\frac{1}{3}$ the thickness, $\frac{1}{3}$ the width, full length; provided that wane not exceed $\frac{2}{3}$ the thickness or $\frac{1}{2}$ the width for $\frac{1}{4}$ the length
Pitch streaks	Not limited	Not limited
Pitch or bark pockets	Not limited	Not limited
Rate of growth	Not limited	Not limited
Stain	Stained sapwood; firm heart stain or firm red heart limited to 10% of pieces	Stained sapwood; firm heart stain or firm red heart not limited
Skips	Hit and miss in 10% of pieces	Hit and miss with a maximum of 5% of the pieces containing hit or miss or heavy skip not longer than 2 ft
Warp	$\frac{1}{2}$ of medium	Light

Source: National Grading Rule.

have the greatest available strength, particularly when loaded in compression, tension, or bending. Thus lumber with a smaller slope of grain (such as 1 in 20) will have the greater strength. For joists and planks, the highest grade is *select structural*. To be placed in this grade a joist or plank must have a slope of grain of 1 in 12, or less. The allowable bending stress is taken as 65% of that allowed for the clear, straight-grained wood. The lowest grade for joists and planks is No. 3, with a slope

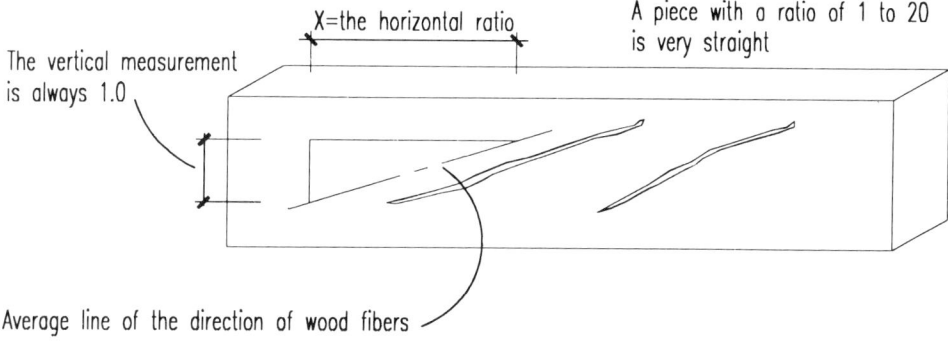

FIGURE 1.5 Slope of grain.

of grain of 1 in 4 or less, and a bending strength of only 26% of that of clear, straight-grained wood. Slope of grain has no effect on compression perpendicular to the grain or on horizontal shear; these two values remain constant for any species regardless of the grade.

Slope of grain can arise from two sources. Most trees grow by laying down the wood cells at a slight angle to the axis of the tree. The cells spiral around the tree, resulting in *spiral grain*. The other possible cause of slope of grain comes from manufacturing, where lumber may be sawn at a slight angle to the axis of the tree. The two forms of slope of grain may coexist. For the purposes of grading and assigning strength, their effects are the same.

A *knot* is a portion of the branch or limb that has become incorporated in a piece of lumber. Knots reduce the strength the wood in two ways. They replace the wood of the trunk with the wood of the knot, which we assume has no strength. This is particularly true for loose knots, but is a good assumption even for knots that are tight. The other consequence is that the grain of the wood winds around a knot, leaving an area with a local deviation of grain. The larger the knot, the larger will be the deviation of grain; as a consequence, the greater will be the reduction of strength.

The National Grading Rule lists the allowable knot sizes for all grades of lumber 2 to 4 in. thick. The rule recognizes that the location of the knot is an important factor in determining the strength reduction. Lumber loaded in bending, such as a joist, will have the maximum stress in the extreme fibers at the top and bottom edges. A knot in this location has a greater impact on the strength than does a similar-size knot in the middle of the piece (see Figure 1.6). Table 1.2 lists the maximum allowable knot sizes for the edge of wide face (see C Figure 1.6) and for the centerline wide face (see B Figure 1.6).

Wood splits for a variety of reasons. Splits are defined and measured depending on their orientation with respect to the growth rings. Figure 1.7 shows three types of splits. Splits and shake occur only at the ends of lumber. Splits, checks, or shake at the ends will reduce the shear capacity of a piece of wood. These defects, particularly checks, are a result of drying and may increase after the lumber has been graded and has left the yard. The NDS therefore assumes that all pieces have the worst possible combination of splits, shake, and checks: namely, a split completely through the piece at the end. The allowable shear strengths listed in the NDS may be adjusted upward if there are smaller splits than assumed. Checks away from the ends have little effect and are not considered.

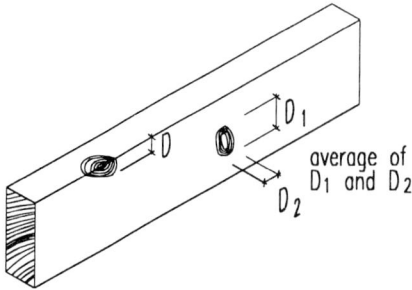

For joists and planks the size of a knot on the wide face is the avg. between the largest and smallest diameters

The size of a knot on the edge of the wide face is its width between lines parallel to the edge of the piece

(a)

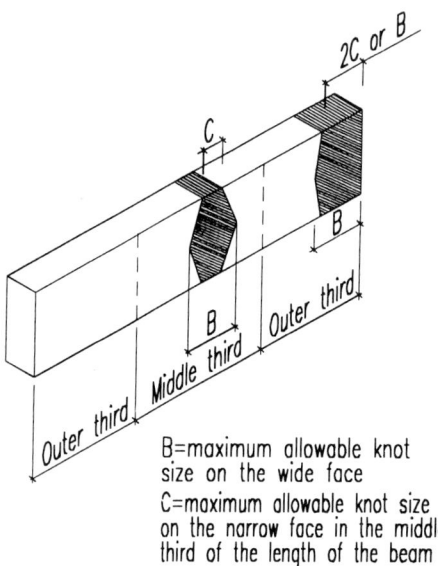

B = maximum allowable knot size on the wide face
C = maximum allowable knot size on the narrow face in the middle third of the length of the beam

Maximum size of knot in the middle third increases from C to B at the center of the wide face
Maximum size of wide face knot in the outer third increases from C to 2C (but not to exceed B) at the ends

(b)

Source: ASTM D 245 section 5.3 and Fig. 2

FIGURE 1.6 Edge knot and center-line knot: (a) measuring knots, and (b) maximum size of knots.

Checks are separations of the wood normally occurring across or through the rings of annual growth and are usually a result of seasoning. *Shake* is a lengthwise separation of the wood which usually occurs between or through the rings of annual growth. Shake between the rings, called *ring shake*, is most common. It is caused principally by the severe stressing of the living tree due to wind storms. *Splits* are separations of the wood due to tearing apart of the wood cells.

Wane is bark, or a lack of wood, on the corner or edge of a board. Wane occurs as a result of the sawing process when a board is cut too close to the edge of the tree. It does not diminish the strength of the wood, but it obviously does result in less wood area than in a full-size piece. Its largest impact is that it destroys the nailing edge, so that nails used to hold on plywood or other sheathing will have less penetration into the wood and less "bite."

Warp is any deviation from a true plane or surface, including *bow*, *crook*, *cup* and *twist*, or any combination. Figure 1.8 shows these types of warp. Warp is a result primarily of unequal shrinking as the material dries. Warp has more of an effect on construction than on strength—warped wood does not fit into straight buildings. Wood that is warped can be stressed while being straightened, leaving less available strength to carry the applied loads. Each grade has limits as to the amount of warp that is allowed.

Compression wood is abnormal wood that forms on the underside of leaning and crooked coniferous trees. In addition to its distinguishing pale color, it is characterized by being hard and brittle and by its relatively lifeless appearance. It is not permitted in readily identifiable and damaging form in stress grades.

Grading also considers a number of defects, such as holes, raised grain, or torn grain, which are put into the wood as an inevitable result of sawing, planing, and

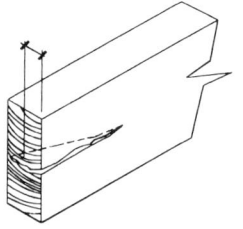
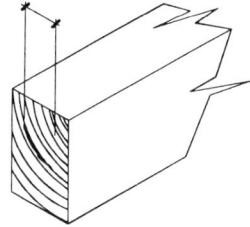
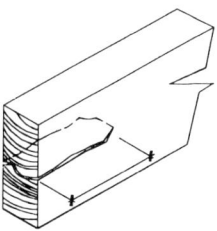

measure average penetration

in joists and planks, and beams and stringers, shake is measured at the end of the piece between lines parallel to the wide faces

splits are measured as the penetration of the split from the end of the piece and parallel to the edges of the piece

(a)   (b)   (c)

FIGURE 1.7 (a) checks, (b) shakes, and (c) splits.

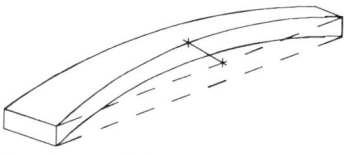

Bow is a deviation flatwise from a straight line drawn from end to end of a piece. It is measured at the point of greatest deviation.

(a)

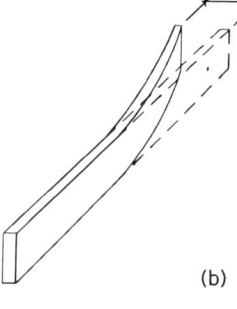

Sweep is a deviation edgewise from a straight line drawn from one end of a piece, parallel to original line of the piece.

(b)

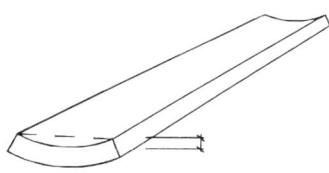

Cup is a deviation in the face of a piece from a straight line drawn edge to edge of a piece. It is measured at the point of greatest deviation.

(c)

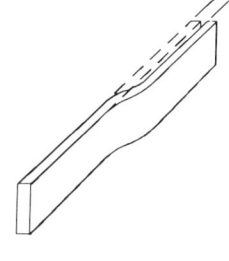

Crook is a deviation edgewise from a straight line drawn from end to end. It is measured at the point of greatest deviation.

(d)

FIGURE 1.8 Warp: (a) bow, (b) cup, (c) sweep, and (d) crook.

handling the wood. These defects usually have little effect on the strength.

MOISTURE CONTENT

After being cut and trimmed, a log is sawed into various pieces of structural lumber. The wood at this point is quite moist, since a live tree is about 35 to 55% water by weight. This *green lumber*, as it is called, can be used on the job, but because it will shrink with time as the moisture works its way out and evaporates, it might give rise to problems with warping or with connections becoming loose. The lumber therefore is usually dried out (seasoned) to become stronger, stiffer, and smaller. Southern Pine is sometimes used in the green state, so its strength values are listed for a specific moisture content. Dried wood has a moisture content of less than 19% and weights about 35–40 pounds per cubic foot, whereas green wood weighs about twice that. Seasoned wood is therefore easier to work with.

There are two methods of drying lumber, air drying and kiln drying. In the first method, the pieces of sawn lumber are laid in a crisscross pattern, in *ricks*, and allowed to remain undisturbed for 3 to 4 months. In the second method the wood is placed in a low-heat oven called a kiln, where it is kept at 70–120° for 4 to 10 days. In both methods it is ready for use when the moisture content (MC) has reached equilibrium, generally about 19% MC. Kilns usually dry wood to a maximum of 15% MC for general use, and may dry them to 5 to 8% for specialized uses such as fabricating glued-laminated beams. The distinction between "dried" (MC < 19%) and "kiln-dried," indicating a moisture content of 15% of less, has been discontinued.

Moisture content is important from a structural point of view for three reasons: (1) shrinkage and swelling, (2) the effect on connection strength, and (3) the effect on material strength. All of the problems associated with moisture content occur as the wood changes its MC. The best solution, if possible, is to select wood with a moisture content as close as possible to what its final MC will be. The effect on connections is discussed in Chapter 4, and the effect on material strength is described in Chapter 2. Here we concentrate on shrinkage.

Green wood has moisture both within the cell wall and in the cell cavity itself. As wood dries, the moisture leaves the cell cavity before leaving the cell wall. The point where the cell cavity is empty but the

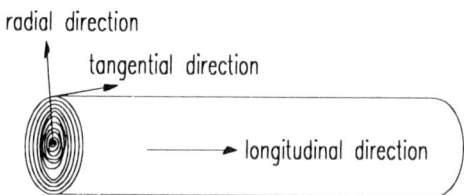

FIGURE 1.9 Radial and tangential directions.

walls are full is called the fiber saturation point (FSP). For most species, this value is usually about 30% MC. Change in moisture content above FSP has no effect on the wood's properties; all shrinkage occurs as the wood dries from 30% to a lower MC value. The increase in strength also occurs in this range as the wood dries.

Wood shrinks at different rates, depending on the direction relative to grain. The most shrinkage is tangential to grain; shrinkage radial to grain is somewhat less; and shrinkage parallel to grain is so small as to be neglected. Figure 1.9 identifies these directions.

If a Southern Pine 6×6 were to dry from 19% at its time of manufacture to an equilibrium MC values of 8%, the radial shrinkage would be only about $\frac{1}{8}$ in. in each direction, not a very significant amount. Shrinkage is not important as a structural concern but is a major consideration in the detailing of wood trim, doors, and windows.

DESIGN VALUES

Whereas in structural steel there is only one modulus of elasticity E regardless of the chemical composition or type or designation of the steel, in timber each species is different, and there can be a large variation in values for the modulus of elasticity within that species, depending on grade and usage. Thus for Douglas Fir–Larch we find values ranging from 1,300,000 to 1,900,000 psi, and in Southern Pine, E

ranges from 1,400,000 to 1,800,000 psi. Furthermore, the allowable stress values do not have any well-defined relationship to the modulus of elasticity. For every allowable stress value of every kind in every grade of any species, reference *must* be made to the appropriate table. All values have been determined by an association of lumber manufacturers, such as the Southern Pine Inspection Bureau or the Western Woods Products Association.

Refer to Tables A.8 and A.9 in Appendix A and consider the design values in pounds per square inch. These are the various allowable stresses that were given as psi in the paragraph above. The first column lists the grade and the second column lists the size classification. To use this table the approximate size must be estimated. See Table 1.1 for a definition of the size classifications. The next column, "bending F_b," lists the allowable bending stress. Moving along in the table, "tension parallel to the grain F_t" and "horizontal shear F_v" are self-explanatory. All allowable stress values have the symbol F with a subscript labeling the type of stress. The allowable compression stresses are of two types, depending on whether the loading condition is applied across the grain, "compression perpendicular to the grain, $F_{c\perp}$," or parallel to the grain, "compression parallel to the grain, F_c." The reason for this difference refers back to the way that trees grow. The trunk grows vertically, of course, and the tubes that carry the sap are all parallel to this growth. The assemblage of all these tubes, which are arranged in rings around the trunk, must be strong enough to carry the superimposed load of all the branches and leaves even when heavily burdened by the weight of rainwater. However, there is no reason for the trunk to have to resist much applied load across itself, so it is weaker in that direction. Since lumber is cut from the trunk parallel to its length (i.e., parallel to the grain), F_c is larger than $F_{c\perp}$. The modulus of elasticity E has already been discussed.

ADJUSTMENT FACTORS

The design values listed in the tables in the Appendix must be adjusted for the actual conditions of use of the member. Not all adjustment factors apply to each type of stress. Table 1.3 lists the applicability of these adjustment factors to the types of stress.

Most of the adjustment factors are defined and discussed in the chapters on member design. Here we discuss two factors in some detail.

Repetitive Member Factor, C_r

When beam or bending type members are installed in a series of not less than three parallel members spaced not more than 24 in. center to center and attached to a floor or roof that can distribute the load to all the members, they are considered to be repetitive. They have higher allowable bending stresses for any one grade in any one species than does a single member of that grade and species, as shown. This is because any flaw or defect in any one member will not occur at exactly the same location along its length as a similar flaw in an adjacent member. The joists must have sheathing or decking over them that distributes all loads equally among all members, so that the burden of carrying any concentrated load is shared among the members. The sheathing or decking actually adds strength to the assembly that is not otherwise credited to the grouping of joists and deck.

Load Duration Factor, C_D

Wood is a rheological material: its strength properties are dependent on the duration

TABLE 1.3 APPLICABILITY OF ADJUSTMENT FACTORS

	Tabulated Design Value Adjustment Factor						
	Bending, F_b	Tension, F_t	Shear F_v	Compression, parallel F_c	Compression perpendicular $F_{c\perp}$	Modulus of Elasticity, E	End Bearing, F_g
Load duration, C_D	×	×	×	×			×
Wet service, C_M	×	×	×	×	×	×	
Temperature, C_t	×	×	×	×	×	×	×
Beam stability, C_L	×[a]						
Size, C_F	×[b]	×		×			
Volume, C_V	×[a,c]						
Flat use, C_{fu}	×[d]						
Repetitive member, C_r	×[e]						
Curvature, C_c	×[f]						
Form factor, C_f	×						
Column stability, C_P				×			
Shear stress, C_H			×[g]				
Buckling stiffness, C_T						×[h]	
Bearing area, C_b					×		

[a] Beam stability factor, C_L, shall not apply simultaneously with the volume factor, C_V (see NDS 5.3.2). The lesser of the adjustment factors apply.
[b] Size factor, C_F, shall apply only to visually graded sawn lumber members and to round timber bending members (see NDS 4.3.2).
[c] Volume factor, C_V, shall apply only to glulam bending members (see NDS 5.3.2).
[d] Flat use factor, C_{fu}, shall apply only to dimension lumber bending members 2 to 4 in. thick (see NDS 4.3.3) and to glulam bending members (see NDS 5.3.3).
[e] Repetitive use factor, C_r, shall apply only to dimension lumber bending members 2 to 4 in. thick (see NDS 4.3.4).
[f] Curvature factor, C_c, shall only apply to curved portions of glulam bending members (see Chapter 5 and NDS 5.3.4).
[g] Shear design values parallel to grain (horizontal shear), F_v, for sawn lumber members shall be permitted to be multiplied by the shear stress factors, C_H, in Table A.6.
[h] Buckling stiffness factor, C_T, shall apply only to 2 × 4 or smaller lumber truss compression chords subjected to combined flexure and axial compression when $\frac{3}{8}$ in. or thicker plywood sheathing is nailed to the narrow face (see NDS 4.4.3).

of time in which the load is in place. Strengths given in Tables A.2, and A.8 through A.11 are based on a "normal" duration of load, which is assumed to be 10 years. If loads are in place for shorter periods, they have less effect on the wood. So we can use higher strengths for shorter loads. Figure 1.10 shows the effects of load duration on wood strength properties. The most common loads, their commonly assumed durations, and the commensurate duration factors are listed.

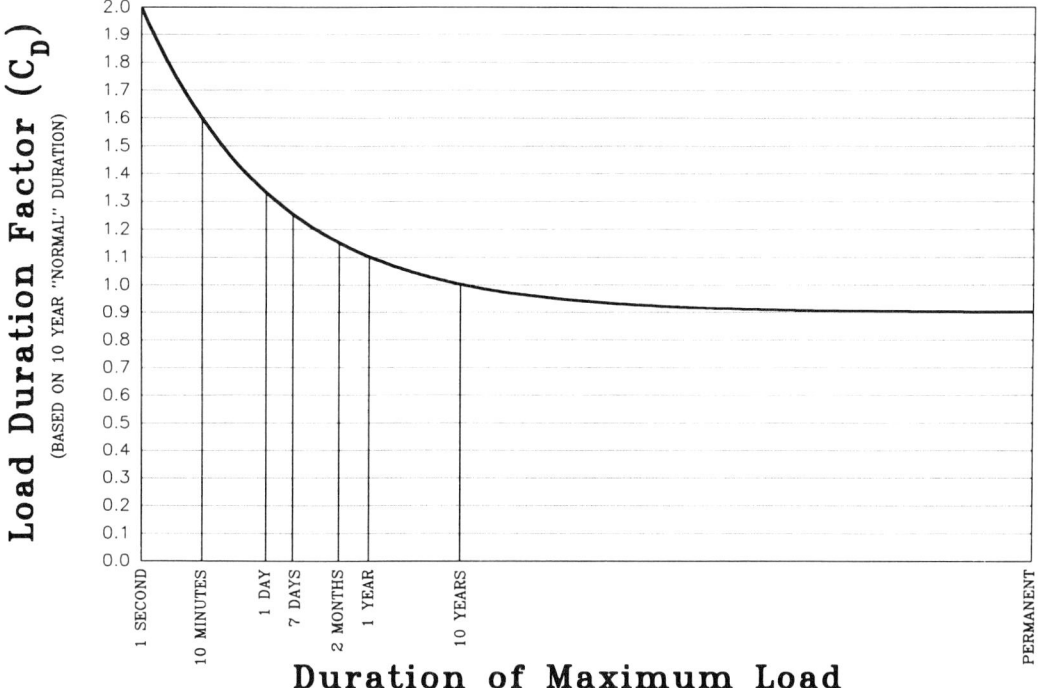

FIGURE 1.10 Load duration factors for various load durations.

Load duration factors (Table 1.4) are used in all portions of wood design, both member design and connection design. They are best thought of as strength modifiers, so the tabulated wood strength or connection strength should be multiplied by the appropriate factor. The factors do not apply to two characteristics, $F_{c\perp}$, perpendicular to the grain, and to E, modulus of elasticity.

In the initial phases of design we must consider a number of load combinations to determine which controls (which load combination is greater). In comparing load

TABLE 1.4 LOAD DURATION FACTORS

Load	Duration	Factor
Dead	Permanent	0.9
Floor live	Normal (10 years)	1.00
Snow	2 months	1.15
Roof live	7 days	1.25
Wind	10 minutes	1.6 or 1.33[a]
Seismic	10 minutes	1.6 or 1.33[a]
Impact	Less than 1 second	2.00

[a] Duration factor of 1.6 applies only to wind loads based on ANSI/ASCE Standard 7-88, specifically loads derived from BOCA, *National Building Code*; CABO, *One- and Two-Family Building Code*; and SBCCI, *Standard Building Code*. For wind loads derived from other sources, such as ICBO, *Uniform Building Code*, use 1.33.

TABLE 1.5 LOAD DURATION CALCULATIONS

Loads	(psf)			
Dead	10			
Snow	30			
Roof live	20			
Wind (uplift)	14			
Wind (downward)	8			
Load Combinations	Design Load (psf)	Load (psf)	Duration Factor	Load Divided by Factor (psf)
Dead + live	10 + 20	30	1.00	30
Dead + snow	10 + 30	40	1.15	34.8
Dead + wind (down)	10 + 8	18	1.6	11.3
Dead + wind (up)	10 − 14	−4	1.6	−2.5
D + S + W (down)	10 + 30 + 8	48	1.6	30

combinations we take into account the load duration factor by dividing the combined load by the appropriate factor. For any combination that includes loads of different durations, the factor is based on the shortest-duration load.

EXAMPLE 1.1—LOAD DURATION

A roof truss must be designed for the loads and combinations shown in Table 1.5. Calculate which load combination controls and what the maximum design load must be.

Results

Dead load plus snow is the most critical, as shown in the final column (34.8). For design use the full combined load is 40 psf, not 34.8 psf. The full load is used and then the duration factor is applied to the wood strength. Using 34.8 psf would invite the error of applying the duration factor twice. The truss would also have to be checked for the net upward force due to dead load plus wind. Use the design force of 4 psf upward. It is unlikely that the truss would require redesigning for this upward load; however, the connections from the truss to the wall would have to be designed for uplift and overturning due to the wind.

2

BEAM DESIGN

At this point the designer should review the basic elements of statics and strength of materials. A knowledge of the effects of external forces and how they are resisted by the internal strength of the members is essential. Hence it is necessary to know where loads really come from, and why building codes specify certain gravity loads for various conditions (live loads) in addition to the weights of all structural members and other materials (dead loads), together with the live loads created by the natural phenomena of wind and earthquake. The effects of these loads are found in *shear*, *bending moment*, and *axial thrust*. The axial thrust can be compressive or tensile. Resistance to these externally applied forces is found in the geometric properties of *area*, *section modulus*, and *moment of inertia*, and in the strength of the material itself. The material strengths are of the several kinds shown in Tables A.2 through A.12 and explained in Chapter 1.

In timber design beams are generally rectangular in cross section and put in place with their long dimension laid horizontally. Beams can be installed on a slant, as with the stringers for a staircase, or they can even be vertical, as with a wall subjected to wind load. Due to the nature and cost of wood it is best to keep beam cross sections relatively small and restrict their lengths to run from 8 to 14 ft. Shorter lengths are certainly acceptable, and using glued-laminated beams, longer lengths are certainly possible, but for dressed lumber these recommendations give the most economical results.

Wood is the most common material used for the structure of residential construction. Due to relatively low absolute strength, even though a good strength-to-weight ratio exists, it is necessary to use many smaller pieces placed close together in residential wood framing. The beam-type pieces, called *joists*, are installed 12, 16, or 24 in. on centers (o.c.). Thus they qualify as repetitive members in bending (explained in Chapter 1). However, as is the case with every beam designed everywhere of any material whatsoever, joists must be designed for *bending moment*, for *shear*, and for *deflection*. This results in determining the required *section modulus S*, the required *area A*, and the required *moment of inertia I*. Perhaps a good mnemonic would be to call this *SAI design*. In addition, in many cases it is necessary to look at the end of the joist to make sure that the *contact area* on the support is sufficient to prevent overstressing the joist in *bearing*, which would utilize the allowable compression perpendicular to the grain.

Design tables containing allowable strength values for stress in all these directions are tabulated by the many lumber grading agencies that operate in the United States and Canada. They are compiled and published by the American Forest and Paper Association in the *National Design Specification*. Each of these values is modified further for member size, load duration, and other conditions of use. Table 1.3 lists the factors to be applied to each direction. In the case of bending, we need to pay particular attention here to understanding the size factor, C_F. There are actually two size factors, one for dimensional lumber (2 to 4 in. thick) and one for beams and stringers (5 in. or thicker and 8 in. or wider) and posts and stringers (5 in. by 5 in. or larger, roughly square). Let us consider here the size factor for dimension lumber. We discuss the factor for beams and stringers in the section "Deep Beams."

Tabulated design values for clear wood are now based on a 13-year-long evaluation of lumber referred to as the "in-grade" testing program. During this time over 1 million board feet of graded lumber was tested nondestructively to determine modulus of elasticity and in some cases modulus of rupture. The purpose of the testing was to determine more exactly the allowable design values associated with the grade of lumber. The 2×6 was chosen as a reference point. Tabulated values for dimensional lumber 2 to 3 in. thick for all species except Southern Pine and Mixed Southern species are given for 2×6's. For Douglas Fir-Larch sizes between 2×4 and 2×14 the allowable bending stress is calculated using values in Table A.2 and multiplying them by the size factor C_F shown in Table A.3. Tables A.8 and A.9 show the allowable strengths incorporating the size factors.

EXAMPLE 2.1—FLOOR JOIST DESIGN

Let us now consider a practical joist design problem. Suppose that in the house we are designing the bedrooms are on the second floor and all other spaces, such as living room, dining room, and kitchen, are on the first floor. We will design the second-floor joists. Building codes specify that the design live load for sleeping spaces shall be 30 pounds per square foot (psf), and that all other residential live loads created by gravity shall be 40 psf, so we will use 30 psf for our second-floor joists. If we have standard construction we will have carpet over plywood floor sheathing on the joists, with a drywall ceiling underneath. It is reasonable to assume that the plywood will be $\frac{5}{8}$ in. thick, since that is fairly standard. Suppose that the carpet is of moderate quality, say 32-ounce, which means that it weighs 32 ounces per square yard. Now the design load can be computed.

Dead load:
Joists	3.0 psf	(a reasonable assumption)
Sheathing	1.8 psf	(plywood weighs about 36 pounds per cubic foot, so $\frac{5}{8} \times \frac{1}{12} \times 36 = 1.8$)
Carpet	0.2 psf	[32 ounces = 2 pounds, and $\frac{2}{9}$ (square foot/square yard)] = 0.2 psf
Ceiling	6.0 psf	(usual for drywall)
Total	11.0 psf	
Live load:	30.0 psf	
Design load:	41.0 psf total	

Now suppose that the clear space from face to face of the studs of the supporting walls underneath our joists is 13′-5″. This dimension might be chosen to fully utilize 14-ft lengths of wood for the joists, since each stud wall is $3\frac{1}{2}$ in. wide, so that $3\frac{1}{2}$ in. + $3\frac{1}{2}$ in. + 13′-5″ = 14′-0″. Even though the clear span is 13′-5″ and the length of each piece is 14′-0″, the *design* length of each joist, considered as a beam, will be 13′-$8\frac{1}{2}$″. This is because the code requires that we take the design length to be from center to center of bearing, or support, for wooden beams such as these. Since each support is $3\frac{1}{2}$ in. wide, we get 13′-5″ + $\frac{1}{2}$ × ($3\frac{1}{2}$ in. + $3\frac{1}{2}$ in.) = 13′-$8\frac{1}{2}$″, or 13.71 ft (see Figure 2.1*a*). Now we can make the necessary preliminary computations for shear, bending moment, and deflection.

If the joists were spaced 12 in. on centers, each joist would carry a strip of floor 1 ft wide. The uniform load per linear foot of length of each joist becomes w = 41 lb/ft² × 1 ft width = 41 lb/linear foot of span length (see Figure 2.1*b*.)

The shear $V = wL/2 = 41 \times 13.71/2$ = 281 lb. *Note*: In this, and in all that follows, the letter l denotes the *length in inches*, whereas the letter L stands for the *length in feet*, unless otherwise noted.

Bending moment $M = \dfrac{wL^2}{8}$

$= 41 \times 13.71 \times \dfrac{13.71}{8}$

$= 963$ ft-lb

The maximum deflection permitted by the code is the span in inches divided by 360, or the span in feet divided by 30, compared with the actual deflection created by the live load only. In our case the maximum permitted deflection would be

$$\Delta = \frac{l}{360} = \frac{13.71 \times 12}{360} = 0.457 \text{ in.}$$

Now we are ready to select the most economical size to fit the design conditions. But first it is necessary to choose a grade and species of wood. For this example, suppose that we use Douglas Fir–Larch. Now the grade of wood has to be chosen. In the case of joists (and studs, too) a rather low strength will generally be satisfactory because the loads for those

20 BEAM DESIGN

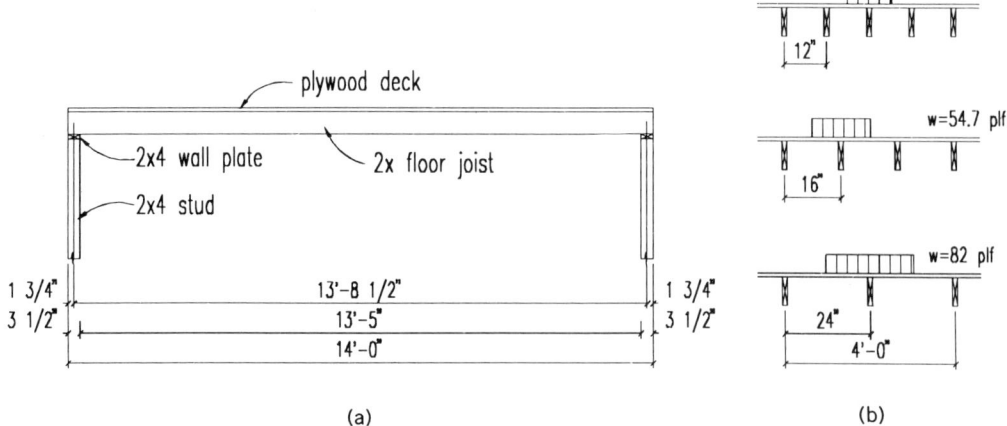

FIGURE 2.1 Joist span diagram.

members are not large. In this case, let us pick No. 2. Joists will usually have a depth of *about* $\frac{3}{4}$ in. for every foot of span. This is not a law or fixed rule—it is simply a guide in approximating a reasonable range of depth so as to save time in using the table. In our case, then, the depth will probably be in the range $\frac{3}{4} \times 13.7 = 9$ or 10 in., more or less. Therefore, use the size classification 2 to 4 in. thick, 2 in. and wider.

The tabulated values are:

F_b, bending strength = 875 psi

F_v, horizontal shear strength = 95 psi

E, modulus of elasticity = 1,600,000 psi

These values must be modified by the following values:

$$F'_b = F_b \times C_D \times C_L \times C_F \times C_r$$
$$F'_v = F_v \times C_D \times C_H$$

E is unchanged

For the factors in the equation, we will use the following values:

C_D = duration factor = 1.0 for floor live loads

C_L = beam stability factor

= 1.0 since the compression edge of the joist is continually supported by the floor decking

C_F = size factor = 1.2 for 2 × 8's
= 1.1 for 2 × 10's

(These values and other applications of size factors are discussed in the "Deep Beams" section of this chapter.)

C_r = repetitive use factor = 1.15, where four or more members, spaces 24 in. o.c. or less, are used to carry the load.

C_h = shear stress factor = 1.0, where there is no control over the splits or shake on the joists.
(This factor is used on larger beams and girders and is explained in more depth in Example 2.10.)

Thus

$$F'_b = F_b \times C_D \times C_L \times C_F \times C_r$$
$$= 875 \times 1.0 \times 1.0 \times 1.2 \times 1.15$$
$$= 1208 \text{ psi for } 2 \times 8\text{'s}$$
$$= 1107 \text{ psi for } 2 \times 10\text{'s}$$
$$F'_v = F_v \times C_D \times C_H$$
$$= 95 \times 1.0 \times 1.0 = 95 \text{ psi}$$
$$E = 1{,}600{,}000 \text{ psi}$$

Using the value for 2×10's, the required section modulus becomes

$$S = \frac{M}{F'_B} = 963 \times \frac{12}{1107} = 10.4 \text{ in}^3$$

The requires cross-sectional area will be

$$A = \frac{1.5V}{F'_v} = \frac{1.5 \times 281}{95} = 4.44 \text{ in}^2$$

Finally, the maximum deflection created by the *live load only* is found from

$$I = \frac{5wL^4}{384E\Delta} = \frac{5 \times 30 \times 13.71^4 \times 1728}{384 \times 1{,}600{,}000 \times 0.457}$$
$$= 32.6 \text{ in}^4$$

The number 1728 is $(12 \text{ in./ft})^3$, and is used to make the units come out right. Now enter Table A.1 and select a 2×8 that has $S = 13.1$ in.^{3}, $A = 10.9$ in.^{2}, and $I = 47.6$ in.^{4}. (The bending calculation was based on the F'_b value of a 2×10, which is smaller than that of a 2×8; thus a 2×8 is satisfactory.) It is easily seen that the 2×8 has the least board feet per lineal foot of any size that satisfies all three criteria, and *all three* must be satisfied or the design is not valid.

For a complete analysis it is now necessary to design for joist spacings of 16 in. center to center, and for 24 in. o.c. If they were spaced on 16-in. centers, each joist would carry a strip of floor 16 in. wide, which is $\frac{16}{12}$ ft. Its total load per lineal foot of span becomes $\frac{16}{12} \times 41 = 54.7$ lb/ft.

The design shear becomes

$$V = \frac{wL}{2} = \frac{54.7 \times 13.71}{2} = 360 \text{ lb}$$

The design bending moment is

$$M = \frac{wL^2}{8} = \frac{54.7 \times 13.71^2}{8} = 1273 \text{ ft-lb}$$

The maximum permitted deflection is independent of the load and thus remains $\Delta = 0.457$ in.

Calculating as before, the required section modulus

$$S = \frac{M}{F'_b} = \frac{1273 \times 12}{1107}$$
$$= 13.8 \text{ in}^3 \text{ to check } 2 \times 10\text{'s}$$

or

$$S = \frac{M}{F'_b} = \frac{1273 \times 12}{1208}$$
$$= 12.65 \text{ in}^3 \text{ to check } 2 \times 8\text{'s}$$

The required cross-sectional area becomes

$$A = \frac{\frac{3}{2}V}{F'_v} = \frac{1.5 \times 360}{95} = 5.68 \text{ in}^2$$

The live load that is carried by the joist at 16-in. spacing becomes $\frac{16}{12} \times 30 = 40$ plf (pounds per lineal foot). The required moment of inertia now becomes

$$I = \frac{5wL^4}{384E\Delta} = \frac{5 \times 40 \times 13.71^4 \times 1728}{384 \times 1{,}600{,}000 \times 0.457}$$
$$= 43.5 \text{ in}^4$$

Again referring to Table A.1, select a 2×8, as before. This size still satisfies all three criteria.

A similar set of computations is made for 24-in. spacing. The load per lineal foot becomes $w = \frac{24}{12} \times 41 = 82$ plf. This creates a design shear of

$$V = \frac{wL}{2} = \frac{82 \times 13.71}{2} = 562 \text{ lb}$$

The maximum bending moment for this 24-in. spacing of joists will be

$$M = \frac{wL^2}{8} = \frac{82 \times 13.71 \times 13.71}{8}$$
$$= 1927 \text{ ft-lb}$$

Now the required section modulus is

$$S = \frac{M}{F'_b} = \frac{1927 \times 12}{1107}$$
$$= 20.89 \text{ in.}^3 \text{ for } 2 \times 10\text{'s}$$

and the cross-sectional area must be at least

$$A = \frac{\frac{3}{2}V}{F'_v} = \frac{1.5 \times 562}{95} = 8.87 \text{ in}^2$$

The largest permitted deflection remains $\Delta = 0.457$ in. since the permitted deflection depends, as before, only on the span. Because the live load increases to $\frac{24}{12} \times 30 = 60$ plf, the moment of inertia cannot be less than

$$I = \frac{5wL^4}{384E\Delta} = \frac{5 \times 60 \times 13.71^4 \times 1728}{384 \times 1,600,000 \times 0.457}$$
$$= 65.2 \text{ in}^4$$

Studying Table A.1, it is apparent that we now need a 2×10 because section modulus now controls. It will be stressed to 98% of its allowable capacity in bending.

We may specify any of the following possibilities:

2×8 joists at 12 in. o.c.
2×8 joists at 16 in. o.c.
2×10 joists at 24 in. o.c.

Which is the cheapest design? Assuming that only board feet of lumber need be considered, obviously 2×8 joists at 16-in. centers use less material than 2×8's at 12-in. centers. Referring to Table A.1, a 2×8 at 12-in. spacing shows that "board feet per lineal foot of piece" will be 1.33. If placed on 16-in. centers, the lumber would reduce to $\frac{12}{16} \times 1.33 = 1.00$ board feet per square foot. On the other hand, the 2×10 at 24-in. centers would yield only $\frac{12}{24} \times 1.67 = 0.84$ board foot per square foot. Hence the solution is to use

2×10's Douglas Fir–Larch, No. 2 grade joists at 24 in. on center.

EXAMPLE 2.2—JOIST DESIGN WITH PARTITION

Now suppose that there is a partition across the joists 3 ft from one end, as shown in Figure 2.2. This wall does not carry any load from above, which is why it is labeled "partition." The weight of the wall itself becomes a concentrated load on the joists. Suppose that the partition has drywall on both sides and is 8'-0" high. Drywall weighs 5 psf. Since it is applied on two sides, each square foot of wall surface will weigh $2 \times 5 = 10$ psf. For the 8-ft height, the wall produces a dead load of $8 \times 10 = 80$ plf. With joists 24 in. on centers, the concentrated load becomes $\frac{24}{12} \times 80 = 160$ lb/joist. As before, with a live load of 30 psf and a dead load for floor construction of 11 psf, the total load on the floor is 41 psf. But with the joists spaced 24 in. on centers, $W = \frac{24}{12} \times 41 = 82$ plf (see Figures 2.1*b* and 2.2).

The maximum shear, V, is 689 lb, and the maximum bending moment, M, is 2160 ft-lb. The maximum permitted deflection remains $\Delta = L/30 = 13.71/30 = 0.457$ in. Then the required geometric properties

JOIST DESIGN   23

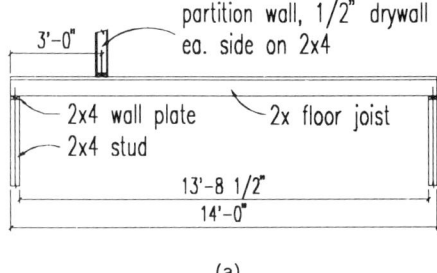

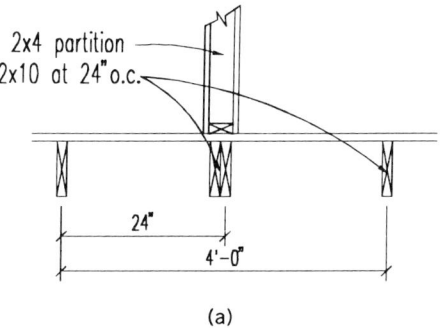

(a)

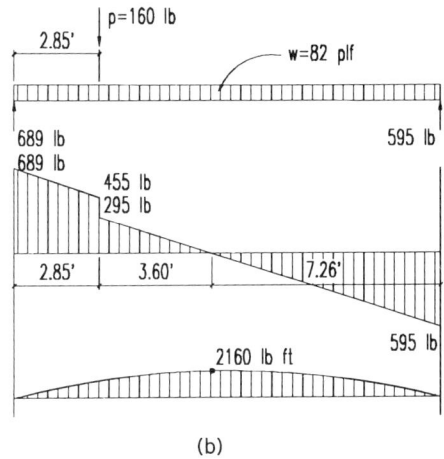

(b)

FIGURE 2.2 Joist with partition wall—perpendicular: (a) diagram and (b) shear and moments.

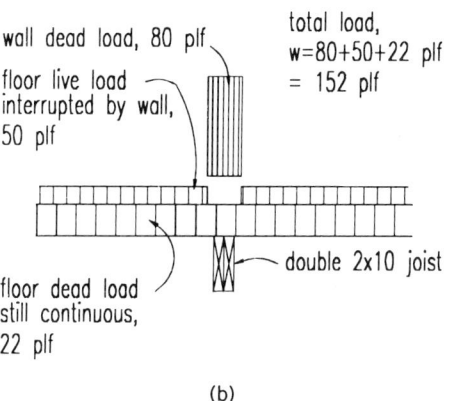

(b)

FIGURE 2.3 Joist with partition wall—parallel: (a) section and (b) loads.

are

$$S = \frac{M}{F'_b} = \frac{2160 \times 12}{1107} = 23.4 \text{ in}^3$$

$$A = \frac{\frac{3}{2}V}{F'_v} = \frac{\frac{3}{2} \times 689}{95} = 10.9 \text{ in}^2$$

$$I = 65.2 \text{ in}^4$$

unchanged, since the required moment of inertia depends only on live load. From Table A.1 try a 2×10, which was successful without the partition loading. A 2×10 joist at 24 in. on centers fails in bending since it provides a section modulus of 21.39 in.³. The beam is stressed by a factor of $23.4/21.4 = 109\%$ of allowable. Reducing the spacing to 16 in. on center reduces the load, moment, and required section modu-

lus by the ratio $\frac{16}{24}$. The required section modulus for this configuration is

$$S = 23.4 \text{ in}^3 \times \tfrac{16}{24} = 15.6 \text{ in}^3$$

So use 2×10's at 16 in. o.c.

EXAMPLE 2.3—JOISTS WITH PARALLEL PARTITION

Next let us see what happens when a partition is placed on the floor parallel to the joists rather than across them (see Figure 2.3). The weight of the wall will still be 82 plf, but since two things cannot occupy the same place at the same time, the live load will be reduced. This is because the wall is

4 in. thick, so the joist under the wall will carry a strip of floor only 20 in. wide, not 24 in. wide. This means that the live load will be 60 plf × $\frac{20}{24}$ = 50 plf, not 60 plf. The rest of the dead load is unchanged, however, and remains $11 \times \frac{24}{12} = 22$ plf. The design load now becomes $w = 80 + 50 + 22 = 152$ plf.

Then $V = wL/2 = 152 \times 13.71/2 = 1042$ lb, and $M = wL^2/8 = 152 \times 13.71 \times 13.71/8 = 3571$ ft-lb. Thus the required geometric properties are

$$S = \frac{M}{F_b} = \frac{3571 \times 12}{1107} = 38.7 \text{ in}^3$$

$$A = \frac{\frac{3}{2}V}{F_V} = \frac{\frac{3}{2} \times 1042}{95} = 16.5 \text{ in}^2$$

$$I = \frac{5wL^4}{384E\Delta} = \frac{5 \times 50 \times 13.71^4 \times 1728}{384 \times 1{,}600{,}000 \times 0.457}$$
$$= 54.4 \text{ in}^4$$

From Table A.1 it is seen that neither a 2×10 nor a 2×12 is satisfactory; the smallest section available is a 2×14. This member is deeper than the rest of the floor joists and will create an unsightly ceiling below them. Rather than changing all the rest of the joists to 2×14's, we will use a double joist in this location, that is, two 2×10's, which have a combined area of 27.8 in.^{2} > 16.5, OK; a total section modulus of 42.8 in.^{3} > 38.7; OK; and a more than adequate moment of inertia. This conforms to standard practice, which is simply to double the joists under a partition running parallel to them (see Figure 2.3a).

JOIST TABLES

Tables are available that greatly facilitate the design of floor joists. These are generally for a live load of 30 psf, as shown in Table B.2a, or 40 psf, as in Table B.2b.

The tables use a dead load of 10 psf, which is commensurate with most residential construction. To learn how to use the joist tables, let us rework Example 2.3, where the clear span was 13'-5", giving a design span of 13'-8$\frac{1}{2}$", and the live load was 30 psf. Once again select Douglas Fir–Larch with a modulus of elasticity E of 1,600,000 psi, an allowable bending stress F_b of 1107 psi for 2×10's, or an F_b value of 1208 psi for 2×6's. Now enter Table B.2a, which is based on a live load of 30 psf, and find the column for E of 1,600,000 psi. Now proceed downward until a span (the upper number in each box) of 13'-8$\frac{1}{2}$" or better is found. The first time this span is exceeded corresponds to 2×8 at 16 in. o.c. joist spacing (read this at the left side of the table), where a span of 14'-1" is permitted. The lower number in each box is the bending stress, F'_b. Thus for a span of 14'-1" the bending strength, F'_b, must be 1215 psi. For a span of 13'-8$\frac{1}{2}$" the actual bending stress is between the two values to the left where the 2×8's at 16 in. span 13'-6" and 13'-9". These values are 1059 psf and 1111 psf, respectively. The actual required strength can be found by linear interpolation. This will not be necessary since the allowable bending stress for a 2×8 is 1208 psi. Thus

2×8 joists at 16 in. o.c. are satisfactory.

Proceeding on down the column of figures under $E = 1{,}600{,}00$ psi, it is seen that a 2×10 at 24-in. spacing will also work in deflection. Scanning to the left the span length of 13'-10" is nearest the required span. This span requires an allowable bending stress of 1083 psi. Our allowable bending stress for 2×10's is 1107 psi, which exceeds the requirement of 1083 psi, and our span is permitted. Obviously, 2×10's at even closer spacing will also work, but they will not be as efficient.

Using Table B.2*a*, we select two options:

2 × 8 joists at 16 in. o.c., 1.0 bd-ft/ft, or
2 × 10 joists at 24 in. o.c. 0.83 bd-ft/ft

Summarizing the results, the table shows that 2 × 10 joists at 24 in. o.c. is the most economical shape available. This is the same result that we got previously with lengthy calculations.

The allowable spans of the joists in these tables do not account for the extra stiffness due to the plywood floor deck. If the deck is attached properly, it can combine with the joist to form a T-shaped beam which will be stiffer than just the joist alone. Such a beam can be made if the decking is glued to the joist in accordance with the APA standards for glued floor systems. This assembly, also referred to as APA Sturd-I-Floor, is described in Table B.3*b*. Its allowable spans are listed in Table B.3. In the earlier example we found that DF-L No. 2, 2 × 8 joists at 16 in. o.c., would span 14'-1" under 30 psf. Under 40 psf they will span 12'-10". When properly glued to the floor deck, they will span 13'-10" under a 40-psf load.

For very long spans dimensional lumber joists become inefficient. Considering a No. 2 grade joist with $E = 1,600,000$ psi, we can read from Table B.2*b* the following maximum spans (live load = 40 psf):

Joist Size	24 in. o.c.	16 in. o.c.	12 in. o.c.
2 × 8	11'-2"	12'-10"	14'-1"
2 × 10	14'-3"	16'-4"	18'-0"
2 × 12	17'-4"	19'-11"	21'-11"
2 × 14	20'-6"	23'-5"	25'-9"

Although longer spans can be effected with higher-grade lumber, these spans, in particular those for the 16-in. o.c. spacing represent the practical limits of standard construction. Both the builder and the architect may choose to memorize these limits. If conditions occur where larger spans (or heavier loads) are necessary, there are a number of newer products available made of wood materials. Some of these products use wood chips or flakes, which are pressed into boards of various thicknesses and then are laminated into beams. Tables B.4 and B.5 list the properties and spans of Micro-Lam[1] and Gang-Lam LVL[2] manufactured lumber.

Another product that is often used for long-span floor systems is a manufactured wood I-beam using plywood or oriented-strand board (OSB) for the web and either solid wood chords or LVL wood chords. Table B.6 shows properties and spans for the Trus Joist wood joist; Table B.7 gives the same information for joists produced by Gang-Nail Systems. These and other producers will supply builders and designers with all pertinent data and specification. As with any other product, the allowable loads and spans shown in the tables depend on accurate installation in accordance with the manufacturers' specifications.

BEARING

All of the joist designs should have been checked for adequacy in bearing at their supports. This involves the area in bearing under the reaction, the allowable compression stress perpendicular to the grain, and the reaction itself. The allowable stress $F_{c\perp}$ is used because the joist is horizontal, with its grain also running horizontally, whereas the reaction is a vertical force, thus acting perpendicular to the grain. The area in bearing is the width of the joist multiplied by the length of contact be-

---

[1] Micro-Lam is a product of Trus Joist Corporation, 9777 West Chinden Boulevard, P.O. Box 60, Boise, ID 83707.
[2] Gang-Lam is a product of Gang-Nail Systems, Inc., P.O. Box 59-2037 AMF, Miami, FL 33159-2037.

tween the end of the joist and its support. The minimum length of bearing required is specified in the building code. Unless the joist is nailed through its side onto a supporting stud, as would be the case in balloon framing, the minimum length of bearing in ordinary frame construction is 1.5 in. on a wood plate or on steel, and 3 in. if resting on masonry. Headers over openings less than 6 ft must have 1.5 in. of bearing on each side, as on a jack stud, or they can be supported on metal hangers, while headers over openings greater than 6 ft must have 3 in of bearing on each side. These requirements will govern almost always, and computations are superfluous.

EXAMPLE 2.4—BEARING

For example, in the case of the cross wall, the maximum end reaction was 689 lb. The length of bearing is 3.5 in., and the width of bearing area is 1.5 in., thus giving an area of bearing of $1.5 \times 3.5 = 5.25$ in.^{2}. From Table A.8 we find an allowable stress in compression perpendicular to the grain to be 625 psi. Then the maximum reaction could be $R = F_{c\perp}A = 625 \times 5.25 = 3281$ lb, which is much, much greater than the actual maximum reaction of 689 lb.

A beam that receives a column or other beam along its span must also be checked for bearing (see Figure 2.4). The concern is the same as bearing at the ends except that the allowable bearing strength can be increased by a bearing area factor, C_b, as defined by Table 2.1. The criteria for using this factor are that (1) the bearing shall not be nearer than 3 in. to the end of the member, and (2) that the length of bearing shall be less than 6 in.

The factors in the formula derive from the equation

$$C_b = (l_b + 0.375 l_b)$$

where l_b is the bearing length measured parallel to the grain, in inches.

EXAMPLE 2.5—BEARING AWAY FROM THE ENDS

Figure 2.4 shows three possible conditions in which the bearing area factor should be considered. For Figure 2.4c, assume that a girder made up of three 2×10's receives an applied load of 1250 lb at a location 4 ft from the end. The load is applied directly from a joist bearing on the girder.

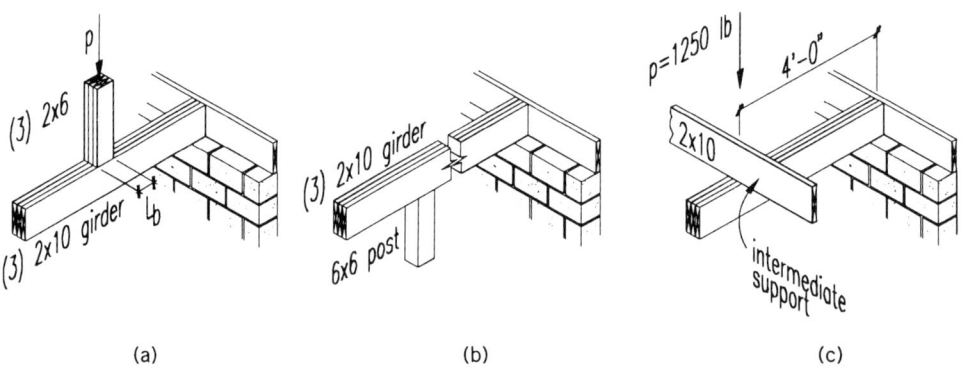

(a) (b) (c)

FIGURE 2.4 Bearing: (a) post above, (b) post below, and (c) beam above.

TABLE 2.1 BEARING AREA FACTORS, C_b

l_b (in.)	0.5	1	1.5	2	3	4	6 or more
C_b	1.75	1.38	1.25	1.19	1.13	1.10	1.00

The bearing area is 1.5 in. by 4.5 in., or 6.75 in.^{2}. The bearing area factor is found from Table 2.1 using a length, l_b, of $1\frac{1}{2}$ in. $C_b = 1.25$. The modified strength perpendicular to the grain is

$$F'_{c\perp} = C_b \times F_{c\perp} = 1.25 \times 625 \text{ psi}$$
$$= 781.25 \text{ psi}$$

and the allowable load in bearing is

$$P = F_{c\perp} \times \text{area} = 781.25 \text{ psi} \times 6.75 \text{ in.}^2$$
$$= 5273 \text{ lb} > 1250 \quad \text{OK}$$

DECKING

Sometimes solid decking is employed instead of framing with joists and sheathing or joists and subfloor. The decking used is not simply planks laid flat next to one another. Instead, it is wood which has each piece configured such that the entire floor acts as a solid unit, and the pieces are all mutually supporting, thus affording the opportunity of cantilevering and random jointing. This is usually accomplished in one of two ways: by using tongue-and-groove or splined pieces. Of course, any piece passing over a beam must be nailed securely to that beam.

There are five types of spans for plank floors: simple, two-span continuous, combination simple and two-span continuous, cantilevered pieces intermixed, and controlled random (Figure 2.5). The last four types are all stiffer, in varying degree, than the simple span because of the continuity over interior supports of the pieces of planking. Notice that in the simple span all planks are of the same length, with end joints over each beam. The plank lengths are also equal for the two-span continuous scheme, with end joints over every other beam. For the combination span, all pieces are two span lengths except for alternating pieces in the end span, and the end joints over interior beams are staggered in adjacent lines of planking. The random arrangement permits the economical use of random lengths of plank. The control requirements for this scheme are that the end joints be well scattered and that each piece of plank bear on at least one beam. This works only because ends and sides of all pieces are both joined by one of the two shapes shown in Figure 2.5b.

The design of a wood deck is straightforward. A width of 12 in. is assumed, and the thickness is then determined by the *SAI* method. Alternatively, the proper thickness may be found by the use of Tables B.1a through B.1c. They give the total uniformly distributed load in psf for the five span types for three deck thicknesses and several span lengths. The maximum permitted load is shown as limited by bending strength, and also as limited by deflection of either $L/20$ or $L/30$, where L is in feet, or $l/240$ and $l/360$, respectively, if l is in inches.

EXAMPLE 2.6—DECKING DESIGN

As an example, suppose that the supporting beams occur at 10'-0" centers, with the planks to be laid in a controlled random pattern, exposed underneath (no finished ceiling underneath). There will be a hardwood finished floor over the planks (allow 3 psf), with a live load of 75 psf. The

BEAM DESIGN

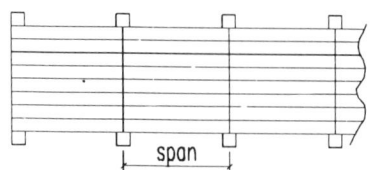

all pieces bear on two supports

(a)

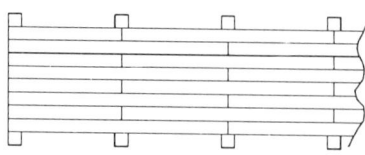

alternate pieces in end spans are simple spans
adjacent pieces are two-span continuous
end joints are staggered in adjacent
courses and occur over supports

(d)

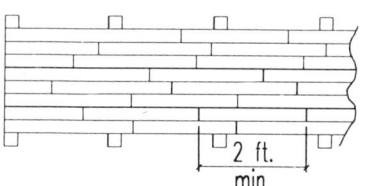

all pieces rest on at least one support
distance between end joints in adjacent
spans must be at least 2 feet

(b)

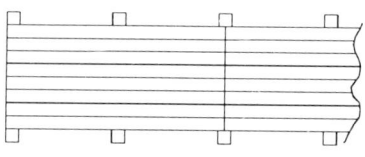

all pieces bear on three supports
all end joints occur in line
on every other support

(e)

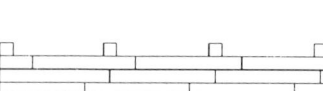

pieces in the starter course and every
third course are simple span
pieces in other courses are cantilevered over
the supports with end joints at alternate
quarter-points or third points.
each piece rests on at least one support

(c)

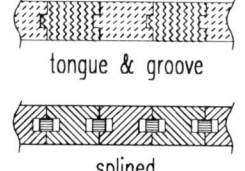

(f)

FIGURE 2.5 Deck layout patterns and profiles: (a) simple span, (b) controlled random lay-up, (c) cantilevered pieces intermixed, (d) combination simple and two-span continuous, (e) two-span continuous, and (f) decking profiles.

deflection may be $L/20$ in this case. Use Commercial grade decking, Douglas Fir–Larch.

Since the loads given in Tables B.1 are based on $F_b = 1000$ psi and $E = 1,000,000$ psi, it is first necessary to find F_b and E for the species and grade used. Table A.10 lists these as $F_b = 1650$ psi and $E = 1,700,000$ psi. The bending strength is listed for 4-in. thick decking; it can be increased by C_F of 1.10 for 2-in. thick decking or by 1.04 for 3-in.-thick decking (see Table A.3). We will apply these factors at the end of the example after we have determined the appropriate deck thickness. Because these values are higher than the corresponding ones in Tables B.1, the allowable loads for Douglas Fir–Larch Commercial decking will be the tabulated load multiplied by the ratio $1650/1000$ for loads limited by bending and by $1,700,000/1,000,000$ for those limited by deflection.

Our design load (the superimposed load on the deck, not including the weight of the deck itself) will be $75 + 3 = 78$ psf. Let us assume that controlled random layup will be the preferred decking pattern. Scanning Table B.1b with our 78-psf superimposed load in mind, try a 3-in.-thick deck on the 10-ft span. The tabular values are

Bending load = 69 psf for controlled random layup

Deflection load = 52 psf for $L/240$

The allowable load limited by bending is now computed for our material as

$$69 \times \frac{1650}{1000} = 114 \text{ psf}$$

The *net* load that may be carried in addition to the dead load of the planking is $114 - 7.6 = 106.4$, say 106 psf.

Similarly, the allowable load limited by deflection is

$$52 \times \frac{1,700,000}{1,000,000} = 88.4 \text{ psf}$$

This yields a net load that may be carried in addition to the weight of the decking of $88.4 - 7.6 = 80.8$, say 81 psf > 78, OK. Since deflection controls, there is no need to apply the size factor to the bending strength.

Use 3-in.-nominal-depth Douglas Fir–Larch Commercial decking.

DEEP BEAMS

As the depth of a beam increases there is a slight decrease in the strength in bending. This means that the allowable bending stress must be decreased accordingly. The size factor, C_F, decreases the allowable bending stress for all members over 12 in. deep. This factor applies to beams and stringers and to posts and timbers. For dimensional lumber, see the notes on adjustment factors that precede Tables 4A and 4B in the *NDS Supplement*.

The size factor, C_F, is determined from the following formula:

$$C_F = \left(\frac{12}{d}\right)^{1/9}$$

in which C_F is the size factor and d is the actual beam depth in inches. Values for C_F for solid timber beams or members in bending having various depths are given in Table 2.2.

First a preliminary design is done in the usual manner. If the beam so designed is deeper than 12 in., its moment capacity must be rechecked using the depth factor. In other words, the resisting bending moment of the beam deeper than 12 in. is found from $M = C_F F_b S$. If the beam is still satisfactory in resisting bending moment, it is accepted since nothing else is changed, and it was designed to be satisfactory in resisting shear and deflection.

TABLE 2.2 SIZE FACTOR, C_F, 5″ × 5″ AND LARGER[a]

If Depth Equals:	C_F Equals:
13.5	0.987
15.5	0.972
17.5	0.959
19.5	0.947
21.5	0.937
23.5	0.928
25.5	0.920
27.5	0.912

[a] Use these factors with lumber that is 5″ × 5″ and larger. For dimensional lumber (2 to 4 in. thick), refer to the NDS Supplement for which species and grades apply.

EXAMPLE 2.7—DEEP BEAM DESIGN

As an example, let us determine the maximum uniformly distributed load that can be carried by a 10 × 20 beam made of Douglas Fir–Larch No. 1 if the design span is 16 ft and the live load is equal to the dead load.

First, from Table A.1, notice that the beam has the geometric properties $S = 602$ in.^{3}, $A = 185$ in.^{2}, and $I = 5870$ in.^{4}. Then, from Table A.11 the properties for Douglas Fir–Larch, entering the section for beams and stringers, read $F_b = 1350$ psi, $F_v = 85$ psi, and $E = 1,600,000$ psi. Then the maximum bending moment that can be carried by this beam is

$$M = C_D \times C_F \times C_L \times F_b \times S$$

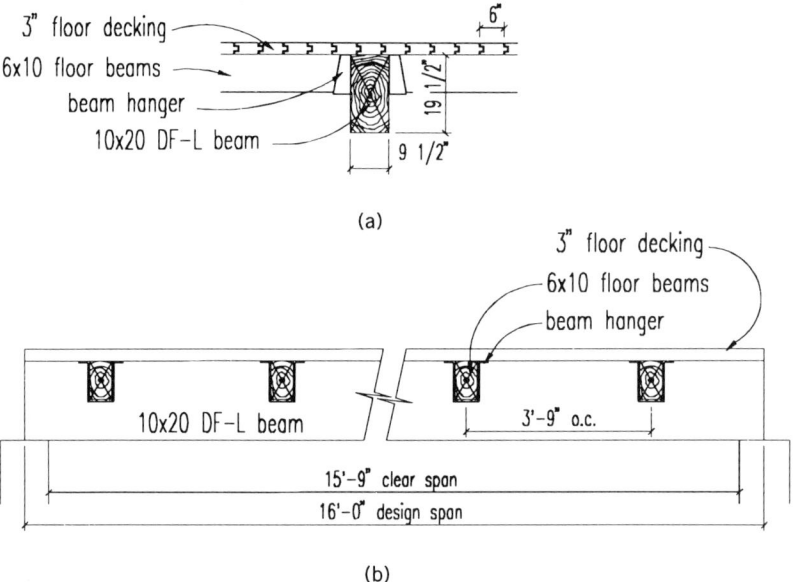

FIGURE 2.6 Decking profiles: (a) beam selection and (b) beam elevation.

Assume that the duration is normal and that C_D equals 1.0. Assume that the beam is braced along its compression edge (see Figure 2.6) and that C_L equals 1.0. Therefore,

$$M = C_F F_b S = 0.947 \times 1350 \times 602$$
$$= 769.9 \text{ in.-kips}$$
$$= 64.1 \text{ ft-kips}$$

Since $M = wL^2/8$, then

$$w = \frac{8M}{L^2} = \frac{8 \times 64.1}{16 \times 16} = 2.0 \text{ kips/ft}$$

The maximum shear that can be carried by this beam is

$$V = \frac{F_v A}{\frac{3}{2}} = \frac{85 \times 185}{1.5}$$
$$= 10,483 \text{ lb or } 10.5 \text{ kips}$$

This shear will be equal to the maximum reaction. For a uniform load both reactions are the same, so each one is equal to half the total load, and the total load is $W = wL$, where w is the load per foot. Then $V = R = W/2 = wL/2$, from which

$$w = \frac{2V}{L} = \frac{2 \times 10.5}{16} = 1.32 \text{ kips/ft}$$

Deflection also depends on the load that is carried. Suppose that the maximum permitted deflection, based on live load only, is $L/30 = 16/30 = 0.533$ in. For a uniform load over the entire beam, $\Delta = 5wL^4/384EI$, which means that

$$w = \frac{384EI\Delta}{5L^4}$$
$$= \frac{384 \times 1,600,000 \times 5870 \times 0.533}{5 \times 16^4 \times 1728}$$
$$= 3394 \text{ lb/ft}$$

Since the deflection criterion is based on live load alone, and since in this problem the live load is equal to the dead load, the total load that can be carried without creating excessive deflection is equal to twice the possible live load, or $2 \times 3394 = 6.79$ kips/ft.

Considering all three criteria, it is apparent that shear controls in this example, and the maximum load that this beam can carry is 1.32 kips/ft.

LATERAL SUPPORT

Whenever a beam is in bending, one edge of it will be in tension while the opposite edge is in compression. Any material subjected to tension is not made unstable thereby, but that same material will attempt to evade compression stress by slipping out away from it. In the case of a simply supported beam, the bottom is in tension but the top is in compression. To escape from the compression, the beam would tend to move either up, down, or sideways. The load pressing down on the top of the beam will prevent an upward motion. If the beam were to move downward, that would simply aggravate the situation by putting even more compression stress into its top edge, so that movement cannot take place. Therefore, the top edge of the beam tries to slip sideways when it is in compression. Since the beam is a solid member, the top cannot move independently of the bottom, so a twisting motion results, with the top sliding one way while the bottom moves in the other direction. The ends, being held in place, do not move at all in this respect (see Figure 2.7).

If the beam is permitted to twist, it gives up some internal strength with which it can resist bending stresses. It can twist only if the beam is not restrained against lateral (i.e., sideways) movement. This restraint, called *lateral support*, is provided in ordinary construction by using the floor or roof deck to prevent the lateral movement of the compression edge. Thus in ordinary construction the beam retains its full strength in regard to bending. However, there is a very complicated reduction in the allowable bending stress whenever there is no lateral support, as in the case of the underside of the haunch of a three-hinged timber arch in some cases. However, even in this case, if the beam is square, or if the breadth is greater than its depth, no lateral support is required to obtain full allowable bending stress.

For rectangular beams, rafters, joists, or other bending members, the following approximate rules should be applied in providing restraint to prevent rotation or lateral displacement. If the ratio of depth to breadth, based on nominal dimensions, is:

1. *2 to 1:*—no lateral support is required. *Example: 2 × 4*
2. *3 to 1 or 4 to 1:*—the ends shall be held in position, as by full-depth solid blocking, bridging, hangers, nailing or bolting to other framing members, or other acceptable means. *Example: 2 × 6 or 2 × 8*
3. *5 to 1:*—one edge shall be held in line for its entire length. *Example: 2 × 10*
4. *6 to 1:*—bridging, full-depth solid blocking, or cross bracing shall be installed at intervals not exceeding 8 ft unless both edges are held in line or unless the compression edge of the member is supported throughout its length to prevent lateral displacement, as by adequate sheathing or subflooring, and the ends at the points of bearing have lateral support to prevent rotation. *Example: 2 × 12*
5. *7 to 1:*—both edges shall be held in line for their entire length, as with 2 × 14 joists having a subfloor above and a ceiling under.

See Figure 2.8 for these conditions.

When a beam is indeed not laterally supported, its allowable bending stress depends on its material and its geometry. We must first consider its unbraced length, which is taken as center to center of lateral supports, or center of support to the free tip in the case of an overhang. This unbraced length is given the symbol L_u. From this actual unbraced length we next compute the effective unbraced length L_e. This is done according to criteria of Table 2.3, where both L_u and the depth d are in inches.

Having computed the effective length we can next determine the reduction in strength given by the lateral support factor, C_L. The allowable bending stress depends, in part, on the slenderness ratio, R_B. It must be understood that the beam has been designed without considering lateral support, so that the cross-sectional dimensions have tentatively been determined. Then the slenderness ratio is found from

$$R_B = \sqrt{\frac{L_e d}{b^2}}$$

where R_B is the slenderness ratio, L_e = the effective length in inches, d the depth of beam in inches, and b the breadth of beam in inches. The allowable bending strength of the wood is calculated as $F_b' = C_L F_b^*$ in which the beam lateral stability factor, C_L, is defined as

$$C_L = \frac{1 + (F_{bE}/F_b^*)}{1.9} - \sqrt{\left(\frac{1 + F_{bE}/F_b^*}{1.9}\right)^2 - \frac{F_{bE}/F_b^*}{0.95}}$$

The value F_b^* is the allowable bending stress, F_b, adjusted for conditions of use.

LATERAL SUPPORT **33**

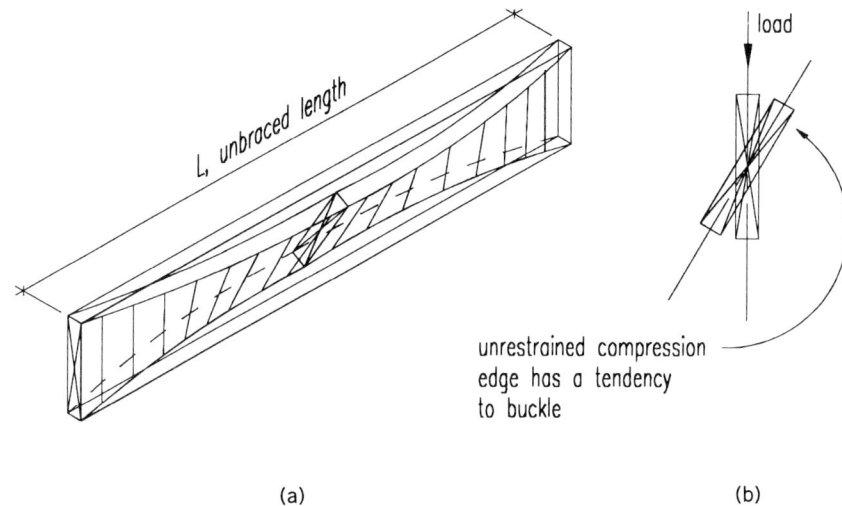

FIGURE 2.7 Compression buckling.

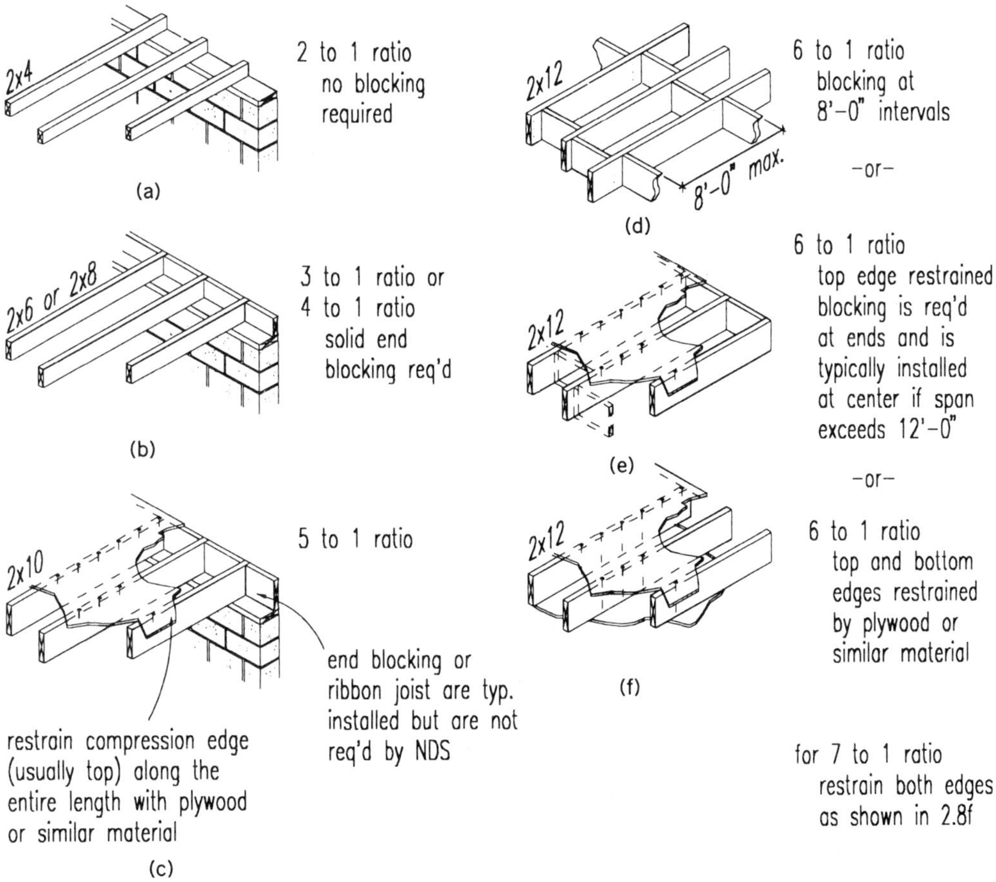

FIGURE 2.8 Lateral support requirements: (a) 2 × 4, (b) 2 × 6 or 2 × 8, (c) 2 × 10, (d) 2 × 12, blocking, (e) 2 × 12, edge restraint, and (f) 2 × 12, two-edge restraint.

34 BEAM DESIGN

TABLE 2.3 EFFECTIVE LENGTH OF A BEAM, $L_e{}^a$

	Value of Effective Length, L_e	
Type of Beam Span and Nature of Load	When $L_u/d < 7$	When $L_u/d \geq 7$
Simple span, load concentrated at center, no intermediate lateral supports	$1.37L_u + 3d$	$1.80L_u$
Simple span, load concentrated at center, with lateral supports at center	$1.11L_u$	$1.11L_u$
Simple span, uniformly distributed load	$1.63L_u + 3d$	$2.06L_u$
Simple span, equal end moments	$1.84L_u$	
Cantilever, load concentrated at free end	$1.44L_u + 3d$	$1.87L_u$
Cantilever, uniformly distributed load	$0.90L_u + 3d$	$1.33L_u$

[a] In all cases, a conservative value for L_e is given by $L_e = 1.84L_u$ if L_u/d is greater than 14.3, and $L_e = 1.63L_u + 3d$ if L_u/d is less than 14.3.

All applicable factors except C_V, C_{fu}, and C_L are used.

$$F_b^* = C_D \times C_M \times C_t \times C_F \times C_c \times C_f \times F_b$$

$$F_{bE} = \frac{K_{bE} E'}{R_b^2}$$

where $K_{bE} = 0.438$ for visually graded lumber and machine-evaluated lumber and 0.609 for MSR (machine stress rated lumber, $COV_E \leq 0.11$)

$$R_B = \sqrt{\frac{L_e d}{b^2}}$$

The procedure for calculating the effect of lateral support is the most complex step in the design of timber members. In 1991 the NDS revised the formulas for this procedure in an attempt to simplify the calculations. In the revision one formula was used to replace three formulas. This revision makes the calculation more systematic and is more adaptable to computer-based calculations. The revised formula was developed carefully to represent the results found by using the three old formulas; the increase in accuracy, if any, is slight. From the point of view of the authors, the new formula masks some of the understanding that was inherent in the old formulas. For this reason we feel that it is necessary to introduce some of the concepts that are now not readily apparent. As the 1991 NDS is adopted by model codes, state codes, and local codes, the designer will be obliged to follow it. Our intent is not to subvert the content of the revised NDS but to elucidate it by referring to the previous version.

Prior to 1991, laterally unsupported beams were classified as short, intermediate, or long, based on the slenderness ratio.

1. If R_B was not more than 10, the beam was *short*.
2. If R_B is more than 10 but less than C_k, the beam was *intermediate*. The value of C_k was found from $C_k = 0.811\sqrt{E/F_b}$ (for most species C_k is appropriately = 30).
3. If R_B is equal to or greater than C_k, the beam was *long*.
4. The value of R_B under the old editions and the 1991 edition is never permitted to be greater than 50.

If the beam was classified as short, there was no change in the design value of the bending stress—C_L was taken as equal to 1.0. It was as if the beam were completely laterally supported.

Notice that a beam with solid decking will have an unbraced length $L_u = 0$, thus

Value of R_B	Classification	Allowable Bending Stress
$0 \leq R_B \leq 10$	Short	Try $F'_b = F_b$; check if critical
$10 < R_B < 30$	Intermediate	Calculate C_L
$30 \leq R_B \leq 50$	Long	Try $F'_b = 0.438E/R_B^2$, check if critical

leading to an effective length $L_e = 0$, so that $R_B = 0$, which is less than 10, and thus a short beam. Under the 1991 revision the formula applies to a beam regardless of the value of R_B, so that C_L must be calculated for a beam with $R_B < 10$. For typical grades and species, short beams will have only a slight reduction in strength, in the range of 5% ($C_L = \pm 0.95$). For short beams, $R_B \leq 10$, it is therefore recommended that C_L would not be calculated unless it is critical (i.e., unless the actual stress is within 5% of that allowable).

For a long beam where $C_k < R_B \leq 50$ (say $30 < R_B \leq 50$) the previous code value for the allowable bending stress was given by the equation

$$F'_b = \frac{0.438E}{R_B^2}$$

Under the revised version this value is referred to as F_{bE}, so that now,

$$F_{bE} = \frac{0.438E}{R_B^2}$$

Since this value is already calculated, it is recommended that if R_B is between 30 and 50, the initial design should be checked using $F'_b = 0.438E/R_B^2$. If the design seems critical (i.e., if the actual stress and the allowable stress are within about 10 to 15% of each other), use the formula for C_L as prescribed by the code.

For intermediate beams, where $10 < R_B \leq C_k$ (say, $10 < R_B \leq 30$), there is no savings of time by using the old formula, so the designer is advised to use the 1991 version.

It is useful to summarize all of the foregoing.

EXAMPLE 2.8—LATERALLY UNSUPPORTED BEAM

As an example, consider a laterally unsupported 24-ft-long beam that is made of Southern Pine No. 1 SR (Structurally Rated) grade. Suppose that the live load is 700 plf, and the dead load is 300 plf, giving a combined uniform load of $w = 1.0$ kip/ft. Assume that the beam will need to be much larger than a 2-in.- or 4-in.-thick member, so use the design values for beams and stringers. The allowable stresses seem to be $F_b = 1350$ psi, $F_v = 110$ psi, and $E = 1{,}500{,}000$ psi. The phrase "seem to be" is used because the allowable bending stress, and only the bending stress, may have to be modified. None of this affects the actual applied bending moment and shear. Then

$$M = \frac{wL^2}{8} = \frac{1.0 \times 24 \times 24}{8}$$
$$= 72.0 \text{ ft-kips}$$
$$V = \frac{wL}{2} = \frac{1.0 \times 24}{2} = 12.0 \text{ kips}$$
$$\Delta_{max} = \frac{L}{30} = \frac{24}{30} = 0.80 \text{ in.}$$

The required cross-sectional area and the required moment of inertia can be found directly and will remain unchanged for this problem, but the required section modulus can only be approximated and is subject to change.

$$S = \frac{M}{F_b} = \frac{72{,}000 \times 12}{1350} = 640 \text{ in}^3$$

$$A = \frac{1.5V}{F_v} = \frac{1.5 \times 12{,}000}{110} = 163.6 \text{ in}^2$$

$$I = \frac{5wL^4}{384E\Delta} \quad (\text{where } w \text{ is live load only})$$

$$= \frac{5 \times 700 \times 24^4 \times 1728}{384 \times 1{,}500{,}000 \times 0.80}$$

$$= 4354 \text{ in}^4$$

Tentatively, try a 10×22 beam, which has $S = 732$ in.^{3}, $A = 204$ in.^{2}, and $I = 7868$ in.^{4}. Now consider the lack of lateral support.

$$L_u = 24 \text{ ft} = 288 \text{ in.}$$

$$L_e = 1.63 L_u + 3d$$

$$\quad (\text{uniform load, simple span})$$

$$= (1.63 \times 288) + (3 \times 21.5) = 534 \text{ in.}$$

$$R_B = \sqrt{\frac{L_e d}{b^2}} = \sqrt{\frac{534 \times 21.5}{9.5 \times 9.5}} = 11.3$$

Since 11.3 is more than 10 but less than 27.8, this beam is *intermediate*. Therefore, the allowable bending stress must be modified. However, since R_B is so close to 10, we should assume that C_L will be close to 1.0, say $C_L = 0.95$. As a quick check, the ratio of actual bending stress to required bending stress can be found from the ratio of the section moduli.

$$\frac{S_{\text{req'd}}}{S_{\text{actual}}} = \frac{640}{732} = 0.87$$

If our guess about C_L is right, the 10×22 will work. Let us proceed with the calculation. The equations we will need are

$$C_L = \frac{1 + F_{bE}/F_b^*}{1.9}$$

$$- \sqrt{\left(\frac{1 + F_{bE}/F_b^*}{1.9}\right)^2 - \frac{F_{bE}/F_b^*}{0.95}}$$

$$F_b^* = C_D \times C_M \times C_t \times C_F \times C_c \times C_f \times F_b$$

(use all applicable factors except C_V, C_{fu}, and C_L)

$$C_F = \left(\frac{12}{d}\right)^{1/9} = \left(\frac{12}{21.5}\right)^{1/9} = 0.937$$

$$F_b^* = 0.937 \times 1350 = 1265 \text{ psi}$$

$$F_{bE} = \frac{K_{bE} E'}{R_B^2} \quad \text{where } K_{bE} = 0.438 \text{ for visually graded lumber}$$

$$R_B = \sqrt{\frac{L_e d}{b^2}}$$

Applying these formulas in the proper order we have

$$R_B = \sqrt{\frac{L_e d}{b^2}} = \sqrt{\frac{534 \times 21.5}{9.5 \times 9.5}} = 11.3$$

$$E' = C_M \times C_t \times E$$

$$= 1.0 \times 1.0 \times 1{,}500{,}000 \text{ psi}$$

$$= 1{,}500{,}000 \text{ psi}$$

$$F_{bE} = \frac{K_{bE} E'}{R_B^2} = \frac{0.438 \times 1{,}500{,}000}{11.3^2}$$

$$= 5145 \text{ psi}$$

$$\frac{F_{bE}}{F_b^*} = \frac{5145}{1265} = 4.067$$

$$C_L = \frac{1 + F_{bE}/F_b^*}{1.9}$$
$$- \sqrt{\left(\frac{1 + F_{bE}/F_b^*}{1.9}\right)^2 - \frac{F_{bE}/F_b^*}{0.95}}$$

$$= \frac{1 + 4.067}{1.9} - \sqrt{\left(\frac{1 + 4.067}{1.9}\right)^2 - \frac{4.067}{0.95}}$$

$$= 0.984$$

$$F_b' = C_L F_b' = 0.984 \times 1265 = 1245 \text{ psi}$$

$$M = F_b' \times S = 1245 \times \frac{732}{12}$$

$$= 75.9 \text{ ft-kips}$$

This moment capacity is greater than the applied moment of 72.0 ft-kips, so the beam as designed is satisfactory. It is rare that we encounter a beam that is laterally unsupported, rarer still when the beam is slender or intermediate to any degree.

As an aid to the designer, Tables B.8a and B.8b list nine common size beams and their allowable uniform load for a range of spans. In these two tables it is assumed that there is no lateral bracing along the compression side. Be careful in using these tables—they do *not* apply to joists that are laterally braced by the floor deck. In addition, you must be aware that Tables B.8a and B.8b show the allowable load based on *bending moments only*. The designer will have to verify the ability of the beam to withstand applied shear loads and to determine the deflection.

There is also a series of tables in the Appendix (Tables B.9a through B.9e) which incorporate all the calculations in this section. With the tables there is an explanation of their use. Where derived values are not listed exactly in the tables, interpolation may be made between tabulated values. Because of the sensitivity of the calculations, approximate linear interpolations are satisfactory. The current example, as derived using the tables, results in a C_L value of 0.986. Compared with the inaccuracies associated with wood design, the two derivations for C_L are identical.

For sawn lumber the size factor, C_F, and the lateral stability factor C_L, are cumulative. This has changed in the 1991 revisions. For glued-laminated timbers the beam lateral stability factor, C_L, and the volume factor, C_V, are not used simultaneously. The lesser factor applies.

NOTCHES

Notches in bending members diminish the strength of the member. The severity of the effect depends on the location of the notch, its length, and its depth. In an ideal world we could say, "notches should be avoided wherever possible," but in the real world, plumbers and other trades are constantly cutting holes and notches in beams. A little knowledge about the forces in the beam will help tremendously in answering the tradesperson with saw in hand. Notches affect wood beams in two ways; one is purely in the removal of wood. The other is in a concentration of stresses at the tip of the crack. Anyone who has seen a small crack in a plaster wall grow has experienced the effect of stress concentrations. Stress concentrations are worst when they align with the weak direction of the wood —tension across the grain. This type of tension occurs anywhere a member is subjected to horizontal shear. Beams loaded on one face have maximum shear at the supports.

For beams notched on the tension side at the ends, the actual shear stress at the end is calculated as

$$f_v = \frac{3V}{2bd_n} \frac{d}{d_n}$$

where d is the depth of the unnotched member, d_n the depth that is left after the section is notched, and V the shear force at the end of the member (see Figure 2.9a). This equation represents both effects of the notch: the first term shows the decrease due to loss of wood. The second term shows the loss due to the stress concentration. The NDS allows the second term to be decreased if the notch is gradually eased where it reenters the beam. In reality, the architect or engineer has no control over how this cut would be made in the field, so that we very rarely consider this increase. The builder would be wise to take extra care to cope the notch or drill the radius before cutting the notch on heavily loaded beams. Examples are shown in Figure 2.9b through d.

It can be seen that if a notch of one-fourth the depth is cut in a piece, the actual shear stress in the beam will increase by 77%. Or, stated conversely, its ability to withstand shear will be reduced to 56% of its original strength. For a notch of one-fourth of d, then d_n equals $\frac{3}{4}$ of d, so that

$$\text{actual shear stress} = f_v = \frac{3V}{2bd_n}\frac{d}{d_n}$$

$$= \frac{3V}{2b(0.75d)}\frac{d}{0.75d}$$

$$= \frac{3V}{2bd}\left(\frac{1}{0.5625}\right)$$

$$= 1.77\left(\frac{3V}{2bd}\right)$$

where $3V/2bd = f_v$ of the full section. Then

$$\text{allowable shear stress, } F_v' = \frac{2bd_n}{3V}\frac{d_n}{d}$$

$$= \frac{2bd}{3V} \times 0.5625$$

For this reason the NDS states that where members are notched for bearing support

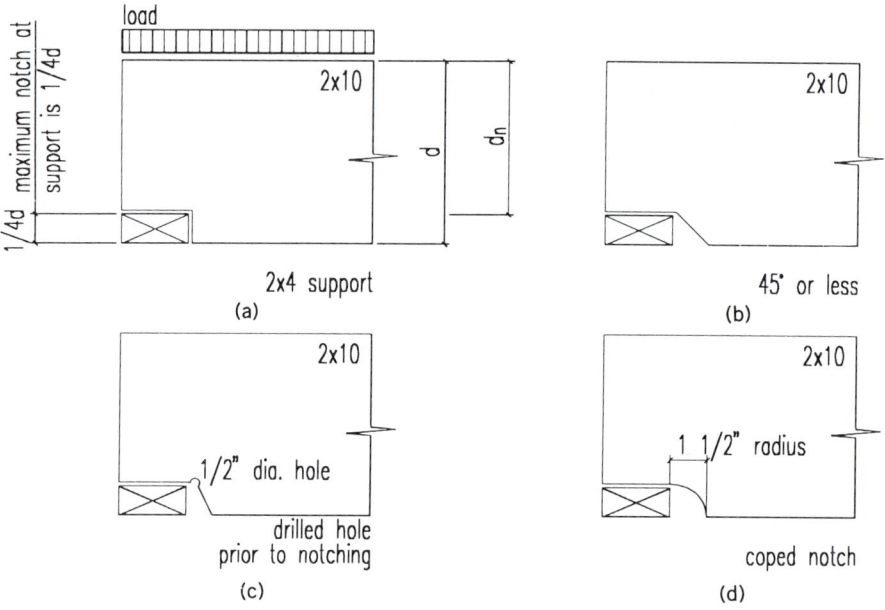

FIGURE 2.9 Notices at ends of beams.

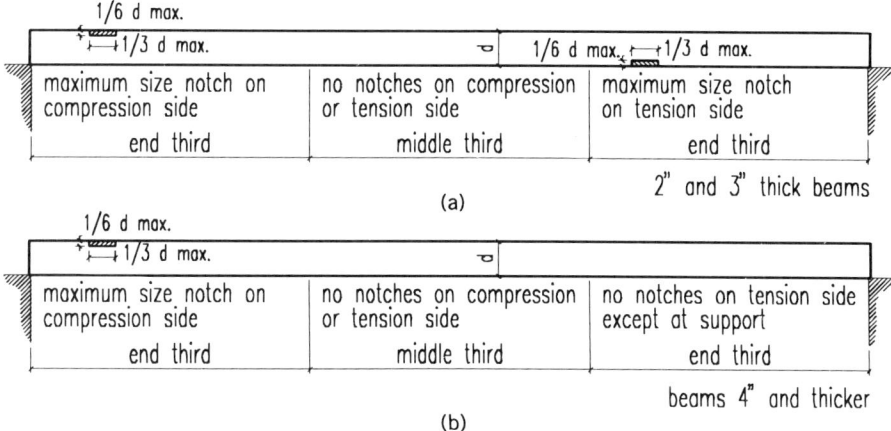

FIGURE 2.10 Allowable notches along a beam: (a) 2″ and 3″ thick beams and (b) beams 4″ and thicker.

at their ends, the notch depth shall not exceed $\frac{1}{4}$ of the beam depth.

Away from the ends a beam is subjected to greater bending stresses, so the effect of notching on bending is considered. The NDS requires that *no* notches be placed in the middle third of a beam (see Figure 2.10). Between the ends and the middle third, the NDS recognizes that the effect of notches on the tension side is much more severe than on the compression side. The NDS requires that no notches be placed in the tension side of any beam or joist thicker than 4 in. nominal. Notches on the compression side are limited to $\frac{1}{6}$ the depth and a length of $\frac{1}{3}$ the depth. For 2-in. or 3-in. thick joists, notches are excluded in the middle third. Away from the ends, both tension face and compression face notches are allowed, but they should be no deeper than $\frac{1}{6}$ the depth and no longer than $\frac{1}{3}$ the depth.

EXAMPLE 2.9—END NOTCHES

Consider the double joists that carried the parallel partition in Example 2.3. If these joists had the maximum notch allowed by NDS on the tension side at the support, would they be satisfactory?

The design load was calculated to be $w = 82 + 50 + 22 = 154$ plf. Then $V = wL/2 = (154 \times 13.71)/2 = 1055$ lb. The shear design strength was taken as $F_V = 95$ psi. Two 2×10's were used, so $b = 3$ in. and $d = 9.25$ in. The maximum depth of the notch $= d/4 = 2\frac{5}{16}$ in., so $d_n = 9\frac{1}{4} - 2\frac{5}{16} = 6\frac{15}{16}$; use $6\frac{7}{8}$ in.

$$f_v = \frac{3V}{2bd_n}\frac{d}{d_n} = \frac{3 \times 1055}{2 \times 3 \times 6.875}\left(\frac{9.25}{6.875}\right)$$
$$= 103.2 \text{ psi}$$
$$f_v = 103.2 \text{ psi} > 95 \text{ psi}$$

Therefore, the notch is too deep. Select a smaller notch. The calculations are left as an exercise. The minimum dimension of d_n is slightly more than $7\frac{1}{8}$ in.; say, the minimum $d_n = 7\frac{1}{4}$ in.

SPECIAL SHEAR DESIGN

Shear often controls for short, heavily loaded beams. When shear becomes a major factor in design, there are a number of factors that should be considered. These

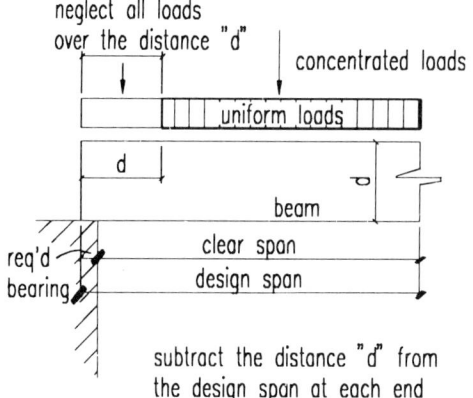

FIGURE 2.11 Shear at ends of beams.

factors will result in a more accurate design for shear. In some cases they will result in a more conservative design and are therefore mandatory.

1. For all beams that are loaded on one face and are fully supported in bearing on the other, any loads within a distance from the support equal to the depth of the beam may be ignored (see Figure 2.11). For a simply supported beam with a uniform load, this results in a design shear less than the maximum shear at the ends. The design shear is given by the equation

$$V_{\text{design}} = \frac{wL}{2} - wd$$

The length L in this equation is the design span, which is the clear span plus half the bearing length from each side (see Figure 2.9). Any concentrated loads within the distance d are also ignored. This consideration is based on the fact that shear is transferred through the beam on a 45° angle; any force applied within the distance d will pass directly into the support and not affect the shear capacity of the beam.

2. For joists and beams with no splits, shakes, or checks on the ends, the design shear value F_V may be increased by a shear stress factor C_H of 2.0. Where shakes, split, and checks exist but are limited in size, the shear stress factor varies from 1.0 to 2.0 as shown in Tables A.7 or A.12. The basis of these factors is the assumption that all joists and beams have checks, splits, or shakes at the end. The tabulated design values for shear have been reduced by a factor of 2.0 to account for these defects. See Figure 2.12 for the definition and measurement of these defects. There is no shear stress factor, C_H, for decking or glued laminated timbers.

If a large shear occurs away from the end of the beam, and the distance to the end exceeds five times the depth of the beam, C_H is taken as 2.0. This condition occurs in beams that are continuous over internal supports. Since shear controls the design in only a few cases, it is reasonable to assume that the designer can specify the allowable checks or splits on the end of a limited number of pieces. We use this approach, for instance, in specifying built-up girders in floor systems. Quite often when the floor girder will be made of four or five 2×10's, a designer can specify that the material for the girder be selected with no checks. This may result in fewer members in the girder.

EXAMPLE 2.10—SHEAR DESIGN

Refer to Figure 2.13 which shows the plan of the floor structure used in Examples 2.1 through 2.4. In this plan let us assume that we require a 6'-0"-wide door in one of the outer bearing walls below the first floor. This header must carry the load of the first floor as well as that of the second floor. In addition the weight of the first floor wall will bear on the header. To accomplish this we will design a wood header than can span the opening. Depending on headroom requirements, the header may fit above the window and below the floor

SPECIAL SHEAR DESIGN   41

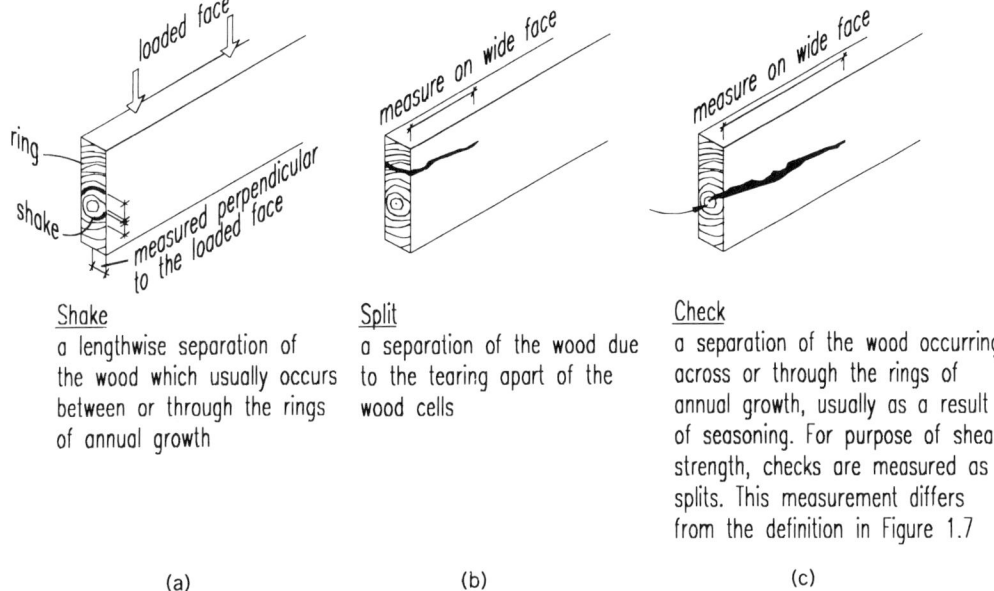

Shake
a lengthwise separation of the wood which usually occurs between or through the rings of annual growth

Split
a separation of the wood due to the tearing apart of the wood cells

Check
a separation of the wood occurring across or through the rings of annual growth, usually as a result of seasoning. For purpose of shear strength, checks are measured as splits. This measurement differs from the definition in Figure 1.7

(a)   (b)   (c)

FIGURE 2.12 (a) Shakes, (b) splits, and (c) checks.

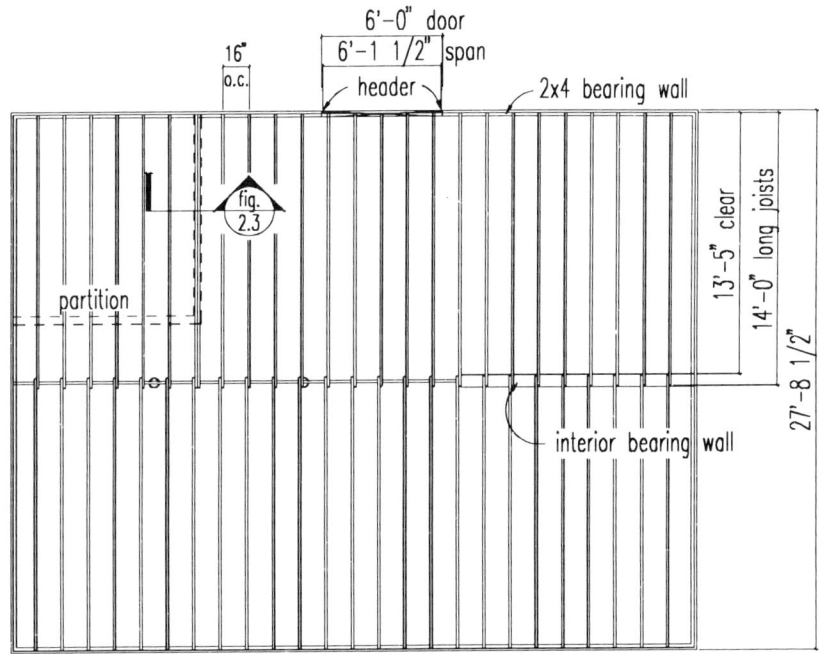

FIGURE 2.13 Floor plan—Example 2.10.

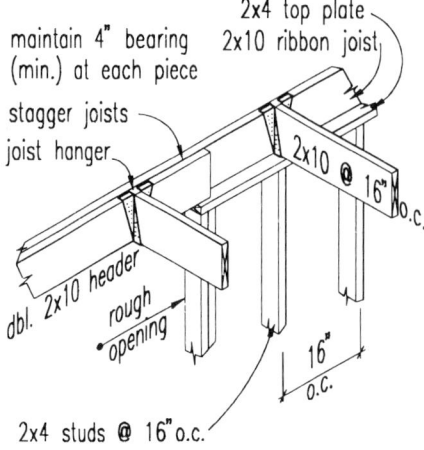

FIGURE 2.14 Header detail.

joists, or it might be necessary to make the header align with the floor system. In this example the joists will be attached to the header with joist hangers (see Figure 2.14). The standard practice is to use the same material for the header as for the joists. These were specified as 2×10 Douglas Fir–Larch No. 2. Let us assume that the rough opening for the door will be slightly wider than the 6-ft door, say $6' - 1\frac{1}{2}''$, and therefore the span of the header is $6' - 3''$ (opening plus half the bearing on each side). We will design for bending, shear, and deflections.

At each floor the load on the joists was calculated earlier as 40 psf live load and 11 psf dead load. The dead load included an allowance for the joists themselves, so this amount still applies. The total load is 51 psf. The joist span was $13' - 8\frac{1}{2}''$. The total load on the header is $w = 51$ psf $\times$ 13.71 ft $\times \frac{1}{2} = 350$ plf. Although the joists act as point loads at 16-in. spacing, it is acceptable to consider them as a uniform load. Whenever there are five or more equal loads equally spaced, they can be treated as a uniform load.

The weight of the wall can be assumed to be 10 psf. For a wall height of 9 ft, the total wall load is 10 psf $\times$ 9 ft or 90 plf.

The total weight on the header is 350 plf + 350 plf + 90 plf = 790 plf.

$$V = \frac{wL}{2} = \frac{790 \text{ plf} \times 6.25 \text{ ft}}{2} = 2470 \text{ lb}$$

The required area to resist the shear is

$$A_{req'd} = \frac{3}{2} \times \frac{V}{F_V} = 1.5 \times \frac{2470 \text{ lb}}{95 \text{ psi}}$$

$$= 39.0 \text{ in}^2$$

Each 2×10 has an area of 13.875 in.^{2}, so the number of 2×10's needed in the header is

$$N = \frac{A_{req'd}}{A_{2 \times 10}} = \frac{39.0}{13.875} = 2.8 \quad \text{use 3}$$

If the support wall is constructed of 2×4 studs, its structural width is $3\frac{1}{2}$ in. Three 2×10's will not fit in this width. We may consider whether it is possible to use any of the special shear considerations to minimize the number of 2×10's in the header. But first we should check bending and deflection criteria to see if two 2×10's are feasible.

For bending,

$$M = \frac{wL^2}{8} = \frac{790 \times (6.25 \text{ ft})^2 \times 12 \text{ in/ft}}{8}$$

$$= 46{,}290 \text{ in-lb}$$

The required section modulus is

$$S_{req'd} = \frac{M}{F_b} = \frac{46{,}290 \text{ in.-lb}}{1107 \text{ psi}} = 41.8 \text{ in}^3$$

Each 2×10 has an S value of 21.39 in.^{3}, so the required number to resist bending is

$$N = \frac{41.8}{21.39} = 1.95 \quad \text{use 2}$$

Checking deflection for two 2 × 10's yields

$$I = (2 \times 98.932 \text{ in}^2) = 197.8 \text{ in}^4$$

$$\Delta = \frac{5wL^4}{384EI} = \frac{5 \times 790 \times 6.25^4 \times 1728}{384 \times 1,600,000 \times 197.8}$$

$$= 0.086 \text{ in.}$$

The relative deflection is

$$\frac{\Delta}{L} = \frac{0.086}{6.25 \times 12} = \frac{1}{870}$$

This clearly meets the deflection criterion of $L/360$.

Since two 2 × 10's will satisfy the requirements for bending and deflection, let us delve a little deeper into the shear requirements. There are two considerations that we might look at: one is the location of the load near the end and the other is the presence of splits and checks at the end. As to the location of loads, the NDS allows us to neglect any load placed less than a distance d from the end of the header, where d equals the depth of the header, in this case $9\frac{1}{4}$ in. As a designer we have no way of knowing clearly where the joists will line up. As a builder, we are concerned about other, more important factors that dictate the location of the joists, such as maintaining the 4-ft module for the plywood. We can conclude that there is no reason for considering this further.

As to the location of splits and checks, we have more control. Good builders are already in the practice of selecting the better-looking pieces for special locations. By making the header of the same material as the joints we have assured that there is a large pile of 2 × 10's to choose from. Since the header is $6'-3''$, a joist with bad-looking ends could be cut to length with the bad ends removed. So it seems reasonable that we may apply this criterion to selected pieces such as headers and girders and expect that it will be built as speicfied.

The design procedure is to determine the maximum allowable splits and checks using two 2 × 10's. For two 2 × 10's the requried shear stress will be

$$F_V = \frac{3}{2} \times \frac{V}{A} = 1.5 \times \frac{2470 \text{ lb}}{2 \times 13.875 \text{ in}^2}$$

$$= 133.5 \text{ psi}$$

This represents a required increase in shear of

$$C_H = \frac{133.5}{95} = 1.4$$

Table A.7 relates the shear stress factor to the size of splits and shakes. An increase of 1.5, which slightly exceeds our needs, can be used if the splits on the wide face are no more than $\frac{3}{4}$ times the width of the wide face, namely $7\frac{1}{2}$ in. long and the size of the shake is no greater than $\frac{1}{4}$ of the narrow face, namely $\frac{1}{2}$ in. To arrive at these sizes use the nominal lumber dimensions. These requirements are shown in Figure 2.15. The complete specification for the header is

Two 2 × 10 Douglas Fir–Larch No. 2 header; limit splits at end to $7\frac{1}{2}$ in. long and shakes on ends to $\frac{1}{2}$ in.

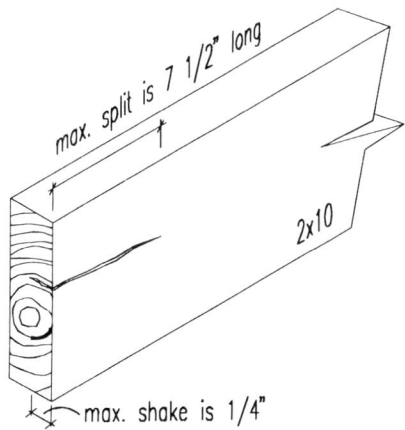

FIGURE 2.15 Maximum allowable checks and splits for $C_D = 1.50$.

44 BEAM DESIGN

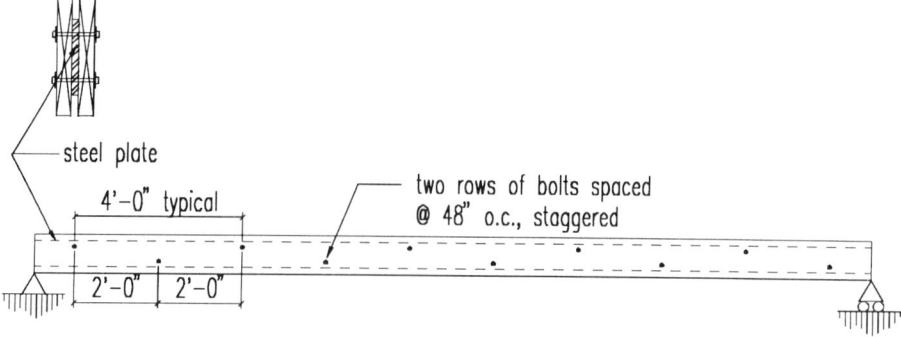

FIGURE 2.16 Flitch beam bolting pattern.

FLITCH BEAM

Under certain conditions it is either impossible or impractical to use a solid timber beam or even to use multiple pieces joined together to form a beam. In such cases a sandwich composed of two exterior pieces of wood together with an internal steel plate is employed. Such a composite built-up beam is called a flitch beam. Generally, the three pieces are bolted together, with the bolts inserted in two parallel horizontal lines, one near the top side of the beam, and the other, of course, near the bottom. The bolts are spaced about 48 in. on centers on each line, with spacing staggered from line to line so that the spacing from bolt to bolt in a horizontal projection is 24 in. (see Figure 2.16).

As to the mathematical analysis of such a built-up beam, the complication of using two different materials can be resolved by theoretically transforming one material into the other. This is an imaginary process that does not actually happen but is a useful device for analysis. It is accomplished by using the ratio of the moduli of elasticity, with the larger divided by the smaller. Call this ratio n. Thus, in our case, $n = E_s/E_w$, where E_s is the modulus of elasticity of steel and E_w is the modulus of elasticity of timber or wood.

EXAMPLE 2.11—FLITCH BEAM

As an example, suppose that we have a clear (rough-to-rough) opening in the first-floor exterior wall of a two-story house of $10'-0''$. The actual *lintel* (a beam placed over an opening in a wall) will be $10'-8''$, allowing 4 inches of bearing on each end. Since the design span is measured from center to center of bearing, the length to be used in computations is not the actual length but will be $L = 10'-4''$. Where we are dealing with a window opening, the deflection must be limited to some value that will not damage the glass or interfere with operation of the window. Suppose in this case that the manufacturer of the window recommends an absolute maximum deflection of $\Delta_{max} = \frac{3}{16}$ in. = 0.188″. Note that if the usual criterion of $L/30$ were used, we would be permitted a deflection of $10.33/30 = 0.344$ in., but we must use 0.188 in. as the maximum. Let us further suppose that we have selected 2 × 10's of Douglas Fir–Larch No. 1 with the properties $F_b = 1.1 \times 1000$ psi = 1100 psi, $F_v = 95$ psi, and $E = 1,700,000$ psi. Since this beam is in an exterior wall of a two-story house, it must carry the wall above it, the portion of the roof resting on that wall, and the load placed on it by the second floor framing into it. Let us suppose that the total of all this is 620 lb/linear foot along the beam.

$$M = \frac{wL^2}{8} = \frac{620 \times 10.33 \times 10.33}{8}$$
$$= 8275 \text{ ft-lb}$$
$$S = \frac{M}{F_b} = \frac{8275 \times 12}{1100} = 90.3 \text{ in}^3$$

At this point, considering only the requirement for section modulus, we would need five 2×10's, or three 2×12's, or two 2×14's. A 2×14 is not commonly stocked in a lumberyard, and a special order would greatly slow down the job. If three 2×12's are used, the combined breadth would be 3×1.5 in. wide = 4.5 in. wide, which would not fit in the nominal 4-in. wall, which is actually only 3.5 in. wide. Therefore, try a flitch beam composed of two 2×10's with a steel plate $\frac{1}{2}$ in. thick $\times$ 9 in. deep.

The computations start by transforming the steel into an equivalent piece of wood and then considering the transformed beam to be one solid beam with a strange shape, somewhat akin to the letter I on its side (see Figure 2.17).

$$n = \frac{E_s}{E_w} = \frac{29{,}000{,}000}{1{,}700{,}000}$$

(The modulus of elasticity of steel is always 29,000,000 psi, for all strengths of structural steel.) Then $n = 17.05$, say 17.

The simplest procedure is to transform the width of the steel plate by converting it into an equivalent width of wood, while retaining its depth. Then the new width becomes $nb = 17 \times \frac{1}{2} = 8.5$ in. The necessary geometric properties are now computed for the transformed beam.

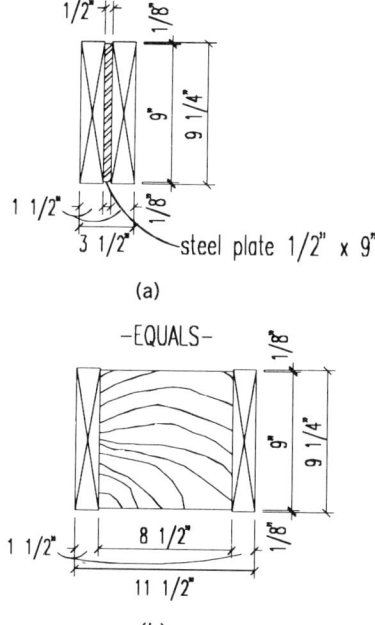

FIGURE 2.17 Transform section: (a) original and (b) transformed.

Item	Area	$I = bh^3/12$
2×10	13.9	98.9
8.5×9	76.5	516.4
2×10	13.9	98.9
	104.3	714.2

Since the overall height or depth of the beam is the depth of a 2×10, which is 9.25″ in., half the height is 4.625 in., and the actual section modulus is

$$S = \frac{I}{c} = \frac{714.2}{4.625} = 154.4 \text{ in}^3 > 76.4$$

$$\text{required } S \quad \text{OK}$$

$$V = \frac{wL}{2} = \frac{620 \times 10.33}{2} = 3202 \text{ lb}$$

The actual shear stress is

$$f_v = \frac{3}{2} \times \frac{V}{A} = \frac{1.5 \times 3202}{137.8} = 46.0 \text{ psi} < 95$$

$$\text{allowable } F_V \quad \text{OK}$$

The actual deflection is found by using the total load, not just the live load in this case, to avoid binding the window. Use the *EI* value for the transformed section.

$$\Delta = \frac{5wL^4}{384EI}$$

$$= \frac{5 \times 620 \times 10.33^4 \times 1728}{384 \times 1{,}700{,}000 \times 714.2}$$

$$= 0.130 \text{ in., } < 0.188 \text{ in.} \qquad \text{OK}$$

The design is accepted.

BUILT-UP BEAMS

Instead of using a steel plate between two pieces of lumber it is possible to build up a beam using several pieces of wood alone. In such cases the usual design employs one or two vertical pieces, called the *web*, and one horizontal piece at the top with another horizontal piece at the bottom, called the *flanges*. The web can be either solid lumber or plywood, with the latter being the most common. An approximate method of design, which gives eminently satisfactory results, will now be explained. See Figure 2.18 for two examples.

The depth of a girder, whether it be of timber or steel, will generally be about one-tenth of the span in order to get the most economical balance of material for the flanges and the web. If the beam is too deep, the flanges will become quite small, with a rather stiff web. The web thickness is needed to prevent the beam from becoming too wobbly, and is wasted as far as strength is concerned. If the beam is too shallow, the flanges will become excessively large. The most economical design is a balance between these two extremes with a depth of one-eighth to one-twelfth of the span, although shallower beams are at times good solutions, with ratios up to 1/22 having been used successfully. The

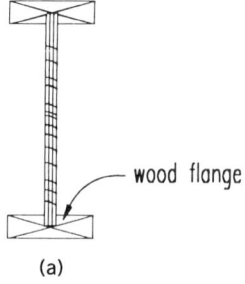

(a)

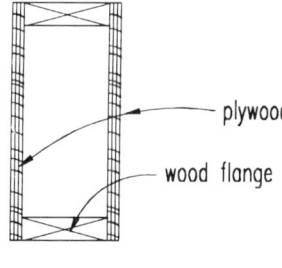

(b)

FIGURE 2.18 Built-up beam: (a) plywood 1-beam and (b) plywood box beam.

depth should be chosen so that the beam may be cut with minimum waste from the standard 4- by 8-ft sheet of plywood.

The total plywood *web thickness* required, t, can be estimated from the formula

$$t = \frac{5V}{4hf_v}$$

where V is the maximum shear in pounds, h the beam depth in inches, and f_v the allowable shear stress in psi through the thickness of the plywood for the grade used. This will be 190 psi for type S-2 using species group 1, and 140 psi for species groups 2 or 3. These values are given in Table 7.6. For a more complete discussion of the strengths of plywood, and for the grading designations, see Chapter 7.

If a box-beam section with two webs is used, each web should be $t/2$ inches thick. The top and bottom flanges will be made of lumber. The required *size of these flanges* can be estimated from the formula

$$A = \frac{M}{F_c h_l}$$

where A is the net area in square inches of *each* lumber flange, M the bending moment in inch-pounds, F_c the allowable lumber compression stress in psi, and h_l the center-to-center distance between flanges in inches. Theoretically, the tension flange could be smaller than the compression flange, but they are both made the same for overall economy and to circumvent the effect of errors in erection.

To have sufficient contact area between the flanges and web if the surfaces are glued, the flange depth should be at least four times the adjoining plywood thickness. If only nailing is used, a greater depth may be required to avoid splitting. After the required area and depth are found, the width can be determined. The flange may be selected as either a single piece or a number of pieces of lumber.

For a box beam that is not glued but rather is nailed only, the *nail spacing* on *each side* of the flange-to-web connection can be found from the formula

$$p = \frac{2rh_l}{V}$$

where p is the nail spacing or pitch in inches, r the allowable nail load for lateral bearing in pounds, and h_l and V are as defined previously. The nail spacing will be closest at the ends where the shear is highest, and the nails may be spaced farther apart toward the midspan in proportion to the shear.

Whenever the beam is longer than 8 ft it will be necessary to *splice the plywood webs*. The best way of doing this is to use glued plywood splice plates at least 6 in. wide on both the inner and outer surfaces of each web. This will permit the web to resist bending as well as shear, even though bending resistance of the web is neglected in the design method presented here. If external splice plates are objectionable because of their appearance, only the glued internal 6-in. plywood splice plates should be used, but solidly backed with a lumber stiffener that is glued or nailed in place.

Vertical lumber *stiffeners* should be placed over the support and at points of concentrated load. Moreover, intermediate stiffeners should be placed near the ends of the beams where the shear stress is high in order to reinforce the webs against buckling. The webs should be glued or well nailed to all stiffeners, and the stiffeners should fit snugly against both flanges. Intermediate stiffener spacing depends on the plywood thickness and the clear distance between flanges as well as the ratio of the actual stress in the plywood to its allowable value.

Generally, *deflection* will not be a problem because deflection of beams with a depth of at least $\frac{1}{12}$ of the span is considerably less than the usual allowable of $\frac{1}{30}$ of the span, especially if it is computed on the basis of live load only. Deflection may be computed on the basis of the standard deflection formulas, but the computed value must be increased 25% to account for shear deformation. The bending-deflection may be multiplied by the following shear-deflection factors to the arrive at an approximation of the total deflection:

Span/Depth Ratio	Shear-Deflection Factor
10	1.5
15	1.2
20	1.0

EXAMPLE 2.12—PLYWOOD BEAM DESIGN

As an example of this design method, suppose that we have a box beam with a 30-ft span with a uniform load of 400 pounds per lineal foot (400 plf). Then

$$M_{max} = \frac{wL^2}{8} = \frac{400 \times 30 \times 30}{8}$$
$$= 45{,}000 \text{ lb-ft}$$
$$V_{max} = \frac{wL}{2} = \frac{400 \times 30}{2} = 6000 \text{ lb}$$

If the depth is, say, $\frac{1}{10}$ of the span, the depth would be $30 \times \frac{12}{10} = 36$ in. Selecting a higher value for the allowable shear stress in the plywood, let $f_v = 190$ psi. To find the *web thickness*,

$$t = \frac{5V}{4hf_v} = \frac{5 \times 6000}{4 \times 36 \times 190} = 1.10 \text{ in.}$$

The stock thickness most nearly meeting this requirement is *two* webs of $\frac{5}{8}$ in. each, giving a total thickness t of $2 \times 0.625 = 1.25$ in.

If other factors are equal, the most economical of the various grades of plywood should be selected. For stress level S-2 plywood, B, C-plugged or D grades are allowed. The B-C and B-B face grades are more expensive than the C-C or C-D face grades and would be chosen only if appearance were important. In fact, for even better appearance, one could select a plywood having an A face on the outside, but this is quite expensive, relatively speaking. It would, though, have a higher stress level (S-1) but the same allowable shear stress. Since it would not lead to a thinner web, there would be no savings in selecting group A. Interior grades are satisfactory for most interior or protected uses. Thus the interior C-D grade will be selected.

Assume that the lumber flanges will have a depth of 4 in. Then the vertical distance from the center of one flange to the center of the other flange will be

$$h_1 = 36 - \left(\tfrac{1}{2} \times 4\right) - \left(\tfrac{1}{2} \times 4\right) = 32 \text{ in.}$$

For lumber with allowable compression parallel to the grain of 1200 psi, the *flange area* required is

$$A = \frac{M}{f_c h_1} = \frac{45{,}000 \times 12}{1200 \times 32} = 14.0 \text{ in}^2$$

Select for each flange two 2×4's with two 1×4's placed between them, giving a gross area of about 16 in.^{2} (see Figure 2.19). The reason for using several pieces instead of just one or two is that they will have to be spliced along their length, and it is necessary to stagger the splices. This arrangement provides a good distribution of flange splices in the central portion of the span. Place the 4-in. faces vertically so as to be parallel with the plywood webs, and arrange butt joints in the 2×4's so that they occur outside the quarter points of the span. The 1×4's may be butt jointed at any convenient place, but all joints, including those of the web, should be staggered at least 24 in. Plywood webs will be spliced every 8 ft with a $\frac{5}{8}$ by 6-in. splice plate glued to the inner side of each web.

The beam will be stiffer and probably stronger if nail-glued, but the same number of nails will be required to set the glue as if it were nailed only. The shear being a maximum at the ends, the nail spacing will be computed at that location. Assuming 16d common nails will be used with an allowable load of 160 lb each, we have

$$p = \frac{2rh_1}{V} = \frac{2 \times 160 \times 32}{6000} = 1.71 \text{ in.}$$

For ease of fabrication it might be best to space the nails only 3 in. apart along the length but staggered in two lines to lessen splitting, and driven into each side, top, and bottom. The spacing can be increased toward the midspan but should not exceed about 6 in. Thus at 5 ft from the end the shear is 4000 lb and the required spacing is 2.57 in. (use $2\tfrac{1}{2}$ in.), and at 10 feet from the end the required spacing is only 5.13 in., so use 5 in. for the central 10 ft of the beam. It is best *never* to use decimals of an inch when giving instructions to the field. If the beam is not glued, the upper flange pieces must be restrained from buckling laterally by using $\frac{3}{8}$-in. bolts placed every 24 in. on the sides, along the length.

Now let us consider the interior stiffeners. Bearing stiffeners over reactions and where other heavy concentrated loads occur are needed to distribute the loads into the beam. Interior stiffeners must be of the same width as the flanges. The total thickness of the flange is $1.5 + 1.5 + 0.75 + 0.75 = 4\frac{1}{2}$ in. They should be at least six times as wide (measured along the length

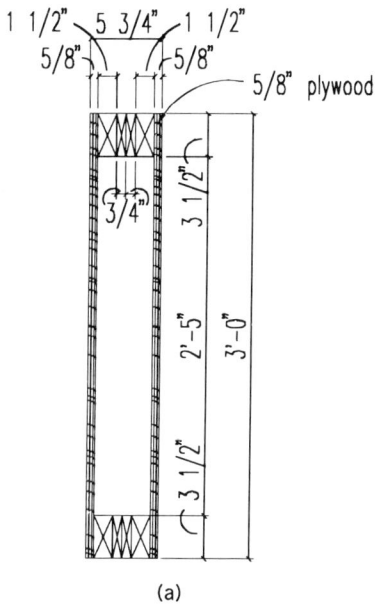

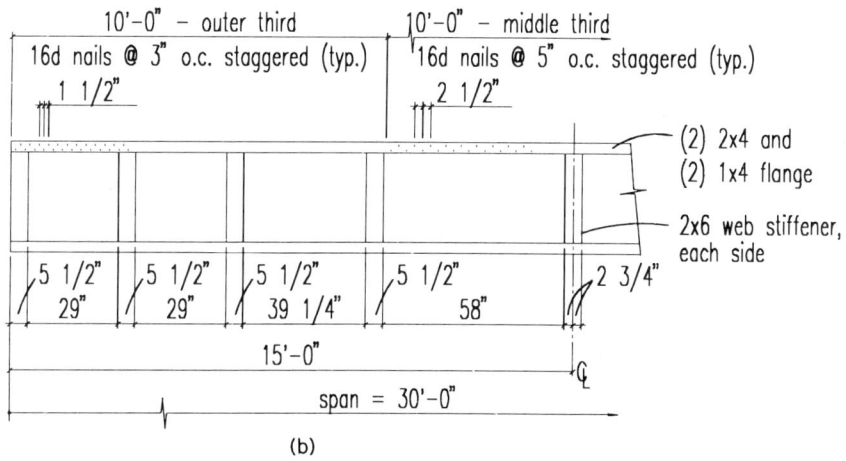

FIGURE 2.19 Built-up beam—details: (a) section and (b) elevation.

of the beam) as web thickness, in this case $6 \times \frac{5}{8}$ in. $3\frac{3}{4}$ in. Make them $4\frac{1}{2}$ in. $\times 5\frac{1}{2}$ in. Since there are no stock sizes to provide this, the stiffeners will be ripped from a 6×6. They will be set over the supports and at intermediate points. The clear distance between stiffeners should be equal to or less than two times the distance between flanges. Since they are to stiffen the webs as they resist shear, we will use a closer spacing at the ends of one times the distance between flanges, 29 in. This is the *clear* distance from the end stiffener to the first intermediate stiffener, so the 2-in. width of stiffener permits the first intermediate stiffener to be placed 31 in. from the end. Thereafter the spacing may be increased from b up to $2b$ as the shear stress decreases from full to one-half the allowable value. Thus the clear spacing of the rest of the stiffeners can be increased uniformly up to 2 times 28 in., or 56 in. (4'-8") at midspan.

As the beam has a depth-to-thickness ratio of about 7:1 it will require good lateral bracing. The provisions for bracing are based on the ratio of the moments of inertia in each direction, I_X/I_Y. I_X is the moment of inertia in the strong direction, I_Y is taken in the weak direction. For a box beam with equal flange depths, the moment of inertia may be found as the I of the outer rectangle minus the I of the inner rectangle. In our example we have

$$I_X = \frac{5.75 \times 36^3}{12} - \frac{4.5 \times 29^3}{12}$$
$$= 22{,}356 - 9146 = 13{,}210 \text{ in}^4$$
$$I_Y = \frac{5.75^3 \times 36}{12} - \frac{4.5^3 \times 29}{12}$$
$$= 570 - 220 = 350 \text{ in}^4$$
$$\frac{I_X}{I_Y} = \frac{13{,}210 \text{ in}^4}{350 \text{ in}^4} = 37.7$$

This beam will need bridging or bracing at intervals of not more than 8 ft. See Table 2.4, which lists the APA lateral bracing provisions. Figure 2.19 details the beam as designed.

As an aid to selecting trial member sizes, Table B.10 in the Appendix has been reproduced from the American Plywood Association, Plywood Design Specification, Supplement 2. This table gives maximum moments for various sizes of plywood beams. In our example the required moment was 45,000 lb-ft. Table B.10 shows that a 36-in.-deep plywood beam with three 2×4 flanges has a capacity of 36,185 lb-ft. The footnotes to this table allow increases based on increased strength of materials. Incorporating these, we arrive at an allowable moment of $1.2 \times 32{,}842 + 1.51 \times 3343 = 44{,}460$ lb-ft. This answer is within 1% of the more exact answer that we derived.

TABLE 2.4 PLYWOOD-LUMBER BEAM LATERAL BRACING PROVISIONS

I_X/I_Y	Provision for Lateral Bracing
Up to 5	None required
5–10	Ends held in position at bottom flanges at support
10–20	Beams held in line at ends (both top and bottom flanges restrained from horizontal movement in planes perpendicular to beam axis)
20–30	One edge (either top or bottom) held in line
30–40	Beam restrained by bridging or other bracing at intervals of not more than 8 ft
More than 40	Compression flange fully restrained and forced to deflect in a vertical plane, as with a well-fastened joist and sheathing or stressed-skin panel sheathing

Source: APA PDS Supplement 2, *Design and Fabrication of Plywood Lumber Beams.*

Obviously, use of this table cannot replace a complete engineered design. As can be seen from this example, the design of a plywood beam is rather complex and includes a number of decisions. It takes some experience with the process to develop confidence in one's results. However, as a preliminary design, Table B.10 can provide a quick means of accurately estimating the size of a plywood beam.

There are available tables that are based on a single plywood web with lumber flanges. Such members are intended to be used as joists, and the use of such tables is similar to those discussed earlier under joist design. Examples from typical manufacturers are shown in Tables B.6 and B.7.

GLUED-LAMINATED BEAMS

Large heavy timbers are getting rather scarce and are certainly expensive. It is possible to build up a beam of substantial size by placing enough small pieces of wood next to each other until the desired size is obtained. These smaller pieces, with a nominal thickness of either 1 or 2 in., are glued together under pressure to form a solid beam that acts as a single unit. Such a beam is called a *glulam* (short for "glued-laminated") beam. The most common arrangement is to place the broad side of the individual pieces horizontally, laminating one piece on top of the next. Thus the total height or depth of such a beam will be some multiple of either $\frac{3}{4}$ in. or of $1\frac{1}{2}$. The width will have started as a nominal 3, 4, 6, and so on, up to 16 in. But after the glue has set, the sides of the built-up beam are sanded down so that the finished width will be $2\frac{1}{4}$, $3\frac{1}{8}$, $5\frac{1}{8}$, $6\frac{3}{4}$, $8\frac{3}{4}$, $10\frac{3}{4}$, $12\frac{1}{4}$, or $14\frac{1}{14}$ in.

When a beam is subjected to bending, the maximum stresses occur only in the top and bottom surfaces of that beam. The central portion is useful in transferring the bending stress from the compression on one edge to the tension on the opposite edge. It does this by means of horizontal shear stress. But in any event, the central portion can certainly be weaker than the outer portions of a beam subjected mainly to bending stress. With this in mind, the laminations are built up of different qualities of wood. Many combinations are possible, but certain of these have been standardized and are listed in the tables.

EXAMPLE 2.13—GLULAM BEAM

As an example, let us design a Southern Pine glulam beam designated by combination symbol "51." The beam is to support a uniform load of 200 plf, of which 165 plf is live load. Limit deflection to $L/30$. The beam will be laterally supported at 6-ft intervals along its 24-ft span.

From the tables, $F_b = 2100$ psi, $F_v = 175$ psi, and $E = 1,700,000$ psi. The design value for bending will have to be modified by the volume factor, C_V, or the lateral support factor, C_L. As with the sawn timber beam we must first establish a preliminary size and then check these factors since they are dependent on the beam size.

In beams such as this, the weight of the beam becomes an important consideration because it adds a considerable amount to the load the beam must carry. Based on experience, let us assume that the glued laminated beam will weigh a little more than 7% of the superimposed load. Since 7% of 200 plf is 14, let us assume a beam weight of 15 plf. Thus $w = 200 + 15 = 215$ plf.

$$M = \frac{wL^2}{8} = \frac{215 \times 24 \times 24}{8}$$
$$= 15{,}480 \text{ ft-lb}$$
$$V = \frac{wL}{2} = \frac{215 \times 24}{2} = 2580 \text{ lb}$$
$$\Delta_{max} = \frac{L}{30} = \frac{24}{30} = 0.80 \text{ in.}$$

$$S_{reqd} = \frac{M}{F_b} = \frac{15{,}480 \times 12}{2100} = 88.5 \text{ in}^3$$

$$A_{reqd} = \frac{3}{2} \times \frac{V}{F_v} = 1.5 \times \frac{2580}{175}$$

$$= 22.1 \text{ in}^2$$

$$I_{reqd} = \frac{5wL^4}{384E\Delta}$$

where w is the live load

$$= \frac{5 \times 165 \times 24^4 \times 1728}{384 \times 1{,}700{,}000 \times 0.80}$$

$$= 906 \text{ in}^4$$

Notice that in all computations so far the span used is the design span, not the unbraced length. Side braces do not prevent downward deflection, nor can they reduce bending or shear in the vertical plane.

To design the beam, first assume a breadth by selecting one of the standard widths. If the ensuing proportions are not satisfactory, a new guess is made on the width and the depth is again calculated. In this case, let us try a breadth $b = 3\frac{1}{8}$ in.

Now we calculate geometric properties that will be at least as great as those required, solving for the unknown depth d. For a solid rectangle, $S = bd^2/6$, and thus

$$S(\text{required}) = S(\text{provided})$$

$$88.5 = \frac{bd^2}{6} = \frac{3.125d^2}{6}$$

$$d^2 = \frac{6 \times 88.5}{3.125}$$

$$d = 13.0 \text{ in.}$$

$$A(\text{required}) = A(\text{provided})$$
$$22.1 = bd = 3.125d$$
$$d = 7.0 \text{ in.}$$
$$I(\text{required}) = I(\text{provided})$$

$$906 = \frac{bd^3}{12} = \frac{3.125d^3}{12}$$

$$d^3 = \frac{12 \times 906}{3.125}$$

$$d = 15.2 \text{ in.}$$

Since every one of the required geometric properties must be satisfied, and the largest of the values encompasses all of the others, the largest required value of d controls. Use $d = 15.2$ in. Each lamination will be either 1.5 or 0.75 in. high. For this straight beam select 1.5 in. The required number of layers will be

$$\frac{15.2}{1.5} = 10.13$$

Since only whole numbers are permissible, use 11. (If $\frac{3}{4}$-in.-thick pieces were used, we would need $15.2/0.75 = 20.3$; use 21.) The actual depth of the beam becomes $d = 11 \times 1.5 = 16.5$ in. The beam is obviously satisfactory for shear ($16.5 > 6.19$) and for deflection ($16.5 > 15.2$), but may not be acceptable for bending stress because it is not laterally supported along its entire length, and it is rather deep. Now let us check the beam for modification of allowable bending stress.

Start with the size factor C_F:

$$C_F = \left(\frac{12}{d}\right)^{1/9} = \left(\frac{12}{16.5}\right)^{1/9} = 0.965$$

Next consider the lateral support factor, C_L.

$$L_u = 6 \text{ ft} = 72 \text{ in.}$$

$$L_e = 1.63L_u + 3d$$

$$= (1.63 \times 72) + (3 \times 16.5)$$

$$= 117 + 49.5 = 166.5$$

$$R_B = \sqrt{\frac{L_e d}{b^2}} = \sqrt{\frac{166.5 \times 16.5}{3.125 \times 3.125}} = 16.8$$

Since $10 < 16.8 < 30$, we have an intermediate beam, and

$$C_F = 0.965$$
$$F_b^* = 0.965 \times 2100 = 2026.5 \text{ psi}$$

$$F_{bE} = \frac{K_{bE}E'}{R_b^2}$$

where $K_{bE} = 0.609$ for glulam, which has a COV_E value of $0.10 < 0.11$.

$$F_{bE} = \frac{0.609 \times 1,700,000}{16.8^2} = 3668 \text{ psi}$$

$$\frac{F_{bE}}{F_b^*} = \frac{3668}{2026.5} = 1.81$$

$$C_L = \frac{1 + F_{bE}/F_b^*}{1.9}$$
$$- \sqrt{\left(\frac{1 + F_{bE}/F_b^*}{1.9}\right)^2 - \frac{F_{bE}/F_b^*}{0.95}}$$

$$= \frac{1 + 1.81}{1.9}$$
$$- \sqrt{\left(\frac{1 + 1.81}{1.9}\right)^2 - \frac{1.81}{0.95}} = 0.948$$

Now calculate the volume factor from the equation

$$C_v = K_L \left(\frac{21}{L}\right)^{1/x} \times \left(\frac{12}{d}\right)^{1/x}$$
$$\times \left(\frac{5.125}{b}\right)^{1/x} \leq 1.0$$

where $x = 20$ for Southern Pine and 10 for all other species. K_L is defined as follows:

Type of Beam Span and Nature of Load	Value of K_L
Simple span, load concentrated at center	1.09
Simple span, uniformly distributed load	1.0
Simple span, two concentrated loads at $\frac{1}{3}$ points	0.96
Cantilever, all load conditions	1.0

$$C_v = 1.0 \times \left(\frac{21}{24}\right)^{1/20} \times \left(\frac{12}{16.5}\right)^{1/20}$$
$$\times \left(\frac{5.125}{3.125}\right)^{1/20}$$
$$= 0.993 \times 0.984 \times 1.025 = 1.002 \geq 1.0$$

Since C_L is less than C_V, C_L controls.

$$F_b' = C_L F_b = 0.948 \times 2026.5$$
$$= 1921 \text{ psi}$$
$$M = F_b' \times S$$
$$= \frac{1921 \times (3.125 \times 16.5^2/6)}{12}$$
$$= 22.7 \text{ ft-kips}$$
$$M_{\text{allowable}} = 22.7 \text{ ft-kips} > 15.48 \text{ ft-kips}$$

OK

The final design is a beam

$3\frac{1}{8}$ in. $\times$ $16\frac{1}{2}$ in. Southern Pine glulam, combination 51.

DESIGN CONSIDERATIONS

Wet Service Factor, C_M

All of the allowable stresses in the tables for Douglas Fir–Larch and most other species are based on the assumptions that the final usage will be in a fairly dry condition, with a moisture content in the wood of 19%. Where the designer knows that the in-use moisture content will exceed 19% for an extended period of time, the design strengths are reduced by applying wet service factors, C_M, which are found in Table A.6. For Southern Pine the C_M factor applies to dimension lumber (2 to 4 in. thick). For Southern Pine and Mixed Pine beams and stringers or posts and timber, the tabulated design values already include reductions based on wet use; do not use further increases for these species. *In-*

creases for dry use are not provided for. For example, if the member being designed is to be used in an outdoor location, exposed to the weather, or used exposed over an indoor swimming pool, its moisture content will become greater than 19%, and the wet service factors, C_M, must be used.

Load Duration Factor, C_D

The design values for all sizes, species, and grades are based on the assumption that the members are fully stressed to the design value by application of the full maximum load for 10 years. All design factors are modified to account for the actual load duration. The duration factors are shown in Table 1.4.

For combinations of loads with different durations, the duration factor of the shortest duration is used. Where more than one combination is considered, the controlling design load must be found by considering the magnitude of the load and its duration factor. Example 1.1 demonstrates this concept.

3

AXIAL LOADS

The design of columns is in some ways more important than the design of beams. If a beam collapses, the rest of the structure might remain in place, but if a column fails, the entire structure will possibly collapse. Furthermore, most beams will sag a great deal before they collapse, thus giving ample warning of danger, but timber columns can collapse suddenly, without warning. Partly for these reasons a large factor of safety is used with columns.

A *column* is a structural member in compression. The major load creating this compression is acting along the long axis of the column, and this long dimension is more than three times either of its shorter dimensions. Since the allowable compression stress parallel to the grain is much larger than the allowable stress perpendicular to the grain, wood columns are built so that the grain runs along the length of the column.

However, long columns are subjected to many uncertainties, and strictly speaking, the formula $F_c = P/A$ does not apply unless modified. If the column is not absolutely plumb, that is, if it is not erected perfectly vertically, the load will force it to move sideways. If there is a flaw within the material of the column, such as a knot or a split, or if the column does not have exactly the same cross section along its entire length, the stress will not be uniformly distributed.

The most important factor, however, is a physical one. Molecules are not happy to be in compression and try to escape what for them is an unhappy state. When the compression stress becomes too great, the column will suddenly slip sideways in the direction of its weakest dimension. If there is lateral support along that dimension, it will surely slip in the other short dimension. This purely elastic phenomenon,

56 AXIAL LOADS

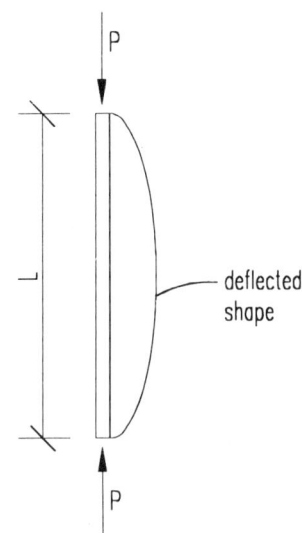

FIGURE 3.1 Euler buckling.

known as *buckling*, is similar to the "pop" of the bottom of an oil can that has been pressed hard enough. Yet whereas an oil can will return to its original shape when the pressure is removed, a column might collapse if it buckles. Actually, it has some postbuckling strength, but the loads might be thrown under the sudden snapping, and the building might collapse. Therefore, a factor of safety must be used so that the applied load is never permitted to become sufficient to cause a buckling failure.

About 200 years ago a Swiss mathematician, Leonard Euler, (pronounced "*Oiler*") set out to find the load that would cause a long, thin column to buckle (Figure 3.1). Purely by mathematics he found that the critical load was given by

$$P = \frac{n^2 \pi^2 EI}{l^2}$$

where n is any integer, E the modulus of elasticity of the material, I the least moment of inertia of the cross section of the two smaller dimensions, and l the long dimension of the column in inches. If the column does not have braces anywhere between its top and bottom, the integer n is unity, and

$$P = \frac{\pi^2 EI}{l^2}$$

For design purposes, it has been found more convenient to replace I by Ar^2, where A is the cross-sectional area and r is the radius of gyration. This leads to

$$P = \frac{\pi^2 EAr^2}{l^2}$$

Dividing both sides of the equation by A yields

$$\frac{P}{A} = \frac{\pi^2 Er^2}{l^2}$$

Then dividing the numerator and the denominator of the right side of the equation by r^2, we have

$$\sigma = \frac{P}{A} = \frac{\pi^2 E}{(l/r)^2}$$

The resulting equation shows that the *critical stress*, σ (P/A) depends on the modulus of elasticity E, a measure of the material of the column, and inversely on the *slenderness ratio* l/r. Notice that the slenderness ratio is not based solely on the length or height of the column, but is the ratio of the height to the *least* radius of gyration. A very slender column will buckle quite easily, and therefore a slender column as a low critical stress—a quite small load will cause it to buckle. A stiff column will not buckle easily and so has a higher critical stress. As an example, a round column only 10 in. high with a diameter of $\frac{1}{4}$ in. will have a much lower critical stress than one 10 ft high but with a diameter of 20 in., because although the second height

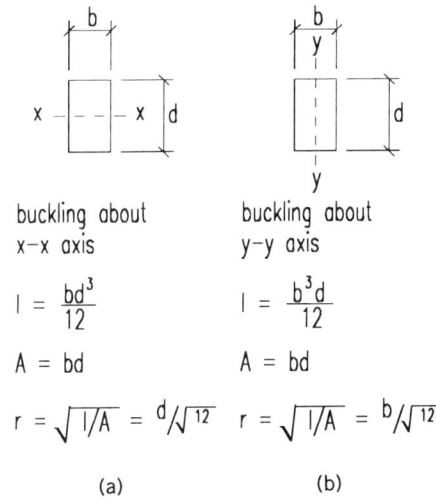

FIGURE 3.2 Moments of inertia and radius of gyration: (a) strong direction and (b) weak direction.

is 12 times as much as the first, the l/r ratio is only one-fifth as much.

For timber columns it is not necessary to compute the radius of gyration. The radius of gyration is found from $r^2 = I/A$. For the rectangular column shown in Figure 3.2, the cross-sectional area $A = b \times d$, and the moment of inertia would be $I = bd^3/12$. Then

$$r^2 = \frac{I}{A} = \frac{bd^3}{12b \times d} = \frac{d^2}{12}$$

Substituting in the Euler formula for critical stress gives

$$\frac{P}{A} = \frac{\pi^2 E}{12(l/d)^2} = \frac{0.82 E}{(l/d)^2}$$

The NDS (*National Design Specification*) requires the high safety factor of 2.7, which takes into account the various defects that can occur in wood, such as knots, splits, and bark streaks. The maximum allowable stress that can be used in design will be the critical stress divided by the safety factor. Thus

$$F_{c(\text{allowable})} = \frac{0.82 E}{2.7(l/d)^2}$$

or

$$F_{cE} = \frac{0.3 E}{(l/d)^2}$$

where F_{cE} is the allowable axial stress and d is the *smallest* dimension of the unbraced column.

All of the foregoing applies to what are called *long*, slender columns. Very short columns are less likely to buckle. For columns with a small slenderness ratio, l/r or l/d, the Euler buckling stress may exceed the crushing or compression stress. For these columns design is based more on compression, and modification of the allowable compression stress is slight.

In Figure 3.1 the first column is shown to be pinned at the top and bottom. For this condition its buckling length equals its actual length ($l_e = l$). In timber construction it is very difficult to build a moment connection, so it is a good assumption that columns will always be pinned at the ends. If, however, the ends of a column are restrained differently, the buckling length will not equal the actual length. In all the formulas the slenderness ratio the proper term is l_e/d, where $l_e = l \times k$ and k is found from the recommendations in Figure 3.3.

As with beam design, the 1991 edition of the *National Design Specification* has revised the formulas used in column design. The revision parallels that in beam design. While maintaining relatively the same values, slight numeric differences have resulted from the change. The authors believe that as with beam design, some of the intuitive understanding of column action has been sacrificed in this change. Therefore, we discuss the effect of the slenderness ratio in terms common to

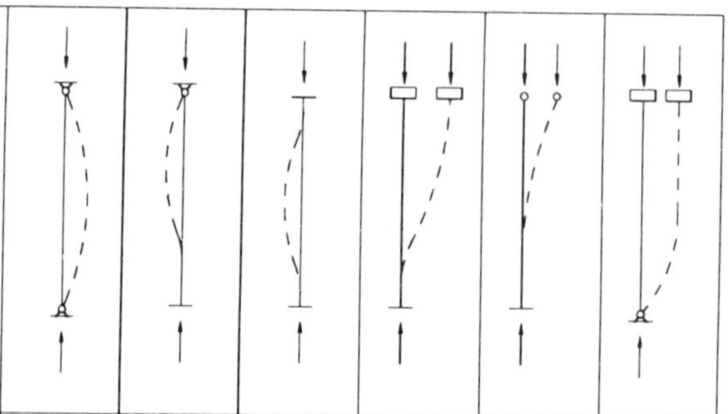

58 AXIAL LOADS

buckling modes						
theoretical K values	1.0	0.7	0.5	1.0	2.0	2.0
recommended K values when ideal conditions approximated	1.0	0.8	0.65	1.2	2.1	2.4
end conditions	rotation fixed, translation fixed rotation free, translation fixed rotation fixed, translation free rotation free, translation free					

FIGURE 3.3 Effective lengths and end conditions.

engineering but now not apparent in the NDS.

In the current edition of the NDS, there is one set of formulas that is to be used for all columns, regardless of their slenderness. Those formulas are

$$F'_c = C_D \times C_M \times C_t \times C_F \times C_P \times F_C$$

where

$$C_P = \frac{1 + F_{cE}/F_c^*}{2c} - \sqrt{\left(\frac{1 + F_{cE}/F_c^*}{2c}\right)^2 - \frac{F_{cE}/F_c^*}{c}}$$

$$F_{cE} = \frac{0.3 \times E'}{(l_e/d)^2}$$

$$F_c^* = C_D \times C_M \times C_t \times C_F \times F_c$$

The adjustment factor C_P is termed the column stability factor. Its effect is to reduce the allowable compression stress as the column becomes more slender. For very short, bulky columns, C_P is close to 1.0, so the allowable stress is not affected much. For very slender columns ($l/d = 50$) the stability factor may approach 0.25. By recognizing the relationship between the slenderness ratio (l/d) and the stability factor (C_P), a designer may make a well-reasoned first estimate of the strength that a column may have.

The procedures outlined here are not intended to replace the use of these formulas. In the final analysis, if the actual capacity of a column is needed, the designer must use these formulas. The procedure that is presented here is a means that is intended to speed the design by making accurate estimates of member sizes and stresses before applying the design formulas.

Like laterally unsupported beams, columns may be classified as short, intermediate, and long (or slender). The classification depends on the l/d ratio. Those columns with a slenderness ratio of 11 or less, (i.e., $l/d \leq 11$) are considered to be *short*. In the first analysis, F_c' may be taken equal to the table value of F_c.

Long columns are defined as those where l/d exceeds a special l/d ratio. This ratio, which varies slightly with wood species and grade, for our purposes may be taken as a value of 30. No column may have an l/d ratio that exceeds 50. (This provision is retained in the 1991 revisions and in previous editions.) For a long column the estimated allowable stress can be taken as

$$F_c' = F_{cE} = \frac{0.3 \times E}{(l_e/d)^2}$$

Between the short columns and the long are the *intermediate* columns. Those columns with an l/d ratio between 11 and 30 are classified as *intermediate*. The allowable axial stress in compression F_c' for an intermediate column must be derived by the equations. Approximate ranges of the stability factor are given for the three classifications.

Summarizing, we have the following:

Slenderness Ratio l/d	Classification	Range of C_P	Estimated Stress
$0 \leq l/d \leq 11$	Short	1.0 to 0.85	F_c (NDS tables)
$11 < l/d < 30$	Intermediate	0.85 to .12	F_c' from equation
$30 \leq l/d \leq 50$	Long	.12 to .10	$F_c' = \dfrac{0.3E}{(l/d)^2}$

The big difficulty in design is that the allowable stress depends on the classification of the column, and the classification depends on the l_e/d ratio. (*Note:* $l_e \times k$. For simplicity we will henceforth refer to l_e/d as l/d.) The l/d ratio is not known until the size is known, since d is one of the size dimensions. So the allowable stress depends on the size, and the size depends on the allowable stress. The only thing to do in such a dilemma is to guess a size and see if it works. It should be neither too large nor too small, but rather, should be just big enough to do the job.

Suppose that we wish to design three solid timber columns carrying an axial load of 24,000 lb each with lengths of 5 ft., 10 ft., and 14 ft.

EXAMPLE 3.1—SHORT COLUMN.

Specify the size of a 5-ft-long column to be built of Douglas Fir–Larch No. 1 to carry a load of 24,000 lb. First go to Table A.11 and find the section for posts and timbers. Notice that the value for F_c is 1000 psi and $E = 1,600,000$ psi.

In this example, let us start by finding the required cross-sectional area. Where $L = 5$ ft, we would guess that we have a *short* column, so assume that C_P is close to 1.0 and the allowable value for allowable axial stress in compression, F_c', will probably be unchanged. Therefore, for the time being, assume that $F_c = 1000$ psi.

$$A = \frac{P}{F} = \frac{24{,}000}{1000} = 24.0 \text{ in}^2$$

The best shape for a timber column is either a square or a circle. In this case, if it were square, each side would be the square root of the area, or $\sqrt{24} = 4.9$ in. The closest stock size that is larger than 4.9 is seen to be a 6×6, which has an actual least dimension of 5.5 in. and a cross-sectional area of 30.25 in². If this column is loaded with 24,000 lb, the actual stress in the column will be

$$f_C = \frac{P}{A} = \frac{24{,}000 \text{ lb}}{30.25 \text{ in}^2} = 793.4 \text{ psi}$$

Thus a 6 × 6 will be satisfactory as long as C_P is less than or equal to 0.793. We have estimated C_P to be close to 1.0. Unless our estimate of C_P is quite poor, this column should work. We should at least have enough confidence at this point to continue the calculation using a 6 × 6. To calculate C_P, we start with

$$\frac{l_e}{d} = \frac{5 \times 12}{5.5}$$

$$= 10.9 < 11 \ (a\ \textit{short}\ \text{column.})$$

$$F_{cE} = \frac{0.3 \times E'}{(l_e/d)^2} = \frac{0.3 \times 1{,}600{,}000}{(10.9)^2}$$

$$= 4040$$

$$F_c^* = C_D \times C_M \times C_t \times C_F \times F_c$$

$$= 1.0 \times 1.0 \times 1.0 \times 1.0 \times 1000 \text{ psi}$$

$$= 1000 \text{ psi}$$

$$\frac{F_{cE}}{F_c^*} = \frac{4040}{1000} = 4.04$$

$$C_P = \frac{1 + F_{cE}/F_c^*}{2c}$$

$$- \sqrt{\left(\frac{1 + F_{cE}/F_c^*}{2c}\right)^2 - \frac{F_{cE}/F_c^*}{c}}$$

$$= \frac{1 + 4.04}{1.6} + \sqrt{\left(\frac{1 + 4.04}{1.6}\right)^2 - \frac{4.04}{0.8}}$$

$$= 3.15 - 2.207 = 0.943$$

$$F_c' = 0.943 \times 1000 \text{ psi} = 943 \text{ psi}$$

$$P = F_c' A = 943 \times 30.25 = 28{,}525 \text{ lb}$$

$$> 24{,}000 \quad \text{OK} \quad (85\% \text{ stress})$$

EXAMPLE 3.2—INTERMEDIATE COLUMN.

The second example column has a length of 10 ft. Let us therefore assume that one of the sides will again be a nominal 6 in. as the least dimension. This gives

$$\frac{l}{d} = \frac{10 \times 12}{5.5} = 21.8$$

We now have an l/d value between 11 and 30, so assume an *intermediate* column. The recommended procedure is to calculate the allowable force by first finding C_P. The only thing that has changed is the l/d ratio, so

$$F_{cE} = \frac{0.3 \times E'}{(l_e/d)^2} = \frac{0.3 \times 1{,}600{,}000}{(21.8)^2}$$

$$= 1010 \text{ psi}$$

$$F_{cE} = C_D \times C_M \times C_t \times C_F \times F_c = 1000 \text{ psi}$$

$$\frac{F_{cE}}{F_c^*} = \frac{1010}{1000} = 1.01$$

At this point we can turn to the column tables in the Appendix, Table C.6, to find the value of C_P.

$$C_P = 0.691$$

$$F_c' = 1000 \times 0.691 = 691 \text{ psi}$$

Proceeding as before, we have

$$A = \frac{P}{F_c'}$$

$$= \frac{24{,}000}{691} = 34.7 \text{ in}^2$$

From Table A.1, try a 6 × 6, $A = 30.3 \text{ in}^2$: No Good. Try a 6 × 8, $A = 41.26 \text{ in}^2$: OK. The actual stress divided by the allowable stress is equal to the ratio of required area

COLUMN DESIGN 61

to actual area. Thus

$$\frac{34.7}{41.25} = 84\% \text{ stress}$$

Since the l/d ratio is based on the least thickness of the column, the value for C_P is the same for any $6 \times$ member (using the same species, grade, and length). Once C_P and F'_c have been found, find the required area as shown above and select any $6 \times$ that can provide that area. Quite often in a building design the columns on any floor will have the same length. If possible, a good strategy is to select one minimum dimension for all columns of that length. Even if the axial loads vary, all the columns can be selected easily by finding the required area ($A = P/F'_c$) and choosing $6 \times$ sizes.

Referring again to Table C.1, we find that Douglas Fir–Larch No. 1 can carry 20,989 lb with an unrestrained length of 10 ft. This is not sufficient. Increasing the area to a 6×8 will provide an increase of

$$\frac{5.5 \times 7.5}{5.5 \times 5.5} = \frac{41.25}{30.25} = 1.364$$

The capacity of the 6×6 in Table C.1 can be used to find the capacity of a 6×8 column for the same condition. Thus for the 6×8, the maximum allowable axial load is

$$20{,}989 \text{ lb} \times 1.364 = 28{,}620 \text{ lb}$$
$$> 24{,}000 \text{ lb} \quad \text{OK}$$

EXAMPLE 3.3—LONG COLUMN.

In the third example the length is 14 ft. Again we will try 6 in. as the least nominal dimension. Start by calculating the slenderness ratio, l/d.

$$\frac{l}{d} = \frac{14 \times 12}{5.5} = 30.55$$

This time l/d is greater than 30 but less than 50 ($30 < 30.55 < 50$), so we have a long column. As a trial size, start with the approximation for F'_c.

$$F'_c = F_{cE} = \frac{0.3E}{(l/d)^2}$$

We may calculate this value as 514 psi. This means that the required cross-sectional area is

$$A = \frac{P}{F'_c} = \frac{24{,}000}{514} = 46.7 \text{ in}^2$$

Entering Table A.1, it seems that we need a 6×10 with an area of 52.25 in^2 to satisfy these conditions. Since this size does not seem excessive, let us continue with the design.

$$F_c^* = 1000 \text{ psi}$$

$$\frac{F_{cE}}{F_c^*} = \frac{514}{1000} = 0.514$$

$$C_P = \frac{1 + F_{cE}/F_c^*}{2c}$$
$$- \sqrt{\left(\frac{1 + F_{cE}/F_c^*}{2c}\right)^2 - \frac{F_{cE}/F_c^*}{c}}$$

$$= \frac{1 + 0.514}{1.6}$$
$$- \sqrt{\left(\frac{1 + 0.514}{1.6}\right)^2 - \frac{0.514}{0.8}}$$

$$= 0.946 - 0.503 = 0.443$$

$$F'_c = 0.443 \times 100 \text{ psi} = 443 \text{ psi}$$

$$A_{\text{req'd}} = \frac{P}{F'_c} = \frac{24{,}000}{443} = 54.2 \text{ in}^2$$

$$> 52.25 \text{ in}^2 \quad \text{No Good}$$

Although this section is overstressed by only 3%, to meet the required safety factors we will choose a larger section. If we select a 6×12, we will be assured that its area of 63.25 in^2 will be sufficient. The only necessary calculation will be to compare the required area with that provided. However, a 6×12 is not even close to a square cross section. A square section would be more economical. In addition, the design values that we took from Table A.11 were for posts and timbers, which by definition must have side dimensions not differing by more than 2 in. A quick check of the design values for beams and stringers shows F_c equal to 925 psi for the same grade, No. 1. The best way out of this morass is to look for a square shape that will do the job.

To this end, assume the least dimension to be a nominal 8 in. instead of 6. If we select an 8×8, with an area of 56.25 in^2, we can see that it will meet the required area. But remember that the required area was derived from the slenderness of a $6 \times$. For an $8 \times$, the slenderness will decrease, the allowable stress will go up, and the required area will actually go down. We can conclude that the 8×8 is safe and satisfactory, but we cannot say quantitatively how safe it is. To do this we would have to recalculate everything (l/d, F_{cE}, C_P, and F'_c) using the least dimension, $d = 7.5$ in. We leave it to the reader to complete this calculation. (The result will show that an 8×8 will be stressed to 63% of its allowable capacity.) For now we may specify

8×8 Douglas Fir–Larch No. 1.

COLUMN DESIGN TABLES

The appendix contains two sets of tables intended as design aids. The most direct set of tables, C.1 and C.2, list allowable axial loads for specific sizes and species and grade of columns. To estimate the allowable load on a particular column, the designer must know the column's effective length and its size, species, and grade. Entering the charts with these values will result in finding the allowable axial load.

To use the charts as a design tool, the designer must estimate in advance the approximate size of the column to select which table to turn to. This can be done by guessing the stability factor, C_P, after having determined the l/d ratio. Alternatively, the designer could scan through the various tables for 4×4's, 6×6's, and 8×8's to find a column that meets the load requirement.

Reworking Example 3.1, we find in Table C.1 that the allowable value for a Douglas Fir–Larch No. 1 6×6 column with a length of 5 ft is 28,511 lb. This is very close to what we found in our calculation. The slight discrepancy between this number and the value calculated above is insignificant and due to rounding off intermediate values in the example.

To find the solution to Example 3.3, look at Table C.1, which shows that a 6×6 with a length of 14 ft can carry 13,421 lb. Although this is insufficient, the allowable load on a 6×8 of the same length can be determined by multiplying 13,421 lb by the ratio of 7.5 in. to 5.5 in. The 6×8 can carry 18,300 lb. Because the lesser dimension remains the same, the allowable stress is the same. The increased capacity of the 6×8 is due solely to the increased area. If the 6×8 were not sufficient, we would probably look for an 8×8.

A comparison of Tables C.1 and C.2 shows that for the 14-ft length, a 6×6 can carry 13,421 lb, whereas the 8×8 can carry 37,995 lb. To reach the required load of 24,000 lb, the $6 \times$ column will need about twice the area; hence a 6×12 could be checked. We could approximate the capacity of a 6×10 or a 6×12 by the same method by which we found the actual capacity of the 6×8. Since the 6×10 is a beam and stringer, not a post and timber, its allowable compression strength, F_c, is different from that of a 6 × 6 or 6 × 8. Therefore, the value derived from Table C.1 is only approximate. It is, however, accurate enough to use to select a column, which can then be checked using the full equations.

The 8 × 8 has an area about twice that of the 6 × 6 but can carry almost three times the load. Therefore, selecting the 8 × 8 will result in a column with a lower level of stress while using slightly less material. The 8 × 8 is the preferred solution. In using Tables C.1 and C.2, remember that $l_e = l \times k$, so for columns that are braced, l_e may not equal the column length, l.

For columns that are not listed in Tables C.1 and C.2, an alternative exists to solving a complex series of equations by calculator. Tables C.3 through C.6 can be used to look up all of the intermediate values and arrive at a final value of C_P. The process is described in the tables. Values from one table will usually need to be rounded off before entering the next table. This results in a slight inaccuracy but should not affect the confidence of the design.

ROUND COLUMNS

Round columns are used in pole buildings and as utility poles. They come in full-inch diameters, so a 6-in. round column will actually have a 6-in. diameter. The slenderness ratio for a round column is not l/d but rather, depends on the radius of gyration. The radius of gyration r is actually one-half of the actual radius of a circular area, or one-fourth of the diameter of that area. To use the column formulas that have been presented so far, instead of using l/d we must use $l/0.866D$, where D is the diameter of a round column. It should also be noted that the design load for a round cross section shall not exceed that for a square column of the same cross-sectional area. All other requirements and formulas remain the same.

EXAMPLE 3.4—ROUND COLUMN.

As an example, let us design a round column for a load of 24,000 lb with an unbranched height of 14 ft, using Douglas Fir–Larch No. 1. From Table A.11 read $F_c = 1000$ psi and $E = 1,600,000$ psi. As a first guess the area required will be

$$A = \frac{P}{F_c} = \frac{24,000}{1000} = 24 \text{ in}^2$$

Solving for the required diameter yields

$$A = \frac{\pi D^2}{4}$$

$$24 = \frac{3.1416 D^2}{4}$$

from which $D^2 = 30.56$, or $D = 5.53$. Since round columns come only in full-inch diameters, try $D = 6$ in. Then

$$\frac{l}{0.866D} = \frac{14 \times 12}{0.866 \times 6} = 32.3$$

Then, since $30 < 32.3 < 50$, we have a *long* column. The estimated allowable axial compressive stress is found from the equations.

Assume that

$$F_c' = F_{cE} = \frac{0.3E}{(l/d)^2} = \frac{0.3E}{(32.3)^2} = 460 \text{ psi}$$

A quick check of the capacity of the 6-in. diameter shows that

$$P_{\text{allowable}} = F_c' \times A = 460 \text{ psi} \times \frac{\pi 6^2}{4} \text{ in}^2$$

$$= 460 \times 28.27 = 13,006 \text{ lb.}$$

This is far less than the actual load, so we need to try a larger column. Try an 8-in. diameter

$$\frac{l}{0.866D} = \frac{14 \times 12}{0.866 \times 8}$$

$$= 24.25 \quad intermediate$$

Use the formulas.

$$F_{cE} = \frac{0.3E}{(l/d)^2} = \frac{0.3E}{(24.25)^2} = 816 \text{ psi}$$

$$\frac{F_{cE}}{F_c^*} = \frac{816}{1000} = 0.816$$

$$C_P = \frac{1 + 0.816}{1.6}$$

$$- \sqrt{\left(\frac{1 + 0.816}{1.6}\right)^2 - \frac{0.816}{0.8}}$$

$$= 1.135 - 0.517 = 0.617$$

$$F_c' = 0.617 \times 1000 \text{ psi} = 617 \text{ psi}$$

$$A = \frac{\pi D^2}{4} = \frac{\pi \times 8^2}{4} = 50.26 \text{ in}^2$$

Then the maximum axial load that this column can carry is

$$P = F_c' A = 617 \times 50.26 = 31{,}018 \text{ lb}$$

The column is stressed to 77% of its allowable load. This must be checked against the capacity of an equivalent square column whose area is also 50.26 in². Each side of such a column would be $\sqrt{50.26} = 7.09$ in. This would lead to $l/d = (14 \times 12)/7.09 = 23.7 < 24.25$. Since a lower l/d yields a higher allowable stress, the load capacity of the equivalent square column would by necessity be higher than that of our round column, so it need not be calculated.

Use an 8-in.-diameter Douglas Fir–Larch No. 1 column.

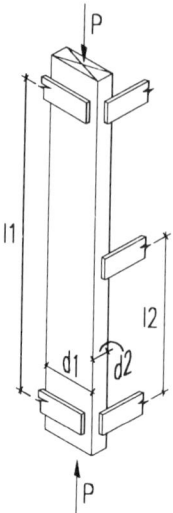

FIGURE 3.4 L/d ratios.

BRACING

One of the major controlling elements in the design of a column is the l/d ratio. In fact, once the material has been selected for species and for grade, the l/d ratio controls the strength of the column. Sometimes the ratio is determined by the conditions of construction or by architectural design, but in some cases it can be reduced by the application of a lateral brace. Braces are also used to resist wind forces. If the two cross-sectional dimensions are not the same, the brace should be attached to the thinner side of the column, the one with the least d, so as to be most effective.

A typical braced column is shown in Figure 3.4. Note that the brace is on the thinner side of the column, leading to two different l/d ratios, l_1/d_1 and l_2/d_2. Both of these must be computed, and since the larger one gives the lower allowable stress, the larger one is used in design so as to make the column as safe as possible. If the column fails in one direction, there is no point in considering the other direction.

EXAMPLE 3.5—BRACED COLUMN.

As an example, consider a column 14 ft high with a brace at midheight. Suppose that the load is 24,000 lb and that we use Douglas Fir–Larch No. 1, which has $F_c = 1000$ psi and $E = 1,600,000$ psi. Based on previous calculations, try a 4×8 for a first guess. We will brace the narrower side at midheight so that the slenderness ratio will be about the same in both directions.

$$\frac{l_1}{d_1} = \frac{14 \times 12}{7.5} = 22.4$$

$$\frac{l_2}{d_2} = \frac{7 \times 12}{3.5} = 24$$

The larger of these controls the design. Use $l/d = 24$. Notice that it is not the longer unbraced length, but rather the larger value of l/d, that controls. Since $11 < 24 < 30$ we have an *intermediate* column. Start with the exact formulas.

$$F_{cE} = \frac{0.3E}{(l/d)^2} = \frac{0.3E}{(24)^2} = 833 \text{ psi}$$

$$F_c^* = 1000 \text{ psi}$$

$$\frac{F_{cE}}{F_c^*} = \frac{833}{1000} = 0.833$$

$$C_P = \frac{1 + F_{cE}/F_c^*}{2c}$$

$$- \sqrt{\left(\frac{1 + F_{cE}/F_c^*}{2c}\right)^2 - \frac{F_{cE}/F_c^*}{c}}$$

$$= \frac{1 + 0.833}{1.6}$$

$$- \sqrt{\left(\frac{1 + 0.833}{1.6}\right)^2 - \frac{0.833}{0.8}}$$

$$= 1.146 - 0.521 = 0.625$$

$$F_c' = 0.625 \times 1000 \text{ psi} = 625 \text{ psi}$$

Since the area of a 4×8 is 25.4 in^2, the load our selection can carry is

$$P = F_c' A = 625 \times 25.4 = 15,875 < 24,000$$

The first guess was too small. It is necessary to increase the cross-sectional area. Since the nominal 4-in. dimension controlled the slenderness ratio, if we keep the 4-in. side the same, the allowable stress will remain unchanged no matter how large the other side becomes. Thus we can solve for the other side directly, without further guesswork.

$$A_{\text{req'd}} = \frac{P}{F_c'} = \frac{24,000 \text{ lb}}{625 \text{ psi}} = 38.4 \text{ in}^2$$

Since $A = bd$ and we have 3.5 as one side,

$$A = 38.4 = 3.5 \times d$$

$$d = 10.95, \text{ at least} \quad \text{use a } 4 \times 12$$

It can be shown that a 6×8 will also work and will be stressed to 86% of its allowable stress. The 4×12 has slightly less area and could be used if there was a severe space requirement. However, the 6×8 would probably be preferable since it is closer to a square section.

BUILT-UP COLUMNS

It is getting increasingly difficult, and certainly expensive, to find structural timbers of large cross section. As a substitute it is quite common in construction to place a number of studs together to form a column. An example of this approach can be found at door jambs where a large header must be supported over an opening. Whenever a column is composed of a number of pieces it is referred to as a *built-up column*. Because each piece may slip with respect to the adjacent pieces, built-up columns have far less capacity than solid columns of the same size. The capac-

66 AXIAL LOADS

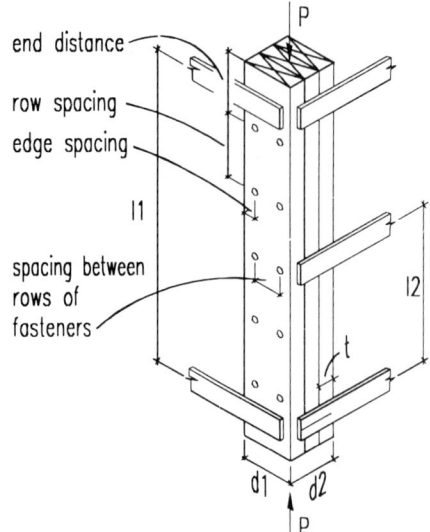

FIGURE 3.5 Built-up columns.

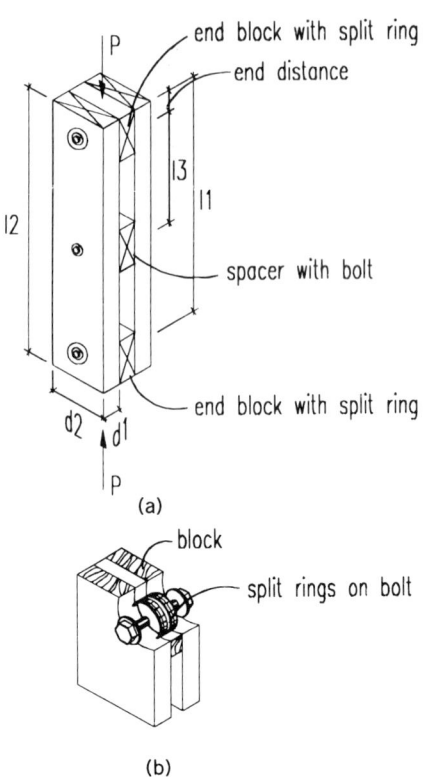

FIGURE 3.6 Spaced columns: (a) full length and (b) end block.

ity is based on the slip of the fasteners, so to build such a column the nails or bolts holding the individual pieces together must be specified. The NDS specifies the fastening requirements for two types of built-up columns: nailed or bolted. When these specifications are met (see Figure 3.5), the axial strength of the column can be calculated using the procedure given above for column design with an additional factor, K_f, used to reduce the stability factor, C_P. The factors for the two types of columns are

	Nailed Column	Bolted Column
$K_f =$	0.6	0.75

In performing these calculations the actual thickness of the total column is used in the l/d ratios. Thus for three 2×6's, d is taken as 4.5 in. in the weak direction, or 5.5 in. in the strong direction. Tables C.1 and C.2 list the allowable axial loads for a series of solid and built-up columns using Douglas Fir–Larch or Southern Pine. In these tables the assumption is made that the weak direction of the column is braced by the wall sheathing. If it is not braced, the designer must calculate the column capacity since the tabulated value will be unconservative.

As can be seen from the calculations or from the tables, there is a rather severe loss of strength due to the connection slip and the K_f factor. As a good alterative, it is possible to build up columns using two pieces that are held a certain distance apart for their entire length. These columns are referred to as *spaced columns* (Figure 3.6). A portion of their increased strength is due to the type of connectors used: *split rings* This technique is also used in building compression members of large trusses not only because of the load, but also to facilitate attaching connecting members, called web members, at the ends and at intermediate points along the length of the main compression member. By using a divided main member, the intermediate member can slip into the open space.

Notice how a spaced column is built with two end blocks and one spacer block. Each of the end blocks can be either a piece of wood or an intermediate member that frames into our spaced column at that point. The same is, of course, also true of the spacer block. Thus the spacer block is not necessarily located at the exact mid-length of the column.

Both of the end blocks are held in place by split rings, a connector that will be discussed more fully later. In the center of such a ring there is always a bolt. The spacer block, if simply a piece of wood and not an intermediate member, is secured by a bolt without a split ring. The distance from the end bolt to the end of the column, and from end bolt to spacer block bolt, is determined by the location of the members framing into the column. The first of these, from end bolt to end of column, cannot be greater than one-tenth of the length of the column. There is no minimum end dimension, it being determined by the conditions of construction.

The overall unbraced length of the column, l, is measured from end to end of the wood. The unbraced length l_1 is used with the smaller thickness dimension, d_1, to get the ratio l_1/d_1, and similarly the unbraced length l_2 is used with the other column thickness, d_2. A third slenderness ratio must also be checked. The length from end bolt to spacer bolt, l_3, is used with the lesser thickness dimension d_1. The maximum slenderness ratios are

$$l_1/d_1 \leq 80$$
$$l_2/d_2 \leq 50$$
$$l_3/d_1 \leq 40$$

Once the controlling slenderness ratio is determined, the design proceeds in the same manner as for a solid timber column, where the column stability factor, C_P, is calculated. As a preliminary design step the designer may approximate the allowable stress by considering the column classifications of *short*, *intermediate*, or *long*. For the spaced column the ranges are somewhat different than they were for solid columns. We may consider them as follows:

Slenderness Ratio, l_1/d_1	Classification	Allowable Stress
$0 \leq l_1/d_1 \leq 11$	Short	F_c (NDS tables)
$11 < l_1/d_1 < 50$	Intermediate	$F_c' = F_c \times C_P$
$50 \leq l_1/d_1 \leq 80$	Long	$F_c' = \dfrac{0.30 K_x E}{(l_1/d_1)^2}$

$K_x = 2.5$, when $a \leq \dfrac{l_1}{20}$

$K_x = 3.0$, when $\dfrac{l_1}{20} \leq a \leq \dfrac{l_1}{10}$

The dividing point between intermediate and long has increased from 30 to 50, and the limit for long columns has increased from 50 to 80. The term K_x depends on the column end condition. The end condition is one of two types, classified as condition a or condition b. Condition a exists when the end distance $a \leq l_1/20$, and condition b exists when $l_1/20 \leq a \leq l_1/10$. If a spaced column cannot be classified as either a or b, such as when $a \geq l_1/10$, it cannot be used. Both cases lead to a fixity factor K_x which changes the value of F_{cE} and modifies the allowable stress in long columns. For condition a, $K_x = 2.5$. For condition b, $K_x = 3.0$ (see Figure 3.6). Note that in the definition of short, intermediate, or long the slenderness ratio that is always used is l_1/d_1.

Regardless of whether a spaced column is classified as short, intermediate, or long, the allowable stress is given as

$$F'_c = C_D \times C_M \times C_t \times C_F \times C_P \times F_c$$

where

$$C_P = \frac{1 + F_{cE}/F^*_c}{2c} - \sqrt{\left(\frac{1 + F_{cE}/F^*_c}{2c}\right)^2 - \frac{F_{cE}/F^*_c}{c}}$$

$$F_{cE} = \frac{K_{cE} \times K_x \times E'}{(l_e/d)^2}$$

$K_{cE} = 0.3$ for visually graded

and machine-evaluated lumber

$= 0.418$ machine stress rated

or glued-laminated lumber

$$F^*_c = C_D \times C_M \times C_t \times C_F \times F_C$$

Note that these equations are the same as those for the solid column except that the value for F_{cE} has been further modified by the end condition factor K_x.

EXAMPLE 3.6—SPACED COLUMN.

As an example, let us design a spaced column that is 14 ft long, carrying an axial load of 24,000 lb. Suppose that Douglas Fir–Larch is to be used. Select grade No. 2, so that $E = 1,600,000$ psi and $F_c = 1300$ psi. We will first find a trial size and then make all necessary modifications based on size (C_F) and stability (C_P). The absolute minimum thickness is found from utilizing the rule $l_1/d_1 \leq 80$, which leads to

$$d_1 \geq \frac{l_1}{80} = \frac{14 \times 12}{80}$$

$$\geq 2.1$$

Try a 3×, which has a thickness of 2.5 in. Then

$$\frac{l_1}{d_1} = \frac{14 \times 12}{2.5} = 67.2$$

Since $50 < 67.2$, we have a *long* column. We will detail the column so that end condition a will apply. K_x will be 2.5, so a trial value for allowable stress is

$$F'_c = F_{cE} = \frac{0.30 K_x E}{(l_1/d_1)^2}$$

$$= \frac{0.30 \times 2.5 \times 1,600,000}{67.2 \times 67.2} = 266 \text{ psi}$$

Solving for the required area, in order to get the required cross-sectional dimensions, gives us

$$A = \frac{P}{F_c} = \frac{24,000}{266} = 90.2 \text{ in}^2$$

This will be made up with two equal pieces. Each piece will have one side of 2.5 in., since we selected 2.5 as our least dimension. Then the other side, d_2, will be found from

$$2 \times 2.5 \times d_2 = 90.2$$

$$d_2 = 18 +$$

which means a 3×20. This would be a poor choice. Let us increase the least dimension by one size. Try a 4×, so that $d_1 = 3.5$ in. Now we must recompute the slenderness ratio and the allowable stress.

$$\frac{l_1}{d_1} = \frac{14 \times 12}{3.5} = 48.0$$

which is approximately 50

Let us continue with the assumption of a *long* column for our preliminary analysis.

$$F'_c = F_{cE} = \frac{0.30 K_x E}{(l_1/d_1)^2}$$

$$= \frac{0.30 \times 2.5 \times 1{,}600{,}000}{48 \times 48}$$

$$= 521 \text{ psi}$$

$$A = \frac{P}{F'_c} = \frac{24{,}000}{521} = 46.1 \text{ in}^2$$

$$2 \times 3.5 \times d_2 = 46.1 \text{ in}^2$$

$$= 6.6 \quad \text{use two } 4 \times 10\text{'s}$$

The final design requires checking the trial size against the l/d ratios and determining the allowable stress.

$$\frac{l_1}{d_1} = 14 \times \frac{12}{3.5} = 48 \le 80 \quad \text{OK}$$

$$\frac{l_2}{d_2} = 14 \times \frac{12}{9.25} = 18.2 \le 50 \quad \text{OK}$$

$$\frac{l_3}{d_1} = 6.5 \times \frac{12}{3.5} = 22.3 \le 40 \quad \text{OK}$$

$$F_c^* = C_D \times C_M \times C_t \times C_F \times F_C$$

$$= 1300 \text{ psi}$$

Note: C_F is dependent on the member size and needs to be checked at this juncture. For the selected 4×10, $C_F = 1.0$.

$$F_{cE} = \frac{K_{cE} \times K_x \times E'}{(l_e/d)^2}$$

$$= \frac{0.3 \times 2.5 \times 1{,}600{,}000}{(48)^2}$$

$$= 521 \text{ psi}$$

$$\frac{F_{cE}}{F_c^*} = \frac{521}{1300} = 0.4$$

$$C_P = \frac{1 + F_{cE}/F_c^*}{2c}$$

$$- \sqrt{\left(\frac{1 + F_{cE}/F_c^*}{2c}\right)^2 - \frac{F_{cE}/F_c^*}{c}}$$

$$= \frac{1 + 0.4}{1.6} - \sqrt{\left(\frac{1 + 0.4}{1.6}\right)^2 - \frac{0.4}{0.8}}$$

$$= 0.359$$

$$F'_c = C_D \times C_M \times C_t \times C_F \times C_P \times F_c$$

$$= 0.359 \times 1300 = 467.5 \text{ psi}$$

The actual stress is

$$f_c = \frac{P}{A} = \frac{24{,}000 \text{ lb}}{2 \times 3.5 \times 9.25 \text{ in}^2}$$

$$= \frac{24{,}000}{64.75} = 370.7 \text{ psi}$$

This column is loaded to 79% of its allowable stress.

AXIAL LOAD WITH BENDING

When a vertical column is on the outside of a building it is subjected to the horizontal load of wind pressure. This uniform load produces a bending moment. In addition, the column must carry a vertical load due to the gravity acting on the weight of the building and its contents.

If we were to consider each of these loads separately, we would see that each load induces compression in parts of the column. The downward-acting load is a concentrated load P acting at the centroid of the shaft, creating a purely compressive stress, P/A, over the entire cross section. The lateral load, which causes the column to act like a beam, produces a moment that has a maximum value at the center ($M = wL^2/8$). This moment produces a bending stress of M/S—tension on the edge farthest from the load, and compressive bending stress on the edge where the load is applied (Figure 3.7).

Thus on the side of the column under the beam there are two compressive stresses, the axial stress P/A and the bending stress M/S. We must consider these two stresses acting together (their interaction). But because the allowable stress in compression parallel to the grain F_c is a different value from the allowable bending stress F_b, we cannot compare the sum thus obtained with any single allowable stress. The only logical approach is to compare each actual stress with it own allowable stress, giving two fractions. The sum of these two fractions cannot exceed 1, or else we would have overstressed the material. In formula form,

$$\left(\frac{f_c}{F_c'}\right)^2 + \frac{M/S}{F_b'(1 - f_c/F_{cE})} \leq 1$$

Notice that if either of the fractions is missing, the formula is still correct. That is, if $P = 0$ or if $M = 0$, the formula still holds. This formula is called the *interaction formula*.

Buckling is a different phenomenon from bending. The column will try to buckle in the easiest way it can, which is against the path of least resistance, which is the weaker side. Bending, however, occurs whenever a bending moment occurs (Figure 3.8) and is created by the positioning of the loads. Bending cannot be prevented, but we must guard against buckling by making the column stiff enough.

With combined loading the column is somewhat bent from the effect of the moment. The axial load will not be acting at the centroid when the column is thus deflected. This means that the axial load will actually be slightly eccentric at the point of maximum deflection and will therefore increase the bending moment somewhat. To account for this extra bending stress we reduce the allowable bending stress by the term in parentheses, $(1 - f_c/F_{cE})$, given in the interaction formula.

EXAMPLE 3.7—INTERACTION.

As an example, suppose that we have a column with an unbraced height of 8'-0", carrying an axial load of 20 kips and a lateral load of 625 plf. Suppose further that the material we have selected, Douglas Fir–Larch Dense No. 1, has the allowable stress values of $F_c = 1100$ psi, $F_b = 1550$ psi, and $E = 1,700,000$ psi found in Table A.11. We use the strength of a beam and stringer in anticipation that a square cross section will not be economical because of the bending.

The bending moment $M = wL^2/8 = 625 \times 8 \times 8 \times 12/8 = 60,000$ in.-lb or 60 in.-kips. Let us try a nominal 6×8, which has an area $A = 41.25$ in^2 and a section modulus $S = 51.56$ in^3. The column is in bending and is unbraced from top to bottom. The effective unbraced length as a beam is $L_e = 1.84L_u = 1.84 \times 96 = 194$ in. Then

$$R_b = \sqrt{\frac{L_e d}{b^2}} = \sqrt{\frac{194 \times 7.5}{5.5 \times 5.5}}$$

$$= 6.93 < 10, \quad \text{a short beam}$$

Try the table value for $F_b = 1550$ psi.

Considering column slenderness, $l/d = 96/5.5 = 17.5$. Since $11 < 17.5 < 30$, we have an intermediate column as to axial load.

$$F_{cE} = \frac{K_{cE} \times E'}{(l_e/d)^2} = \frac{0.3 \times 1,700,000}{(17.5)^2}$$

$$= 1665 \text{ psi}$$

$$\frac{F_{cE}}{F_c^*} = \frac{1665}{1100} = 1.514$$

AXIAL LOAD WITH BENDING 71

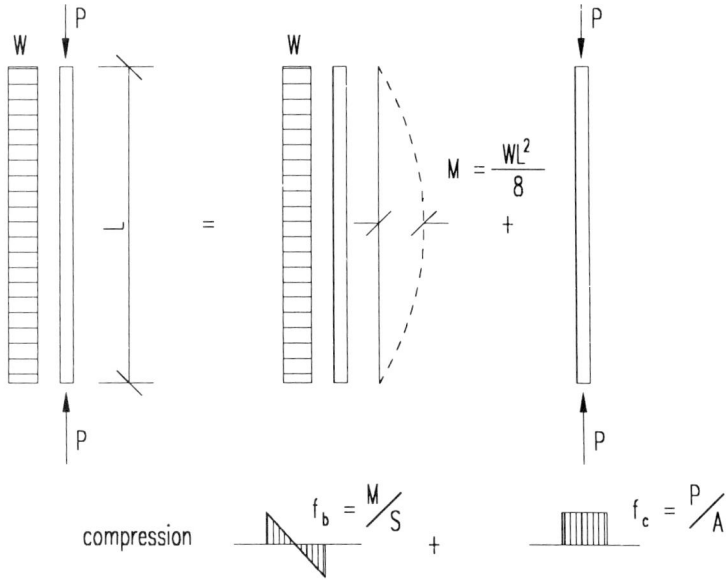

FIGURE 3.7 Interaction—axial and bending stresses combined.

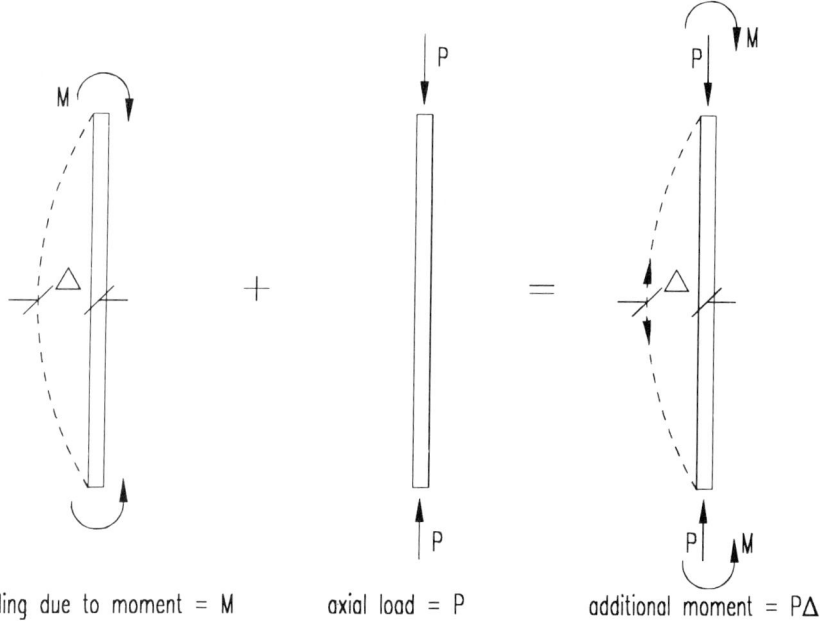

bending due to moment = M   axial load = P   additional moment = PΔ

FIGURE 3.8 P-Delta, moment due to deflection.

AXIAL LOADS

$$C_P = \frac{1 + F_{cE}/F_c^*}{2c}$$
$$- \sqrt{\left(\frac{1 + F_{cE}/F_c^*}{2c}\right)^2 - \frac{F_{cE}/F_c^*}{c}}$$

$$= \frac{1 + 1.514}{1.6}$$
$$- \sqrt{\left(\frac{1 + 1.514}{1.6}\right)^2 - \frac{1.514}{0.8}} = 0.812$$

$$F_c' = C_D \times C_M \times C_t \times C_F \times C_P \times F_c$$
$$= 0.812 \times 1100 = 893.3 \text{ psi}$$

The actual axial stress,

$$f_c = \frac{P}{A} = \frac{20,000 \text{ lb}}{41.25 \text{ in}^2} = 484.8 \text{ psi}$$

The actual bending stress,

$$f_b = \frac{M}{S} = \frac{60,000 \text{ lb-in.}}{51.56 \text{ in}^3} = 1164 \text{ psi}$$

The interaction formula states that

$$\left(\frac{f_c}{F_c'}\right)^2 + \frac{M/S}{F_b'(1 - f_c/F_{cE})} \leq 1$$

$$\left(\frac{484.8}{893.3}\right)^2 + \frac{1164}{1550 \times (1 - 484.8/1665)}$$

$$= (0.543)^2 + \frac{1164}{1550 \times (1 - 0.291)}$$

$$= 0.295 + \frac{1164}{1099}$$

$$= 0.295 + 1.06 \quad \text{greater than 1.0}$$
$$\text{No Good}$$

Since the sum of the fractions is greater than unity, the section selected is too small. It is overstressed by 0.94 or 94%. For a second guess we try the next larger section, an 8×8, with $A = 56.25 \text{ in}^2$ and $S = 70.31 \text{ in}^3$. Now

$$R_b = \sqrt{\frac{L_e d}{b^2}} = \sqrt{\frac{194 \times 7.5}{7.5 \times 7.5}}$$
$$= 5.09 < 10 \quad \text{a short beam}$$

$F_b = 1550 \text{ psi}$

$$\frac{l}{d} = \frac{96}{7.5} = 12.8 \quad 11 < 12.8 < 30,$$
$$\text{intermediate}$$

$$F_{cE} = \frac{0.3 \times ,700,000}{(12.8)^2} = 3113 \text{ psi}$$

$$\frac{F_{cE}}{F_c^*} = \frac{3113}{1100} = 2.83$$

$$C_P = \frac{1 + 2.83}{1.6}$$
$$- \sqrt{\left(\frac{1 + 2.83}{1.6}\right)^2 - \frac{2.83}{0.8}}$$
$$= 0.913$$

$F_c' = 0.913 \times 1100 = 1004.3 \text{ psi}$

The actual axial stress,

$$f_c = \frac{P}{A} = \frac{20,000 \text{ lb}}{56.25 \text{ in}^2} = 355.6 \text{ psi}$$

The actual bending stress,

$$f_b = \frac{M}{S} = \frac{60,000 \text{ lb-in.}}{70.31 \text{ in}^3} = 583.4 \text{ psi}$$

The interaction formula becomes

$$\left(\frac{355.6}{1004.3}\right)^2 + \frac{853.4}{1550 \times (1 - 0.114)}$$
$$= 0.125 + 0.621 = 0.746,$$
$$\text{less than 1.0} \quad \text{OK}$$

Use 8×8 Douglas Fir–Larch No. 1

Generally, the sum of the two fractions should be close to 1. Although this is not always possible using standard timber sizes, it should be attempted. We have used slightly less than 100% (i.e., 0.746 = 74.6%) of the strength of our section. This is ideal. The design is accepted.

ECCENTRIC AXIAL LOADS

The load on a column is not always directed downward along the centroid of the shaft, so as to produce compression only. Sometimes the load is such that it creates bending and compression simultaneously. For example, if the architectural design calls for a beam attached to the side of a column (Figure 3.9), the load on the column is not at its centroid but is actually half a column width away. This is called an *eccentric* load, and the amount of eccentricity is $e = t/2$.

Let us cut a thin slice across the column somewhere below the top (see Figure 3.9c). If the reaction of the beam has a magnitude of P pounds, it will transfer to the column as a concentrated load acting downward on the face of the column. Let us now add two equal loads P in opposite directions at the centroid of the column. This is permitted because we have not changed any of the conditions involved in static equilibrium, nor have we added any load to the column. These two new theoretical loads can be recombined to produce two loading conditions. The first condition, induced by the downward-acting load, is that of a concentrated load P acting at the centroid of the shaft, creating a purely compressive stress, P/A, over the entire cross section. The upward-acting load, however, does not act independently, but works with the original load P to produce a moment couple equal to Pe. This moment produces a bending stress that has the maximum values of M/S—tension on the edge farthest from the load, and compressive bending stress on the edge to which the beam is attached. It can be seen that this condition is very similar to that of the column with combined axial and bending stresses.

For columns with eccentric loads the NDS uses a modified version of the interaction formula, which is intended to slightly improve the accuracy in determining the effects of eccentric loads. The eccentricity equation is more conservative than the interaction formula and will result in higher stresses in the column.

$$\left(\frac{f_c}{F'_c}\right)^2 + \frac{f_c(6e_1/d_1) \times [1 + 0.234(f_c/F_{cE})]}{F'_b(1 - f_c/F_{cE})} \leq 1$$

EXAMPLE 3.8—ECCENTRICITY.

As an example, suppose that we have a column with an unbraced height of 8'-0", carrying a load of 20 kips, which acts at an eccentricity of 3 in. The same material will be selected, Douglas Fir–Larch Dense No. 1, with allowable stress values of $F_c = 1100$ psi, $F_b = 1550$ psi, and $E = 1,700,000$ psi.

The moment will be the same as before, $M = Pe = 20,000$ lb $\times$ 3 in. = 60,000 lb-in. Let us try the section that was successful in Example 3.7, an 8×8, with $A = 56.25$ in^2 and $S = 70.31$ in^3. All values remain the same going in to the interaction formula, so

$$\frac{f_c}{F'_c} = \frac{355.6}{1004.3} = 0.354$$

$$(0.354)^2 + \frac{355.6(6e_1/d_1) \times [1 + 0.234(355.6/3113)]}{F'_b(1 - 355.6/3113)}$$

$$= (0.354)^2 + \frac{355.6 \times (6 \times 3/7.5)[1.027]}{1550(1 - 0.114)}$$

$$= 0.125 + \frac{876.3}{1373}$$

$$= 0.125 + 0.638 = 0.763$$

This column is stressed to about 76% and is acceptable. Note that for the same load, the eccentric load produced an increase in stress of about 2% (0.763/0.746).

74 AXIAL LOADS

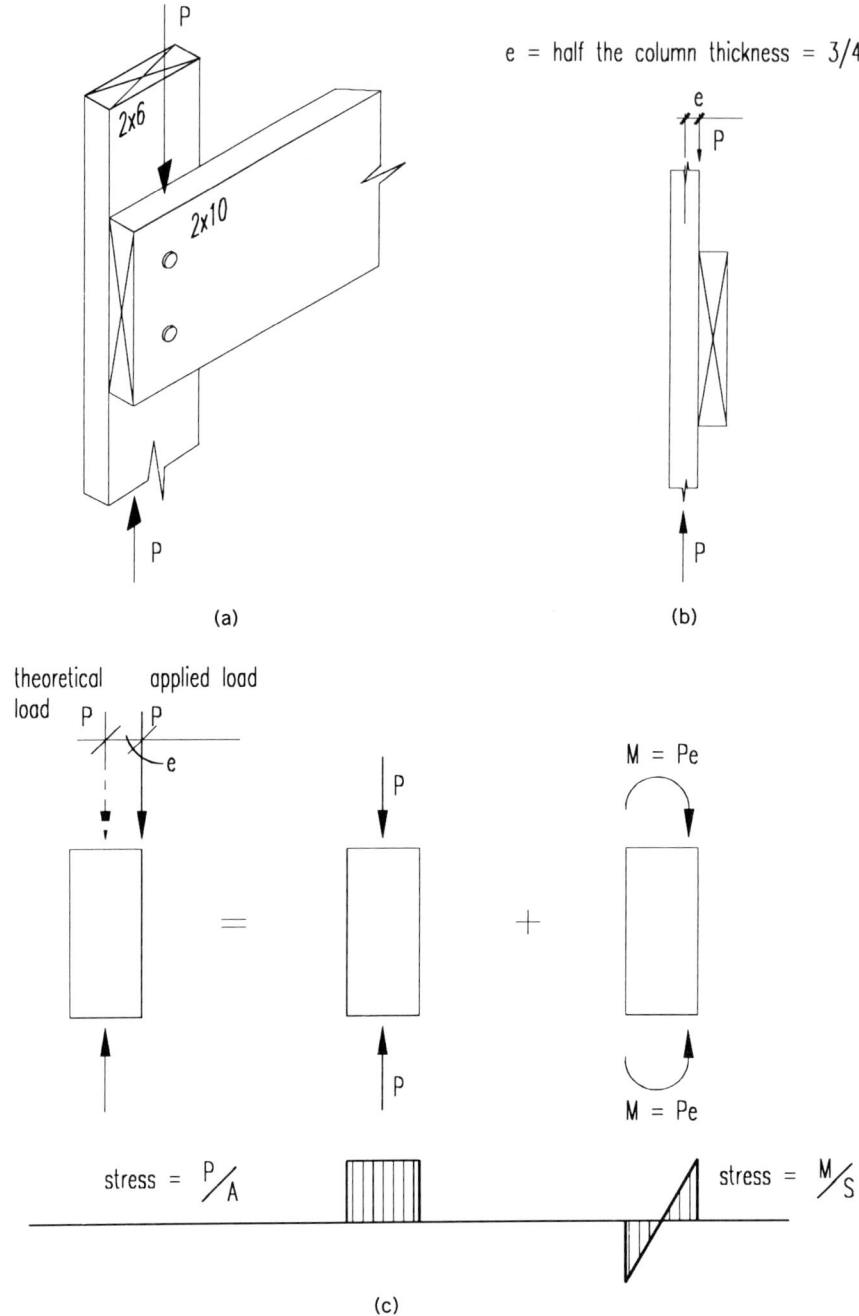

FIGURE 3.9 Eccentric loading: (a) application, (b) measurement, and (c) effect.

Generally, if the simpler interaction formula results in a column that is stressed to less than about 90% of its allowable, there is no need to use the more exact calculation.

Other cases of combined loading are fairly common. When a vertical column is subjected to the horizontal load of wind pressure and, in addition, the vertical load is eccentric, the two moments are added together. The eccentric–load interaction formula is expanded in this case to read

$$\left(\frac{f_c}{F'_c}\right)^2 + \frac{f_b + f_c(6e_1/d_1)[1 + 0.234(f_c/F_{cE})]}{F'_b(1 - f_c/F_{cE})} \leq 1$$

where f_b is the bending stress due to side loads.

If wind load is used, *all* allowable stresses are increased by the duration factor for wind. In such a case it is necessary to design the column for the eccentric load acting alone with normal allowable stresses and then check the design for the combined effect of the eccentric load and the wind load acting simultaneously, with the increased allowable stresses from the wind duration factor. If the vertical load is concentric, with no eccentricity, the column should be designed for the axial load with normal allowable F'_c and then checked for wind interaction with all allowable stresses increased by the duration factor for wind.

In Chapter 5 combined bending and axial loading in trusses is considered. Finally, the case of interaction of axial load and bending moment is found always to be present in three-hinged arches. These are discussed in Chapter 6.

Important Note: As was the case with allowable bending stress in beams, all allowable stresses F'_c and F'_b are subject to adjustment for load duration and moisture conditions. These adjustments are the same as with beams, 0.90 for dead load only, 1.33 for wind or earthquake (UBC), 1.6 for wind or seismic (BOCA or SBCCI), and so on.

NET SECTION

It is always best not to cut out any part of a structural member, but sometimes the removal of some material is unavoidable. For example, when a piece is bolted to the side of a column, a hole slightly larger than the bolt is drilled through the column. In such a case, and whenever material is removed by drilling, boring, dapping, grooving, notching, or other means, the projected area of all material removed is subtracted from the gross section to get a *net section*. It is this net section that is used in calculating the load-carrying capacity of compression members. The material thus removed will also adversely affect the section modulus.

These deductions must also be computed whenever they occur in the middle third of the length of intermediate and long columns because the middle third is the part most subject to potential buckling. Notice that only the area and the section modulus are affected by the net section. Typically, cuts are so short that they do not affect the average moment of inertia. As opposed to strength calculations, where the smallest cross section is the most critical, deflections are influenced by more of an average moment of inertia. Any material removed might also reduce the least thickness d when computing the slenderness ratio l/d.

TENSION

Certain structural members resist a tensile load. Such a load is one that acts to stretch or pull the member apart. Examples can be found in trusses and timbers used as hangers.

The internal resisting tension strength must be great enough to safely resist the applied tension load. This internal strength is measured in terms of allowable tension stress parallel to the grain. Since the allowable tension stress perpendicular to the grain is unknown, loads should not be applied in that manner. When tension perpendicular to the grain stresses cannot be avoided, the load should be held by a mechanical device, such as a steel strap, and not attached to the bare member. An example of a design that induces critical tension perpendicular-to-grain stresses is found in the practice of hanging loads below the neutral axis of a beam. This design should be avoided by using a U-strap that fits over the top of the beam.

Whenever a tension member is held in place by steel plates, as with residential trusses, or by nails, screws, or lag screws, the load is resisted by the entire cross section, called the *gross area* A_g. However, when bolts or split rings are employed, it is absolutely essential to use the *net section* A_n in computing the actual tension stress, and thus the load-carrying capacity of the member. The net section is found by taking a section through the holes on a plane perpendicular to the line of the force acting on the member.

EXAMPLE 3.9—TENSION ANALYSIS.

Let us determine the maximum tension load P that can be applied to a 2×6 made of Douglas Fir–Larch No. 2 that is held in place by six bolts of $\frac{5}{8}$-in. diameter, as shown in Figure 3.10.

The net section is found by cutting a typical section A–A. Notice that only two bolt holes are found in the section, since it is cut orthogonally to the load direction. Although for small bolts the hole may be drilled only $\frac{1}{32}$ in. larger than the bolt, most holes are made $\frac{1}{16}$ in. larger than their bolts. Therefore, let us take the hole to be the bolt diameter plus $\frac{1}{16}$ in., or $\frac{5}{8}$

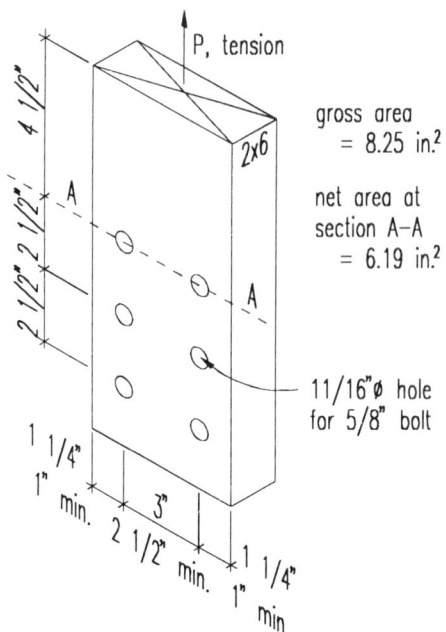

FIGURE 3.10 Net section.

$+ \frac{1}{16} = \frac{11}{16}$ in. The amount of wood removed for each hole is the thickness of the piece times the diameter of the hole, $1\frac{1}{2} \times \frac{11}{16} = 1.03$ in^2. Since there are two holes it is necessary to subtract 2.06 in^2. The gross area A_g of a 2×6 is 8.25 in^2. The net area is thus $A_n = 8.25 - 2.06 = 6.19$ in^2. For the wood selected the allowable tension stress parallel to the grain is $F_t = 575$ psi $\times 1.3 = 748$ psi ($C_F = 1.3$ for 6-in.-wide members in tension). The maximum load

$$P = F_t \times A_n$$
$$= 748 \times 6.19$$
$$= 4630 \text{ lb}$$

Design is a mater of trial and error. After the species and grade of wood are selected, it is necessary to assume the bolt size and the number of rows of bolts that will be used. It is also necessary to select a trial thickness of timber. All of these may be adjusted after the first set of computations is made if the answer warrants it because either the piece is too large or it is too small for the assumed number of rows of bolts or rings.

EXAMPLE 3.10—TENSION DESIGN.

For example, suppose that a tension load of 7500 lb is to be carried by No. 2 Southern Pine using $\frac{1}{2}$-in.-diameter bolts. Let us assume, first, that a $2\times$ will do the job nicely, and second, that two rows of bolts will be needed.

Entering Table A.9 for allowable stresses, it is seen that F_t might be either 650 (for a 2×8) or 575 psi (for a 2×10 or 2×12) of this grade. Assuming the worst case, try $F_t = 575$ psi. The required net section will be

$$A_n = \frac{P}{F_t}$$
$$= \frac{7500}{575} = 13.04 \text{ in}^2$$

With $\frac{1}{2}$-in.-diameter bolts the hole diameter will be $\frac{1}{2} + \frac{1}{16} = \frac{9}{16}$ in. The projected area of the material that is cut out is thus

$$\frac{9}{16} \times 1\frac{1}{2} \text{ (material thickness)} = 0.844 \text{ in}^2$$

With two rows of holes we get $2 \times 0.844 = 1.69$ in^2. The gross area less the hole area yields the net area. $A_g - 1.69 = 13.04$ in^2. Thus the required gross area is found to be

$$A_g = 13.04 + 1.69 = 14.73 \text{ in}^2$$

as a minimum

Select a 2×12 with $A_g = 16.88$ in$^2 > 14.73$ in^2, OK.

Use 2×12 Southern Pine No. 2.

4
CONNECTIONS

In designing wood structures an engineer is responsible not only for the design of the various members but also for the connections. On a typical project it is not surprising to find that the design of the connections may comprise half of the work. Furthermore, it is estimated that as many as 90% of the structural failures experienced in wood-frame buildings originate at the connections. It is common practice in wood stud construction to detail connections so that the gravity loads are carried in bearing through the wood. There are some noticeable exceptions, such as joist hangers, that violate this principle, but it is a principle that should be adhered to as much as possible. What this means is that connectors such as nails, screws, and bolts are primarily the paths available to carry wind and seismic loads. For most people a gravity force is easier to understand; we constantly feel its effects on our bodies. Since very large wind and seismic loads occur less frequently, we are less accustomed to thinking about them; however, we must always work in structural design to recognize the effects of these types of loads. In this chapter and in Chapter 7 we discuss the effects of lateral loads on the elements and connections in timber construction.

In many instances the connection detail will require drilling or cutting away a portion of the wood. Obviously, any cut will decrease the strength of the wood, often at the location that needs the most strength. The importance of accurately calculating, specifying, detailing, and installing timber connections should never be underestimated. Through this chapter we discuss the principles behind the various types of connectors used and discuss the important concerns and precautions that must be considered in connection design and detailing.

Throughout the history of wood construction a number of types of fasteners have been developed for use with wood, ranging from intricate timber-frame connections using wooden dowels to the more modern steel connectors currently available. The most common connections used currently include nails, staples, screws, bolts, and special connectors such as split rings, shear plates, and truss plates. In this chapter we discuss these connectors in detail.

RECOMMENDED FASTENING SCHEDULE

We start by looking at the connection schedule that is used in common practice. Where light framing is used under light to moderate loads, it is often sufficient to specify nails by using the recommended fastening schedule from the model code that prevails in your area. Table 4.1 lists the recommended fasteners as specified by the CABO *One- and Two-Family Dwelling Code*. This model code has been endorsed by all the major code agencies and is applicable throughout most of the United States. (In some instances, recommended fasteners vary from code to code.) Figure 4.1 shows the section through a wall of a typical wood-stud residential building and indicates the location of some of the connections that are referred to in the fastening schedule.

The principal requirement of the connections in Figure 4.1 is to resist wind or seismic loads. In Example 4.1 we calculate the forces that are exerted on some of these connections to determine their capacity to resist wind loads. Wind loads on a building can have any of three effects: they may cause the building, or pieces of it, to slide sideways; they may cause the building, or more often the roof, to lift up; or they may cause overturning about some point that does not move. Figure 4.2 diagrams these conditions.

There are two categories of forces available to resist these loads: one is the weight of the structure itself (commonly referred to as the dead load) and the other is the connectors. The standard engineering principle is to assume that the weight of the building and the connections can combine to resist the effects of the wind. However, to assure adequate safety, there are two provisions that are always required. First, with regard to sliding, the wind produces a horizontal force, whereas the building weight is a vertical force. The dead load can resist a lateral load only through developing friction between the building and its foundation. Friction forces are very dependent on the conditions at the plane of contact: the two materials in contact and their condition, such as wet or dry, clean or dirty. Since there is very little quality control on a job to assure these conditions, the designer does not have a quantifiable sense of the factor of safety related to this portion of the design. It is best to use a rather conservative estimate of the friction factor. In the example that follows we take the friction factor between two wood surfaces as 0.25 and that between wood and concrete as 0.50. The actual factors are substantially greater. The other concern with friction is whether the dead load will actually be in place when the lateral load is applied. Under wind loads the uplift force can be calculated. The total downward load should be taken as the dead load less the wind uplift force. Under seismic loads there is a possibility that the earth movement will have a vertical as well as a horizontal component. If the earth essentially "falls away" below the building, there will be an interval of time in which the building is "weightless"—there will be no dead load. For this reason, *friction is never considered in calculating the effects of sliding due to seismic loads*. The major building codes differ in their wording because of this. In the western states the *Uniform Building Code* (ICBO) prohibits the use of friction to resist sliding for any type of load. In the northeast the BOCA *National Building Code* allows friction to resist sliding arising from wind loads. With regard to seismic loads, BOCA is not specific, saying, "Consideration shall be given to minimum gravity loads acting in combination with lateral loads." Even under

TABLE 4.1 RECOMMENDED FASTENING SCHEDULE

Key	Building Element	Connection Size and Type	Number Location	Notes
a	Sill plate to foundation	$\frac{1}{2}$-in-diameter bolts	6'-0" o.c.	Seismic zones 1 and 2
			4'-0" o.c.	Zones 3 and 4
b	Floor joist to sill plate	8d common	3 toenail ea.	
c	Rim joist to floor joist	10d common	2 at each joist	Rim is 6 in. or less
			3 at each joist	Rim is 6 in. or more
d	Sole plate to joist or blocking	16d common	16 in. o.c.	
e	Stud to sole plate	8d common	4 toenail	
		or		
		16d common	3 direct nail	
f	Double studs	10d common	12 in. o.c. direct	
g	Corner studs	16d common	24 in. o.c. direct	
h	Stud to top plate	16d common	2 toenail	
			or	
			2 direct nail	
i	Diagonal bracing	8d common	2 each direct bearing	
j	Double top plate	10d common	16 in. o.c.	
k	Splices at top plate	10d common	2 direct nail	
l	Floor joist to stud	10d common	5 direct	Balloon frame, no ceiling joist
		or		
		16d common	3 direct	
m	Floor joist to stud	10d common	2 direct	Balloon frame with ceiling joist
n	Ceiling joist to top plate	16d common	3 toenail	
o	Ceiling joists	10d common	3 direct	Laps over partition
p	Ceiling joists	10d common	3 direct	Parallel to rafter
q	Roof rafter to plate	16d common	3 toe nail	
r	Roof rafter to ridge	16d common	2 direct or 2 toenail	
s	Built-up girder or beam	20d common	32 in. o.c.	[a]
t	Continuous header beam	16d common	16 in. o.c.	Along each edge
u	Header to trimmers	20d common	1 each end per 8 ft^2 floor area	Where nailing permitted
v	Tail beam to header	20d common	1 each end per 4 ft^2 floor area	Where nailing permitted

[a] At top and bottom, staggered. Two at ends and at each splice.
Source: CABO *One- and Two-Family Dwelling Code.*

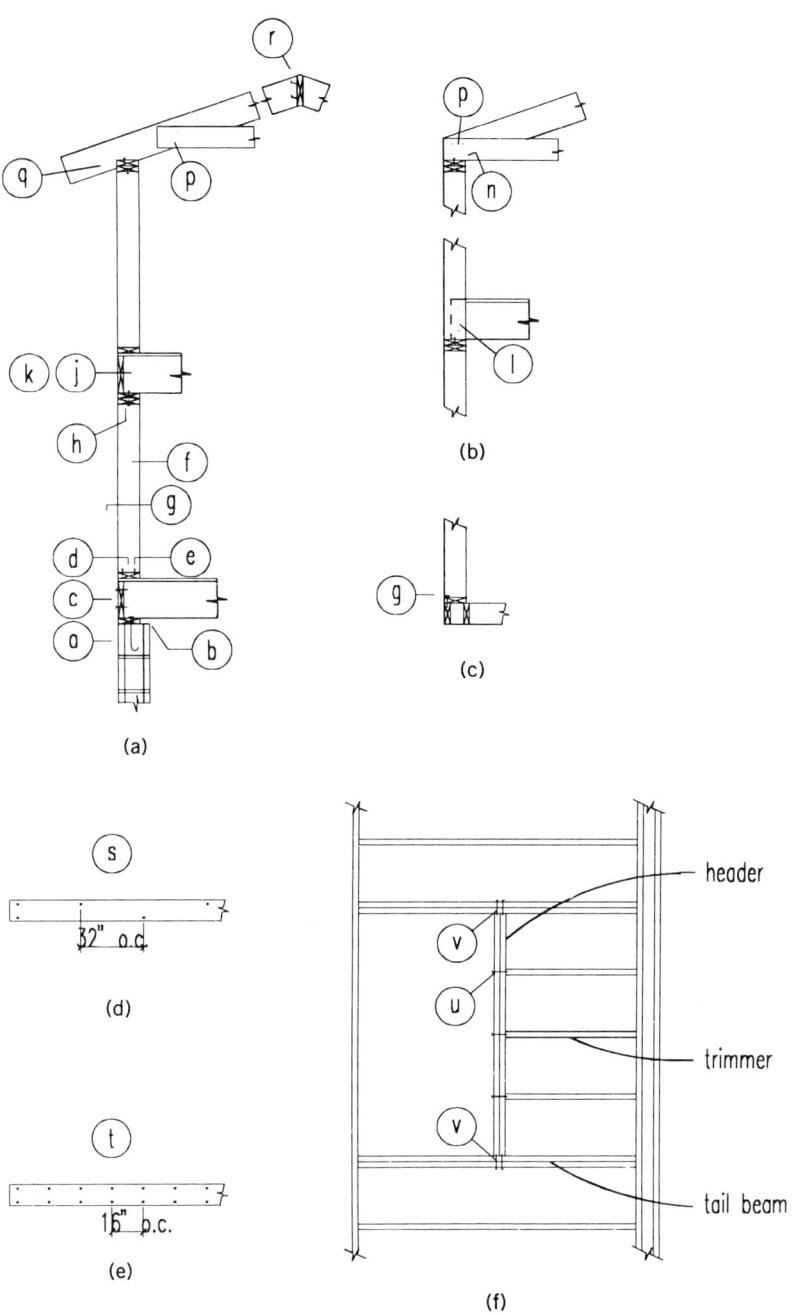

FIGURE 4.1 Recommended fastener schedule: (a) wall section, (b) section at balloon frame, (c) plan of corner studs, (d) built-up beam, (e) header, and (f) plan of header at floor opening.

82 CONNECTIONS

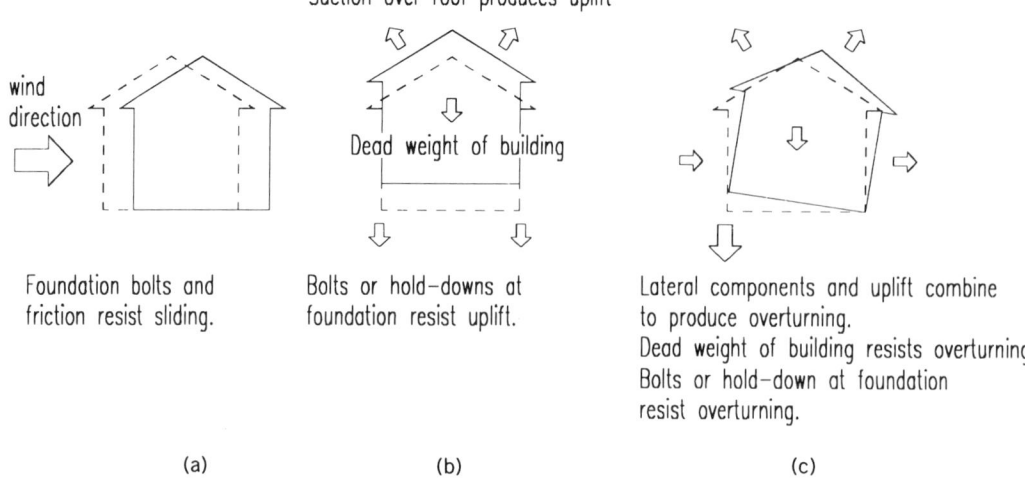

FIGURE 4.2 (a) Sliding, (b) uplift, and (c) overturning.

BOCA conditions friction should not be considered to resist sliding.

The second provision occurs in the design to resist overturning. It is assumed that in this case an additional factor of safety is necessary to account for the variability of the dead load. Standard practice is to use only two-thirds of the effective dead load. This provision can be considered to represent the case when one-third of the dead load, say a portion of the roof, has blown away. If the remaining dead load (two-thirds of the total dead load) can resist the wind load, no mechanical resistance (connectors) are required. If however, the dead-load resistance is not sufficient, connectors will be needed to overcome the difference between the wind-induced overturning and the dead-load-induced resistance.

Mathematically, this provision can be written as

If $\frac{2}{3}$ R.M. ≥ O.T.M, no mechanical resistance is needed.

If $\frac{2}{3}$ R.M. < O.T.M, M.R. must be ≥ O.T.M − $\frac{2}{3}$ R.M.

where $\frac{2}{3}$ R.M. is two-thirds of the resisting moment due to dead loads, O.T.M. is the overturning moment due to wind, and M.R. is the moment resistance due to the connectors. In calculating the resistance to uplift, the two-thirds reduction of dead loads is not used; the full dead load is assumed to be in place in resisting uplift.

EXAMPLE 4.1—WIND UPLIFT AND OVERTURNING

Consider the wood-frame building shown in Figure 4.3. The building is a typical two-story suburban home measuring 55 ft long and 28 ft wide. The elevation of the top plate below the roof is 18 ft and the elevation of the ridge is 25 ft. The roof slope is 6 in 12. Assuming that the standard connection schedule is used, will the connections be able to withstand winds of 70 mph? Use wind loads and coefficients from the BOCA *National Basic Building Code*.

To investigate this problem, let us select three locations to check: (1) the connection between the roof trusses and the wall plate, (2) the connection that holds the second floor wall to the floor, and (3) the connection that holds the entire house to the foundation. For each connection we

RECOMMENDED FASTENING SCHEDULE   83

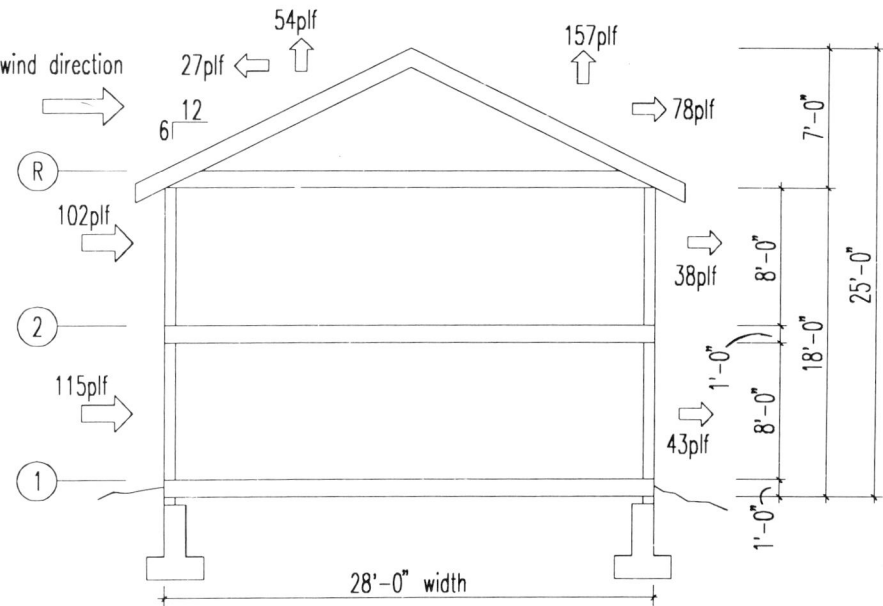

FIGURE 4.3 Section through typical residence—Example 4.1.

will first consider the effects of uplift, overturning, and sliding. Both uplift and overturning will produce a vertical force at the joint. The connection will have to resist the portion of the wind load not resisted by the dead load. We will take the larger of those two forces in selecting the connector. Sliding requires a horizontal resistance. Each connection must carry both the vertical force and the horizontal force.

Figure 4.4 shows the calculations of uplift, overturning, and sliding at these points. It also computes the dead load that resists these forces and the required force at the connection. These values are summarized in Table 4.2a. Under the heading of 70 mph wind speed you will find the required lateral resistance of 28 lbs. This value is based on stud spacing of 16 in. o.c. For studs 12 in. o.c. the required value would be 75% of 28 lbs. or 21 lbs. This is the same value (21 plf) calculated in Figure 4.4c for overturning. All solutions in Figure 4.4 are computed in units of pounds per foot. In Table 4.2 they are converted to 16 ft stud spacing by multiplying them by 16/12.

Figure 4.5 shows a section through the wall of the residence and calls out all of the CABO code standard nailed connections and their calculated capacities. The procedure for calculating these values is explained in detail in Examples 4.6 through 4.9. In the figure you will see that there are two numbers listed for the vertical capacity. The number in parentheses is the capacity calculated with a wind duration factor of 1.33 according to ICBO. The other figure is the capacity derived with a duration factor of 1.6, according to BOCA and SBCCI. The same is true for the horizontal capacities. In the example we use the larger capacities ($C_D = 1.6$) since we are designing for the BOCA *National Building Code*.

Figure 4.5 may be compared with Figures 4.6 and 4.7, which show the same wall using metal connectors, which are rated for higher loads than the standard nailed connections. In these figures the metal connectors are rated based on tests and calculations. The loads shown use the duration factor of 1.33. Since these ratings include the possibility of failure in the

(*text continues on page* 88)

84 CONNECTIONS

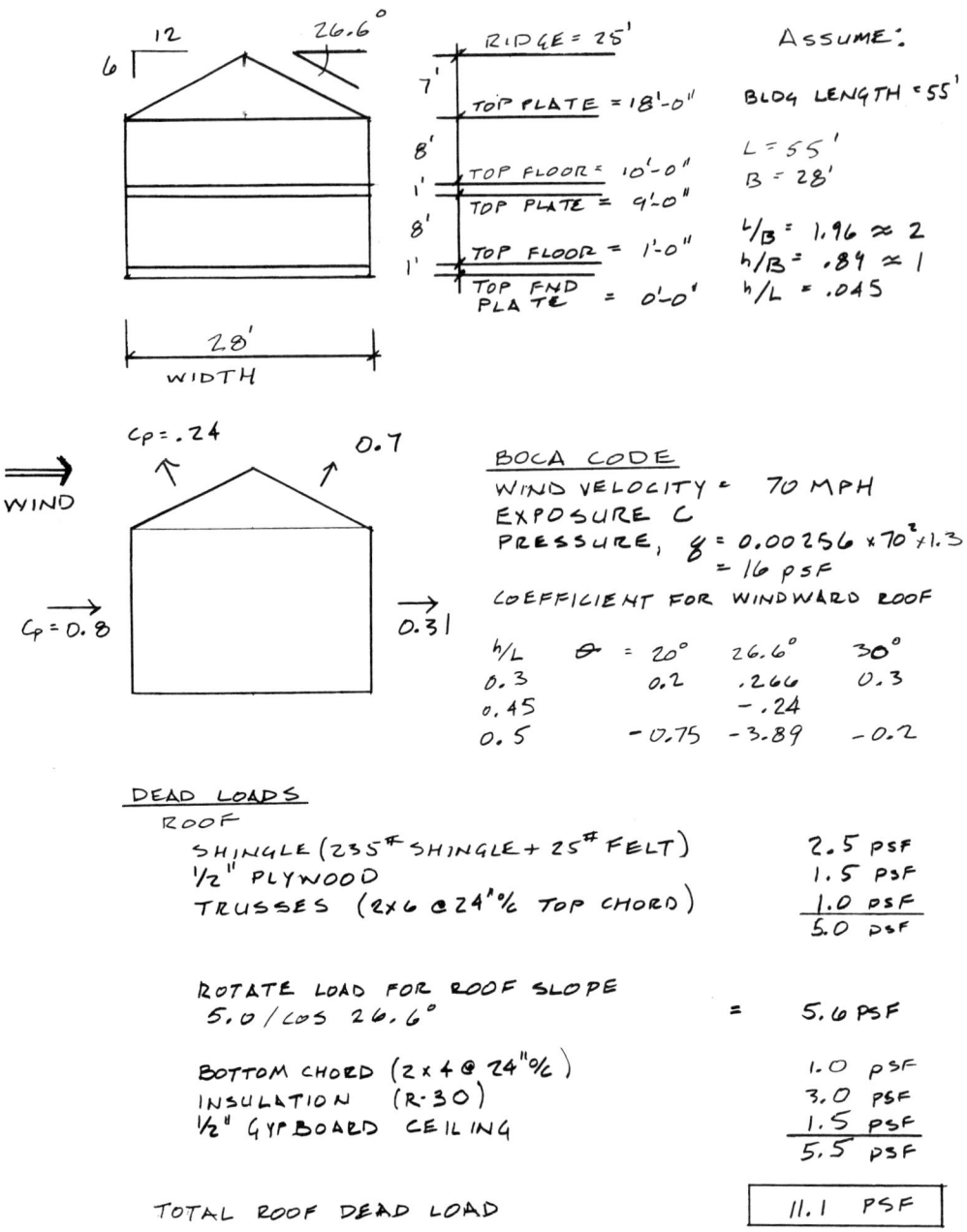

FIGURE 4.4 Calculations for uplift, overturning, and sliding.

DEAD LOADS, CONT.

    EXTERIOR WALLS

3/4" WOOD SIDING	2.5 PSF
2x4 @ 16" O/C STUDS	1.0
INSULATION (R-11)	1.5
1/2" GYPBOARD	1.5
	6.5 PSF

    INTERIOR WALLS

1/2" GYPBOARD	1.5 PSF
2x4 @ 16" O/C STUDS	1.0
1/2" GYPBOARD	1.5
	4.0 PSF

AS A CONSERVATIVE ESTIMATE, ASSUME THERE IS AT LEAST ONE INTERIOR WALL IN ANY SECTION THROUGH THE BUILDING. PLACE IT AT CENTERLINE OF SECTION.

    FLOORS

CARPET AND PAD	1.75 PSF
3/4" PLYWOOD	2.25
2x10'S @ 16" O/C JOISTS	2.5
	6.5 PSF

REQUIRED MECHANICAL CONNECTIONS

UPLIFT
    IF UPLIFT > DEAD LOAD

$$\boxed{R = UPLIFT - DL}$$

OVERTURNING
    IF OVERTURNING MOMENT > 2/3 DEAD LOAD MOMENT

$$RM = OTM - \tfrac{2}{3} DLM$$

$$\boxed{R = \frac{OTM - \tfrac{2}{3} DLM}{MOMENT\ ARM}}$$

SLIDING
    IF SLIDING FORCE > FRICTION RESISTING FORCE
    THEN R_H = SLIDING - FRICTION

AS CONSERVATIVE ESTIMATE OF FRICTION COEFF
TAKE $f = 0.25$ WOOD-TO-WOOD
    OR $f = 0.50$ WOOD-TO-CONCRETE

$$\boxed{R = S - .25\,(DL - UPLIFT)}$$

(b)

FIGURE 4.4 (*Continued*)

CONSIDER EFFECTS AT ROOF PLATE

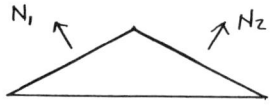
WINDS NORMAL

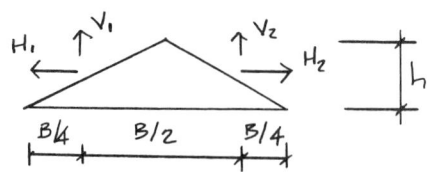
WIND COMPONENTS

THE WIND PRESSURE ACTS NORMAL (PERPENDICULAR) TO THE SURFACE. THIS SYSTEM IS EQUIVALENT TO ONE IN WHICH WIND COMPONENTS ACT UPON THE PROJECTED AREA OF THE ROOF. IN THE COMPONENT SYSTEM, USE THE SAME PRESSURE COEFFICIENTS.

SO $H_1 = 0.24 \times 16$ PSF $\times 7'$ = 26.88 PLF
    $V_1 = 0.24 \times 16$ PSF $\times 14'$ = 53.76 PLF
    $V_2 = 0.7 \times 16$ PSF $\times 14'$ = 156.8 PLF
    $H_2 = 0.7 \times 16$ PSF $\times 7'$ = 78.4 PLF

UPLIFT

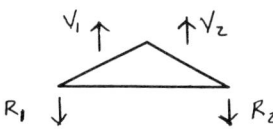

$R_1 \times 28' = 53.76 \times (3/4 \times 28') + 156.8 \times (1/4 \times 28')$
$R_1 = 79.5$ plf
$R_2 = 156.8 + 53.76 - 79.5 = 131$ plf
$DL = 11.1$ psf $\times 28' \times 1/2 = 155.4$ plf
$R_2 - DL = 131 - 155.4 = -24.4$ plf
CONVERT TO 16" STUD SPACING
-24.4 plf $\times 16/12 = \boxed{-32.5 \text{ lb/STUD}}$
NEGATIVE MEANS NO CONN. REQ'D.

OVERTURNING

$R_1 \times 28' = 53.76 \times 21' + 156.8 \times 7' - 26.88 \times 3.5'$
$\qquad + 78.4 \times 3.5'$
$R_1 = 86$ plf
$R_2 \times 28' = 53.76 \times 7' + 156.8 \times 21' + 26.88 \times 3.5'$
$\qquad - 78.4 \times 3.5'$
$R_2 = 124.6$ plf
$DL = 155.4$ plf
$R_2 - 2/3 DL = 124.6 - 2/3 \times 155.4 = 21.0$ plf
$21 \times 16/12 = \boxed{28 \text{ lb/STUD}}$

SLIDING

$DL - UPLIFT = 155.4 - 131$
$= 24.4$ plf
FRICTION $= .25 \times 24.4$ plf
$= 6.1$ plf

$(-H_1 + H_2) = -26.88 + 78.4 = 51.5$ plf
APPLY 1/2 TO EACH SIDE $= 25.76$ plf
$25.76 - 6.1 = 19.66$ plf
$19.66 \times 16/12 = \boxed{26.2 \text{ lb/STUD}}$
$=$ UNRESTRAINED SLIDING

(c)

FIGURE 4.4 (*Continued*)

CONSIDER EFFECTS AT SECOND FLOOR

$V_1 = 53.76$ $V_2 = 156.8$
$H_1 = 26.88$ $H_2 = 78.4$

AT TOP OF FLOOR $h_w = 8'$ $h_r = 7'$ $h = 15'$
$W_1 = 0.8 \times 16\,psf \times 8' = 102.4\,plf$
$W_2 = .31 \times 16 \times 8' = 39.7\,plf$

UPLIFT $DL = 11.1\,psf \times 28' \times \frac{1}{2}$ (roof) $= 155.4\,plf$
$ + 6.5 \times 8'$ (wall, ext) $= 52$
$ + 4.0 \times 8 \times \frac{1}{2}$ (wall, int) $= \underline{16}$
$\phantom{DL = + 4.0 \times 8 \times \frac{1}{2} \text{ (wall, int)} =\;} 223.4\,plf$

$R_2 = 131\,plf$ (same as at roof)
$R_2 - DL = 131 - 223.4\,plf = -92.4\,plf \Rightarrow \boxed{-123.2\,lb/STUD}$
 NO CONNECTION REQ'D

OVERTURNING $R_1 \times 28' = 102.4 \times 4' - 26.88 \times 11.5 + 53.76 \times 21'$
$ + 156.8 \times 7' + 78.4 \times 11.5 + 39.7 \times 4'$
$R_1 = 121\,plf$
$R_2 \times 28' = -102.4 \times 4' + 26.88 \times 11.5 + 53.76 \times 7$
$ + 156.8 \times 21 - 78.4 \times 11.5 - 39.7 \times 4$
$R_2 = 89.6\,plf$
$OTM - \frac{2}{3} DL = 121 - \frac{2}{3} \times 223.4 = \boxed{-28.9\,plf \Rightarrow -37.2\,lb/STUD}$

SLIDING
$102.4 - 26.88 + 78.4 + 39.7 = 193.6\,plf$
half each side = $96.8\,plf$
$96.8 - \frac{1}{4} \times (223.4 - 131) = 73.7\,plf \Rightarrow \boxed{+98.2\,lb/STUD}$

AT TOP OF WALL PLATE
$h_w = 9'$ $h = 16'$ $W_1 = 115.2\,plf$ $W_2 = 44.64\,plf$
$DL = 223.4 + 6.5\,psf \times 28' \times \frac{1}{2} = 314.4\,plf$

UPLIFT $R_2 - DL = 131 - 314.4 = -183.4\,plf \Rightarrow \boxed{-244.5\,lb/STUD}$

OVERTURNING $R_1 - \frac{2}{3} DL = 128.2 - \frac{2}{3} \times 314.4 = -81.4\,plf \Rightarrow \boxed{-108.5\,lb/STUD}$

SLIDING   SLIDING FORCE = $211.4\,plf / 2 = 105.7\,plf$ @ each end
FRICTION = $\frac{1}{4}(314.4 - 131) = 45.85$
CONNECTION FORCE = $105.7 - 45.85 = 59.8\,plf$
$59.8 \times 16/12 = \boxed{79.8\,lb/STUD}$

(d)

FIGURE 4.4 (*Continued*)

CONSIDER EFFECTS AT FIRST FLOOR

AT TOP OF FLOOR $h_w = 17'$ $h = 24'$
$W_1 = 217.6 \text{ plf}$
$W_2 = 84.3 \text{ plf}$
$DL = 314.4 + 52 + 16 = 382.4 \text{ plf}$

UPLIFT $131 - 382.4 = -251.4 \text{ plf} \Rightarrow \boxed{-335.2 \text{ lb/STUD} \quad \text{NO UPLIFT}}$

OVERTURNING $R_1 = 208.9 \text{ plf}$ $R_2 = 1.7 \text{ plf}$
$R_1 - 2/3 \, DL = 208.9 - 254.9 = -46 \text{ plf} \Rightarrow \boxed{-61.8 \text{ lb/STUD}}$

SLIDING $\frac{1}{2} \times 353.4 - \frac{1}{4}(382.4 - 131) = 113.9 \text{ plf} \Rightarrow \boxed{151.8 \text{ lb/STUD}}$

AT TOP OF FOUNDATION $h_w = 18'$ $h = 25'$
$W_1 = 230.4 \text{ plf}$
$W_2 = 89.3 \text{ plf}$
$DL = 382.4 + 91 = 473.4 \text{ plf}$

UPLIFT $131 - 473.4 = -342.4 \text{ plf} \Rightarrow \boxed{-456.5 \text{ lb/STUD}}$

OVERTURN $R_1 = 221.8 \text{ plf}$ $R_2 = -11.28$
$R_1 - 2/3 \, DL = -93.8 \text{ plf} \Rightarrow \boxed{-125.0 \text{ lb/STUD}}$

SLIDING $\frac{1}{2} \times 371.2 - \frac{1}{2}(473.4 - 131) = 14.4 \text{ plf} \Rightarrow \boxed{19.2 \text{ lb/STUD}}$

(e)

FIGURE 4.4 (*Continued*)

metal parts, *an increase to 1.6 is not allowed under any condition*. In many cases the connectors are rated for vertical forces only. It is recommended that the standard nailing be used to resist the sliding forces. In Figures 4.6 and 4.7 connection capacities shown in parentheses are based on the horizontal capacity of the CABO recommended nailing schedule.

Table 4.2*a* summarizes the results for the 70-mph wind as well as for higher wind speeds. Refer to the first group of numbers under the column labeled "70 mph." This column shows the difference between the uplift and the dead load. If a number is shown in parentheses, the dead load is sufficient to resist the uplift without any mechanical connectors, such as nails, being needed. This condition is true, for 70 mph, at all locations except at the connections between the roof trusses and wall plates and between the wall plates and the second floor studs. The connection that holds the truss to the wall plate must resist 28 lb uplift. Looking down to the lower section of the table, titled "Sliding," we find the comparison of the sliding force due to wind and the friction force due to the gravity loads. In this section none of the values are written with parentheses, so mechanical fasteners are needed at all locations. At the roof-to-wall plate location, the nails must hold 30.3 lb, laterally. To provide that much resistance only one 16d toenail, capable of carrying 187 lb, is needed. Comparing the capacity of all the recommended connectors with the required capacity shows that under the 70-

(*text continues on page 96*)

RECOMMENDED FASTENING SCHEDULE

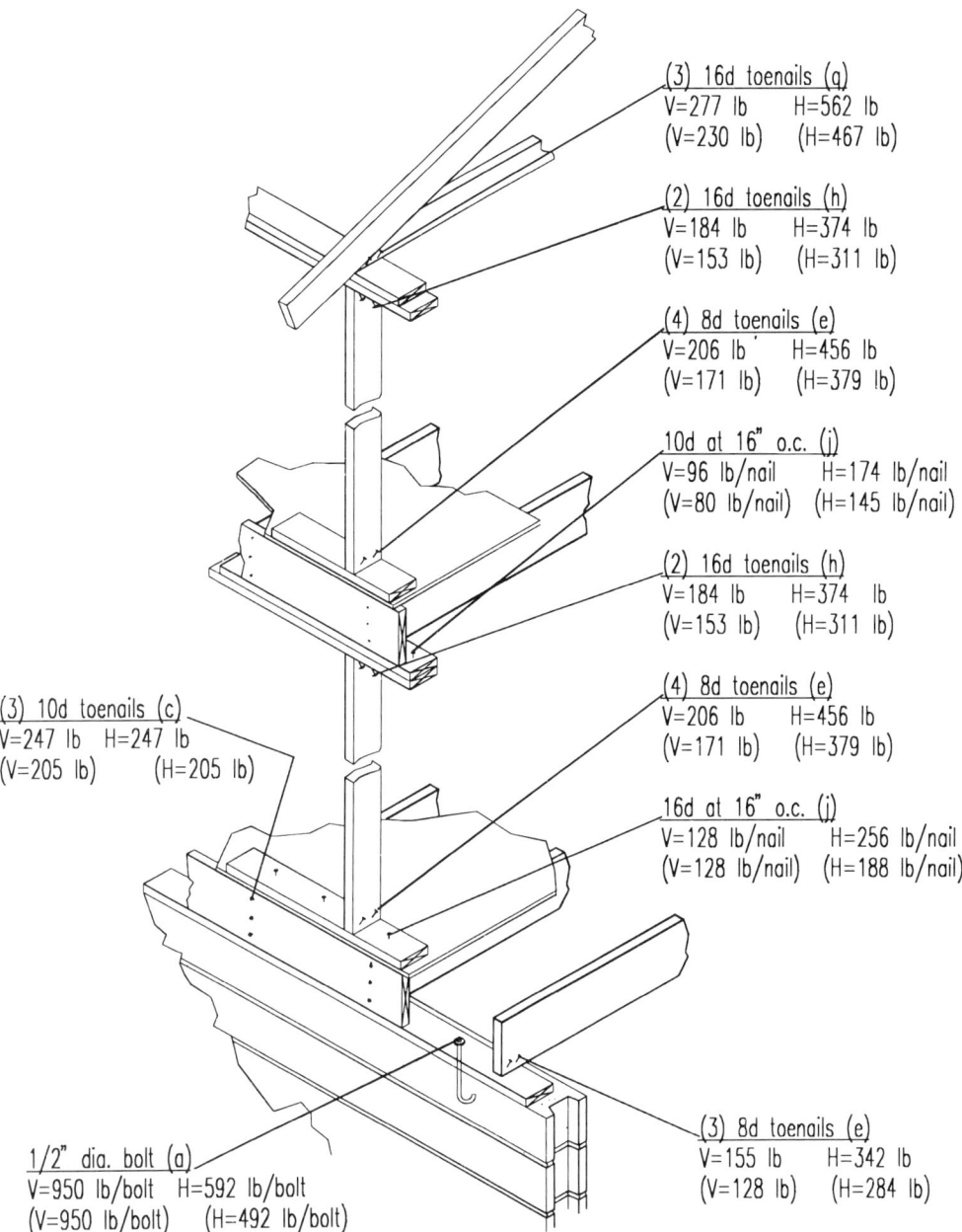

note: values in parentheses use a duration factor of 1.33 as per UBC. Values without parentheses use a duration factor of 1.60 as per SBC and BBC

FIGURE 4.5 Connection capacities — nails.

90 CONNECTIONS

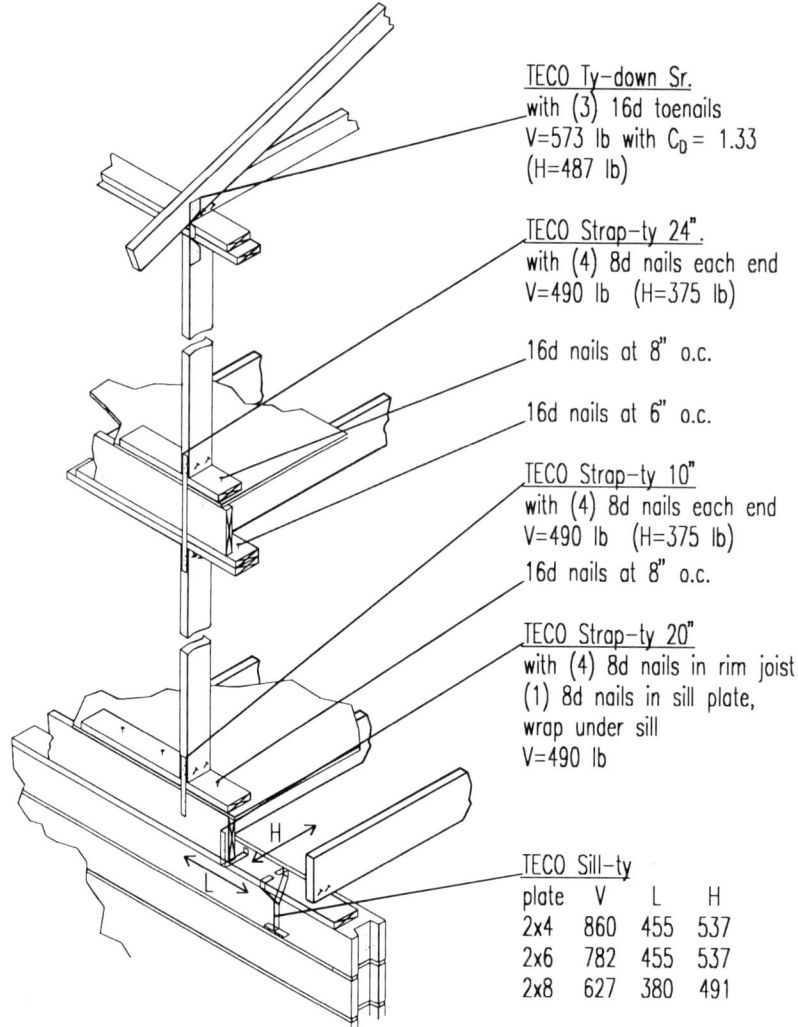

notes:
a. Use recommended CABO fastening schedule at all locations.
b. Sliding strength is developed by CABO toenails unless noted. Values in parentheses are due to CABO recommended nailing schedule.
c. All values use a duration factor of 1.33 as per UBC. Values may not be increased to use a duration factor of 1.60.
d. Values are for DF-L or SP framing. Multiply values by 0.82 for group III framing or by 0.65 for group IV framing.

Source: TECO Publication No. 141, Storm Resistant Construction

FIGURE 4.6 Connection capacities—moderate wind loads.

RECOMMENDED FASTENING SCHEDULE   91

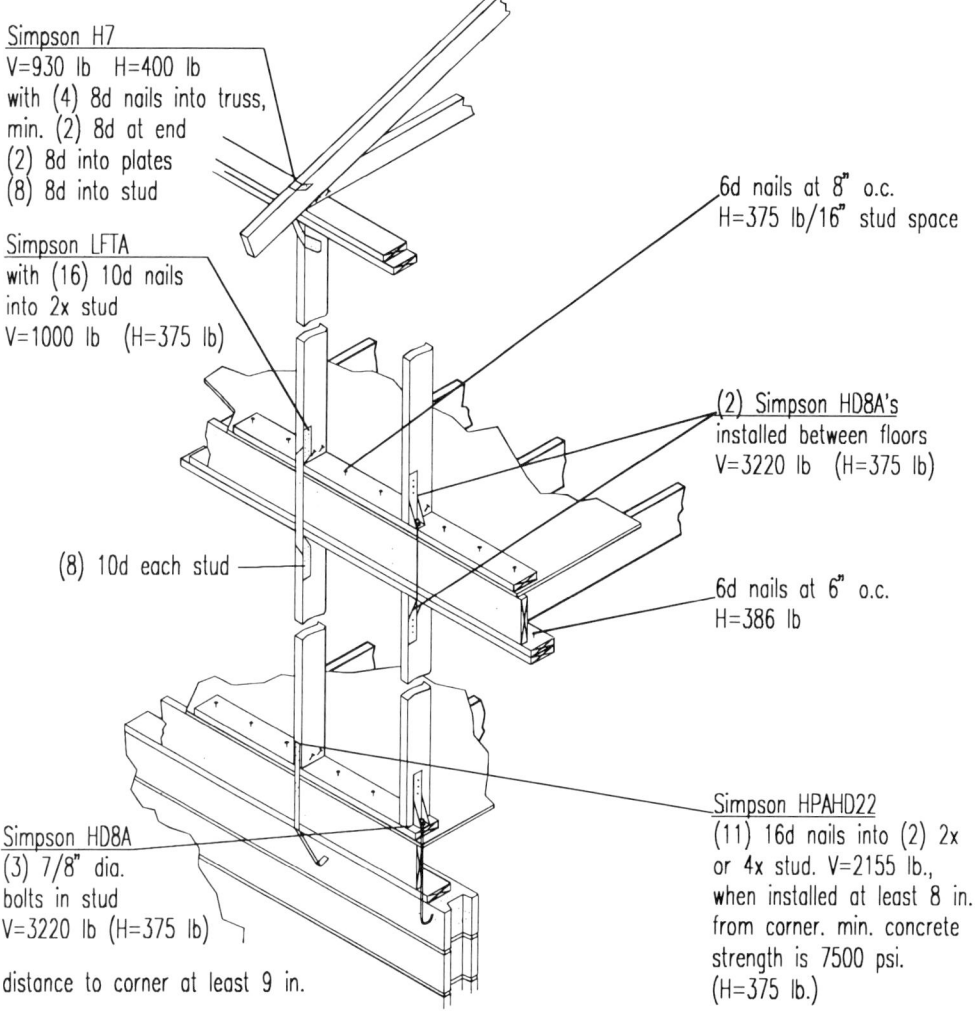

Simpson H7
V=930 lb   H=400 lb
with (4) 8d nails into truss,
min. (2) 8d at end
(2) 8d into plates
(8) 8d into stud

Simpson LFTA
with (16) 10d nails
into 2x stud
V=1000 lb   (H=375 lb)

(8) 10d each stud

Simpson HD8A
(3) 7/8" dia.
bolts in stud
V=3220 lb (H=375 lb)

distance to corner at least 9 in.

6d nails at 8" o.c.
H=375 lb/16" stud space

(2) Simpson HD8A's
installed between floors
V=3220 lb   (H=375 lb)

6d nails at 6" o.c.
H=386 lb

Simpson HPAHD22
(11) 16d nails into (2) 2x
or 4x stud. V=2155 lb.,
when installed at least 8 in.
from corner. min. concrete
strength is 7500 psi.
(H=375 lb.)

cast into concrete (2000 psi or higher) with 10 in. embedment

notes:
a. All values use a duration factor of 1.33 as per UBC. Values may not be increased to use a duration factor of 1.60.
b. Values are for DF-L or SP framing. Multiply values by 0.82 for group III framing or by 0.65 for group IV framing.
c. Sliding strength is developed by CABO toenails unless noted. Values in parentheses are due to CABO recommended nailing schedule.

Source: Simpson Strong-Tie Connectors, product and instruction manual, C-94H-1

FIGURE 4.7 Connection capacities—high wind loads.

TABLE 4.2a REQUIRED CONNECTIONS FOR 28-FT-WIDE BUILDING WITH 6-IN-12 ROOF PITCH (EXTERIOR ONLY), FRAMING MEMBERS AT 16 IN. O.C.

Wind velocity (mph)	70	80	90	100	110	120
Wind pressure (psf)	16	21	27	33	40	48

Uplift/Overturning (pounds per connector)

Roof to wall plate	28.0	79.9	142.2	204.5	277.2	360.3
Wall plate to stud	28.0	79.9	142.2	204.5	277.2	360.3
Sill plate to 2nd floor	(37.3)	13.1	73.6	134.1	204.7	285.3
2nd floor to wall plate	(108.5)	(55.1)	9.0	73.1	147.9	233.4
Sill plate to 1st floor	(61.4)	38.5	195.2	358.1	534.6	743.5
1st floor to foundation	(125.0)	(32.6)	78.3	189.2	318.6	466.5

Location	Recom'd			Required Connector		
Roof to wall plate	3-16d toenail	1-16d toenail	2-16d toenail	3-16d toenail	TYDOWN SR	TYDOWN SR
Wall plate to stud	2-16d toenail	1-16d toenail	2-16d toenail	3-16d toenail	TYDOWN SR	TYDOWN SR
Stud to 2nd sill plate	4-8d toenail	None needed	2-8d toenail	3-8d toenail	4-8d toenail	24" STRAP
Sill plate to 2nd floor	16d @ 16" o.c.	16d @ 13' o.c.	16d @ 28" o.c.	16d @ 15" o.c.	16d @ 10" o.c.	16d @ 7" o.c.
1st floor stud to wall plate	2-16d toenail	None needed	1-16d toenail	1-16d toenail	2-16d toenail	3-16d toenail
Stud to 1st sill plate	4-8d toenail	None needed	4-8d toenail	24" STRAP	Simpson LFTA	Simpson LFTA
Sill plate to 1st floor	16d @ 16" o.c.	None needed	16d @ 26" o.c.	16d @ 11" o.c.	16d @ 6" o.c.	Addt'l conn.
1st floor to foundation	1/2" bolt @ 6'	None needed	1/2" bolt @ 16'	1/2" bolt @ 7'	1/2" bolt @ 4'	Addt'l conn.
	Recom'd OK	Recom'd OK	Recom'd OK			

Wind velocity (mph)		70	80	90	100	110	120
Wind pressure (psf)		16	21	27	33	40	48

Sliding (pounds per connector; friction factor = 0.25)

		70	80	90	100	110	120
Roof to wall plate		30.3	47.8	68.9	90.0	114.6	142.7
Wall plate to stud		30.3	47.8	68.9	90.0	114.6	142.7
Sill plate to 2nd floor		113.7	160.8	217.4	274.0	340.0	415.5
2nd floor to wall plate		110.3	161.2	222.2	283.3	354.5	435.8
Sill plate to 1st floor		193.7	274.2	370.7	467.3	579.9	708.7
1st floor to foundation		133.3	224.3	333.5	442.7	570.1	715.6

Location	Recom'd			Required Connector			
Roof to wall plate	3-16d toenail	1-16d toenail	1-16d toenail	1-16d toenail	1-16d toenail	1-16d toenail	1-16d toenail
Wall plate to stud	2-16d toenail	1-16d toenail	1-16d toenail	1-16d toenail	1-16d toenail	1-16d toenail	1-16d toenail
Stud to 2nd sill plate	4-8d toenail	1-8d toenail	2-8d toenail	2-8d toenail	3-8d toenail	3-8d toenail	Addt'l conn.
Sill plate to 2nd floor	16d @ 16″ o.c.	16d @ 36″ o.c.	16d @ 25″ o.c.	16d @ 19″ o.c.	16d @ 15″ o.c.	16d @ 12″ o.c.	16d @ 10″ o.c.
1st floor stud to wall plate	2-16d toenail	1-16d toenail	1-16d toenail	2-16d toenail	2-16d toenail	3-16d toenail	
Stud to 1st sill plate	4-8d toenail	2-8d toenail	3-8d toenail	4-8d toenail	Addt'l conn.	Addt'l conn.	Addt'l conn.
Sill plate to 1st floor	16d @ 16″ o.c.	16d @ 31″o.c.	16d @ 18″o.c.	16d @ 12″o.c.	16d @ 9″o.c.	16d @ 7″o.c.	Addt'l conn.
1st floor to foundation	1/2″ bolt @ 6′	1/2″ bolt @ 6′	1/2″ bolt @ 4′	Addt'l conn.	Addt'l conn.	Addt'l conn.	Addt'l conn.

Recom'd OK

Notes:
1. Compare uplift/overturning with sliding. Use the larger connector.
2. Required forces are shown as pounds per connector. Only one metal connector per wood member is assumed. To use two connectors on a member, see manufacturer's requirements on member sizes.
3. Negative values (shown in parentheses) show that no connector is required.
4. At a minimum, *always use the CABO-recommended fastening schedule.*

TABLE 4.2b REQUIRED CONNECTIONS FOR 28-FT-WIDE BUILDING WITH 6-IN-12 ROOF PITCH (EXTERIOR PLUS INTERIOR WIND PRESSURE), FRAMING MEMBERS AT 16 IN. O.C.

Wind velocity (mph)		70	80	90	100	110	120
Wind pressure (psf)		16	21	27	33	40	48

Uplift/Overturning (pounds per connector)

Roof to wall plate		180.2	279.7	399.1	518.4	657.7	816.9
Wall plate to stud		180.2	279.7	399.1	518.4	657.7	816.9
Sill plate to 2nd floor		121.8	222.0	342.1	462.3	602.5	762.7
2nd floor to wall plate		48.8	151.4	274.5	397.5	541.2	705.3
Sill plate to 1st floor		77.2	311.3	545.9	780.5	1,054.2	1,367.0
1st floor to foundation		10.7	145.3	307.3	469.1	657.8	873.6

Location	Recom'd				Required Connector		
Roof to wall plate	3-16d toenail	2-16d toenail	TYDOWN SR	TYDOWN SR	TYDOWN SR	SIMPSON H7	SIMPSON H7
Wall plate to stud	2-16d toenail	2-16d toenail	TYDOWN SR	TYDOWN SR	TYDOWN SR	SIMPSON H7	SIMPSON H7
Stud to 2nd sill plate	4-8d toenail	3-8d toenail	24" STRAP	24" STRAP	24" STRAP	Simpson LFTA	Simpson LFTA
Sill plate to 2nd floor	16d @ 16" o.c.	16d @ 17" o.c.	16d @ 9" o.c.	addt'l conn.	addt' l conn.	addt'l conn.	addt'l conn.
1st floor stud to wall plate	2-16d toenail	1-16d toenail	2-16d toenail	3-16d toenail	TYDOWN SR	TYDOWN SR	SIMPSON H7
Stud to 1st sill plate	4-8d toenail	2-8d toenail	24" STRAP	Simpson LFTA	Simpson LFTA	Simpson LFTA	addt'l conn.
Sill plate to 1st floor	16d @ 16" o.c.	16d @ 16' o.c.	16d @ 14" o.c.	16d @ 7" o.c.	addt'l conn.	addt'l conn.	addt'l conn.
1st floor to foundation	1/2" bolt @ 6'	1/2" bolt @ 7'4"	1/2" bolt @ 9'	1/2" bolt @ 4'	addt'l conn.	addt'l conn.	addt'l conn.

Recom'd OK

Wind velocity (mph)		70	80	90	100	110	120
Wind pressure (psf)		16	21	27	33	40	48
Sliding (pounds per connector; friction factor = 0.25)							
Roof to wall plate		52.1	84.5	123.5	162.4	207.9	259.8
Wall plate to stud		52.1	84.5	123.5	162.4	207.9	259.8
Sill plate to 2nd floor		111.0	169.0	238.5	308.1	389.2	481.9
2nd floor to wall plate		90.9	152.0	225.4	298.8	384.4	482.2
Sill plate to 1st floor		149.8	236.4	340.4	444.4	565.7	704.3
1st floor to foundation		52.9	168.0	306.2	444.4	605.6	789.8

Location	Recom'd			Required Connector			
Roof to wall plate	3-16d toenail	1-16d toenail	1-16d toenail	1-16d toenail	1-16d toenail	2-16d toenail	2-16d toenail
Wall plate to stud	2-16d toenail	1-16d toenail	1-16d toenail	1-16d toenail	1-16d toenail	2-16d toenail	2-16d toenail
Stud to 2nd sill plate	4-8d toenail	1-8d toenail	2-8d toenail	3-8d toenail	3-8d toenail	3-8d toenail	addt'l conn.
Sill plate to 2nd floor	16d @ 16" o.c.	16d @ 37" o.c.	16d @ 24" o.c.	16d @ 17" o.c.	16d @ 13" o.c.	16d @ 11" o.c.	16d @ 8" o.c.
1st floor stud to wall plate	2-16d toenail	1-16d toenail	1-16d toenail	2-16d toenail	2-16d toenail	3-16d toenail	addt'l conn.
Stud to 1st sill plate	4-8d toenail	2-8d toenail	3-8d toenail	3-8d toenail	addt'l conn.	addt'l conn.	addt'l conn.
Sill plate to 1st floor	16d @ 16" o.c.	16d @ 6" o.c.	16d @ 24" o.c.	16d @ 13" o.c.	16d @ 9" o.c.	16d @ 7" o.c.	addt'l conn.
1st floor to foundation	1/2" bolt @ 6'	1/2" bolt @ 15'	1/2" bolt @ 5'	addt'l conn.	addt'l conn.	addt'l conn.	addt'l conn.

Recom'd OK   Recom'd OK

Notes:
1. Compare uplift/overturning with sliding. Use the larger connector.
2. Required forces are shown as pounds per connector. Only one metal connector per wood member is assumed. To use two connectors on a member, see manufacturer's requirements on member sizes.
3. Negative values (shown in parentheses) show that no connector is required.
4. At a minimum, *always use the CABO-recommended fastening schedule.*

mph wind the recommended connectors are sufficient at all locations to resist uplift, overturning, and sliding.

EXAMPLE 4.2—WIND UPLIFT AND OVERTURNING UNDER DIFFERENT WIND SPEEDS

Consider the building described in Example 4.1. Using the connections listed in the recommended fastening schedule, determine the maximum wind speed that each connection can resist. Table 4.2a shows the uplift, overturning, and sliding forces for the same building with wind speeds ranging from 70 mph to 120 mph. We can check each connection in turn to find when the wind load will exceed the capacity of the connector.

The trusses are held to the wall plate using three 16d nails driven at an angle—called *toenails*. The uplift capacity of the three toenails is 276.6 lb (see Figure 4.5). Table 4.2a lists the required capacity for each truss spaced 16 in. on center. For a wind velocity of 100 mph the connection must be able to carry 204.5 lb, so three 16d toenails are adequate for 100 mph in uplift and overturning. For sliding the three nails can resist 562 lb per connection. On Table 4.2a the maximum sliding force listed is 143 lb. Thus the roof connection is satisfactory.

At the connection between the roof trusses and the top plate, three 16d toenails are adequate for winds of 100 mph or less.

Moving down the wall, we find that the resistance to uplift becomes less important since the dead load of the building increases. However, the sliding force increases rapidly because more wall area is exposed to the wind pressure. This increase in wind load is not generally offset by the increased dead load. Following the procedure of comparing both uplift and sliding requirements, we find that between the bottom of the second floor stud and the sill plate the recommended four 8d toenails are adequate in uplift and sliding for 110 mph but will not suffice for a 120 mph wind.

At the connection between the wall stud and the second floor, four 8d toenails are adequate for winds of 110 mph or less.

The same connection is recommended between the bottom of the first floor stud and the sill plate below. At this point the connection, which can carry 206 lb in uplift and 456 lb laterally, will be satisfactory for 90 mph under both uplift and sliding but that with a wind of 100 mph it will no longer suffice to carry the uplift force of 358 lb per stud.

At the connection between the wall stud and the first floor, four 8 toenails are satisfactory for winds of 90 mph or less.

For the two-story, 28-foot wide building as described, the CABO Recommended Fastening Schedule is satisfactory for winds of 90 mph or less. Additional tables showing other building widths and roof pitches are found in Appendix E.

EXAMPLE 4.3—FLOOR CONSTRUCTION NAILING

Look at Figure 4.5 and trace the load path between the first-floor stud and the second-floor stud. This path is made up of five sets of connectors. The path is only as strong as the weakest link, in this case the 10d nails at 16 in. on center that secure the two top wall plates together. This connection can carry 174 lb per nail laterally and 96 lb in withdrawal (uplift). Rather than limiting the capacity of the entire structure on this connection, it seems reasonable to revise the nailing so that this

BALANCING CONNECTIONS AT THE FLOOR

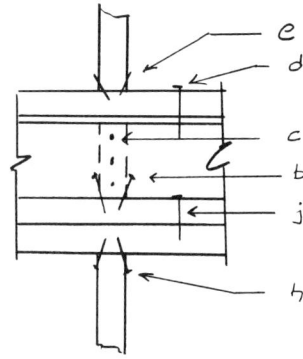

ASSUME JOIST SPACING EQUALS STUD SPACING. CALL STUD SPACING "S"

CALCULATE CAPACITIES OF ALL CONNECTIONS. FIND MINIMUM OF THOSE SPACED AT 'S'. IE, STUDS AND JOISTS.

FIND SPACING REQ'D FOR OTHER CONNECTORS (d, j) TO MEET MINIMUM

Designation	Connection	spacing	V	H
e	(4) 8d toe	S	206#	456#
c	(3) 10d (end)	S	246.6#	246.6#
b	(3) 8d toe	S	154.5#	342#
h	(2) 16d toe	S	184.9#	374.4#
	MINIMUM		154.5# (b)	246.6# (c)
d	16 d		V= 128#/nail	H= 255.6#/nail

req'd spacing = $\frac{128}{154.5}$ S   or $\frac{255.6}{246.6}$ S

= .83 S   1.04 S

| j | 10 d | | V = 96#/nail | H= 174.4#/nail |
| | | S = .62 S | | S = .707 S |

10d's should be spaced at 75% of 16d's ($\frac{.62}{.83} \cong .75$)

RECOMMENDED SPACING FOR CONNECTIONS			STUD SPACING O.C.		
			12"	16"	24"
d	16 d SILL TO FLOOR	S=.83	10"	12"	20"
j	10 d DOUBLE PLATES	S=.62	7"	10"	14"

FIGURE 4.8 Calculations for nailing at floor plate.

98 CONNECTIONS

TABLE 4.3 RECOMMENDED SPACING FOR PLATE CONNECTORS

		For Stud Spacing of:		
		12 in. o.c.	16 in. o.c.	24 in. o.c.
d	16d nails from sill plate to floor	10 in. o.c.	12 in. o.c.	20 in. o.c.
j	10d nails in double top plate	7 in. o.c.	10 in. o.c.	14 in. o.c.

portion of the connection is at least as strong as the rest of the load path. The same is true with the 16d nails at 16 in. o.c. securing the wall sill plate to the floor. This connection can carry 128 lb per nail vertically and 255.6 lb per nail horizontally. Looking through the other connections in this area we see that the next smallest vertical capacity is 154.5 lb, due to the three 8d toenails holding the joist to the wall plates. The next-smallest horizontal force is 246.6 lb, due to the three 10d nails end nailed from the rim to the joist. Figure 4.8 shows the engineering calculation used to find the required spacing of the plate nailing. In this figure the required spacing for the plate nails is calculated as a function of the stud spacing. Table 4.3 summarizes these results. In effect, you can see that the CABO *Recommended Fastening Schedule* spaces the plate nailing farther apart than recommended here.

In Table 4.2*a* you will find the required spacing of 16d nails from sill plate to floor. At the first-floor level the spacing is called out as 11 in. o.c. for 100 mph and 6 in. o.c. for 110 mph to resist uplift. For a wind velocity of 120 mph or greater, the nails spacing will have to be less than 6 in. o.c., which is not recommended because it will tend to split the plate. The entry for this connection under sliding due to a 120-mph wind advises you to use an "additional connector." Figures 4.7 and 4.8 show two options. Both of these options have the advantage that they will bypass the sill plate-to-floor plate connection. With the metal plate in place the original specification of 16d at 16 in. o.c. should be followed for the sill plate-to-rim joist connection.

This analysis indicates that the CABO *Recommended Fastening Schedule* would be made more consistent by revising the recommendations for plate connections. Table 4.3 represents the authors' recommendation for tighter nail spacing for wall plate-to-floor connections.

Figure 4.6 shows a series of connectors made by the TECO company for use in light residential construction. The capacities for each of these connections were derived through testing and standard engineering. These connectors have been accepted for use by CABO and most state and local building codes. Figure 4.7 shows the same information for Simpson Strong-Tie connectors, which are also approved by CABO and most local and state codes. The connectors in Figure 4.6 were selected to resist moderate loads; those in Figure 4.7 were selected for severe loads. TECO, Simpson Strong-Tie, Silver, and others all make connectors that can carry both moderate and severe loads.

EXAMPLE 4.4—WIND UPLIFT AND OVERTURNING WITH METAL CONNECTORS

For the building described in Example 4.1, use the connectors and their capacities given in Figure 4.6 to determine the maximum wind speed that this set of connections can resist. The TECO series in Figure 4.6 shows two options for rafter or truss connections. When the truss is over the stud, the connector can carry 650 lb in uplift, which exceeds all the required values on Table 4.2*a*. The lateral value for the connector is not listed; instead, the note on the drawing requires that toenails be used to resist the lateral load. The TECO connector is not used laterally. Three 16d toenails at this location will be able to produce 562 lb, which is more than enough to carry the largest sliding load of 143 lb. One 16d toenail carrying 187 lb would suffice, but the recommended three 16d toenails will be specified. As we move down through the building we see that all of the other connectors are rated for 520 lb of uplift. This will be sufficient to resist uplift for a 100-mph wind.

Regarding sliding loads, the principal line of resistance is through the recommended nailing schedule. Many of the metal connectors are not rated for horizontal loads. Scanning down Table 4.2a in the sliding section we see that under the 100-mph load the connection from the bottom of the stud to the plate will not suffice at the first-floor level. An additional connection, such as a Simpson Stud Plate Tie, is needed.

Tables F.1 through F.6 in the Appendix show the same information for a variety of building widths and two roof pitches. These tables are intended to give an understanding of the relation of building width and height to connection capacity. These tables also list the effects and required connectors for wind loads, ranging from 70 mph, which is the standard velocity for most of the United States, to 120 mph,

TABLE 4.4 ABILITY OF RECOMMENDED FASTENERS TO RESIST WIND LOADS

	External Pressure Only					External + Internal Pressure			
Uplift	Maximum Wind Velocity (mph)				Uplift	Maximum Wind Velocity (mph)			
Roof pitch Bldg. width	4:12	6:12	8:12	10:12	Roof Pitch Bldg. width	4:12	6:12	8:12	10:12
24 ft	80	80	80	80	24 ft	—	70	70	70
28 ft	80	90	90	80	28 ft	—	70	70	70
32 ft	80	90	90	90	32 ft	—	—	70	70
Sliding	Required Spacing for $\frac{1}{2}$-in. Anchor Bolts				Sliding	Required Spacing for $\frac{1}{2}$-in. Anchor Bolts			
Bldg. width					Bldg. width				
24 ft	6' o.c.	6' o.c.	6' o.c.	6' o.c.	24 ft	—	6' o.c.	6' o.c.	6' o.c.
28 ft	4' o.c.	2' o.c.	2' o.c. stud[b] plate first floor	2' o.c. stud[b] plate first floor	28 ft	—	6' o.c.	6' o.c.	6' o.c.
32 ft	4' o.c.	2' o.c.	2' o.c. stud[b] plate first floor	2' o.c. stud[b] plate both levels	32 ft.	—	—	6' o.c.	6' o.c.

[a] Maximum wind velocity allowed for CABO-recommended fastener schedule (see Figure 4.1).
[b] Secure stud to plate with strap or floor tie.

Procedure: 1. Identify applicable wind velocity from local building code. If wind velocity exceeds 90 mph, use "External + Internal Pressure" in table.
2. Identify building width and roof pitch.
3. Refer to the top portion of the table marked "Uplift." Find the maximum allowable wind velocity based on building configuration.
4. If the maximum allowed velocity exceeds the actual wind velocity, the recommended fastening schedule is adequate. If not, additional metal fasteners are needed. Turn to Tables F.1 through F.6 to find where fasteners are needed.

which is the maximum hurricane load specified by code. The results of these calculations are summarized in Table 4.4. By referring to these tables you will find that the recommended fasteners perform satisfactorily under low to moderate loads (generally 70 mph to 90 mph), but are not sufficient for larger loads. In these cases special fasteners, referred to generically as "hurricane clips," are needed. The connections in Figure 4.6 were selected to carry moderate to high wind loads, 90 to 100 mph, for the lower-pitched roof. Those in Figure 4.7 are designed to carry severe wind loads. In the example they are sufficient to carry the 120-mph winds at all

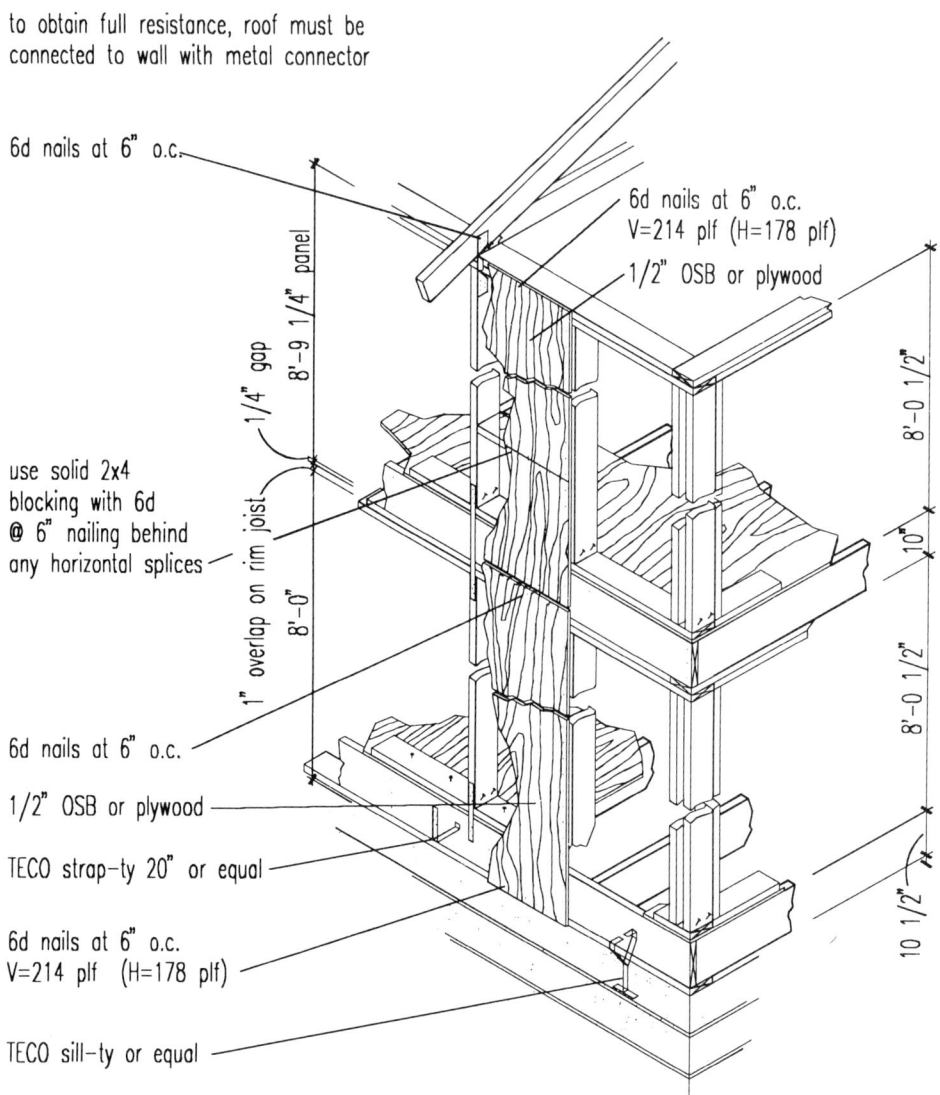

FIGURE 4.9 Plywood sheathing to resist sliding and overturning.

locations except the floor-to-foundation connection and the stud-to-sill plate connection.

Another possible connection detail (Figure 4.9) makes use of the sheathing—plywood or oriented-strand board—to tie the building together. This solution is satisfactory only if the entire wall is sheathed in plywood and if the plywood is detailed so that adjoining pieces are nailed into the same member. Typically, in low-wind-load areas, plywood is used only in a 4-ft width at the corners. For this application additional connectors are needed at the remaining studs.

CONNECTOR DESIGN GUIDELINES

The previous examples are intended to indicate generally the safe range of wind velocities for which the recommended fastener schedule was developed. Obviously, one example will not establish a limit but will provide a sense of what the normal conditions are and where the recommended fastener schedule can be used without calculation. It should be remembered that the CABO-recommended fastening schedule is not intended to replace engineering judgment or calculations for all buildings and all conditions. The guidelines listed here, as all aspect of a building code, represent the *minimum* requirements for most standard conditions. A designer who is not familiar with wood connections will need to design the majority of the connections; a more experienced designer will check the critical connectors for a design when any of the following conditions occur.

1. **Large Loads.** Typical fastening schedules are based on moderate loads. Where snow loads exceed 30 psf or where wind loads exceed 80 mph, it is important to identify the principal connectors and check their capacity. In Examples 4.1 and 4.2 we investigated the capacity of the recommended connections to determine the maximum wind load that they can carry for a particular-size building. As the building size varies, some connectors may become more critical.

2. **Building Configuration.** In the examples we looked at a building of typical width and height. As the height increases, the total wind load and the length of the moment arm will increase, so the effects of overturning will increase at an even greater rate. As the building width decreases, both the amount of dead load (as a function of the floor area) and the resisting moment arm decrease, so the resisting moment decreases, again at a rate greater than the decrease of width. For exceptionally tall or narrow buildings the standard connections will carry significantly smaller wind loads.

3. **Roof Pitch.** For low-pitched roofs (4 in 12 or less) the uplift will be far more important than roof sliding or overturning. For steep-pitched roofs (9 in 12 or greater) both sliding and overturning can be critical.

4. **Uplift.** Uplift can occur whenever the wind loads exceed the gravity dead loads. Uplift will occur wherever the roof pitch is relatively flat (30° or less), wherever there are large overhangs (greater than 18 in.), or wherever the wind pressure is approximately equal to the dead load. The effects of uplift need to be considered at all locations, from the roof connections all the way down to the foundation. Example 4.1 demonstrates wind uplift calculations.

5. **All or Most of the Load Carried by the Connector.** Ideally, all wood connections are detailed so that the primary forces are carried in bearing of wood against wood. In reality, this is not always possible. The connector must carry the load in joist or beam hangers, deck connections, and uplift on roofs and walls. Wherever

wood members lap over each other, the load will be carried entirely by the connector.

6. **Internal Pressure.** Table 4.2a and 4.2b show the difference between a building under external pressure and one under combined external and internal pressure. The internal pressure is due to openings in the building. Under any large wind load of 100 mph or more, it is reasonable to assume that portions of the roofing may be torn away or windows may break. When this happens, an internal pressure occurs that adds to the external pressure. In Table 4.2a the moderate level of hurricane strapping was generally sufficient for a 100-mph wind. In Table 4.2b, where internal pressure is also considered, heavier strapping is needed in many locations to resist the uplift forces. Protecting the windows with sturdy shutters, adequately nailing all portions of the roof—shingles, felt paper, and sheathing—and using $\frac{3}{4}$-in.-thick exterior plywood siding will go a long way to protecting the building and the property within it.

CONNECTION DESIGN

The primary source for design of wood connections in the United States is the *National Design Specification*,[1] published by the American Forest and Paper Association. In December 1991 the specifications on connections were revised considerably from the edition of 1986. The new connection design is based on a procedure referred to as the European yield theory. All the major code agencies have reviewed this information and it is anticipated that the revisions will be adopted by the agencies by late 1993. States and local building departments usually adopt specifications within a year of their approval by the major agencies. Until the revised specifications are adopted within a local jurisdiction, the designer should contact the building official to determine what provisions apply. For the purposes of this book, all examples will use the revised, 1991 approach and values.

To aid the engineer in designing connections, the NDS has tabulated *allowable* values for typical connection types, wood species, and typical connection sizes which are derived from the yield theory equations. Occasionally, an engineer will need to design a connection that exceeds the bounds of the NDS tables. When that occurs the designer may use the yield formulas that are included in the NDS. A sample of this calculation is shown in Example 4.20.

The connection design formulas and tables incorporated in the NDS are all based on engineering theory and substantiated by a number of tests conducted on the connectors. The theory identifies a number of critical variables that the designer must consider during the design of each connection. In beginning the design of a connection, it is useful to start by listing all the conditions that are known or that will be needed. The following list shows the basic variables. The list starts with what is most commonly known and ends with what we most commonly solve for: number and size of connectors.

1. Connection type
2. Wood species
3. Thickness of each wood or steel piece
4. Moisture content (conditions of wood at fabrication and in service)
5. Direction of load (load parallel to grain, load perpendicular to grain, connector in side grain or end grain)
6. Spacing of connectors
7. Size of connectors
8. Number of connectors within a group

---

[1] *National Design Specification for Wood Construction.* American Forest and Paper Association, Washington, DC, 1991.

The primary wood characteristic that affects the strength of a connection is the density of the wood. For convenience the density is usually referred to as the specific gravity (the density of an oven-dried sample of wood divided by the density of water). Each of the commonly used species is categorized by its specific gravity. Within the NDS tables for each type of connector list allowable strengths based on the wood's specific gravity. Table E.1 in the Appendix also lists the species by group. This approach, used prior to the 1991 NDS, specified connection strengths by four wood species group. Although the recent changes in the specifications provide more accuracy, the groups are a valuable framework for understanding the relation of connection strengths to species; all species within the same group will tend to have similar strengths. If additional strength is needed, changing to the next-higher group will be

TABLE 4.5 NAIL DIAMETERS, LENGTHS, AND PENETRATIONS

Box Nails

	6d	8d	10d	12d	16d	20d	30d	40d				
Diameter (in.)	0.099	0.113	0.128	0.128	0.135	0.148	0.148	0.162				
Length (in.)	2	$2\frac{1}{2}$	3	$3\frac{1}{4}$	$3\frac{1}{2}$	4	$4\frac{1}{2}$	5				
Minimum penetration, $6D$	0.594	0.678	0.768	0.768	0.810	0.888	0.888	0.972				
Recommended penetration, $12D$	1.188	1.356	1.536	1.536	1.620	1.776	1.776	1.944				

Common Wire Nails

	6d	8d	10d	12d	16d	20d	30d	40d	50d	60d		
Diameter (in.)	0.113	0.131	0.148	0.148	0.162	0.192	0.207	0.225	0.244	0.263		
Length (in.)	2	$2\frac{1}{2}$	3	$3\frac{1}{4}$	$3\frac{1}{2}$	4	$4\frac{1}{2}$	5	$5\frac{1}{2}$	6		
Minimum penetration, $6D$	0.678	0.678	0.888	0.888	0.972	1.152	1.242	1.350	1.464	1.578		
Recommended penetration, $12D$	1.356	1.356	1.776	1.776	1.944	2.304	2.484	2.700	2.928	3.156		

Threaded Hardened-Steel Nails

	6d	8d	10d	12d	16d	20d	30d	40d	50d	60d	70d	80d	90d
Diameter (in.)	0.120	0.120	0.135	0.135	0.148	0.177	0.177	0.177	0.177	0.177	0.207	0.207	0.207
Length (in.)	2	$2\frac{1}{2}$	3	$3\frac{1}{4}$	$3\frac{1}{2}$	4	$4\frac{1}{2}$	5	$5\frac{1}{2}$	6	7	8	9
Minimum penetration, $6D$	0.720	0.720	0.810	0.810	0.888	1.062	1.062	1.062	1.062	1.062	1.242	1.242	1.242
Recommended penetration, $12D$	1.440	1.440	1.620	1.620	1.776	2.124	2.124	2.124	2.124	2.124	2.484	2.484	2.484

Common Wire Spikes

	10d	12d	16d	20d	30d	40d	50d	60d	$\frac{5}{16}$	$\frac{3}{8}$
Diameter (in.)	0.192	0.192	0.207	0.225	0.244	0.263	0.283	0.283	0.312	0.375
Length (in.)	3	$3\frac{1}{4}$	$3\frac{1}{2}$	4	$4\frac{1}{2}$	5	$5\frac{1}{2}$	6	7	$8\frac{1}{2}$
Minimum penetration, $6D$	1.152	1.152	1.242	1.350	1.464	1.578	1.698	1.698	1.872	2.250
Recommended penetration, $12D$	2.304	2.304	2.484	2.700	2.928	3.156	3.396	3.396	3.744	4.500

Note: The length given is the most common. Most nails are available in other lengths as well.
Source: NDS Tables 12.2A through 12.2D.

Nail Connections

Nails are the most commonly used connectors for light timber framing. Since nails offer less strength than do other connection types, they are used where small forces are applied. There are a wide variety of types of nails currently available. The NDS specifies values for four types of nails (Table 4.5):

1. Box nails
2. Common wire nails
3. Common wire spikes
4. Threaded hardened steel nails and spikes

The first three nail types are made from the same material, a low-carbon steel, and vary from each other primarily in thickness. Their size is designated by *pennyweight* and marked with a "d." A 16d nail is referred to as a *16 penny nail*. The most commonly used nails are common wire nails; examples and questions in this book will use common wire nails unless otherwise noted. Threaded hardened steel nails and spikes are made of high-carbon steel and are tempered to have more strength. Because they are threaded they are much more resistant to withdrawal than are smooth-shank nails. Currently, the NDS gives withdrawal values about 9% higher for threaded nails than for smooth-shank nails. The threading also improves their holding power in wood subject to wetting and drying; therefore, the reduction factors due to moisture conditions in Table 4.6 do not apply to threaded hardened-steel nails and spikes.

Nails are generally used to connect two pieces of wood together directly or by means of a metal plate that is attached to the two pieces. Nails can be loaded along their own axis, which is referred to as withdrawal loading, or they may be loaded laterally, where the load is perpendicular to the axis of the nail. Tabulated values for each condition are given in the NDS in Tables 12.2A for withdrawal, in Tables 12.3A through 12.3D for lateral loads on joints with wooden side pieces, and in Tables 12.3E through 12.3H for lateral loads on joints with metal side pieces. Factors of safety in withdrawal generally exceed 5.0. In lateral resistance the factor of safety ranges from about 3.5 for softwoods to 7 for hardwoods.

Design Modification Factors. The values listed in the NDS must be modified to account for the actual condition of use. The specific requirements, modifications, and recommendations are listed in the NDS and summarized here in Table 4.6.

Withdrawal. Nails loaded in withdrawal are effective only when the nail is driven into the side grain of the holding piece (the piece that contains the point of the nail). Because the holding power of the nail in end grain is small, *the use of nails in withdrawal in end grain is not allowed.* Example 4.2 showed a common condition where this may occur—the connection between studs and plates—and a solution for it—toenails.

When the moisture content of the wood changes from the time of fabrication throughout the service life of the structure, the holding power in withdrawal can be greatly reduced—in some cases by as much as 400%. The tabulated values in NDS Table 12.2A must be modified by the values shown in Table 4.6 to account for the actual conditions of use.

Lateral Resistance. Nails loaded laterally are also most effective when the nail

TABLE 4.6 NAIL MODIFICATION FACTORS

Condition		Factors	
Direction of Load		Lateral	Withdrawal
Direction of grain (of holding piece)	Side grain	1.0	1.0
	End grain	0.67	Not allowed
Orientation of nail	Direct or face nail	1.0	1.0
	Toenail	0.83	0.67^a
Penetration	Less than $6D$	Not allowed	Not recommended
	Less than $12D$	$C_d = \frac{\text{actual penetration}}{12D}$	b
	Greater than $12D$	$C_d = 1.0$	b
Number of Members in Connection		Lateral	Withdrawal
Single shear		1.0	Not applicable
Double shear, clinched (three member joints)	Side members $\frac{3}{8}$ in. thick or greater, nails 12d or smaller and extend at least $3D$ beyond side member and are clinched	2.0 times C_d, where $C_d = 1.0$	Not applicable
Double shear, not clinched	Not clinched or side members less than $\frac{3}{8}$ in. thick, or nails are greater than 12d, or nails extend less than $3D$ beyond side member	2.0 times $C_d^{\,c}$	Not applicable
Side plates	Wood	Tables 12.3A through 12.3D	Table 12.2A
	Metal	Tables 12.3E through 12.3H	Table 12.2A
	Side member thickness	Tables 12.3A through 12.3D	Table 12.2A

	Moisture Content			
	Condition of Wood			
Nail Type	At Fabrication	In Service	Lateral	Withdrawal
Common wire nails, box nails, and common wire spikes	Dry	Dry	$C_M = 1.0$	$C_M = 1.0$
	Partially seasoned or wet	Wet	0.75	1.0
	Partially seasoned or wet	Dry	0.75	0.25
	Dry	Subject to wetting and drying	$(0.75)^d$	0.25
Threaded, hardened steel nails	Dry or wet	Dry or wet	1.0	1.0
Metal connector plates	Dry	Dry	1.0	1.0
	Partially seasoned or wet	Dry or wet	0.8	$(0.8)^e$

[a] Values for toenail in withdrawal need not be further reduced for conditions of wood or service conditions (i.e., moisture content).
[b] Multiply values in Table 12.2A by actual penetration.
[c] C_d is derived from the penetration of the nail into the third member.
[d] Not specifically listed in NDS. Interpretation from values in rows above.
[e] Not normally loaded in withdrawal.

Source: National Design Specification, 1991 Edition.

is perpendicular to the grain in the holding piece. The common condition of toenailing (driving the nail at an angle into the holding piece) has an allowable load capacity of $\frac{5}{6}$ (83%) of the tabulated value. When a nail is driven into the end grain of the holding piece and loaded laterally, the allowable load capacity is taken as $\frac{2}{3}$ (67%) of the tabulated value.

Penetration. The allowable strengths for lateral loads on nails given in NDS Tables 12.3A through 12.3H are based on a required nail penetration. Penetration is the length of nail in the holding piece (the piece that has the point end of the nail). Specified penetration is given as 12 times the nail diameter ($12D$). The penetration may be less than that specified but cannot be less than $6D$, that is, 6 times the nail diameter. If the penetration is less than $12D$, the capacity of the nail will be reduced by a factor C_d, taken as the ratio of the actual penetration divided by the specified penetration:

$$C_d = \frac{\text{actual penetration}}{12D}$$

No increase is taken if the penetration exceeds the amount specified. Thus the reduction factor for penetration will range from a maximum of 1.0 to a minimum of 0.50.

As with withdrawal loads, changes in the moisture content of the wood between the time of fabrication throughout the service life of the structure affect the strength of the connection, but the effect is not so dramatic. The moisture content factors are defined in Table 4.6.

Metal Side Plates. The allowable strengths listed in NDS Tables 12.3E through 12.3H are for nails joining one piece of wood with a metal side plate. These connections are assumed always to be in single shear. To use metal side plates, the plate must be of sufficient strength to carry the shear. In addition, the nail should fit tightly into the holes in the plate. The NDS specifies the type of metal grade A steel (ASTM A446). The table lists thicknesses ranging from 20 gage (0.036 in.) up to 10 gage (0.134 in.).

Double Shear. Double shear may occur where more than two pieces are connected, for example where two side pieces pull against a central, main piece (see Figure 4.10). The allowable capacity of a nail connection in double shear can be increased to twice the capacity of the single-shear value, provided that the middle member is thicker than six times the nail diameter ($6D$). If the two side pieces are of the same thickness and the same wood species, the capacity of the double shear connection is twice that of the single-shear connection. If the side pieces are different, you must look at each set of two members independently. The three-member connection is divided into two two-member con-

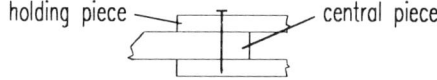

if the thickness of the center piece is at least six times the nail dia., the strength of the connection equals two times the strength of a nail in single shear

(a)

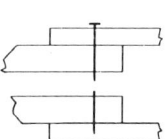

two connections in single shear

the total allowable shear is twice that of the weaker connection

(b)

FIGURE 4.10 (a) Nails in double shear and (b) equivalent connection.

nections, each in single shear. The allowable lateral load for each of those is found in the NDS tables. The capacity of the total connection is two times the strength of the *weaker* single-shear connection.

In calculating the capacity of the double-shear connection the penetration depth factor, C_d, must be found for the third piece (the piece holding the point of the nail). The actual value of C_d must be used when the nails are not clinched. Furthermore, if the nails are clinched but they are large (greater than 12d), or if the side members are less than $\frac{3}{8}$ in. thick, the actual value of C_d must be used. In any of these cases it is possible that the penetration factor, C_d, will be less than 1.0, so the capacity of the double-shear connection may be less than two times that of the weaker single-shear capacity.

If the nail is clinched, the penetration factor, C_d, is taken as 1.0 and there is no need to calculate the actual penetration factor. The double-shear connection will have a strength of 2.0 times the weaker single-shear capacity. To meet the criteria for clinching, the nail must be 12d or smaller, the side members must be $\frac{3}{8}$ in. thick or more, the nail must extend at least three times its diameter ($3D$) beyond the third piece, and it must be bent over (i.e., clinched). Threaded, hardened steel nails should not be clinched, but if all the other conditions apply, the factor of 2.0 may still be used.

Nail Slip. When nailed joints are tested to failure there is a decidedly nonlinear relationship between load and deformation (nail slip). To control the deformation of nailed structures the allowable load values of nail connections were established as the loads that produce a slip of 0.015 in. In the tests the ultimate loads often range as high as 5 to 7 times the allowable loads, with ultimate deflections in the range of 0.5 in. This implies a factor of safety based on strength of approximately 5 to 7; although this seems rather high, it should be remembered that deformation, not strength, is the critical criterion in setting the allowable loads. As a consequence, nailed connections are not dependent on the direction of load relative to the wood grain. (When we look at lag screws and bolted connections, we will see that the strength varies markedly between parallel-to-grain and perpendicular-to-grain directions.)

If two nails of different diameters are loaded to their maximum values, the larger-diameter nail will slip more than the smaller nail. Therefore, it is usually preferable to use a greater number of small nails than a lesser number of large nails. In choosing the best combination, however, the designer must also remember that smaller nails might not have sufficient penetration and will therefore produce less capacity than will a slightly larger nail. A rule of thumb that attempts to balance these two conflicting positions is to choose a nail so that its length will fully penetrate all the members of the joint.

Nail Spacing. The NDS specifies, rather generally, that nail spacing must be adequate "to avoid unusual splitting of the wood." More specific information is given in the *Uniform Building Code* (UBC), which states that in the direction of the load the minimum center-to-center spacing between nails should be equal to the required penetration (i.e., 12 diameters). The UBC also requires that the end and edge distances parallel to the load not be less than one-half the center-to-center spacing. No requirements for distances perpendicular to the load are given.

Number of Nails in a Joint. The NDS states, "When more than one nail or spike is used in a joint, the total design value for the joint in withdrawal or lateral resistance is the sum of the design values for the individual nails or spikes." No reductions are needed as the number increases. No recommendations are given for situations in which a nail is loaded in a combination of lateral load and withdrawal; this situation is rare and should be avoided.

Diaphragms. Where nails are used in diaphragm construction, the values listed in NDS Tables 12.3A through 12.3D can be increased 10%. Typically, diaphragms are specified using standard nail patterns which have specified capacities so that the engineer does not have to calculate diaphragm strengths. Diaphragms are covered in Chapter 7.

Load Duration Factor. The load duration factors shown in the NDS Table 2.3.3 apply to all connections. The only exception is the impact factor ($C_D = 2.0$), which cannot be used in connection design.

A number of example problems are given to show the range of conditions.

EXAMPLE 4.5—NAILS IN SINGLE SHEAR

Design a nail connection between the roof rafter and the ceiling tie shown in Figure 4.11. The roof spans 24 ft. Its pitch is 6 on 12 (26.6°). The roof system is designed to carry a dead load of 10 psf and a snow load of 30 psf. The rafters are 2 × 8's and the ceiling ties are 2 × 6's; they are spaced 2'-0" o.c. Assume that both the ceiling tie and the rafter are Southern Pine.

To calculate the load on each rafter, the dead load must be increased to account for the slope of the roof.

$$w_{tot} = \frac{10 \text{ psf}}{\text{cosine } 26.6°} + 30 \text{ psf} = 41.2 \text{ psf}$$

$$w_{tot} \text{ per rafter} = w_{tot} \times \text{spacing}$$
$$= 41.2 \text{ psf} \times 2 \text{ ft}$$
$$= 82.4 \text{ plf}$$

This load acts vertically.

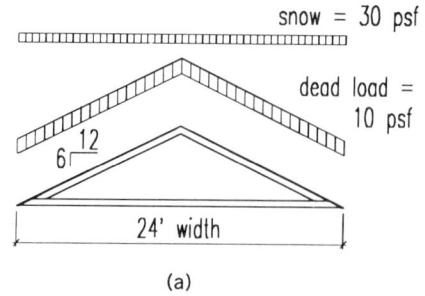

(a)

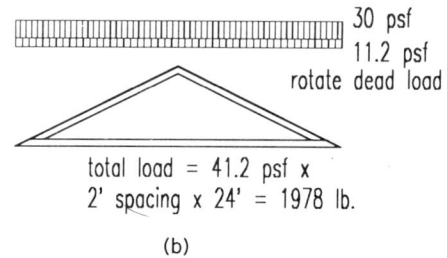

(b)

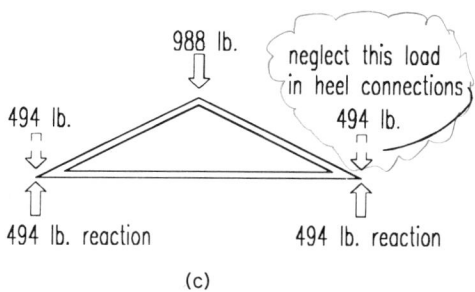

(c)

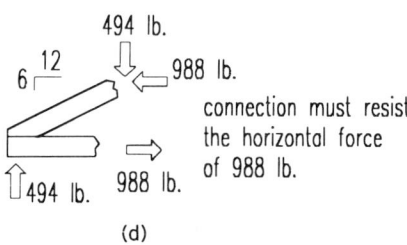

(d)

FIGURE 4.11 Roof framing and heel connection: (a) snow load plus dead load, (b) rotate dead load, (c) concentrate at ridge, and (d) forces in heel connection.

The ceiling tie overcomes the thrust and therefore is loaded in tension equal to the thrust:

$$T = \frac{wL^2}{8h} = \frac{82.4 \text{ plf} \times (24 \text{ ft})^2}{8 \times 6 \text{ ft}} = 988 \text{ lb}$$

Try lapping the ceiling tie and rafter and use direct nailing.

Since the total thickness of the joint will be 2×1.5 in. $= 3$ in., we need to use at least a 12d common nail, which is $3\frac{1}{4}$ in. long. Typically, a 16d nail is used when the side member is $1\frac{1}{2}$ in. thick. A 16d nail is $3\frac{1}{2}$ in. long and has a diameter of 0.162 in. The required penetration is 12 diameters or 1.94 in. and the allowable strength of the 16d nail, given in Table 12.3B, is 154 lb. Since the actual penetration is 1.5 in., less than the required 1.94 in., the strength must be reduced by the ratio $C_d = 1.5/1.94$. Thus

$$Z = \frac{1.5}{1.94} \times 154 \text{ lb}$$
$$= 0.773 \times 154 \text{ lb} = 119.0 \text{ lb}$$

Since the connection carries a snow load, the strength may be increased by the duration factor for snow, 1.15.

$$Z = 1.15 \times 119.0 \text{ lb} = 136.8 \text{ lb per nail}$$

The total number of nails needed in this connection is

$$N = \frac{\text{force on connection}}{\text{capacity per nail}}$$
$$= \frac{988 \text{ lb}}{136.8 \text{ lb/nail}} = 7.22$$

which we round up.

Use eight 16d nails.

To check the practicality of this connection, we need to sketch it to scale and try placing the nails with the recommended nail spacing. Let us use the UBC requirements so that the spacing parallel to the direction of the load equal 12 diameters or approximately 2 in., and the end and edge distances (between the last nail and the end or edge of each piece) are half the penetration requirement, or in this case,

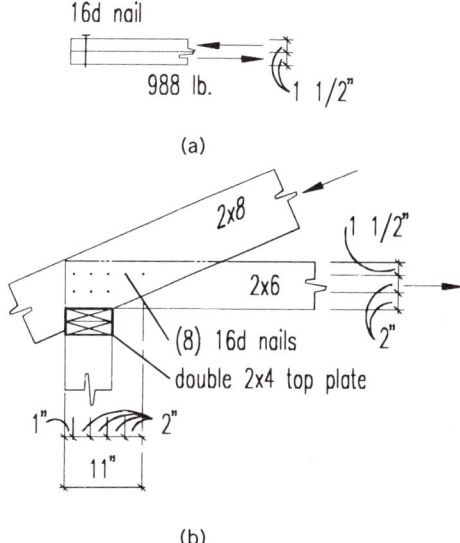

FIGURE 4.12 Heel connection—nails in single shear: (a) plan view and (b) elevation.

1 in. By referring to Figure 4.12, we can see that one row of five nails and one row of three nails will result in a nail pattern that is 9 in. long, which fits within the overlap area of the wood. The connection as designed is acceptable. We should also note that more nails could not be added and remain within the spacing guidelines. This connection can therefore only carry 8×136.8 lb, or 1094 lb.

The next three examples explore variations of the heel connection. In them we look at the possibility of using steel side plates, plywood side plates, or doubling the rafter and using much larger nails. In actual construction there are a number of factors that impinge on the selection of the best design: economy of materials (number and cost of nails), economy of labor (speed and ease of construction), and reliability (assurance that the connection can be built as detailed). Not the least to consider is the common practice in the area. In our estimate the solution in Example 4.5, using eight 16d nails, or the solution in Example 4.6, using steel side plates, are the best solutions.

EXAMPLE 4.6—NAILS WITH A STEEL SIDE PLATE

Let us try a revised detail that uses a metal side plate. The top and bottom chords of the truss will align, and the plate will lap over them both. Select a plate with a thickness of 16 gage for the initial trial. The plate size must be checked at the end of the design. This connection will be easier to construct and therefore seems more likely to be built the way it is specified.

A smaller nail must be used since the width of the wood is only $1\frac{1}{2}$ in. The smallest nail that is shown in Table 4.5 is a 6d nail with a length of 2 in. Most suppliers of metal plates for this application also provide nails with the same diameters as 8d or 10d nails that are $1\frac{1}{2}$ long. Therefore, let us try an 8d nail. The tabulated value, found in Table 12.3F, for a 16-gage plate is 101 lb. The actual penetration of 1.5 in. is less than the required penetration of $12D$ ($12 \times 0.131 = 1.57$ in.). There will be a decrease for penetration. Accounting for the snow-load duration, the nail capacity becomes

$$Z = \frac{1.5}{1.57} \times 1.15 \times 101 \text{ lb} = 111 \text{ lb}$$

The number of nails needed is

$$N = \frac{\text{force on connection}}{\text{capacity per nail}}$$
$$= \frac{988 \text{ lb}}{111 \text{ lb/nail}} = 8.9$$

Use nine 8d nails through each chord with 16-gage steel side plates.

In the connection with overlapping wood members, all the nails penetrated both pieces of wood. With the steel side plates, nine nails will be needed to penetrate the rafter and another nine to penetrate the ceiling tie; thus 18 nails will be

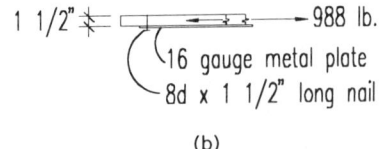

(b)

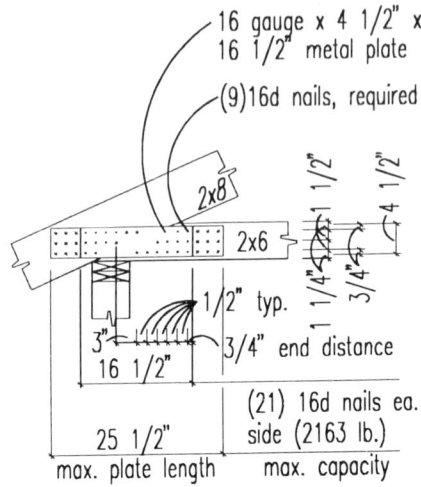

plate can be extended 4 1/2" more on either end to increase capacity to 2163 lb. if needed

(a)

FIGURE 4.13 Heel connection—nails with steel side plates: (a) plan view and (b) elevation.

needed. Reference to Figure 4.13 shows that there is ample room to place nine nails and meet the spacing criteria. Most plates come with prepunched nail holes spaced either $\frac{3}{4}$ in. apart or $1\frac{1}{2}$ in. apart.

This connection is somewhat preferable, although it uses more nails. Its main advantages are that the nails can be placed farther from the end of the tension member and that the rafter and ceiling-tie members can be placed in one plane. Although nail slip is usually not critical, this connection will probably result in less nail slip because the side plate is stiffer and because smaller nails are used. Figure 4.13 shows that a longer plate can be used and that more nails could be added to each side. With a total of 21 nails on each side, this connection will carry a maximum of 2331 lb.

EXAMPLE 4.7—NAILS IN DOUBLE SHEAR

As another example, consider designing the truss with a double top chord and a single bottom chord between the top chords. (This design is common for heavy trusses but would rarely be used for a truss as small as this residential one. It is included here for the sake of demonstrating nails in double shear.) The nails at the heel connection of this truss are loaded in double shear. By referring to Figure 4.14, we can see that the width of the side pieces equals the width of the main piece, $1\frac{1}{2}$ in., and that the total joint width is $4\frac{1}{2}$ in. To use the fewest number of nails, we choose a nail that will give maximum penetration in each piece. A 30d nail is $4\frac{1}{2}$ in. long and has a diameter of 0.207 in. The capacity of this connection will be twice that of either of the single-shear connections. We need not concern ourselves with clinching since the nail is larger than a 12d nail. We will have to consider penetration and the penetration factor, C_d, for the final shear plane.

The required penetration is $12D = 12 \times 0.207 = 2.48$ in.

$$C_d = \frac{1.5}{2.48} = 0.60$$

For the single-shear value, refer to NDS Table 12.3B.

$$Z = 203 \text{ lb}$$
$$Z'_{\text{single shear}} = 0.6 \times 203 \text{ lb} = 121.8 \text{ lb}$$

Including the duration factor and the double shear factor gives

$$Z'_{\text{double shear}} = 1.15 \times 2.0 \times 121.8 \text{ lb} = 280 \text{ lb}$$

The number of nails needed is

$$N = \frac{988 \text{ lb}}{280 \text{ lb/nail}} = 3.5 \text{ nails}$$

Use four 30d nails

The recommended spacing is $12D$ or $2\frac{1}{2}$ in. For larger nails the end distances and loaded edge distances become more critical. In Figure 4.14 a nailing pattern was chosen such that the distance to the end of the bottom chords was kept at $2\frac{1}{2}$ in. as well as the distance to the bottom edge of the top chord, this being the edge that the nails pull toward (the loaded edge). To do this, the spacing between nails was reduced slightly, to $2\frac{1}{4}$ in. Since the spacing distances for nails are only recommendations, a slight reduction in spacing (say, 10%) can be taken without a commensurate loss of strength. This nail pattern covers considerably less area than that used in the first example for eight 16d

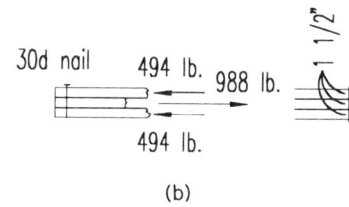

(b)

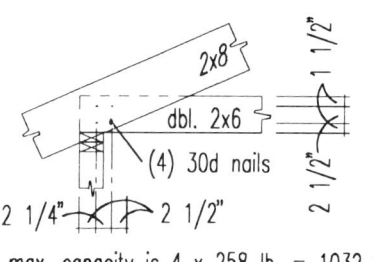

max. capacity is 4 × 258 lb. = 1032 lb.

(a)

FIGURE 4.14 Heel connection—nails in double shear: (a) plan view and (b) elevation.

nails and is also acceptable. The maximum load for this connection is 4×280 lb, or 1120 lb.

EXAMPLE 4.8—NAILS WITH A PLYWOOD GUSSET PLATE

For the final example in this group, consider replacing the metal plates with plywood gussets. The plywood plates will lap both sides of the truss members. The nails will be loaded in double shear. With this type of connection it is common to choose nails that are long enough to drive through all the pieces and clinch on the far side. To make use of the highest values available, we need to specify that the plywood be a minimum of $\frac{3}{8}$ in. thick, that the nail be no greater than a 12d, and that the nail length extends at least three diameters beyond the far piece of plywood. To check this consider, as a trial, using a 12d nail and $\frac{1}{2}$-in.-thick plywood. The $\frac{1}{2}$-in. thickness is more commonly available on the construction site than the $\frac{3}{8}$-in. thickness. Scraps could be cut to make the gussets.

A 12d nail is 3.25 in. long and has a diameter of 0.148 in.

Three nail diameters = 3×0.148 in.
$= 0.444$ in.

The total width of the connection is

0.5 in. + 1.5 in. + 0.5 in. = 2.5 in.

The required length of the nail is

2.5 in. + 0.444 in. = 2.944 in.

The actual nail length exceeds the required length,

3.25 in. > 2.944 in.

The middle piece must have a thickness greater than $6D$ or 0.89 in. Since all the requirements for clinched nails are met, the design can proceed using a factor of 2.0 times the allowable capacity for a 12d nail. The penetration factor can be taken as $C_d = 1.0$. The tabulated value is found in NDS Table 12.3B using a 12d nail with a $\frac{1}{2}$-in.-thick side piece. $Z = 101$ lb.

When the double-shear factor with clinching and the snow duration factor are included,

$$Z = 2.00 \times 1.15 \times 101 \text{ lb} = 232.3 \text{ lb}$$

The number of nails needed is

$$N = \frac{988 \text{ lb}}{232.3 \text{ lb/nail}} = 4.25 \quad \text{say 5 nails}$$

As with the steel side plates, each nail will penetrate only one truss chord, so the total number of nails needed is 2×5 or 10 nails.

From Table 12.3B we see that a 10d nail has the same diameter and therefore the same allowable load, $Z = 101$ lb. Before we could use the 10d nails, we would have to check if there is sufficient length to clinch them. 10d nails are 3 in. long, so they will also meet the clinching requirement. We can then specify:

Use five 10d nails with $\frac{1}{2}$ in. plywood gussets each side.

Figure 4.15 shows this connection. By extending the plate along the chords, a great number of nails can be added. The maximum capacity of the connection will be limited by the allowable tension of the bottom chord. If the gusset plates extend more than 2 or 3 in. past the intersection of the chords, a triangular piece of solid wood blocking needs to be added between the gussets. The blocking, when nailed to the gussets, will brace the gussets against buckling.

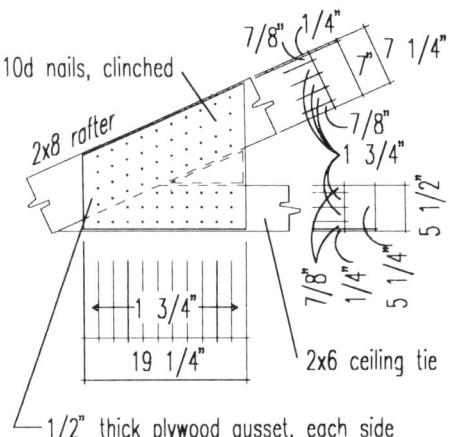

FIGURE 4.15 Heel connection—nails with a plywood gusset.

The purpose of these examples has been to lead the reader through a variety of exercises. Normally, a designer will select one type of connection based on experience or what is conventional within an area and then calculate the number of nails. You should note that these designs fell into two categories: those with gussets and those without gussets. For those without gussets the choice of connector did not greatly affect the outcome—approximately the same maximum strength of 1100 lb was developed using eight 16d nails or four 30d nails. For the two examples with gussets the maximum capacity was based on the allowable tension in the bottom chord of approximately 7000 lb. The gusset plates in either case could be extended to add enough nails to reach the maximum strength.

Regarding spacing recommendations, you should note that although larger nails provide greater strength, they also require more room. Generally, these two factors balance each other. If there is not enough space in the overlap to meet the spacing requirements for small nails, increasing the nail size will not help matters. The size of the connection pattern will remain about the same. One effective solution will be to use side plates, either steel or plywood, which result in a larger area through which to spread out the nails. This logic will also apply to bolted connections, split rings, and shear plates. The other solution is to switch to a different type of connection, such as lag screws or bolts, which we consider later in this chapter.

EXAMPLE 4.9—TOENAILS

Determine the capacity of the roof rafter connection to the top wall plate that is shown in Figure 4.16. This connection is shown in the recommended fastening schedule in Table 4.1 and Figure 4.1. In Example 4.1 the capacity of the connection was given, but not derived. In this example we will find the capacity of the connection to resist wind uplift through withdrawal and to resist sliding through lateral loads. We will use Douglas Fir–Larch for all members.

The recommended fastening schedule calls for three 16d toenails from the roof rafter to the plate. To calculate the capacity of each nail we must first determine the length of penetration in the holding piece. The NDS recommends "that toenails be driven at an angle of approximately 30° with the piece and started approximately one-third the length of the nail from the end of the piece" (see Figure 4.16). A 16d nail is 3.5 in. long. The length of the nail

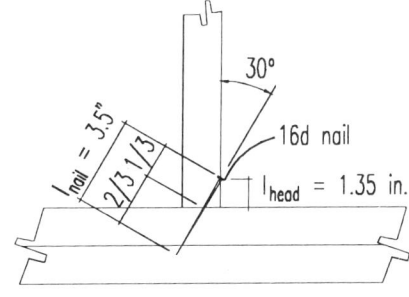

FIGURE 4.16 Toenails.

in the top piece then is

$$l_{head} = \frac{l_{nail} \times 0.333}{\cosine 30°}$$

$$= \frac{0.333 \times 3.5 \text{ in.}}{0.866} = 1.35 \text{ in.}$$

penetration = 3.50 in. − 1.35 in. = 2.15 in.

To find the nail capacity for withdrawal, turn to Table E.1 to find the specific gravity of Douglas Fir–Larch as 0.50 and to Table 4.5 to find the diameter of a 16d common wire nail as 0.162 in. Next turn to NDS Table 12.2A to find the withdrawal capacity for a 0.162-in.-diameter nail (16d) in Douglas Fir–Larch ($G = 0.50$).

$$P_w = 40 \text{ lb/in.}$$

To find the capacity for each nail, multiple the capacity per inch times the length of penetration times all applicable modifications times the duration factor for wind.

$$P'_w = P_w \times \text{penetration} \times C_{tn} \times C_D$$
$$= (40 \text{ lb/in.}) \times (2.15 \text{ in.}) \times 0.67 \times 1.6$$
$$= 92.2 \text{ lb/nail}$$

The factor 0.67 is the reduction for toenails, and the factor 1.6 is the duration factor for wind loads under BOCA or SBCCI codes. For the group of three nails, $P'_w = 3 \times 92.2 = 276.6$ lb.

To resist sliding due to wind, the connection must be able to develop lateral loads. The lateral-load for this connection is the full tabulated value since the actual penetration, 2.15 in., exceeds the required penetration of $12D$, or 1.94 in. The tabulated value of Z is 141 lb, found in NDS Table 12.3B, opposite t_s, the side member thickness, of 1 in.

$$Z' = Z \times C_{tn} \times C_D$$
$$= 141 \times 0.83 \text{ (toenail factor)} \times 1.6$$
$$= 187.2 \text{ lb/nail}$$

For the group of three nails,

$$Z = 3 \times 187.2 = 561.6 \text{ lb.}$$

Staples and Pneumatically Driven Connectors

The use of staples and other pneumatically driven connectors has increased dramatically as the technology of the tools has improved. Except for "do-it-yourself" projects, it is unlikely that you will visit a job site where these connectors aren't used for the majority of the framing. Where nails are properly driven using a pneumatic gun, the design values are no different than the ones given.

Staples are used primarily to attach sheet materials such as plywood and oriented-strand board which operate as horizontal diaphragms or vertical shearwalls. The capacities of both nails and staples in these applications are discussed in Chapter 7. For other applications, the same modifiers that apply to nails can be used for staples. The design strengths for a variety of staples have been determined by the International Staple, Nail and Tool Association (ISANTA) and have been accepted by the major national codes. These values are published in the CABO *National Evaluation Report NER-272*. Selected values have been excerpted from Table 1 of the report and are included here in Table 4.7.

Following the severe damage caused by Hurricane Andrew in south Florida in August 1992, the Dade County building department banned the use of staples in any portion of roof construction. Staples had failed in holding roofing down as well as in holding down OSB panels. The failures were due primarily to errors in construction. Staples that were overdriven cut through the shingle. Overdriven staples also cut through the upper layers of the OSB board. Due to the failure of the roofing materials, the OSB got wet and

TABLE 4.7 STAPLES: DIMENSIONS AND ALLOWABLE LOADS

Fastener Description[a]	Wire Dia. (in.)	Wire Ga.	Penetration Required for Lateral Strength (in.) into Main Member[b]	Allowable Load[c] (lb)		
				Lateral Strength[d]	Withdrawal Strength	
					Southern Pine	Douglas Fir–Larch
6d Cooler nail	0.0915	13	1	46	27	23
6d Box nail	0.099					
P-Nail	0.097	12½	1⅛	51	29	25
Staple	0.0625	16	1	51	36	32
6d Casing nail	0.099					
Finish T-nail	0.097	12½	1⅛	51	—	—
6d Common nail						
8d Cooler nail						
8d Box nail	0.113	11½	1¼	63	35	29
P-nail						
6d Ring shank nail						
6d Screw shank nail	0.120	11	1⅜	78	41	34
8d Casing nail						
Finish T-nail	0.113	11½	1¼	63	—	—
Staple	0.072	15	1	64	42	37
10d Cooler nail	0.1205	11	1⅜	69	36	31
Staple	0.080	14	1	75	46	41
10d Box nail						
12d Box nail	0.128	10½	1½	76	38	33
10d Casing nail						
Finish T-nail	0.128	10½	1½	76	—	—
8d Common nail	0.131	10¼	1½			
P-Nail				78	41	34
8d Ring Shank nail						
8d Screw shank nail	0.120	11	1⅜			
16d Box nail	0.1350	10	1½	81	41	—

swelled. As the layers of wood fibers pushed up, they were cut by the staples until they eventually failed and were blown off. Nailed roofs fared better, either because there was less tendency to overdrive the nails or because the nail head did not cut the wood fibers as much. Other instances of staple failure resulted from staples not hitting the framing or from the staples being driven at an angle to the roof deck so that one leg of the staples was not fully embedded. These errors did not occur so much with nailed construction, either when hand-nailed or when nails were pneumatically driven. It is the opinion of the authors that the failure of the staples

TABLE 4.7 *Continued*

Fastener Description[a]	Wire Dia. (in.)	Wire Ga.	Penetration Required for Lateral Strength (in.) into Main Member[b]	Allowable Load[c] (lb)		
				Lateral Strength[d]	Withdrawal Strength	
					Southern Pine	Douglas Fir–Larch
Staple	0.0915	13	1	92	53	49
10d Common nail						
P-nail	0.148	9				
10d Ring shank nail						
10d Screw shank nail	0.135	10				
12d Common nail			$1\frac{5}{8}$	94	46	38
16 Sinker nail	0.148	9				
12d Ring shank nail						
12d Screw shank nail	0.135	10				
Staple	0.1055	12	$1\frac{1}{8}$	113	62	54
16d Common nail	0.162	8				
16d Ring shank nail			$1\frac{3}{4}$	108	50	42
16d Screw shank nail	0.148	9				

[a] Special-length P-nails or nails with shank diameters as noted above having smooth, barbed, screw, or ring shanks may be used, provided that the total length of fastener provides the penetration into the receiving member plus the thickness of the attachment material. These fasteners will have the same values as tabulated above. Staples shall have a $\frac{7}{10}$-in. minimum o.d. crown width.

[b] The tabulated penetrations are for fasteners installed in group I or II species. Penetration shall be increased to 13 diameters for group III and 14 diameters for group IV species.

[c] Allowable values shall be adjusted for duration of load in accordance with standard engineering practices. Where metal side plates are used, lateral strength values may be increased 25%. Withdrawal values are for fasteners inserted perpendicular to the grain in pounds per lineal inch of penetration into the main member of Douglas Fir–Larch or Southern Pine. Withdrawal strength for fasteners shall not be increased by a factor greater than 2, regardless of increased penetration. Loads for 8d, 10d, and 16d threaded nails (ring and screw shank) are the same as for common nails of the same pennyweight (8d, 10d, and 16d).

[d] For wood diaphragms resisting wind or seismic loading these values may be increased 30% in addition to the $33\frac{1}{3}$% increase permitted for duration of load. The tabulated allowable lateral values are for fasteners installed in Douglas Fir–Larch or Southern Pine (group II species). To determine the allowable values when both the attached wood member and the supporting (main) wood member are in the same group, but not group II, multiply the values listed in the above table by the following conversion factors: I-1.23, III-.82, IV-.65. If the attached and supporting members are in different groups, use the conversion factor for the wood in the higher group. See Table D.1 in the Appendix for a list of the wood species included in each group.

Source: National Evaluation Report NER-272.

resulted primarily from the combination of very high winds and heavy rains, in association with the construction errors mentioned. We believe that in areas without hurricanes and with only moderate winds (90 mph or less), staples are a satisfactory type of connection.

Wood Screws

Wood screws have roughly the same lateral resistance as nails but are superior to nails in withdrawal. Wood screws can therefore be used instead of nails in withdrawal. The NDS recommends against using either wood screws or nails in with-

TABLE 4.8 WOOD SCREWS: RECOMMENDED LEAD HOLE SIZES

Specific Gravity of Wood	Recommended Lead Hole Sizes		
	Loaded in Withdrawal	Loaded Laterally	
		Smooth Part	Threaded Part
$G > 0.6$	90% shank	100% shank	100% root
$0.5 < G \leq 0.6$	70% shank	$\frac{7}{8}$ shank	$\frac{7}{8}$ root
$G \leq 0.5$	None	$\frac{7}{8}$ shank	$\frac{7}{8}$ root

Source: National Design Specification, 1991 Edition.

drawal from end grain. Screws are often used in attaching sheetrock to wood structures but are not often used in other structural connections. Tabulated values for wood screws loaded laterally or in withdrawal are given in NDS Tables 11.2A, 11.3A, and 11.3B. Factors of safety in withdrawal generally exceed 5.0; those in lateral resistance are somewhat less.

Proper installation of wood screws is essential to developing their potential strength. Wood screws must be turned into a drilled lead hole with a screwdriver, not driven with a hammer, as this will destroy the wood fiber. The screw or lead hole may be lubricated with soap, wax, or other lubricant before insertion. The size of the lead hole depends on the direction of load, either withdrawal or lateral, and on the density of the wood. The recommended sizes of the lead holes are given in Table 4.8 as a ratio of the screw diameter. Table 4.9 converts these values into drill-bit sizes for the most common sizes, or gages, of wood screws. The shank diameter is the width of the smooth portion of the screw; the root diameter is the smallest diameter of the threaded part. It is usually two-thirds of the shank diameter. Each gage of screw is available in a number of lengths. The lengths given in Table 4.9 are the most common. The design val-

TABLE 4.9 WOOD SCREW DIMENSIONS

Wood Screw Gage	Shank Diameter (in.)	Root Diameter (in.)	Lead Holes Sizes				Minimum Length (in.)	Maximum Length (in.)
			90% Root Dia.	70% Root Dia.	$\frac{7}{8}$ Shank Dia.	$\frac{7}{8}$ Root (in.)		
6 g	0.138	0.092	$\frac{1}{16}$	$\frac{1}{16}$	$\frac{1}{8}$	$\frac{1}{16}$	$\frac{1}{2}$	$1\text{-}\frac{1}{2}$
7 g	0.151	0.100	$\frac{3}{32}$	$\frac{1}{16}$	$\frac{1}{8}$	$\frac{3}{32}$	$\frac{3}{4}$	2
8 g	0.164	0.109	$\frac{3}{32}$	$\frac{1}{16}$	$\frac{1}{8}$	$\frac{3}{32}$	$\frac{3}{4}$	2
9 g	0.177	0.118	$\frac{3}{32}$	$\frac{3}{32}$	$\frac{5}{32}$	$\frac{3}{32}$	$\frac{3}{4}$	$2\text{-}\frac{1}{2}$
10 g	0.190	0.127	$\frac{3}{32}$	$\frac{3}{32}$	$\frac{5}{32}$	$\frac{3}{32}$	$\frac{3}{4}$	$2\text{-}\frac{1}{2}$
12 g	0.216	0.047	$\frac{1}{8}$	$\frac{3}{32}$	$\frac{3}{16}$	$\frac{1}{8}$	1	3
14 g	0.242	0.161	$\frac{1}{8}$	$\frac{3}{32}$	$\frac{3}{16}$	$\frac{1}{8}$	$1\frac{1}{2}$	3
16 g	0.268	0.179	$\frac{5}{32}$	$\frac{1}{8}$	$\frac{7}{32}$	$\frac{5}{32}$	2	3
18 g	0.294	0.196	$\frac{5}{32}$	$\frac{1}{8}$	$\frac{1}{4}$	$\frac{5}{32}$	$2\frac{1}{2}$	3
20 g	0.320	0.213	$\frac{3}{16}$	$\frac{1}{8}$	$\frac{9}{32}$	$\frac{3}{16}$	3	—
24 g	0.372	0.248	$\frac{7}{32}$	$\frac{5}{32}$	$\frac{5}{16}$	$\frac{7}{32}$	3	—

Source: National Design Specification, 1991 Edition.

TABLE 4.10 WOOD SCREW MODIFICATION FACTORS

Condition		Factor	
Direction of grain		Lateral	Withdrawal
Side grain		1.0	1.0
End grain		$C_{eg} = 0.67$	Not allowed
Penetration			
Recommended $7D$		$C_d = 1.0$	No minimum
Minimum $4D$		$\frac{4}{7}$	
Between $4D$ and $7D$		$p/7D$	
Moisture Content			
At Fabrication	In Service		
Dry or wet	Dry	$C_m = 1.0$	$C_m = 1.0$
Dry or wet	Exposed to weather	0.75	0.75
Dry or wet	Wet	0.67	0.67

Source: National Design Specification, 1991 Edition.

ues in the NDS are accurate for lengths shown in Table 4.9. For lengths other than those, the yield equations should be used.

Modifications affecting the design strength are given in Table 4.10. These factors are multiplied by the base design values (NDS Tables 11.2A, 11.3A, and 11.3B) to arrive at design values that take into account conditions of use.

Penetration. The tabulated values for wood screws loaded laterally are given in NDS Tables 11.3A and 11.3B. These values are based on an assumed screw penetration of seven times the screw diameter. The penetration may be less than that specified but can not be less than four times the screw diameter. If the penetration is between $7D$ and $4D$, a linear reduction is allowed. The reduction factor is taken as the ratio of the actual penetration divided by the specified penetration of $7D$. No increase is taken if the penetration exceeds $7D$. Thus the reduction factor for

TABLE 4.11 WOOD SCREWS: RECOMMENDED PENETRATIONS

Wood Screw Gage	Shank Diameter (in.)	Recommended Penetration $7D$ (in.)	Minimum Penetration $4D$ (in.)
6 g	0.138	0.966	0.552
7 g	0.151	1.057	0.604
8 g	0.164	1.148	0.656
9 g	0.177	1.239	0.708
10 g	0.190	1.330	0.760
12 g	0.216	1.512	0.864
14 g	0.242	1.694	0.968
16 g	0.268	1.876	1.072
18 g	0.294	2.058	1.176
20 g	0.320	2.240	1.280
24 g	0.372	2.604	1.488

Source: National Design Specification, 1991 Edition.

penetration will range from a maximum of 1.0 to a minimum of $\frac{4}{7}$ or 0.57. Recommended penetrations for each screw size are shown in Table 4.11.

No limitations on penetration are given for screws loaded in withdrawal. The minimum of $4D$ is a good rule of thumb in this case as well.

The NDS does not specify a minimum thickness of the wood side piece next to the head. Most wood screws are made so that one-third the length is smooth and two-thirds is threaded. It is recommended that the side piece be as close to one-third the screw length as is possible.

End Grain. Design values for wood screws loaded laterally where the point is inserted into the end grain shall be taken as two-thirds the value given in NDS Tables 11.3A and 11.3B.

Withdrawal. The allowable strengths for wood screws in withdrawal are given in NDS Table 11.2A. Modifications in Table 4.10 should be applied. In addition, the designer must check that the root area of the screw will provide enough tensile strength to carry the load. Wood screws shall not be loaded in withdrawal from the end grain of wood.

Metal Side Plates. The values listed in NDS Table 11.3B are for screws connecting a main member of wood with a steel side plate. To use metal side plates the plate must be of sufficient strength to carry the shear.

EXAMPLE 4.10—WOOD SCREW WITH PLYWOOD GUSSET PLATES

Design the rafter-to-ceiling tie connection from Example 4.5 using wood screws and plywood gusset plates. Assume that a gusset plate will be used on each side. Assume that the rafters and collar ties are Southern Pine and are spaced 2 ft on center.

Assume that plywood plates are $\frac{1}{2}$ in. thick. A screw that is $1\frac{1}{4}$ in. long will provide a penetration of $\frac{3}{4}$ in. To meet the penetration requirement of $4D$, the diameter must not exceed $\frac{3}{4} \div 4$, or $\frac{3}{16}$ in. To meet the recommendation of $7D$, the diameter must not exceed $\frac{3}{4}$ divided by 7, or approximately $\frac{1}{8}$ in. We can try a screw with diameters near the average of these two extremes.

Try an 8-gage screw. The diameter equals 0.164 in. and the length ranges from $\frac{3}{8}$ to $2\frac{1}{2}$ in. Select a length of $1\frac{1}{2}$ in.

Penetration equals 1.5 in. − 0.5 in. = 1.0 in.

Recommended penetration = $7D$ = 7 × 0.164 in. = 1.148 in.

Minimum penetration = $4D$ = 4 × 0.164 in. = 0.656 in.

The actual penetration is greater than the minimum value, $4D$, but less than the recommended value, $7D$, so the penetration factor, C_d, is $1/1.148 = .87$. As always, the duration factor is used; in this case the factor for snow-load duration is 1.15.

$$Z = 1.15 \times .87 \times 108 \text{ lb} = 108 \text{ lb}$$

The number of screws needed is

$$N = \frac{988 \text{ lb}}{108 \text{ lb/screw}} = 9.14 \quad \text{say 10 screws}$$

We will need five screws on each side into the rafter and five screws on each side into the ceiling tie, for a total of

20 8-gage by $1\frac{1}{2}$ in.-long wood screws with $\frac{1}{2}$-in.-thick plywood gusset plates each side.

Let us also consider a larger-diameter screw—try a 14-gage screw. The diameter equals 0.242 in. and the length ranges from $\frac{3}{4}$ to 5 in.

Recommended penetration = $7D$
= 7×0.242 in. = 1.694 in.
Minimum penetration = $4D$
= 4×0.242 in. = 0.968 in.

Because the minimum penetration exceeds half the thickness of the rafter, it will be easier to apply the plates to only one side. Therefore, choose a screw length that will give the largest penetration possible. As screws are available in $\frac{1}{4}$-in. increments for lengths between $1\frac{1}{4}$ and 3 in., select a screw length of 2 in. The resulting penetration is

Penetration equals 2 in. − 0.5 in.
= 1.5 in.

The design value will be reduced by the ratio 1.5 divided by 1.694 = 0.885. The duration factor for snow loads, 1.15, is used.

$$Z = 0.885 \times 1.15 \times 146 \text{ lb} = 149 \text{ lb}$$

The number of screws needed is

$$N = \frac{988 \text{ lb}}{149 \text{ lb/screw}} = 6.6 \quad \text{say 7 screws}$$

We need

fourteen 14-gage by 2-in.-long wood screws with $\frac{1}{2}$-in.-thick plywood gusset plate on one side only.

This design seems a bit easier to construct and is preferable to the previous one. Both use the same-size gusset plates, although the first could be shortened substantially by using the recommended spacing of $\frac{3}{4}$ in. between screws in a row. As a general rule, though, it is always preferable to use fewer, larger connectors where possible.

If we choose to use steel side plates instead of plywood, the capacity of each 14-gage screw will increase. Select a steel plate with a 12-gage thickness. In NDS Table 11.3B we find the capacity of one 14-gage screw to be 192 lb, about a 30% increase.

$$Z = 0.885 \times 1.15 \times 192 \text{ lb} = 195.4 \text{ lb}$$

The number of screws needed is

$$N = \frac{988 \text{ lb}}{195.4 \text{ lb/screw}} = 5.05 \quad \text{say 6 screws}$$

Use six screws each in rafter and in ceiling tie. The total number of screws is

twelve 14-gage by 2-in.-long wood screws with 12-gage steel plate on one side.

These options are all shown in Figure 4.17a. Although wood screws are not typically used in this application, Figure 4.17b—using the 14-gage screws with the plywood gussets—is the easiest and most economical of these options to construct.

Lag Screws

Lag screws, which are also called lag bolts, have much more strength than nails or wood screws and therefore are used commonly where higher-strength connections are needed. Lag screws can be loaded either laterally or in withdrawal and can be used with either wood side pieces or metal side plates. Lag screws are used exclusively in single shear.

Lag screws differ from wood screws by having larger diameters and by having hex heads instead of slotted heads. The shank is smooth for approximately one-fourth the length of the screw for small diameters and up to half the length for larger diameters. Screw sizes, ranging from $\frac{1}{4}$ to $1\frac{1}{4}$ in. in diameter and from 1 to 12 in. long, are generally available. Standard sizes are

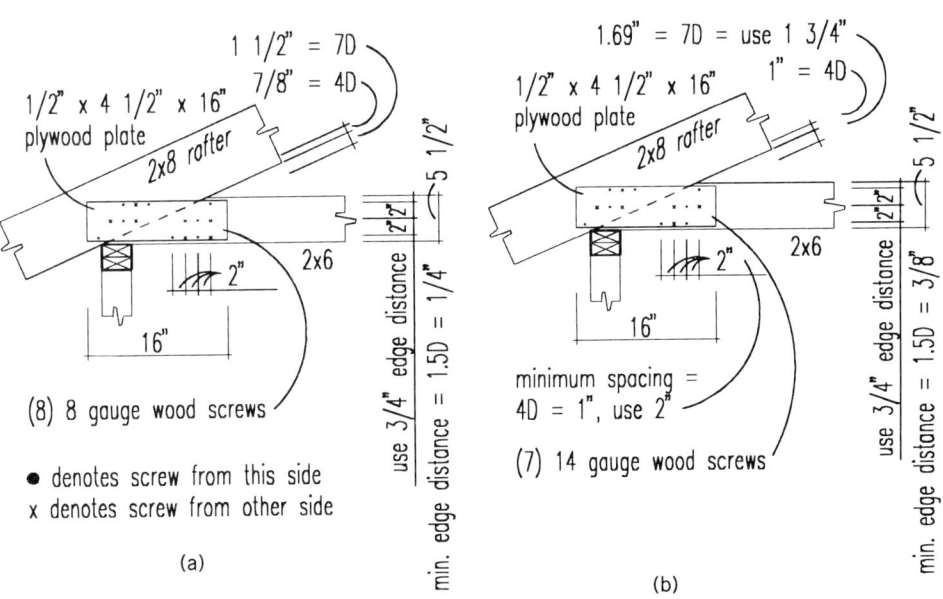

FIGURE 4.17 Heel connection—wood screws with plywood gussets.

TABLE 4.12 DIMENSIONS OF STANDARD LAG BOLTS OR LAG SCREWS[a]

		Dimensions of Lag Bolts with Nominal Diameter, D (in.)									
		$\frac{1}{4}$	$\frac{5}{16}$	$\frac{3}{8}$	$\frac{7}{16}$	$\frac{1}{2}$	$\frac{9}{16}$	$\frac{5}{8}$	$\frac{3}{4}$	$\frac{7}{8}$	1
Nominal	$D_s = D$	0.250	0.3125	0.375	0.4375	0.500	0.5625	0.625	0.750	0.875	1.000
Length	D_r	0.173	0.2270	0.265	0.3280	0.371	0.4350	0.471	0.5790	0.683	0.780
of	E	$\frac{3}{16}$	$\frac{1}{4}$	$\frac{1}{4}$	$\frac{19}{32}$	$\frac{15}{16}$	$\frac{13}{8}$	$\frac{13}{8}$	$\frac{17}{16}$	$\frac{1}{2}$	$\frac{9}{16}$
Lag Bolt,	H	$\frac{11}{64}$	$\frac{13}{64}$	$\frac{1}{4}$	$\frac{19}{64}$	$\frac{21}{64}$	$\frac{3}{8}$	$\frac{27}{64}$	$\frac{1}{2}$	$\frac{19}{32}$	$\frac{21}{32}$
L	W	$\frac{3}{8}$	$\frac{1}{2}$	$\frac{9}{16}$	$\frac{5}{8}$	$\frac{3}{4}$	$\frac{7}{8}$	$\frac{15}{16}$	$1\frac{1}{8}$	$1\frac{5}{16}$	$1\frac{1}{2}$
(in.)	N	10	9	7	7	6	6	5	$4\frac{1}{2}$	4	$3\frac{1}{2}$
1	S	$\frac{1}{4}$	$\frac{1}{4}$	$\frac{1}{4}$	$\frac{1}{4}$	$\frac{1}{4}$	—	—	—	—	—
	T	$\frac{3}{4}$	$\frac{3}{4}$	$\frac{3}{4}$	$\frac{3}{4}$	$\frac{3}{4}$	—	—	—	—	—
	$T - E$	$\frac{9}{16}$	$\frac{1}{2}$	$\frac{1}{2}$	$\frac{15}{32}$	$\frac{7}{16}$	—	—	—	—	—
$1\frac{1}{2}$	S	$\frac{3}{8}$	$\frac{3}{8}$	$\frac{3}{8}$	$\frac{3}{8}$	$\frac{3}{8}$	—	—	—	—	—
	T	$1\frac{1}{8}$	$1\frac{1}{8}$	$1\frac{1}{8}$	$1\frac{1}{8}$	$1\frac{1}{8}$	—	—	—	—	—
	$T - E$	$\frac{15}{16}$	$\frac{7}{8}$	$\frac{7}{8}$	$\frac{27}{32}$	$\frac{13}{16}$	—	—	—	—	—
2	S	$\frac{1}{2}$	$\frac{1}{2}$	$\frac{1}{2}$	$\frac{1}{2}$	$\frac{1}{2}$	$\frac{1}{2}$	$\frac{1}{2}$	—	—	—
	T	$1\frac{1}{2}$	$1\frac{1}{2}$	$1\frac{1}{2}$	$1\frac{1}{2}$	$1\frac{1}{2}$	$1\frac{1}{2}$	$1\frac{1}{2}$	—	—	—
	$T - E$	$1\frac{5}{16}$	$1\frac{1}{4}$	$1\frac{1}{4}$	$1\frac{7}{32}$	$1\frac{3}{16}$	$1\frac{1}{8}$	$1\frac{1}{8}$	—	—	—
$2\frac{1}{2}$	S	1	$\frac{7}{8}$	$\frac{7}{8}$	$\frac{3}{4}$	$\frac{3}{4}$	$\frac{3}{4}$	$\frac{3}{4}$	—	—	—
	T	$1\frac{1}{2}$	$1\frac{5}{8}$	$1\frac{5}{8}$	$1\frac{3}{4}$	$1\frac{3}{4}$	$1\frac{3}{4}$	$1\frac{3}{4}$	—	—	—
	$T - E$	$1\frac{5}{16}$	$1\frac{3}{8}$	$1\frac{3}{8}$	$1\frac{15}{32}$	$1\frac{7}{16}$	$1\frac{3}{8}$	$1\frac{3}{8}$	—	—	—
3	S	1	1	1	1	1	1	1	1	1	1
	T	2	2	2	2	2	2	2	2	2	2
	$T - E$	$1\frac{13}{16}$	$1\frac{3}{4}$	$1\frac{3}{4}$	$1\frac{23}{32}$	$1\frac{11}{16}$	$1\frac{5}{8}$	$1\frac{5}{8}$	$1\frac{9}{16}$	$1\frac{1}{2}$	$1\frac{7}{16}$
4	S	$1\frac{1}{2}$	$1\frac{1}{2}$	$1\frac{1}{2}$	$1\frac{1}{2}$	$1\frac{1}{2}$	$1\frac{1}{2}$	$1\frac{1}{2}$	$1\frac{1}{2}$	$1\frac{1}{2}$	$1\frac{1}{2}$
	T	$2\frac{1}{2}$	$2\frac{1}{2}$	$2\frac{1}{2}$	$2\frac{1}{2}$	$2\frac{1}{2}$	$2\frac{1}{2}$	$2\frac{1}{2}$	$2\frac{1}{2}$	$2\frac{1}{2}$	$2\frac{1}{2}$
	$T - E$	$2\frac{15}{16}$	$2\frac{1}{4}$	$2\frac{1}{4}$	$2\frac{7}{32}$	$2\frac{3}{16}$	$2\frac{1}{8}$	$2\frac{1}{8}$	$2\frac{1}{16}$	2	$1\frac{15}{16}$

TABLE 4.12 *Continued*

| | | \multicolumn{10}{c}{Dimensions of Lag Bolts with Nominal Diameter, D (in.)} |
		$\frac{1}{4}$	$\frac{5}{16}$	$\frac{3}{8}$	$\frac{7}{16}$	$\frac{1}{2}$	$\frac{9}{16}$	$\frac{5}{8}$	$\frac{3}{4}$	$\frac{7}{8}$	1
Nominal	$D_s = D$	0.250	0.3125	0.375	0.4375	0.500	0.5625	0.625	0.750	0.875	1.000
Length	D_r	0.173	0.2270	0.265	0.3280	0.371	0.4350	0.471	0.5790	0.683	0.780
of	E	$\frac{3}{16}$	$\frac{1}{4}$	$\frac{1}{4}$	$\frac{19}{32}$	$\frac{15}{16}$	$\frac{13}{8}$	$\frac{13}{8}$	$\frac{17}{16}$	$\frac{1}{2}$	$\frac{9}{16}$
Lag Bolt,	H	$\frac{11}{64}$	$\frac{13}{64}$	$\frac{1}{4}$	$\frac{19}{64}$	$\frac{21}{64}$	$\frac{3}{8}$	$\frac{27}{64}$	$\frac{1}{2}$	$\frac{19}{32}$	$\frac{21}{32}$
L	W	$\frac{3}{8}$	$\frac{1}{2}$	$\frac{9}{16}$	$\frac{5}{8}$	$\frac{3}{4}$	$\frac{7}{8}$	$\frac{15}{16}$	$1\frac{1}{8}$	$1\frac{5}{16}$	$1\frac{1}{2}$
(in.)	N	10	9	7	7	6	6	5	$4\frac{1}{2}$	4	$3\frac{1}{2}$
5	S	2	2	2	2	2	2	2	2	2	2
	T	3	3	3	3	3	3	3	3	3	3
	$T - E$	$2\frac{13}{16}$	$2\frac{3}{4}$	$2\frac{3}{4}$	$2\frac{23}{32}$	$2\frac{11}{16}$	$2\frac{5}{8}$	$2\frac{5}{8}$	$2\frac{9}{16}$	$2\frac{1}{2}$	$2\frac{7}{16}$
6	S	$2\frac{1}{2}$	$2\frac{1}{2}$	$2\frac{1}{2}$	$2\frac{1}{2}$	$2\frac{1}{2}$	$2\frac{1}{2}$	$2\frac{1}{2}$	$2\frac{1}{2}$	$2\frac{1}{2}$	$2\frac{1}{2}$
	T	$3\frac{1}{2}$	$3\frac{1}{2}$	$3\frac{1}{2}$	$3\frac{1}{2}$	$3\frac{1}{2}$	$3\frac{1}{2}$	$3\frac{1}{2}$	$3\frac{1}{2}$	$3\frac{1}{2}$	$3\frac{1}{2}$
	$T - E$	$3\frac{5}{16}$	$3\frac{1}{4}$	$3\frac{1}{4}$	$3\frac{7}{32}$	$3\frac{3}{16}$	$3\frac{1}{8}$	$3\frac{1}{8}$	$3\frac{1}{16}$	3	$2\frac{15}{16}$
7	S	3	3	3	3	3	3	3	3	3	3
	T	4	4	4	4	4	4	4	4	4	4
	$T - E$	$3\frac{3}{16}$	$3\frac{3}{4}$	$3\frac{3}{4}$	$3\frac{23}{32}$	$3\frac{11}{16}$	$3\frac{5}{8}$	$3\frac{5}{8}$	$3\frac{9}{16}$	$3\frac{1}{2}$	$3\frac{7}{16}$
8	S	$3\frac{1}{2}$	$3\frac{1}{2}$	$3\frac{1}{2}$	$3\frac{1}{2}$	$3\frac{1}{2}$	$3\frac{1}{2}$	$3\frac{1}{2}$	$3\frac{1}{2}$	$3\frac{1}{2}$	$3\frac{1}{2}$
	T	$4\frac{1}{2}$	$4\frac{1}{2}$	$4\frac{1}{2}$	$4\frac{1}{2}$	$4\frac{1}{2}$	$4\frac{1}{2}$	$4\frac{1}{2}$	$4\frac{1}{2}$	$4\frac{1}{2}$	$4\frac{1}{2}$
	$T - E$	$4\frac{5}{16}$	$4\frac{1}{4}$	$4\frac{1}{4}$	$4\frac{7}{32}$	$4\frac{3}{16}$	$4\frac{1}{8}$	$4\frac{1}{8}$	$4\frac{1}{16}$	4	$3\frac{15}{16}$
9	S	4	4	4	4	4	4	4	4	4	4
	T	5	5	5	5	5	5	5	5	5	5
	$T - E$	$4\frac{13}{16}$	$4\frac{3}{4}$	$4\frac{3}{4}$	$4\frac{23}{32}$	$4\frac{11}{16}$	$4\frac{5}{8}$	$4\frac{5}{8}$	$4\frac{9}{16}$	$4\frac{1}{2}$	$4\frac{7}{16}$
10	S	$4\frac{3}{4}$	$4\frac{3}{4}$	$4\frac{3}{4}$	$4\frac{3}{4}$	$4\frac{3}{4}$	$4\frac{3}{4}$	$4\frac{3}{4}$	$4\frac{3}{4}$	$4\frac{3}{4}$	$4\frac{3}{4}$
	T	$5\frac{1}{4}$	$5\frac{1}{4}$	$5\frac{1}{4}$	$5\frac{1}{4}$	$5\frac{1}{4}$	$5\frac{1}{4}$	$5\frac{1}{4}$	$5\frac{1}{4}$	$5\frac{1}{4}$	$5\frac{1}{4}$
	$T - E$	$5\frac{1}{16}$	5	5	$4\frac{31}{32}$	$4\frac{15}{16}$	$4\frac{7}{8}$	$4\frac{7}{8}$	$4\frac{13}{16}$	$4\frac{3}{4}$	$4\frac{11}{16}$
11	S	$5\frac{1}{2}$	$5\frac{1}{2}$	$5\frac{1}{2}$	$5\frac{1}{2}$	$5\frac{1}{2}$	$5\frac{1}{2}$	$5\frac{1}{2}$	$5\frac{1}{2}$	$5\frac{1}{2}$	$5\frac{1}{2}$
	T	$5\frac{1}{2}$	$5\frac{1}{2}$	$5\frac{1}{2}$	$5\frac{1}{2}$	$5\frac{1}{2}$	$5\frac{1}{2}$	$5\frac{1}{2}$	$5\frac{1}{2}$	$5\frac{1}{2}$	$5\frac{1}{2}$
	$T - E$	$5\frac{9}{32}$	$5\frac{1}{4}$	$5\frac{1}{4}$	$5\frac{7}{32}$	$5\frac{3}{16}$	$5\frac{1}{8}$	$5\frac{1}{8}$	$5\frac{1}{16}$	5	$4\frac{15}{16}$
12	S	6	6	6	6	6	6	6	6	6	6
	T	6	6	6	6	6	6	6	6	6	6
	$T - E$	$5\frac{13}{16}$	$5\frac{3}{4}$	$5\frac{3}{4}$	$5\frac{23}{32}$	$5\frac{11}{16}$	$5\frac{5}{8}$	$5\frac{5}{8}$	$5\frac{9}{16}$	$5\frac{1}{2}$	$5\frac{7}{16}$

[a]Length of thread (T) on intervening bolt lengths is the same as that of the next-shorter bolt length listed. The length of thread (T) on standard bolt lengths (L) in excess of 12 in. is equal to $\frac{1}{2}$ the bolt length ($L/2$). $D_s = D$, diameter of shank; D_r, diameter at root of thread; H, height of bolt head; W, width of bolt head across flats; N, number of threads per inch; S, length of shank; T, length of thread; E, length of tapered tip.

Source: National Design Specification, 1991 Edition.

TABLE 4.13 LAG SCREWS: RECOMMENDED LEAD HOLE SIZES

Specific Gravity of Wood	Recommended Lead Hole Sizes, Loaded Laterally or in Withdrawal[a]
$G > 0.6$	65–85% shank diameter
$0.5 < G \leq 0.6$	60–75% shank diameter
$G \leq 0.5$	40–70% shank diameter

[a]The larger percentage is used with larger-diameter lag screws. The length of the drilled hole should be at least equal to the threaded length of the lag screw.

Source: National Design Specifications, 1991 Edition, Section 9.2.1.

shown in Table 4.12. Lag screws are made of low-carbon steel.

For proper installation a lag screw must be inserted into a predrilled lead hole. The hole has a diameter that equals that of the smooth shank for a length equal to the smooth length; the threaded portion of the hole is between 40 and 85% of the diameter of the screw (see Table 4.13). Drill bits with the proper diameters are readily available. Wherever possible the screw length should be chosen so that the full length of the threaded portion extends into the holding piece. The screw must be turned into the hole, not driven. The screw or hole should be lubricated with soap or other lubricant to avoid damaging the screw or ripping the wood grain. Particularly when loaded in withdrawal, the holding strength is severely degraded if the wood grain is destroyed.

Withdrawal. Tabulated values for lag screws in withdrawal are given in NDS Table 9.2A. These values were derived by applying a factor of safety of about 5 to ultimate values determined by tests. Lag screws may be loaded in withdrawal from end grain, although this condition should be avoided. Where loaded in end grain the design strength shall be taken as three-fourths of the tabulated strength. As with screws or bolts in tension, the designer must check the root area of the screw for tension capacity. The NDS advises that the tension capacity of the screw is approximately equal to the withdrawal capacity of the connector when the lag screw has a penetration of 8 diameters. In lieu of a more exact analysis, the designer may use the withdrawal capacity based on these penetrations as a lower limit even if the screw has a greater penetration.

Lag Screws Loaded Laterally. Design values for lateral loads are given in NDS Table 9.3A for wood-to-wood connections and in NDS Table 9.3B for connections with steel side plates. These tables both show values for the screw inserted into the side grain of the wood. When a lag screw is loaded with a lateral load in the end grain of the holding piece, two-thirds of the tabulated value shall be used. In Table 9.3A we see that there are three values listed. The first value, $Z_\parallel$ (called "Z parallel"), is to be used when both main and side members are loaded parallel to grain. The main member is the member holding the point. $Z_{s\perp}$ is used when the side member is loaded perpendicular to grain. $Z_{m\perp}$ is used when the main member is loaded perpendicular to grain. If both side and main members are loaded perpendicular to grain, the lesser of $Z_{s\perp}$ and $Z_{m\perp}$ is used (see Figure 4.18). Table 9.3B for lag screws used with steel plates lists two values, $Z_\parallel$ and $Z_\perp$. These values refer to the direction of the load relative to the grain of the wood piece.

Hankinson's Formula. Where the direction of the load on the wood piece is at an angle other than 0° or 90°, the design value can be determined using Hankinson's formula. Hankinson's formula is to be used for both wood-to-wood connections and wood-to-steel connections.

To understand Hankinson's formula, refer to Figure 4.19. In the figure you see that the direction of the load is not parallel to the direction of grain of the wood piece. The least angle (acute angle) between the direction of the load and the direction of grain is used in the formula; this angle in the figure is marked α (alpha). The allowable force on the lag screw is given by the formula

$$P_\alpha = \frac{Z_\parallel \times Z_\perp}{Z_\parallel \times \sin^2 \alpha + Z_\perp \times \cos^2 \alpha}$$

If in Figure 4.19a, the angle, α, is given as 30°, and the forces $Z_\parallel$, and $Z_\perp$ are

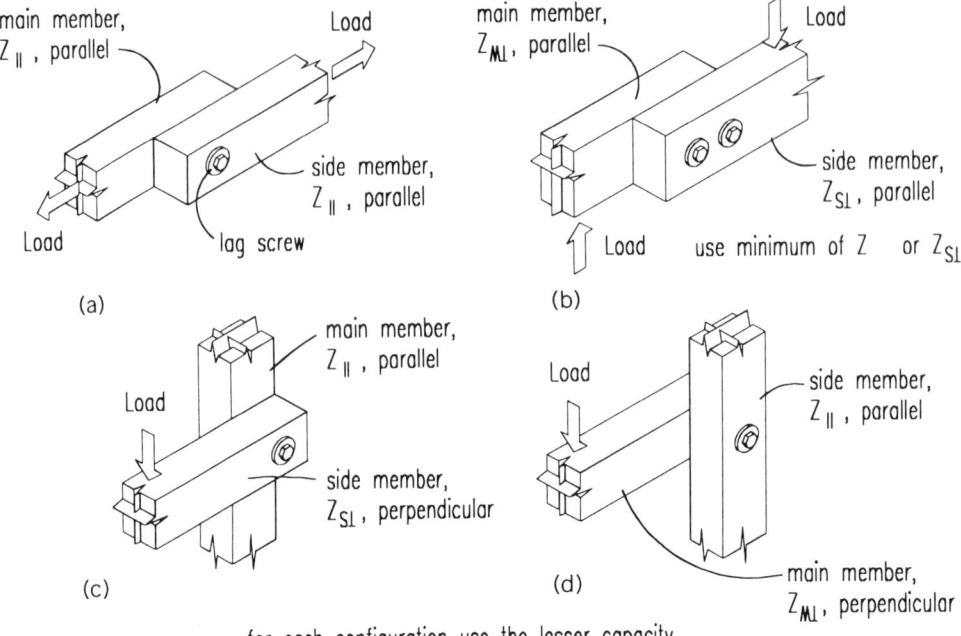

FIGURE 4.18 Definition of load directions.

given as

$Z_\parallel$ = allowable force parallel to grain

$$= 660 \text{ lb}$$

$Z_\perp$ = allowable force perpendicular to grain

$$= 380 \text{ lb}$$

then

$$Z_{30°} = \frac{660 \times 380}{660 \times \sin^2(30) + 380 \times \cos^2(30)}$$

$$= \frac{250{,}800}{(660 \times 0.25) + (380 \times 0.75)}$$

$$= \frac{250{,}800}{165 + 285}$$

$$= \frac{250{,}800}{450} = 557.3 \text{ lb}$$

Alternatively, this value can be found from Figure 4.20. To solve the previous problem using Figure 4.20, first select the portion of the figure labeled "Design Values in Units of 100." Read along the $A - B$ axis (0°) to find the force parallel to grain, 660 lb. Mark this point P. Also, along the $A - B$ line, find the perpendicular to grain force of 380 lb. Follow the horizontal line (perpendicular to $A - B$) from 380 lb until it intersects the 90° axis $A - C$. Mark this point Q. Next, draw a line between the two points, P and Q. Where this line crosses the 30° axis, find the allowable force Z_α. Extend a horizontal line from Z_α to the 0° axis $(A - B)$ to read the value, which in this case appears to be about 560 lb. Since the original values are rounded to the nearest 10 lb, this value is sufficiently accurate.

Once the values of P and Q are established for a particular connector, the chart can easily be used to find the allowable

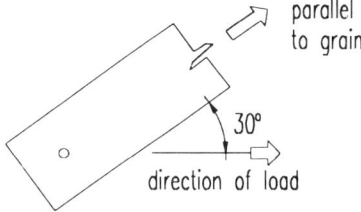

$$Z_\alpha = \frac{Z_\parallel \times Z_\perp}{Z_\parallel \times \sin^2(\alpha) + Z \times \cos^2(\alpha)}$$

$$= \frac{Z_\parallel \times Z_\perp}{.25Z_\parallel + .75Z_\perp}$$

(a)

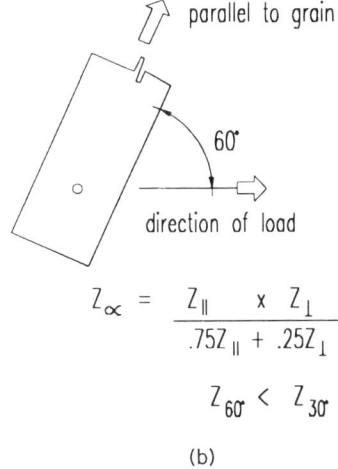

$$Z_\alpha = \frac{Z_\parallel \times Z_\perp}{.75Z_\parallel + .25Z_\perp}$$

$$Z_{60°} < Z_{30°}$$

(b)

FIGURE 4.19 Connectors with load at an angle to grain.

force for any other angle. Thus, in this example, Z_α for a load 60° from the grain is read as about 430 lb. This is much easier than calculating Hankinson's formula each time. For the rest of this chapter, we will use the nomograph in Figure 4.20 to find connection capacities when loads are applied at an angle to the wood grain. The procedure is also exactly the same as that used to find the allowable compression stress at an angle to the grain.

In Figure 4.19*b* the angle between the direction of the force and the grain of the wood is 60°. Note that the angle is always measured from the line of the load to a line parallel to the grain. Applying Hankinson's formula to this will give

$$Z_{60°} = \frac{660 \times 380}{660 \times \sin^2(60) + 380 \times \cos^2(60)}$$

$$= \frac{250,800}{(660 \times 0.75) + (380 \times 0.25)}$$

$$= \frac{250,800}{495 + 95}$$

$$= \frac{250,800}{590} = 425 \text{ lb}$$

When the direction of force is closer to parallel to the grain, the magnitude of the allowable force will be closer to that for parallel to the grain, and therefore relatively large. When the direction of force is closer to perpendicular to the grain, the magnitude of the allowable force will be closer to that for perpendicular to the grain, and therefore smaller.

Number of Screws. The design for nails and wood screws was based on each connector carrying the same amount of load, so that the effect of a number of connectors was additive. With lag screws, as with through bolts, there is a diminishing effect when a number of connectors are used together, due to the destruction of the wood. NDS Table 7.3.6A shows the group action factor, C_g, used in through-bolt and lag screw design.

Combined Lateral and Withdrawal Loads. In the case where a lag screw is loaded both in withdrawal and laterally at the same time, the allowable design load is based on the interaction of the two loads and is calculated using *Hankinson's formula*:

$$Z_\alpha = \frac{W'pZ}{W'p \sin^2 \alpha + Z' \cos^2 \alpha}$$

126 CONNECTIONS

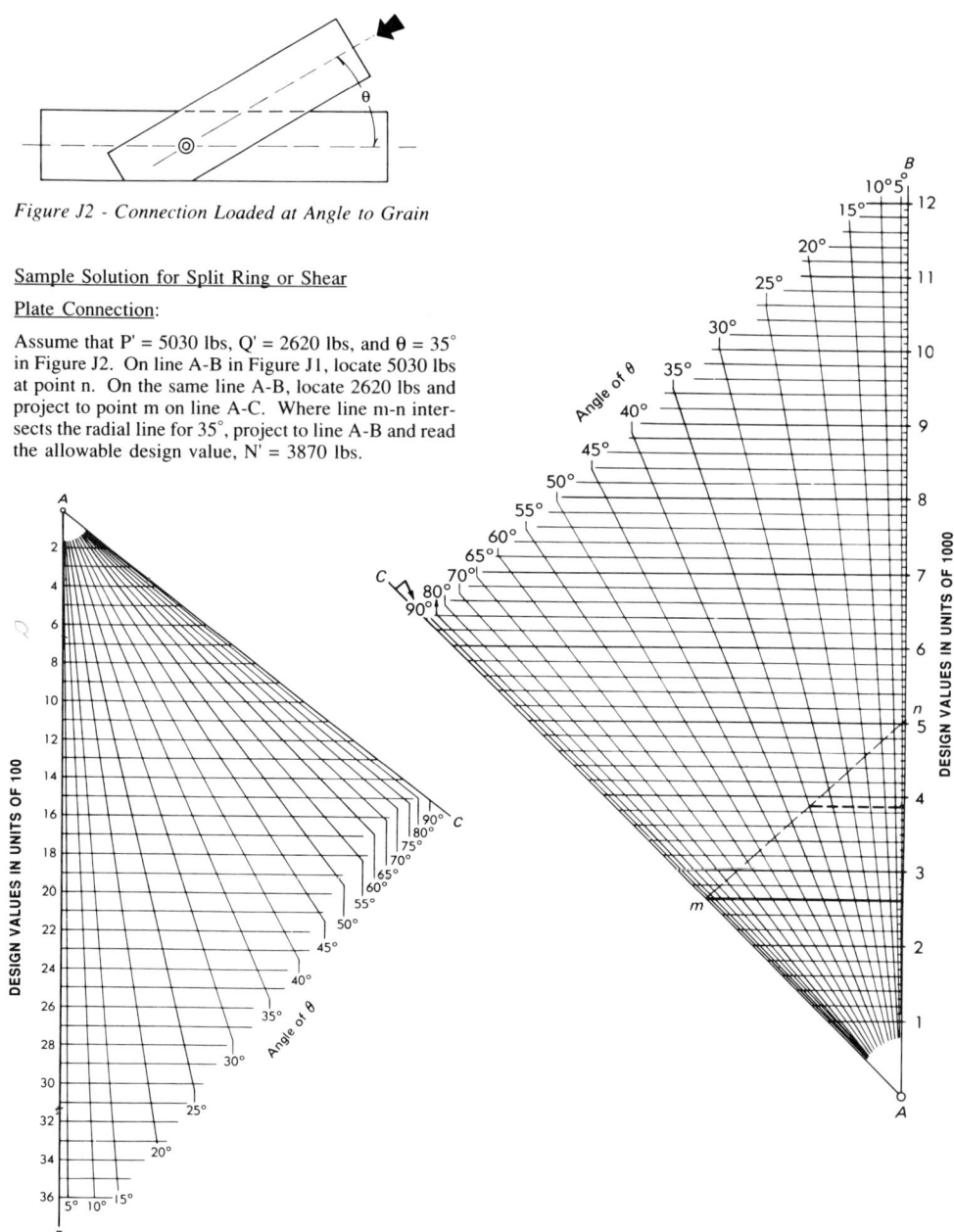

Figure J2 - Connection Loaded at Angle to Grain

<u>Sample Solution for Split Ring or Shear Plate Connection</u>:

Assume that P' = 5030 lbs, Q' = 2620 lbs, and θ = 35° in Figure J2. On line A-B in Figure J1, locate 5030 lbs at point n. On the same line A-B, locate 2620 lbs and project to point m on line A-C. Where line m-n intersects the radial line for 35°, project to line A-B and read the allowable design value, N' = 3870 lbs.

Figure J1 - Solution of Hankinson Formula

FIGURE 4.20 Lag screw connection between beam and post.

where W' is the allowable withdrawal load, p the penetration in inches, Z' the allowable lateral load, and α the angle between the wood surface and the direction of applied load.

These values can also be derived from the nomograph in Figure 4.20. Follow the procedure described, replacing P with W'_p and Q with Z'.

End and Edge Distances and Spacing Requirements. The strength of lag screw connections is dependent on the distances from the middle of the screw to the edge or end of each piece of wood connected by the screw. If these distances, or if the spacing between screws, falls below recommended distances, the calculated strength of the connection must be re-

TABLE 4.14 LAG SCREW MODIFICATION FACTORS

Condition			Factor	
Direction of Load			Lateral	Withdrawal
Direction of grain		Side grain	1.0	1.0
		End grain	$C_{eg} = 0.67$	$C_{eg} = 0.75$
		Angle to grain	Hankinson's	See factor for combined lateral and withdrawal loads
Penetration	Recommended	$8D$	$C_d = 1.0$	No minimum
	Minimum	$4D$	$C_d = 0.5$	
	Between	$4D$ and $8D$	$C_d = p/8D$	
Side piece material		Wood to wood	NDS Table 9.3A	NDS Table 9.2A
		Wood to steel	NDS Table 9.3B	NDS Table 9.2A
Combined lateral and withdrawal loads			Hankinson's formula: $Z_\alpha = \dfrac{WpZ}{Wp \sin^2 \alpha + Z \cos^2 \alpha}$	
Multiple lag screws			See NDS Table 7.3.6A	See NDS Table 7.3.6A
Moisture content	At fabrication:	In service:		
	Dry	Dry	$C_M = 1.0$	$C_M = 1.0$
	Partially seasoned or wet	Dry		
	One screw only, or two or more screws in a single row parallel to grain, or screws in two or more rows with separate splice plates for each row		1.0	1.0
	All other arrangements		0.4	0.4
	Dry or wet	Exposed to weather	0.75	0.75
	Dry or wet	Wet	0.67	0.67

Source: National Design Specification, 1991 Edition.

duced. If the distances fall below certain minimums, the connection is assumed to have no strength. These smaller distances then are minimum allowable distances which are given by the NDS. The minimum and recommended spacings and distances for lag screws are the same as those for through bolts. These spacings are defined and discussed in the next section of this chapter, which deals with through bolts.

Modification Factors. As with all connection types, lag screws are subject to modification factors related to the conditions of the wood at fabrication and in service. The applicable factors are summarized in Table 4.14.

EXAMPLE 4.11—LAG SCREWS: DECK BEAM TO POST

Design the connection between the edge beam and post for the wooden deck shown in Figure 4.21. Assume that the wood is Douglas Fir–Larch and that the connection will use lag screws. Take the live load as 60 psf, which is specified for residential balconies, and the dead load as 5 psf. Assume that the beam is a 4×10 and that the posts are 6×6. All material is S4S.

Find the force at the corner connection.

$$F = \text{load} \times \text{tributary area}$$

$$= \frac{(60 + 5) \text{ psf} \times (8 \text{ ft} \times 16 \text{ ft})}{4}$$

$$= 2080 \text{ lb}$$

Let us start by guessing the number of screws that we would like to use. Assume that there are to be two screws. The required capacity of each screw is

$$F = \frac{2080 \text{ lb}}{2 \text{ screws}} = 1040 \text{ lb/screw}$$

The next step is to find the necessary screw diameter from NDS Table 9.3A to meet the required load. We will need a screw that is approximately 7 in. long so that the threaded portion will be about $3\frac{1}{2}$ in. and be fully within the holding piece, the post. A width of $3\frac{1}{2}$ in. is not given in the table. Use the tabulated value for a $2\frac{1}{2}$-in.-thick side piece. If this proves satisfactory, we can check it using the yield theory formulas.

For a 1-in.- diameter screw, the capacities are given as $Z_{\parallel} = 2390$ lb, $Z_{s\perp} = 1130$ lb, and $Z_{m\perp} = 1530$ lb, where $Z_{\parallel}$ refers to the capacity parallel to grain, $Z_{s\perp}$ refers to the side piece when loaded perpendicular to grain, and $Z_{m\perp}$ refers to the main piece loaded perpendicular to grain. The main piece—the post—is loaded parallel to grain, so its capacity is given by $Z_{\parallel} = 2390$ lb. The side piece is loaded perpendicular to grain so its capacity is given by $Z_{s\perp} = 1130$ lb. This being the least value, the capacity of the connection is 1130 lb, controlled by the side piece.

The penetration requirement to use full design values is $8D$, eight times the screw diameter. Under no circumstances may the actual penetration be less than $4D$. Since the post is only $5\frac{1}{2}$ in. thick, the allowable strength must be reduced by the penetration factor, C_d. Select a lag screw that is 8 in. long; its penetration will be 8 minus the beam width of $3\frac{1}{2}$ in., or $4\frac{1}{2}$ in.

$$C_d = \frac{4.5}{8} = 0.5625$$

The allowable value, $Z_{s\perp}$, must be modified by all appropriate factors.

$$Z' = Z_{\perp} \times C_d \times C_D$$

$$= 1130 \text{ lb} \times 0.5625 \times 1.0 \text{ for live loads}$$

$$= 636 \text{ lb}$$

CONNECTION DESIGN 129

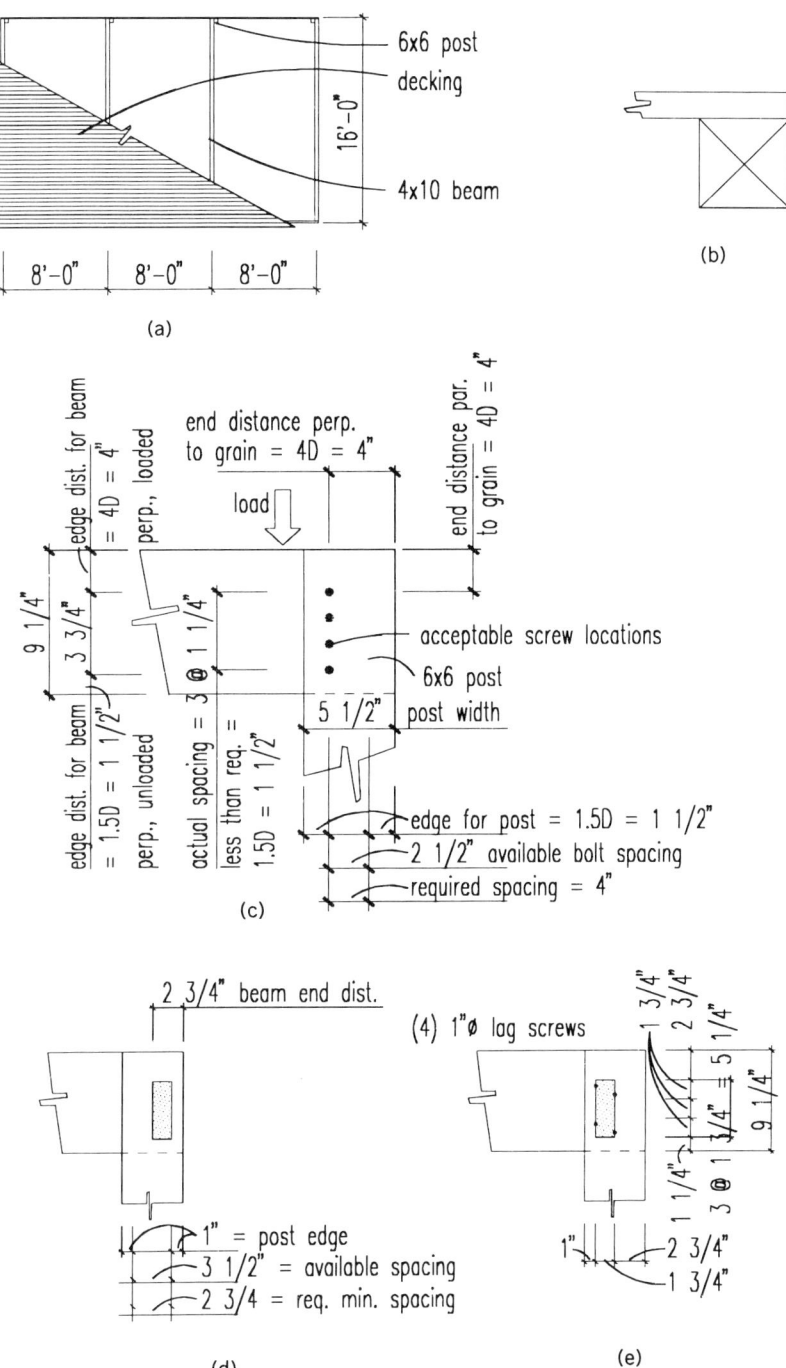

FIGURE 4.21 Lag screw connection between beam and post.

The trial number of screws needed is

$$N = \frac{2080 \text{ lb}}{636 \text{ lb/screw}} = 3.27 \quad \text{use 4 screws}$$

(level of stress in lag screws = 3.27/4 = 82%).

We would suspect that the capacity of the connection with a $3\frac{1}{2}$-in.-thick side piece might be greater than that with a $2\frac{1}{2}$-in.-thick side piece. A check of the yield equations shows that the capacity of the connection is 1332 lb, controlled by mode III$_s$, yield of the screw at the plane between the two members. The actual level of stress within the connection is 2080/(4 × 0.5625 × 1332), or 69%.

Use four 1-in.-diameter lag screws from the beam into the post.

Figure 4.21c shows the layout of the four lag screws, the required spacings for full loads, the required spacings for the actual load (69% of full load), and the actual spacings. You may note in Figure 4.21c that four screws could be put in one vertical line with a spacing of $1\frac{1}{4}$ in., which is somewhat less than the required spacing of $1\frac{1}{2}$ in. All other requirements are met in this arrangement. The total capacity of this connection is

$$Z' = 4 \text{ screws} \times \frac{1.25}{1.5} \times 0.5625$$
$$\times 1332 \text{ lb/screw} = 2500 \text{ lb}$$

This is the maximum capacity for any connection in this location using a 1-in.-diameter by 8-in.-long screw. The capacity could be increased by increasing the length of the screw.

EXAMPLE 4.12—LAG SCREWS: DECK BEAM TO POST WITH BEARING BLOCK

Continue the design of the connection started in Example 4.11 using a bearing

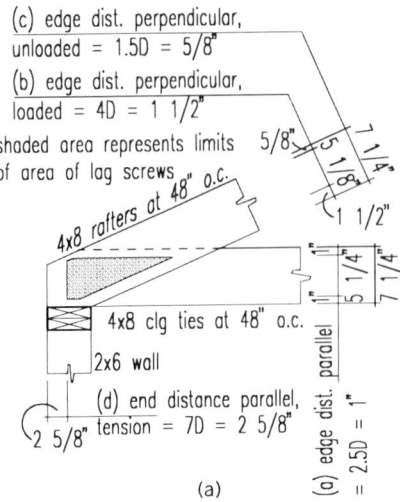

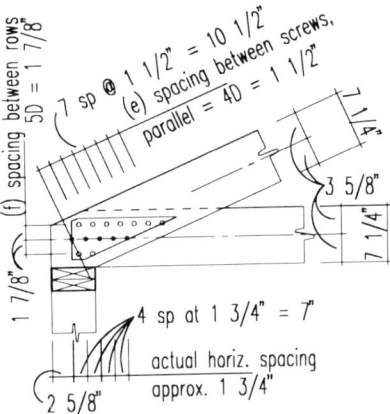

FIGURE 4.22 Lag screw connection with bearing block.

block (see Figure 4.22). In this connection the beam needs to be fixed to the post or the block. Consider first the uplift forces which will be resolved by the lag screws that connects the beam to the post.

Two screws can be used to connect the beam to the post to carry uplift. The amount of uplift might be significant if the deck is exposed to the wind from the underside. Find the maximum wind uplift that can be carried if two 1-in.-diameter lag screws are used to connect the beam to the post.

$$Z_\| = 2390 \text{ lb} \quad Z_\perp = 1130$$
$$F = N \times Q = 2 \times 1130 \text{ lb} = 2260 \text{ lb}$$

Since this is a wind load, apply the duration factor for wind,

$$F = 2260 \text{ lb} \times 1.6 = 3616 \text{ lb}$$

The penetration is found as 8 in. $- 3\frac{1}{2}$ in. $= 4\frac{1}{2}$ in., which is less than the required $8D$. The penetration factor is found.

$$C_p = \frac{4.5}{8} = 0.5625$$

$$Z' = 0.5625 \times 3616$$
$$= 2034 \text{ lb for two 1-in.-diameter screws}$$

To calculate the effect of overturning and the connector resistance, use the formula

$$M_{resisting} = 1.5 \times M_{uplift} - M_{dead}$$

Taking moments around the far edge of the deck yields

$$M_{resisting} = F_{conn} \times 16 \text{ ft}$$

$$M_{uplift} = w_{wind} \times 16 \text{ ft} \times 8 \text{ ft} \times \frac{16 \text{ ft}}{2}$$
$$= w_{wind} \times 1024 \text{ ft}^3$$

The dead load is taken as 5 psf $\times$ (8 ft $\times$ 16 ft) $\div$ 4 = 160 lb. So

$$M_{dead} = 160 \text{ lb} \times \frac{16 \text{ ft}}{2} = 1280 \text{ lb-ft}$$

and

$$F_{conn} \times 16 \text{ ft} = 1.5 \times (W_{wind} \times 1024 \text{ ft}^3)$$
$$- 1280 \text{ lb-ft}$$

or

$$2034 \text{ lb} \times 16 \text{ ft} = 1.5 \times (w_{wind} \times 1024 \text{ ft}^3)$$
$$- 1280 \text{ lb-ft}$$

or

$$w_{wind} \times 1536 \text{ ft}^3 = 2034 \text{ lb} \times 16 \text{ ft}$$
$$+ 1280 \text{ lb-ft}$$
$$w_{wind} = 22.0 \text{ psf}$$

If the deck is exposed on three sides to the wind and high enough up (at least one story) so that the wind can get underneath, there could be a strong uplift force. Using a component factor of 2.0 is often realistic in these cases. Therefore, a wind with a constant pressure of 11.0 psf could result in an uplift of 2.0×11.0 psf $= 22.0$ psf. 11.0 psf corresponds to a wind velocity of about 65 mph.

Two 1-in.-diameter by 8-in.-long lag screws from the beam into the post are sufficient to carry the uplift due to a 65-mph wind.

Wind pressure varies as the velocity squared. A pressure of 22.0 psf is caused by a wind velocity of about 90 mph.

Four 1-in.-diameter by 8-in.-long lag screws from the beam into the post are sufficient to carry the uplift due to a 90-mph wind.

The next step is to determine the number of lag screws between the bearing block and the post. Although the screws from the beam into the post will be able to carry a portion of the downward gravity loads, it is better not to reduce the number of screws in the block. Since the capacity of a connection is based on a number of variables, such as member size, species direction of grain, and connection diameter, if any of these conditions vary, the connections will slip or deform at different rates.

It is conceivable that the less stiff connection will yield, leaving the full load to be carried by the remaining connectors. For this reason it is normally not acceptable to add the capacities of different types of connectors. Therefore, the screws in the block will be designed to carry the full gravity load of 2080 lb.

In Example 4.11 the design was controlled by the allowable load perpendicular to the grain in the beam. Here, however, both the block and the post are loaded parallel to the grain, so the capacity of the connection is determined by the allowable strength in the direction parallel to the grain.

$Z_{\parallel} = 2390$ lb

$Z' = Z_{\parallel} \times C_d \times C_D$

$= 2390$ lb $\times 0.5625 \times 1.0$ for live loads

$= 1344$ lb

The trail number of screws needed is

$N = \dfrac{2080 \text{ lb}}{1344 \text{ lb/screw}} = 1.55$ use 2 screws

(level of stress in lag screws = 1.55/2 = 77%.)

Two 1-in.-diameter by 8-in.-long lag screws from the block into the post are sufficient to carry the live load of 60 psf.

For this example the two designs result in about the same number of lag screws. In situations where the dead and live loads are larger than these, the design with the block will utilize fewer screws since the number needed to resist uplift can be decreased.

Heel Connection

In Example 4.5 the connection between a rafter and a ceiling tie was designed for a roof spanning 24 ft. The rafters were assumed to be spaced 2 ft on center and were to carry a snow load of 30 psf. In that example eight 16d nails were used with a total capacity of 1091 lb. If the spacing of the rafters increased, we would easily see that a nailed connection would not be satisfactory. Example 4.13 explores a simi-

TABLE 4.15   COMPARISON OF RAFTER-TO-CEILING TIE CONNECTIONS

Spacing (ft)	Vertical Reaction (lb)	Horizontal Thrust (lb)	Axial Force (lb)	Top Chord			
				Size	Grade	CSI[a]	OK?
2	988	988	1,397	2×8	1	0.93	Yes
4	1,976	1,976	2,795	4×8	1	0.8	Yes
8	3,952	3,952	5,589	4×10	1	1.1	No
				(3) 2×10	1	0.75	Yes
12	5,928	5,928	8,384	6×10	1 SR	1.04	No
				6×10	1 Dense	0.75	Yes
16	7,904	7,904	11,178	6×12	1 SR	0.59	Yes

Spacing (ft)	Bottom Chord					Connection	
	Size	Grade	CSI	Tension Index	OK?	Number and Size	Type
2	2×8	2	0.91	0.12	Yes	(8) 16d	Nails
4	4×8	2	0.52	0.10	Yes	(5) $\frac{3}{8}$ in.	Lag screw
8	(2)2×8	2	—	0.28	Yes	(2) $\frac{5}{8}$ in.	Through bolt
12	(2)3×10	2	—	0.21	Yes	(3) $\frac{5}{8}$ in.	Through bolt
16	(2)3×10	2	—	0.28	Yes	(2) $2\frac{1}{2}$ in.	Split ring

[a]CSI, the combined stress index, is the level of stress in the chord due to interaction of axial and bending force. CSI must be ≤ 1.0.

lar roof design where the rafters are spaced 4 ft on center. Lag screws are used satisfactorily in this solution. Two other examples, Examples 4.14 and 4.19, demonstrate the design of similar connections using through bolts and split-ring connectors. These examples are intended to help the reader understand the relative strength of connectors and to develop a sense for when nails, lag screws, through bolts, or split rings might be appropriate.

Table 4.15 shows the forces and required member sizes for the four examples, with rafters spaced at 2, 4, 8, 12, and 16 ft on center. The table also summarizes the required connections for each of these rafter spacings. In all of the examples the snow load and dead load remained the same. In actuality, we would assume that the dead load, in psf, would increase slightly as the spacing increases, but this will not be of enough significance to consider in this example.

EXAMPLE 4.13—LAG SCREWS AS HEEL CONNECTION

Redesign the heel connection from Example 4.5 using lag screws. Assume that the spacing of the rafters and ceiling ties has doubled, from 2 ft o.c. to 4 ft o.c. All loads and materials remain the same, but the forces and member sizes are doubled. Therefore, the ceiling tie must carry twice 988 lb in tension, or 1976 lb.

Assume that both the rafter and the ceiling tie are 4×8's. Choose a 7-in.-long lag screw. Refer to NDS Table 9.3A, using Southern Pine. The maximum thickness of side piece in the table is $2\frac{1}{2}$ in., whereas the actual thickness is $3\frac{1}{2}$ in. To calculate this connection accurately will require referring to the yield formulas (Equations 9.3-1 through 9.3-3 in the NDS). Assume as an initial trial a $\frac{3}{8}$-in.-diameter lag screw. The results derived from these equations are $Z_\parallel = 421$ lb, $Z_{s\perp} = 278$ lb, and $Z_{m\perp} = 278$ lb. The tabulated values for $2\frac{1}{2}$-in.-thick pieces are only about 8% less. Those values are $Z_\parallel = 420$ lb, $Z_{s\perp} = 300$ lb, and $Z_{m\perp} = 300$ lb. We can feel fairly confident using the tabulated values, particularly if the screws are not stressed to their full capacity.

Using the tabulated values, apply Hankinson's formula (or Figure 4.20) to find the allowable bolt capacity at an angle of 26.6° from the parallel-to-grain direction. For snow load duration, increase the capacity by $C_D = 1.15$.

$$Z_\alpha = 390 \text{ lb/bolt}$$
$$Z' = 1.15 \times 390 = 448 \text{ lb/bolt}$$
$$N = \frac{1976 \text{ lb}}{448 \text{ lb/bolt}}$$
$$= 4.4 \text{ bolts} \quad \text{use 5} \quad (88\% \text{ stress})$$

To check the spacing, refer to Figure 4.23a. The shaded area represents the area in which screws can be located. These limits are based on edge and end distance requirements for full design loads.

a = edge distance, parallel to grain

Check

$$\frac{L}{D} = \frac{3.5}{\frac{3}{8}} = 9.3 > 6$$

Use the greater of $1.5D$ or half the spacing between rows ($5D$).

$a = 2.5D = 0.9375$ in. use 1 in.
b = edge distance, perpendicular to grain
  at the loaded edge
  $= 4D = 1.5$ in.
c = edge distance, perpendicular to grain
  at the unloaded edge
  $= 1.5D = \frac{5}{8}$ in. $= 0.625$ in.
d = end distance, parallel to grain, loaded
  in tension
  $= 7D = 2.625$ in.
e = spacing between screws in a row $= 4D$
  $= 1.5$ in.
f = spacing between rows, parallel to grain
  $= 5D = 1.875$ in.

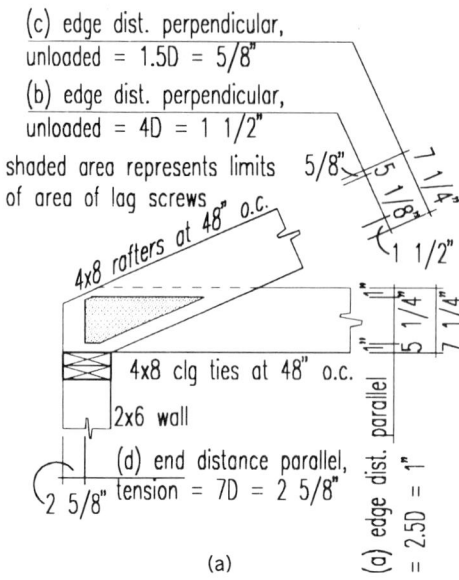

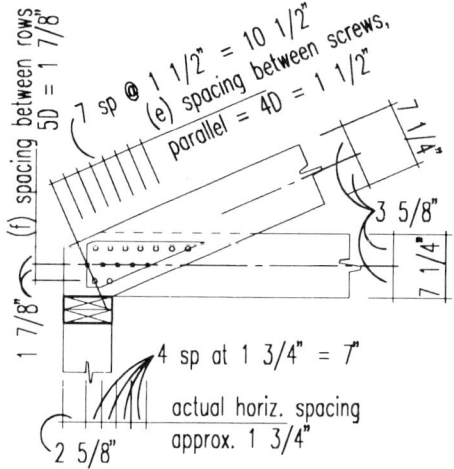

use (5) 3/8"⌀ lag screws – capacity = 2240 lb.
• indicates screw location
maximum number of screws = 14
(14) 3/8"⌀ lag screws – capacity = 5854 lb.
(includes group action factor of 0.933)
○ indicates possible screw location

(b)

FIGURE 4.23 Heel connection—lag screws.

Figure 4.23b shows a screw spacing that meets all of these requirements. To minimize the eccentricity of the connection, the center of the screw group should coincide with the intersection of the centerlines of the two chords. We choose to locate the lag screws in one line to minimize splitting that may result from drying and shrinkage. The actual horizontal spacing is $1\frac{3}{4}$ in., which comes from rotating the spacing from the angle of the rafter.

Use five $\frac{3}{8}$-in.-diameter by 7-in.-long lag screws.

The figure indicates that there is space for 14 lag screws. The total capacity of this connection is 5854 lb. In calculating this value, the group action factor was determined to be 0.933.

In this example we arbitrarily selected a lag screw of $\frac{3}{8}$ in. diameter, which is probably the smallest diameter that should be considered. We might have selected any size larger than that. Table 4.16 tabulates the number of lag screws that would be needed for various other diameters. Design values are based on wood thickness of $2\frac{1}{2}$ in., so they are slightly unconservative. Because the percentage of allowable load generally goes down for larger screws, the required spacing for the connection group decreases. Generally, the fewer number of screws is preferred, since it will reduce labor.

EXAMPLE 4.14—LAG SCREWS WITH STEEL SIDE PLATES

Redesign the heel connection from Example 4.5 using lag screws and steel side plates. Assume that the spacing of the rafters and ceiling ties are 4 ft o.c. All loads and materials remain the same as in

TABLE 4.16 NUMBER OF LAG SCREWS REQUIRED IN HEEL CONNECTION

Lag Screw Diameter (in.)	$Z_\parallel$	$Z_\perp$	Z_α	$Z' = 1.15 \times Z_\alpha$	Number of Screws Required	Number of Screws Used	Percent of Allowable Load
$\frac{3}{8}$	420	300	389	447	4.4	5	88
$\frac{1}{2}$	710	420	624	717	2.75	3	92
$\frac{5}{8}$	980	570	857	985	2.00	2	100
$\frac{3}{4}$	1320	660	1100	1265	1.56	2	78
$\frac{7}{8}$	1710	720	1341	1542	1.28	2	64

Example 4.13. The rafter and ceiling tie are both 4 × 8's and are in the same plane. The ceiling tie will carry 1976 lb in tension.

Figure 4.24 shows three options for the design of the rafter-to-ceiling tie joint. In Figure 4.24a the ceiling tie bears on the wall plate and is cut on an angle to receive the rafter. The arrows indicate the direction of the forces at the joint. Because the rafter can slide past the ceiling tie, the metal plates must carry the complete axial force in the rafter. The force diagram show that the force is 2209 lb and that it acts parallel to grain along the rafter. The connection at the ceiling tie must carry only the horizontal thrust which is 1976 lb in tension, parallel to the grain along the ceiling tie. The connection to the rafter carries 1976 lb acting at 26.6° to the grain. The plate must be long enough to assure adequate end distance from the end of the ceiling tie to the first lag screw. Because the center of the plate is offset to the right of the intersection of the forces, there will be an induced moment in the connection that will require additional design and more screws than shown in option (b).

Figure 4.24b shows the rafter bearing on the wall plate. The ceiling tie is connected via a metal plate to the rafter. This configuration would most likely be constructed on the ground or off-site and swung into place with a crane. Here the ceiling tie connection must also carry 1976 lb in tension, parallel to grain. The plate transfers 1976 lb into the rafter at an angle of 26.6° from the parallel. The vertical force in the rafter is carried in bearing on the wall plate, so the connection to the rafter will not need to carry the full axial force of 2209 lb. For low-pitched roofs (approximately 30° or less) this configuration will result in slightly fewer screws in the rafter. For steeper-pitched roofs this will actually require more screws in the rafter since the force on the screw becomes more perpendicular to grain.

Figure 4.24c is the most difficult of the three to build because of the angled cuts in the collar tie. However, if designed properly, the joint can carry the full gravity loads in bearing, wood to wood. The plate and screws are needed only to counter wind uplift forces. Borrowing a provision from manufactured truss design, we will adopt a conservative approach and design the plates to carry 50% of the load. In the figure you see a purlin on top of the rafter. The purlin, which runs parallel to the ridge and spans between the rafters, is used only where the rafter spacing is very large, in the range 8 to 16 ft on centers.

Table 4.17 summarizes the design for each of these options. The rafter and the ceiling tie are both 4 × 8's. Both are 3.5 in. thick, so choose a 7-in.-long lag screw.

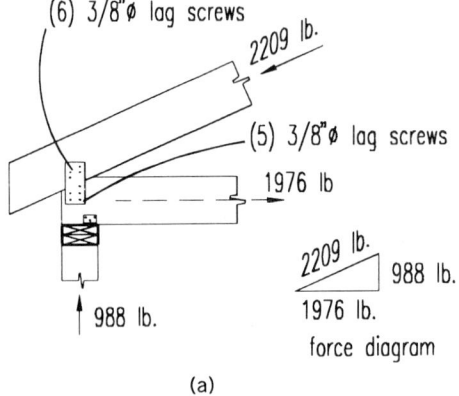

(a)

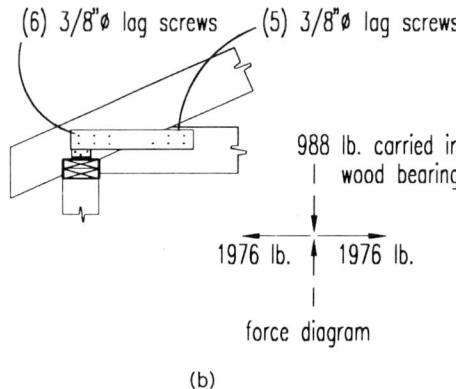

(b)

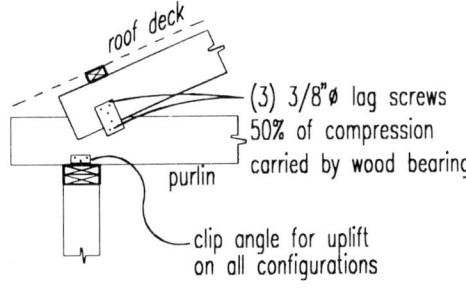

(c)

FIGURE 4.24 Heel connection—lag screws with steel side plates.

Refer to NDS Table 9.3B, using Southern Pine.

$\frac{3}{8}$ in. diameter × 7-in.-long lag screws in Southern Pine framing.

12-gage steel side plates.

Snow-load duration factor of 1.15.

$$Z_\parallel = 400 \text{ lb} \times 1.15 = 460 \text{ lb}$$
$$Z_\perp = 230 \text{ lb} \times 1.15 = 264.5 \text{ lb}$$
$$Z_{26.6°} = 348.5 \text{ lb} \times 1.15 = 400 \text{ lb}$$

Before we decide if this is acceptable, we need to check spacing requirements and also the group factor, C_g. The group action factor, C_g, was calculated for three screws in a row using NDS Table 7.3.6A. The result of this calculation is that C_g equals 0.92. If the actual stress in the connection (shown in table 4.17) is less than 92%, the connection is satisfactory. For connections with three screws in a row, if the actual stress exceeds 92%, the connection is not acceptable.

In connections like this where metal plates are used it is very easy to induce moments into the connections. If the connection is detailed as shown in Figure 4.24a, the effect of the moment would have to be considered. The centroid of the top screw is displaced from the centroid of the bottom screws by about 3 in. (A moment calculation using through bolts is covered in Example D.1 in the Appendix.) If it is detailed as shown in Figure 4.24b, where the line of action of all the forces are coincident (meet at one point), there is no moment.

Through Bolts

Our discussion of connectors has proceeded from those with the lowest capacity toward those with higher capacities. Through bolts are the most frequently used

TABLE 4.17 SUMMARY: LAG SCREWS WITH STEEL SIDE PLATES

Option	Member	Force (lb)	Direction	Screw Capacity (lb)	Required Number of Screws[a]	Use	Percentage of Stress[b]	Required Number for Group Action
a	Rafter	2209	Parallel	460	4.8	5	96	6
	Ceiling tie	1976	Parallel	460	4.3	5	86	5
b	Rafter	1976	26.6°	400	4.94	5	99	6
	Ceiling tie	1976	Parallel	460	4.3	5	86	5
c	Rafter	2209	Parallel	460	4.8 × 0.5	3	(80)	3
	Ceiling tie	1976	Parallel	460	4.3 × 0.5	3	(72)	3

[a] For option c it is assumed that the wood-to-wood connection carries only 50% of the force. The screws are sized to carry the remaining force; hence the number required is less than in the other options.

[b] Where three screws are in a row, there is a group action factor of 0.92. If the actual percentage of stress in less than 0.92, the connection is satisfactory. If not additional screws are needed.

connectors where higher capacities are required. Bolt capacities are high enough that they are most commonly used with heavier timbers, 4 in. and thicker. Bolts, along with split-ring connectors, are also used extensively to join multiple light members together. As the capacity of the connectors increases, edge and end distances and spacing between connectors becomes more critical. They often become the most important criteria in the design: we will see cases where we will not be able to fit all the bolts that are needed in a joint.

A through-bolt connection consists of a bolt with a head, a nut, and a washer on each side. Bolts must be of low-carbon steel meeting ASTM A307 specifications. Standard-size washers are to be used, or if steel plates or straps are used, the washers may be omitted. The bolts are to be installed into drilled holes. The holes should be no more than $\frac{1}{16}$ in. larger than the bolt diameter to ensure that the bolt will bear against the full surface of the wood. Similarly, the hole should not be less than $\frac{1}{32}$ in. smaller than the bolt so that the bolt will not have to be driven into the wood. As with lag screws, the strength of the connection comes from keeping the wood fibers intact. The design values are to be used for conditions where the nut and washers are tight against the wood. The designer needs to be aware of conditions that will cause the wood to shrink and loosen the nut. In these cases the design values must be modified by wet service factors, C_M, but the designer should also assure that there is a procedure to check for and tighten loose connections.

Through bolts are used most often to carry lateral loads. Bolts are used to connect two or more wood pieces or to connect wood pieces with metal side plates. The most common arrangements use the bolts in double shear, although single-shear connections are frequently used.

It is possible to use a through bolt so that the bolt will be loaded axially, but usually axial loads are only components of the total load on the bolt. When bolts are loaded axially, the wood under the washer will be loaded in compression perpendicular to grain (not a very strong property) and the bolt will be loaded in tension.

The *National Design Specification* (NDS) contains eight tables for bolt strengths. Tables identify values for bolts through two-piece connections, referred to as single shear, and for three-piece connections, referred to as double shear. Each of these categories is then divided into four tables that address the composition of the material: either sawn wood, sawn wood main piece with steel side piece(s), glued-laminated wood, or glued laminated with

TABLE 4.18 THROUGH-BOLT MODIFICATION FACTORS

Condition	Factor	
Direction of load		
Parallel to grain	$Z_\parallel$	
Perpendicular to grain	Minimum of $Z_{s\perp}$ or $Z_{m\perp}$	
Load at angle to bolt axis	Component of load perpendicular to bolt axis $< Z_\parallel$	
	Component of load parallel to bolt axis; design washers for wood bearing	
angle to grain	Hankinson's formula	
Side piece material	Single shear:	Double shear:
Wood-to-wood	NDS Table 8.2A	NDS Table 8.3A
Wood-to-steel	NDS Table 8.2B	NDS Table 8.3B
Glulam-to-glulam	NDS Table 8.2C	NDS Table 8.3C
Glulam-to-steel	NDS Table 8.2D	NDS Table 8.3D
Wood-to-concrete or masonry	Compare NDS Table 8.2A values with Table 4.19	
Multiple shear planes	Multiply least value for any single-shear plane by the number of shear planes	
Multiple bolts	C_g, group action factor; see NDS Table 7.3.6A	
End and edge distance and spacing	Reduced design value:	Full design value
	C_Δ = minimum of C_Δ factor in any direction; applies to all bolts in group	$C_\Delta = 1.0$
Moisture content		

At fabrication:	In service:	
Dry	Dry	$C_M = 1.0$
Partially seasoned or wet	Dry	One bolt only, or two or more bolts in a single row parallel to grain, or bolts in two or more rows with separate splice plates for each row $C_M = 1.0$
		All other arrangements $C_M = 0.4$
Dry or wet	Exposed to weather	$C_M = 0.75$
Dry or wet	Wet	$C_M = 0.67$

Source: National Design Specification, 1991 Edition.

steel side piece(s). The values in these tables must be modified by the factors shown in Table 4.18. Although the group of tables in the NDS is quite extensive, there are some considerations that are not specifically covered which require additional investigation.

Side Member Material. The NDS bolt tables are based on wood side plates of the same species as the main member. When a different species is used, both the side plates and the main member must be checked. The design value is taken as the lesser of:

(a) The tabulated value for a comparable joint having all the material the same as the *side plate's* material, or
(b) The tabulated value for a comparable joint having all the material the same as the *main member's* material.

Double Shear with Unequal Side Plates. In double-shear connections, if the side pieces are of differing species or of differing thickness, the procedure is to break the three-member joint into two single-shear connections. The capacity of each connection is found as shown above. The capacity of the total connection is the sum of the two single-shear values. If the side pieces are of slightly different thickness, you may simply double the strength of the single shear found in the tables for the thinner side piece. This approach is simpler and conservative.

Steel Main Member. When the central or main member is steel and the side plates are wood, simply double the design value for the single-shear connection using a steel side plate and the same thickness wood member. If you compare Figure 4.25*a* and *b*, you can see that the connection with the steel side plates is stronger. The difference is very significant for nar-

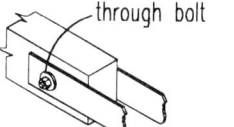
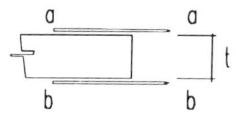

wood main members with steel side piece

capacity equals tabulated value for double shear for double shear of wood with thickness = t

(a)

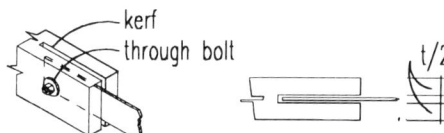

steel main member with wood side plates

capacity equals 2x single shear value for wood thickness = t/2

(b)

Example:

for 3" thick Southern Pine w/ 1"⌀ bolt and 1/4" steel side plate(s):

a) Z = 4610 lb. – double shear
    (NDS Table 8.3B)

b) Z = 2 x 1140 lb. = 2280 lb. – single shear
    2 x single shear for t = 1 1/2"
    (NDS Table 8.2B)
    use 2280 lb.

compare with 8" thick Southern Pine (7 1/2") with 1"⌀ bolt and 1/4" steel side plate(s):

a) Z = 5750 lb. – double shear
    (NDS Table 8.3B)

b) Z = 2 x 2420 lb. = 4840 lb. – single shear
    2 x single shear for t = 3 1/2"
    (NDS Table 8.2B)
    use 4840 lb.

FIGURE 4.25 Through bolts—compare steel side plate with steel centered plate.

row pieces of wood, but becomes less dramatic for thicker pieces of wood. In cases where there is an objection to seeing the metal side plates in a connection exposed to view, the detail shown in Figure 4.25a is often the best option available.

Bolts to Concrete or Masonry. The most common way to connect a wood piece to concrete or masonry is to use a through bolt. An example is the connection between the mud plate of a house and the block or concrete foundation below. The common specification for this connection is $\frac{1}{2}$-in.-diameter bolts at 6'-0" on center. In the higher seismic zones, 3 and 4, or in high wind load areas, the required spacing is 4'-0" on center. In unusual circumstances this connection may require checking. The strength of the bolt in both the wood and the concrete or masonry will need checking. To determine the capacity of the connection relative to the wood, take a design value from NDS Table 8.2A (single shear, wood-to-wood). Enter the table with t_m equal to twice the thickness of the actual wood piece. If the wood is loaded parallel to grain, as in the case of a shearwall, find $Z_\parallel$. If the load is perpendicular to the wood grain, for example in checking the wind transverse to the wall, use $Z_\perp$. The concrete values and masonry values can be found in the UBC. Those

TABLE 4.19 THROUGH BOLTS:
ALLOWABLE SHEAR AND TENSION (POUNDS) ON BOLTS IN CONCRETE AND MASONRY[a]

Bolt Diameter (in.)	Concrete				Masonry[b]		
	Bolt Embedment[c] (in.)	Minimum Concrete Strength (psi)			Bolt Embedment (in.)	Solid Shear	Grouted, Shear
		2000, Shear[d]	3000, Shear[d]	2000 to 5000, Tension[e]			
$\frac{1}{4}$	$2\frac{1}{2}$	500	500	200			
$\frac{3}{8}$	3	1100	1100	500			
$\frac{1}{2}$	4	2000	2000	950	4	350	550
$\frac{5}{8}$	4	2750	3000	1500	4	500	750
$\frac{3}{4}$	5	2940	3560	2250	5	750	1100
$\frac{7}{8}$	6	3580	4150	3200	6	1000	1500
1	7	3580	4150	3200	7	1250	1850
$1\frac{1}{8}$	8	3580	4500	3200	8	1500	2250
$1\frac{1}{4}$	9	3580	5300	3200			

[a]Values are for natural stone aggregate concrete and bolts of at least ASTM A307 quality. Bolts shall have a standard head or an equal deformity in the embedded portion. Values are based on a bolt spacing of 12 diameters with a minimum edge distance of 6 diameters. Such spacing and edge distance may be reduced 50% with an equal reduction in value. Use linear interpolation for intermediate spacings and edge margins.
[b]Masonry values are for all types of masonry except gypsum and unburned clay units. Masonry values exceeding 1500 lb permitted only when f_m^1 is not less than 2500 psi.
[c]An addition 2 in. of embedment shall be provided for anchor bolts located in the top of columns for buildings located in UBC seismic zones 2, 3, and 4.
[d]Values shown are for concrete with or without special inspection.
[e]Values shown are for concrete without special inspection. Where special inspection is provided values may be increased by 100%.
Sources: UBC Table 24-M for Masonry and UBC Table 26-E for Concrete.

values are represented here in Table 4.19. All applicable modification values, including duration of load, shall be applied to the wood capacity but not to the concrete capacities. The capacity of the connection is the lesser of the wood or concrete values.

Direction of Load. As with lag screws, the strength of the bolt connection is greatly affected by the direction of the load with respect to the grain of the wood. A quick review of NDS Tables 8.2 and 8.3 will show that strengths perpendicular to the grain are about half to two-thirds of those parallel to grain. If a joint is loaded so that the force is at a direction between 0 and 90°, the value can be determined using Hankinson's formula.

When the side member(s) are loaded at an angle from the grain, the design value is derived using Hankinson's formula with $Z_\parallel$ and $Z_{s\perp}$: thus,

$$Z_\alpha = \frac{Z_\parallel \times Z_{s\perp}}{\sin^2 \alpha \times Z_\parallel + \cos^2 \alpha \times Z_{s\perp}}$$

When the main member is loaded at an angle from the grain, the design value is derived using Hankinson's formula with $Z_\parallel$ and $Z_{m\perp}$: thus,

$$Z_\alpha = \frac{Z_\parallel \times Z_{m\perp}}{\sin^2 \alpha \times Z_\parallel + \cos^2 \alpha \times Z_{m\perp}}$$

The NDS prescribes a slightly different approach to finding bolt capacities for loads not parallel or perpendicular to grain. The NDS approach asks you to derive the dowel bearing strength, $F_{e\perp}$, using Hankinson's formula and then calculate the allowable strength using the yield mode equations. The difference between these two approaches is in the range 2 to 5%; the approach recommended here is usually but not always conservative. It is felt by the authors that converting the tabulated values is a simpler procedure that still maintains sufficient engineering accuracy.

Connections with More Than Three Members. Connections with more than three members contain multiple shear planes. To determine the capacity of such a joint, break the joint into a series of two-member joints and investigate each one separately. (An example of the following procedure is given in Example 4.16. Since the procedure is somewhat involved it would be a good idea to follow along in the example as you read the procedure.)

If the various members have different thicknesses or species, the following procedure is used:

1. Resolve the multimember connection into the maximum number of contiguous two-member joints.
2. For each such two-member joint determine the appropriate design load using standard procedures.
3. For assemblies in which the load is shared equally among all the members, or in which the distribution of loads among members is indeterminate, the design value for the multimember joint shall be the least design value for any one shear plane times the number of shear planes in the joint.
4. For assemblies in which the load on each member is known, the bolt design value for any member in the joint shall be the sum of the individual bolt design values for each of the shear planes acting on that member. The total design value for the connection shall be the sum of the design values for all of the members in the joint.

This procedure is also to be used on three-member joints (double shear) with side pieces of different thicknesses. For multimember connections where all the pieces are of the same thickness and species, the design value for each shear plane is found in NDS Table 8.2A or 8.2C. The total connection capacity is calculated by summing the capacity of each shear plane. Thus for a connection with four members, there will be three shear planes. The capacity of the total connection will be three times the tabulated value for single shear. It can be seen that the detailed procedure for unequal thicknesses yields the same result when applied to members with equal thicknesses.

Spacing, Edge, and End Distances. Tabulated values for bolts are based on having an adequate amount of wood to carry the load applied by the bolt. Specific requirements for spacing between bolts

TABLE 4.20 THROUGH BOLTS AND LAG SCREWS: SPACING, EDGE, AND END DISTANCE REQUIREMENTS[a]

Condition		Direction of Load			
		Parallel to Grain		Perpendicular to Grain	
		Minimum Distance		Minimum Distance	
		For Reduced Design Value	For Full Design Value	For Reduced Design Value	For Full Design Value
End Distance	Tension Softwood	$3.5D$	$7D$		
	Hardwood	$2.5D$	$5D$		
	Compression Any direction	$2D$	$4D$	$2D$	$4D$
Spacing between bolts in a row		$3D$	$4D$	$3D$	Required Spacing for attached members
Edge distance	When $L/D < 6$	$1.5D$			
	When $L/D > 6$	$1.5D$ or half the spacing between rows, whichever is greater			
	Loaded edge			$4D$	
	Unloaded edge			$1.5D$	
Spacing between rows of bolts	All ratios of L/D	$1.5D$			
	When $L/D < 2$			$2.5D$	
	$2 < L/D < 6$			$\dfrac{5L + 10D}{8}$	
	$L/D > 6$			$5D$	

[a] Definition of terms used in bolt distances: "Row of bolts" means two or more bolts placed in a line parallel to the direction of the load. "End distance" is the distance from the end of the timber to the center of the bolt hole nearest the end. "Edge distance" is the distance from edge of the timber to the center of the nearest bolt. All bolt spacings and distances are measured from the center of the bolt.

Source: National Design Specification, 1991 Edition.

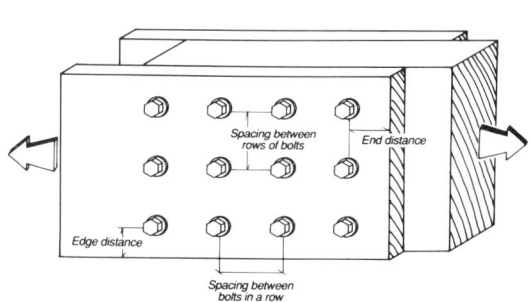

Parallel to grain loading   Perpendicular to grain loading

TABLE 8.5.3 - EDGE DISTANCE REQUIREMENTS FOR BOLTS

Direction of Loading	Minimum Edge Distance
Parallel to Grain:	
when $\ell/D \leq 6$	1.5D
when $\ell/D > 6$	1.5D or 1/2 the spacing between rows, whichever is greater
Perpendicular to Grain:	
loaded edge	4D
unloaded edge	1.5D

TABLE 8.5.5 - SPACING REQUIREMENTS FOR BOLTS IN A ROW

	Minimum Spacing	
Direction of Loading	for Reduced Design Value	for Full Design Value
Parallel to Grain	3D	4D
Perpendicular to Grain	3D	required spacing for attached member(s)

TABLE 8.5.4 - END DISTANCE REQUIREMENTS FOR BOLTS

	Minimum End Distances	
Direction of Loading	for Reduced Design Value	for Full Design Value
Perpendicular to Grain	2D	4D
Parallel to Grain, Compression: (bolt bearing away from member end)	2D	4D
Parallel to Grain, Tension: (bolt bearing toward member end)		
for softwoods	3.5D	7D
for hardwoods	2.5D	5D

TABLE 8.5.6 - SPACING REQUIREMENTS BETWEEN ROWS OF BOLTS

Direction of Loading	Minimum Spacing
Parallel to Grain	1.5D
Perpendicular to Grain:	
when $\ell/D \leq 2$	2.5D
when $2 < \ell/D < 6$	$(5\ell + 10D)/8$
when $\ell/D \geq 6$	5D

FIGURE 4.26 Spacing, edge, and end distance requirements for through bolts and lag screws connections.

and for distances from bolts to the end or edge of the wood must be maintained. All bolt spacings and distances are measured from the center of the bolt. These requirements are listed in Table 4.20. The various terms are defined in Figure 4.26.

Geometry Factor, C_Δ. For end distances and spacing between bolts in a row, two set of values are given: reduced design values and full design values. The NDS tables are based on bolt spacing and end distances meeting or exceeding the full design values. For these cases the full tabulated value may be used. If the bolt spacing or end distance is less than those in the full design column, but not less than those in the reduced design column, the design value must be reduced by the geometry factor, C_Δ, where

$$C_\Delta = \frac{\text{actual spacing}}{\text{``full design'' spacing}} \quad \text{or}$$

$$\frac{\text{actual end distance}}{\text{``full design'' end distance}}$$

If both the end distance and the spacing are less than the full design requirements, C_Δ is taken as the lesser value.

The reduced values are the minimum possible values. This is true for the edge distances and the spacing between rows of bolts as well. Any spacing or end distance that is less than the minimums is a violation of the NDS and should be avoided.

Optimal Design. The approach to calculating bolted connection strengths taken in the 1991 NDS is called yield theory. In this approach the strength of all the possible failure modes is considered; whatever mode has the lowest strength will fail first and determine the strength of the connector. The lowest strength is the allowable load. By reviewing the equations for the various modes we can determine the controlling parameters, which include the thickness of each wood piece in the connection and the diameter of the bolt.

A quick review of the bolt capacity tables in NDS shows that in all cases a larger bolt will carry a greater load. The first rule of bolt design then is simply to

> Use the largest bolt possible.

The bolt strength will be affected, however, by spacing requirements, which are a function of the bolt diameter. So our first rule holds only if the increase in strength with the larger bolt is not offset by decreases due to spacing requirements. Generally, if a bolt of 1 in. diameter has more than twice the strength of a $\frac{1}{2}$-in.-diameter bolt, the larger bolt will be more effective. Scanning through NDS tables for double shear (Tables 8.3A through 8.3C), we can see that this is generally true for all values of $Z_\parallel$ and $Z_{s\perp}$ but not for $Z_{m\perp}$. Therefore, whenever the main member is loaded parallel to the grain, use the largest bolt that meets the spacing requirements. Otherwise, the spacing requirements must be considered. As a rule, though, when a number of bolt sizes will result in the same count of bolts, the smallest bolt will be preferable. Example 4.16 demonstrates this rule. The first rule can then be rewritten to include these conditions.

> (a) For main members loaded primarily parallel to the grain, use the largest bolt possible.
> (b) For main members loaded primarily perpendicular to the grain, use the least number of bolts possible; for this number, use the smallest bolt possible.

First Rule of Bolt Design

When the bolt size is very large in comparison to the thickness of the wood pieces, the bolt will be very stiff. This bolt stiffness is measured by the L/D ratio. A low L/D ratio refers to a stiff bolt. In a three-member connection (double shear) the length of the bolt through the main piece, which equals the thickness of the piece, is taken as L. The diameter of the bolt is D. In a two-member (single-shear) connection L is the minimum thickness of the two pieces. In cases where the bolt is very stiff (a small L/D ratio), the failure mode is pure compression uniformly distributed against the

TABLE 4.21 OPTIMUM SIDE PIECE THICKNESSES

Condition	Direction of Load	Minimum Thickness of Side Piece[a]	Example: Use Southern Pine, $t_m = 3$ in. Bolt Diameter = $\frac{3}{4}$ in.
Single shear	Both pieces parallel to grain	$t_s \leq t_m$	$t_s \leq t_m = 3$ in.
	Main piece parallel to grain, side piece perpendicular	$t_s \leq \dfrac{F_{em}}{F_{es}} \times t_m$	$t_s \leq \dfrac{6150}{2950} \times 3 = 6$ in.
Double shear	Both pieces parallel to grain	$t_s \leq \dfrac{t_m}{2}$	$t_s \leq \dfrac{t_m}{2} = 1.5$ in.
	Main piece parallel to grain, side piece perpendicular	$t_s \leq \dfrac{F_{em}}{F_{es}} \times \dfrac{t_m}{2}$	$t_s \leq \dfrac{6150}{2950} \times \dfrac{3}{2} = 3$ in.

[a] t_s, Thickness of the side piece; t_m, thickness of the main piece; F_{es}, dowel bearing strength in side piece (see NDS Table 8A); F_{em}, dowel bearing strength in main piece (see NDS Table 8A).

Source: Equations from the *National Design Specification*, 1991 Edition.

entire length of the side piece or the main piece. This mode of failure occurs when the L/D ratio generally exceeds 8. Obviously, if in a three-member connection, the side pieces are too thin, they will fail first through crushing. To assure that all members are of equal strength, the following guidelines should be applied. For cases when the L/D ratio > 8, use minimum side piece thickness as shown in Table 4.21.

Thicker side pieces will result in greater capacities throughout the table, regardless of the L/D ratio. Therefore, the second rule of bolt connection design is

> Use side pieces as thick as possible.
> When the L/D ratio > 8,
> (a) In double-shear connections side pieces should be at least one-half the thickness of the main piece.
> (b) In single-shear connections, the side piece should equal the thickness of the main piece.

Second Rule of Bolt Design

Bolts Used in Moment Connections. Wherever possible, wood connections should be detailed so that moments are not induced in the connections. This is accomplished by laying out the connection so that the centerlines of each piece intersect at a common point. Where it is impractical to avoid moment connections, the design must consider the effect of the moment in addition to the loads applied. See Figure 4.27 for a comparison of bolting patterns—one with significant eccentricity and one with a minimum. Appendix D contains examples of moment connections and the calculation of their capacity.

EXAMPLE 4.15—THROUGH BOLTS: TENSION CONNECTION

Design the tension connection for the bottom chord of a truss as shown in Figure 4.28. The connection must carry 30,000 lb in tension under normal duration. The members will not be subjected to any other

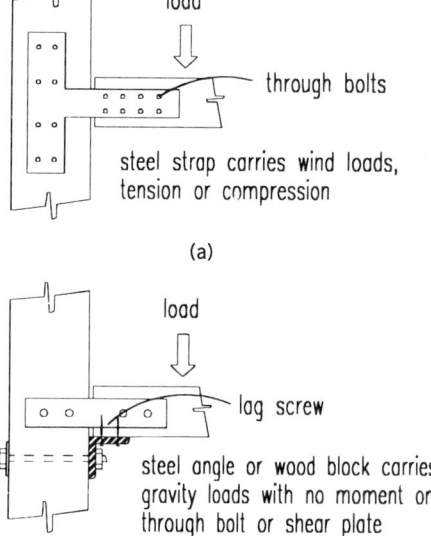

FIGURE 4.27 Comparison of connections—moments.

loads. Assume that side pieces are wood and that all pieces will be Douglas Fir–Larch No. 1.

For a preliminary design of the main member, neglect the bolt holes and determine the gross cross-sectional area required. Assume that the main member will be 5 in. thick or greater and use design values for Douglas Fir–Larch No. 1, beams and stringers.

$$F_t = 675 \text{ psi}$$

$$A_{\text{req'd}} = \frac{30{,}000 \text{ lb}}{675 \text{ psi}} = 44.44 \text{ in}^2$$

Try a 6×10 with an area of 52.25 in^2. The two side pieces combined must have an equal area, so try two 4×10's.

For a trial bolt size, select a 1-in.-diameter bolt, the largest that is shown in NDS Table 8.3A. The ratio L/D is 5.5/1 and is

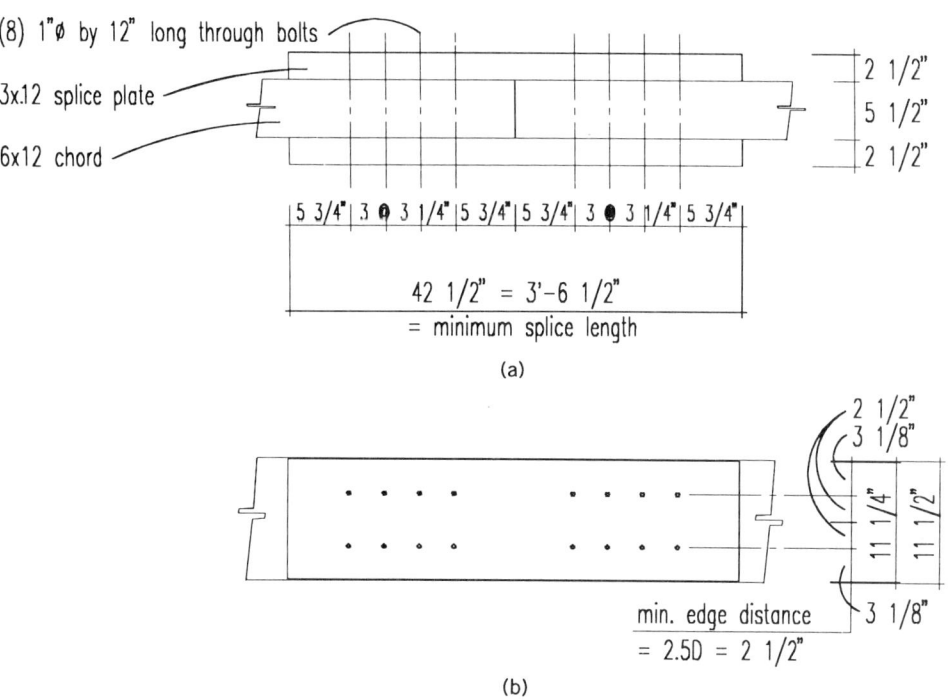

FIGURE 4.28 Through bolts—tension connection.

approximately equal to 6. Referring to NDS Table 8.3A, a 1-in.-diameter bolt, in double shear in a $5\frac{1}{2}$-in.-thick main piece of Douglas Fir–Larch and side pieces of $3\frac{1}{2}$-in. thickness will carry $Z_\parallel = 5330$ lb. The trial number of bolts is

$$N = \frac{30{,}000 \text{ lb}}{5330 \text{ lb/bolt}} = 5.63 \text{ bolts}$$

Consider six bolts in two rows of three bolts each. The required spacing between rows of bolts is $5D = 5$ in. (use $L/D = 6$). The required edge distances are $\frac{1}{2} \times 5D = 2\frac{1}{2}$ in. The trial assembly will fit within a depth of 2×5 in. or 10 in. The initial choice is not satisfactory. Note that neither the edge distance nor the spacing between rows of bolts is affected by the percentage of stress in the bolt. Both of these are minimum standards. We will need to increase the width of the main piece to 12 in. nominal ($11\frac{1}{2}$ in. actual). See Figure 4.28.

The trial member size was also chosen without regard to the bolt holes. Calculate the net tension area by subtracting the two bolt holes that align in the cross section, and determine the capacity of the main member. Consider the 6×12.

$$A_{\text{tot}} = 63.25 \text{ in}^2 - 2 \times (1 \text{ in.} \times 5.5 \text{ in.})$$
$$= 52.25 \text{ in}^2$$
$$T_{\text{allow}} = 675 \text{ psi} \times 52.25 \text{ in}^2$$
$$= 35{,}269 \text{ lb} > 30{,}000 \text{ lb}$$

The 6×12 main member is satisfactory. Select side members with an area of approximately one-half the main member and check their capacity in tension considering the net section. Try two 3×12's for side members.

$$A_{\text{tot}} = 2 \times 28.125 \text{ in}^2 - 2 \times (1 \text{ in.} \times 5.5 \text{ in.})$$
$$= 45.25 \text{ in}^2$$
$$T_{\text{allow}} = 675 \text{ psi} \times 45.25 \text{ in}^2$$
$$= 30{,}540 \text{ lb} > 30{,}000 \text{ lb}$$

Two 3×12 side plates are satisfactory. Note that from NDS Table 4A a value of F_t of 675 psi is allowed for Douglas Fir–Larch No. 1, 3×12's.

The thickness of the side members is less than the $3\frac{1}{2}$ in. used for the side members in deriving the bolt capacity. We must recalculate the capacity by using linear interpolation between the two side thicknesses given in the NDS Table 8.3A.

For $t_s = 1\frac{1}{2}$ in., $\quad Z_\parallel = 4090$ lb

For $t_s = 3\frac{1}{2}$ in., $\quad Z_\parallel = 5330$ lb

For $t_s = 2\frac{1}{2}$ in., $\quad Z_\parallel = 5330 - \dfrac{3.5 - 2.5}{3.5 - 1.5}$

$$\times (5330 - 4090) = 4710 \text{ lb}$$

$$N = \frac{30{,}000 \text{ lb}}{4710 \text{ lb/bolt}} = 6.37 \text{ bolts}$$

The use of linear interpolation between tabular values is permitted in these tables. The exact design value for the $2\frac{1}{2}$-in.-thick side piece from the yield mode formulas is $Z_\parallel = 4560$. The interpolated value, although higher and therefore unconservative is only 3% higher. This is an acceptable difference. Where member thicknesses or bolt diameters exceed the bounds of the tables, extrapolation is not allowed. The yield mode equations will have to be used.

We could either increase the thickness of the side members to 4×12's or try using eight 1-in.-diameter bolts. Let us try eight bolts in two rows of four bolts each. Where there are more than two bolts, reductions for the number of bolts must be considered. These reduction factors are found in NDS Table 7.3.6A for wood side plates.

A_m is the gross cross section of the main member:

$$A_m = 5.5 \text{ in.} \times 11.5 \text{ in.} = 63.25 \text{ in}^2$$

A_s is the sum of the cross-section areas of the side plates,

$$A_s = 2 \times 2.5 \text{ in.} \times 11.25 \text{ in.} = 56.25 \text{ in}^2$$

$$\frac{A_m}{A_s} = \frac{63.25}{56.25} = 1.12,$$

so use the inverse $\frac{1}{1.12} = 0.89$

We will also switch A_m for A_s. Use $A_s = 63.25 \text{ in}^2$

From NDS Table 7.3.6A read the factors for four bolts in a row and A_s equal to 64 in², then interpolate.

For $\frac{A_s}{A_m} = 0.5$, $C_g = 0.98$

For $\frac{A_s}{A_m} = 1$, $C_g = 0.99$

These values are so close together that interpolation is meaningless; use

$$C_g = 0.98$$

The allowable load on each bolt will be

$$P = 0.98 \times 4710 = 4616 \text{ lb}$$

The number of bolts required will equal

$$N = \frac{30{,}000 \text{ lb}}{4616 \text{ lb/bolt}} = 6.5 \text{ bolts}$$

So our choice to use eight bolts is accurate. Each bolt is stressed to 6.5/8 = 81%.

The final step is to check spacing requirements for the 1-in.-diameter bolt.

End distance, parallel to grain in tension = $7D = 7$ in.

Spacing between bolts in a row, parallel to grain = $4D = 4$ in.

End distance for the side plates is also 7 in., so the total length of the side plates, for full design load, is 50 in. Because the bolt load is 81% of the full design value, each of the end distances and spacing for bolts in a row could be reduced. The allowable reduced end distance 0.81×7 in. $= 5\frac{3}{4}$ in. The reduced spacing between bolts is 0.81×4 in. $= 3\frac{1}{4}$ in. The total reduced length of the splice is $42\frac{1}{2}$ in. For distances perpendicular to the grain, spacing between rows of bolts with $L/D = 6$; use $5D = 5$ in.

Edge distances are the same top and bottom; use the greater of $1\frac{1}{2}$ times the bolt diameter or one-half the distance between rows of bolts. Half the spacing between rows of bolts is $\frac{1}{2} \times 5D = \frac{1}{2} \times 5$ in. $= 2\frac{1}{2}$ in. The minimum width of either main piece or side piece then is 2×5 in. = 10 in. This connection is satisfactory.

Use eight 1-in.-diameter bolts.

As an exercise to the reader, try the option of using metal side pieces. The capacity of each bolt will increase due to using the metal plates, which load the wood parallel to grain. Design the side plates assuming A36 steel with an F_t of $0.4 \times 36{,}000$ psi, or 14,400 psi. If you try six bolts, you should find that the reduction factor for steel side plates is about 0.98. The overall strength of the connection will increase enough that only six bolts will be needed.

EXAMPLE 4.16—THROUGH BOLTS AS HEEL CONNECTION

Determine the capacity of the connection shown in Figure 4.29. The figure shows a possible rafter–ceiling tie assembly designed for the conditions used in Example 4.5, except that the rafters and ceiling ties are now spaced at 8'-0" o.c. rather that 2'-0" o.c. Design of the truss results in a top chord made of three pieces, each 2×10's, and a bottom chord of two 2×8's.

CONNECTION DESIGN 149

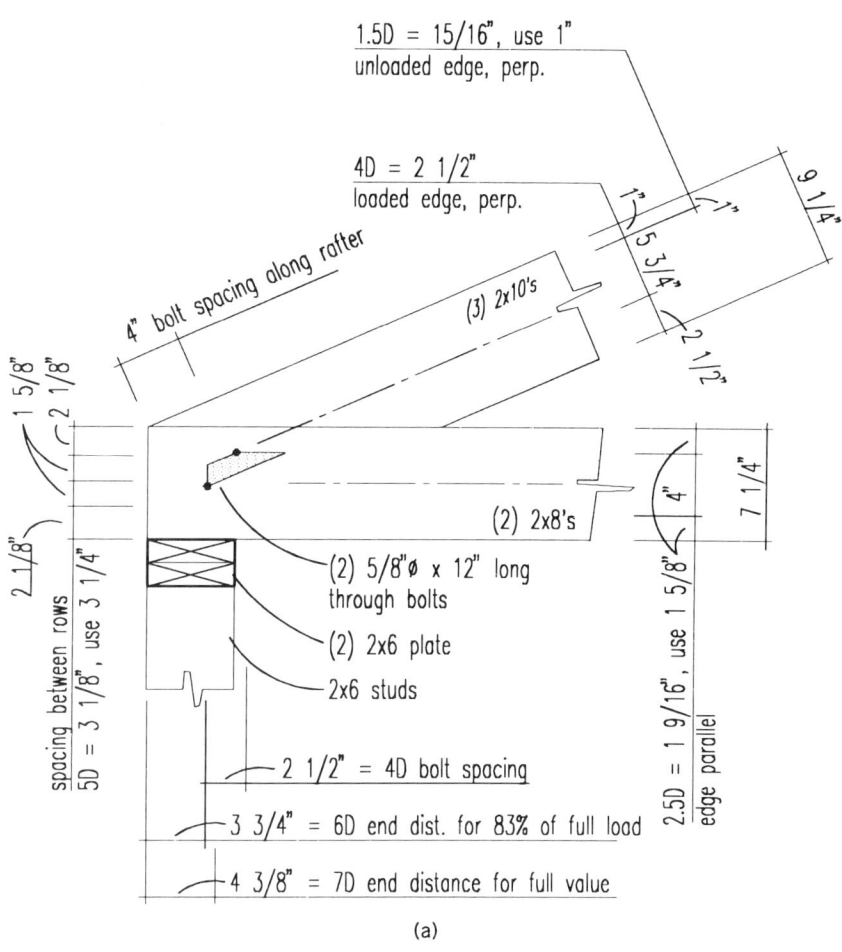

(a)

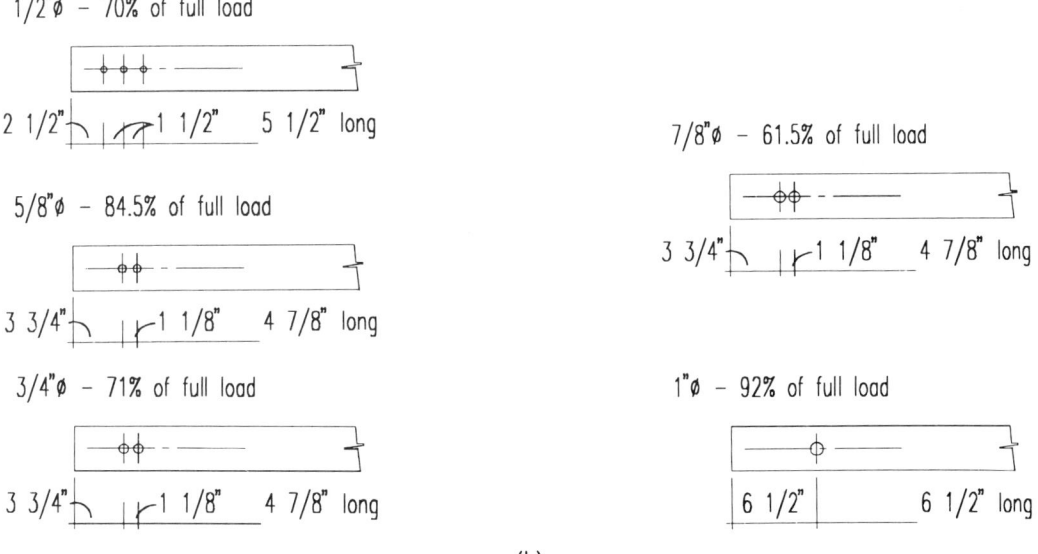

(b)

FIGURE 4.29 Heel connection—through bolts in a sinlge shear.

The members alternate. The tension in the bottom chord is 4 × 988 lb = 3952 lb.

Assume that all the material is Southern Pine and that a 1-in.-diameter bolt is to be used. The roof pitch is 6 in 12, or 26.56°. This connection is made up of five members of equal thicknesses. For a joint with four or more members of equal thickness, the design value is determined by taking the number of shear planes times the single shear value for each shear plane. For a 1-in.-diameter bolt through a $1\frac{1}{2}$-in.-thick main member and a $1\frac{1}{2}$-in.-thick side member using Southern Pine, the values in NDS Table 8.2A are $Z_\parallel = 1060$ lb, $Z_{s\perp} = 580$ lb, and $Z_{m\perp} = 580$ lb.

The bottom chord must transfer the tension force of 3952 lb into the bolts. The bottom chord is in tension, so the bolt capacity is $Z_\parallel$, parallel to grain. The same force, 3952 lb, is then transferred into the rafter, which is loaded at an angle to grain. The allowable bolt capacity is given by Z_α as derived by Hankinson's formula. Because Z_α is less than $Z_\parallel$, the design of the bolt in the rafter will control.

Looking at the rafter, we apply Hankinson's formula or the nomograph in Figure 4.3.

$$Z_\alpha = 910 \text{ lb}$$

The capacity of the complete joint is the number of shear planes times the single-shear capacity.

$$P = 4 \times 910 \text{ lb} = 3640 \text{ lb}$$

Adjusting for duration of a snow load, we obtain

$$P = 1.15 \times 3640 \text{ lb} = 4186 \text{ lb}$$

This capacity is greater than the applied load of 3952 lb. One 1-in.-diameter bolt will suffice.

We might choose to check the capacity of a smaller bolt. Table 4.22 gives the strengths and number of bolts required for all bolt diameters in the NDS tables.

The one 1-in.-diameter bolt may be preferable since it is the easiest connection to construct and because it will not induce any moment into the connection. If a smaller bolt is chosen, the best selection would be using two $\frac{5}{8}$-in.-diameter bolts. Figure 4.29 demonstrates the distance requirements for this connection.

Use one 1-in.-diameter through bolt

or

Use two $\frac{5}{8}$-in.-diameter through bolts.

In this figure only the end distance and the spacing between bolts can be reduced to account for the reduced level of stress in the bolts. You can see by comparing the five conditions in Figure 4.29b that the actual end distances remain about the same and that the net area of the connection is about 5 in. long for all the smaller bolts. This is because the bolt strength is

TABLE 4.22 COMPARISON OF BOLT STRENGTHS: EXAMPLE 4.16

Bolt Diameter (in.)	$Z_\parallel$	$Z_{s\perp}$	$Z_{25.6°}$	4 × $Z_{25.6°}$	Number of Bolts, N		Percent Stress, and C_Δ
					Required	Used	
$\frac{1}{2}$	530	330	472.7	1891	2.09	3	70
$\frac{5}{8}$	660	400	584	2336	1.69	2	84.5
$\frac{3}{4}$	800	460	697	2788	1.42	2	71
$\frac{7}{8}$	930	520	803.3	3213.2	1.23	2	61.5
1	1060	580	909.5	3638	0.92	1	92

proportional to the bolt diameter; thus an increase in diameter will increase the distance requirements at the same rate that the geometry factor, C_Δ, is reduced. Since edge distance to the loaded edge of the top chord is not reduced by C_Δ, it becomes the critical dimension in comparing the designs. Bolts of each of the diameters will fit; however, for the smallest diameter the bolts can be placed nearer the centerline of the bottom chord. This is beneficial for two reasons: it encroaches least on the edge distance on the top of the bottom chord, and most important, it will reduce the amount of eccentricity and moment in the joint. The ideal design is one in which the centroid of the bolt group coincides with the intersection of the centerlines of the wood members.

If the 1-in.-diameter bolt is not practical, another possible solution is to increase the thickness of some of the members in the truss. What is the effect of using three 4×8's in the top chord? The reader may try this exercise to find the smallest single bolt that can be used.

Timber Connectors: Shear Plates and Split Rings

Of the connections that we have described so far, the strongest has been the through bolt. In cases where the L/D ratio stays fairly small (i.e., does not exceed 4 or so), the capacity of the connection is controlled by crushing of the wood under the bolt. The best way to increase the capacity of the connection is to increase the bearing area between the wood and the bolt. The purpose of both split ring connectors and shear plates is essentially to increase the diameter of the connector. Figures 4.25 and 4.26 show examples of these connectors.

Shear Plates. A shear plate is either a $2\frac{5}{8}$-in.-diameter steel plate with a depressed lip or a cast malleable iron plate 4 in. in diameter. Both of them are set into the face of the wood piece, which has been drilled with a special drill that cuts the groove for the lip, daps down the face to receive the thickness of the plate, and drills a center hole. The $2\frac{5}{8}$-in.-diameter plate is used with a $\frac{3}{4}$-in.-diameter bolt and the 4-in. plate can be used with a $\frac{3}{4}$-in. bolt or a $\frac{7}{8}$-in. bolt.

The shear plate connection consists of one or two plates and a bolt. This grouping is referred to as the *connection unit*. Shear plates can be used to connect wood to wood, in which case two plates are used, one in each piece of wood. They can also be used to connect to a steel side plate, in which case only the wood face receives a shear plate. In either case the load is transferred from the wood into the plate, from there to the bolt, and from the bolt into the other half of the connection. Regardless of whether the connection joins wood to wood or wood to metal, when two pieces are connected, the bolt will be loaded in single shear. The design values given in NDS Table 10.2A for split rings and Table 10.2B for shear plates list strengths for single shear only.

Split Rings. A split ring is a hot-rolled carbon steel ring that is thicker in the center than at the edges. It is slotted and joined with a tongue and groove. These modifications are designed to assure that the ring will fit tightly into the wood slot when properly installed. The split ring is installed at the interface of two pieces of wood so that it extends half of its depth into each piece. A bolt is installed through the two pieces, with the axis of the bolt aligned with the axis of the ring. The bolt is drawn up tight to maintain contact between the wood pieces, but unlike with the shear plate connection, the bolt carries no lateral load. All load is transferred from one wood piece directly into the ring, through the bearing, and then directly into the other piece of wood. The split ring must be installed into two pieces of wood, so it cannot be used in a wood-to-metal connection.

Both shear plates and split rings may be installed with lag screws rather than through bolts. Lag screws can be used in conjunction with wood or metal side plates. The diameter of the lag screw will be the same as the diameter of the bolt that it is replacing. Shear plates are currently used more extensively than split rings, in part because they provide slightly higher strengths, in part because they are somewhat easier to install. Their other advantage is that they may be used with steel side plates.

Design Assumptions and Modifications. The design values for split rings are shown in NDS Table 10.2A and for shear plates in Table 10.2B. There are a number of precautions that must be followed in using these tables, and modifications must be applied to the design values.

1. The net thickness of the wood piece cannot be less than the minimum thickness listed. Net thickness refers to the actual thickness before the dap is cut for the plate.

2. If the wood thickness is between the minimum and the thicker width given on the table, the design value can be found by interpolating between given values except that for shear plates, the value may not exceed the maximum allowable value given in footnote 2 to Table 10.2B.

3. If you look at Table 10.2B you will see some values marked with an asterisk; the asterisk indicates that the value tabulated is actually greater than the allowable value, which is controlled by the bolt strength. The higher values are placed in the table only to allow interpolation for intermediate thicknesses of wood. When using these bolt configurations you must use the lesser of the two values—the interpolated value or the bolt value in footnote 2. Furthermore, since steel design does not use duration factors, the bolt strength values in footnote 2 may not be increased by the standard duration factors, such as $C_D = 1.6$ for wind or seismic. Instead, you are allowed to use an increase of 1.33 for load combinations, which include wind or seismic loads in combination with other dead or live loads. Example 4.18 demonstrates this procedure. The procedure and the footnote apply only to shear plates.

4. When either split rings or shear plates are connected with lag screws, minimum penetration of the lag screw into the holding piece is required. Minimum penetration depths for such lag screws are provided in NDS Table 10.2.3. The table lists minimum penetration lengths for full design values. If the lag screw penetration meets or exceeds this penetration, the full design value can be used. The table also lists minimum penetration for reduced design values. The actual penetration may not be less than this minimum for reduced design values. If the actual penetration equals the minimum for reduced values, the tabulated design strength (found in NDS Table 10.2A or 10.2B) must be multiplied by a penetration factor, $C_d = 0.75$, which reduces the allowable strength. If the actual penetration is between the two minimums (greater than the minimum for reduced design values but less than the minimum for full design values), the penetration factor is found through linear interpolation. The penetration factor C_d is used in combination with all the other pertinent modification factors.

5. When metal side plates are used with the 4-in.-diameter shear plate, the design values for parallel-to-grain loading may be increased by the factor C_{st} shown in NDS Table 10.2.4. No increases for any species are allowed for $2\frac{5}{8}$-in. plates, and no in-

creases are allowed for perpendicular-to-grain loading.

6. Required end distances and spacing between connectors are given in NDS Table 10.3 along with the geometry reduction factors, C_Δ. These apply to shear plates as well as split rings. NDS Table 10.3 gives the required distances for the full load and the minimum distances for reduced loads. The actual spacing and edge distances may not be less than the reduced-load values. If they equal the reduced-load distances, the design value will be less than the full-load value (between 50 and 83%). If the distance is between the two listed, the geometry reduction factor, C_Δ, may be found through linear interpolation. This concept is the same as that used with through bolts. All the end and edge distances and spacings must be checked for any connector and a geometry factor, C_Δ, calculated for each distance. The least geometry factor is then used to find the allowable load. Example 4.19 shows the normal approach taken in using the geometry factor, C_Δ.

Figure 4.30 demonstrates the correct measurements for end and edge distances and connector spacings used in NDS Table 10.3. In NDS Table 10.3 we can see that for loading parallel to grain the minimum edge distance for the full design value equals that for reduced design value. Therefore, the minimum edge distance must be maintained regardless of the amount of load in the connection. Table

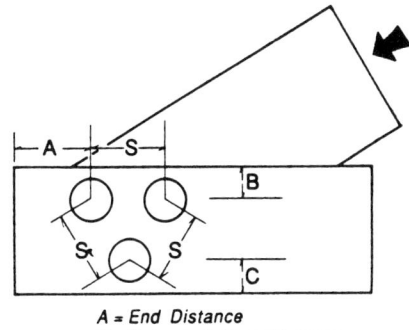

A = End Distance
B = Unloaded Edge Distance
C = Loaded Edge Distance
S = Spacing

(a)

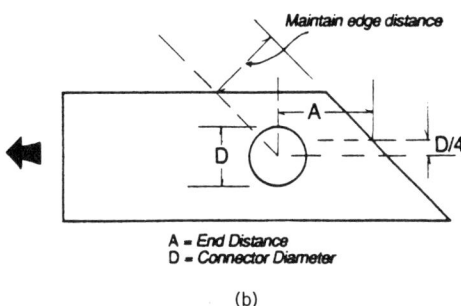

A = End Distance
D = Connector Diameter

(b)

FIGURE 4.30 Timber Connectors — Connection Geometry.

4.23 lists those edge distances and spacings for which the reduced distance equals the full distance.

[The discussion above refers to loads that are either parallel to the grain (0°) or perpendicular to the grain (90°). For loads

TABLE 4.23 TIMBER CONNECTORS: CRITICAL SPACING DIRECTIONS

	Parallel-to-Grain Loading	Perpendicular-to-Grain Loading
Edge distance		
Unloaded edge	No reduction	No reduction
Loaded edge	No reduction	
Spacing		
Spacing parallel to grain		No reduction
Spacing perpendicular to grain	No reduction	

Source: National Design Specification, 1991 Edition.

at angles other than these, or for connectors installed in the end grain, specific requirements are listed in the NDS, Chapter 10.]

7. When two pieces are connected, the bolt will be loaded in single shear and each wood piece will have one face with a connector on it. The design value for that connector is given in NDS Table 10.2A or 10.2B in the row designated "1 face." If the connection consists of three members joined (a main member and two side members), the bolt will be in double shear and there will be two shear planes. The capacity for the connector will be the sum of the capacity of each shear plane. The procedure for determining the capacity of such a connection is as follows:

(a) For the side members find the capacity from the row marked "1 face"; use the thickness of the side piece.

(b) The main member will have two faces with connectors. Find its ca-

TABLE 4.24 SHEAR PLATES AND SPLIT RINGS—MODIFICATION FACTORS

Condition			Factor[a]			
Direction of load Parallel to grain			P Table 10.2A or 10.2B			
Perpendicular to grain			Q Table 10.2A or 10.2B			
Angle to grain			Hankinson's formula $N_\alpha = \dfrac{P \times Q}{P \times \sin^2 \alpha + Q \times \sin^2 \alpha}$			
Thickness of wood member Less than minimum			No allowable design value			
Between min and larger listed			User linear interpolation			
greater than larger listed			1.0			
Penetration depth factor C_d			Length of penetration of lag screw into main member (number of shank diameters)			
Connector	Side member	Minimum penetration	Group A	Group B	Group C	Group D	C_d
$2\frac{1}{2}$ in. Split Ring or 4 in. split ring or 4 in. shear plate	Wood or metal	Full design value	7	8	10	11	1.0
		Reduced design value	3	$3\frac{1}{2}$	4	$4\frac{1}{2}$	0.75
$2\frac{5}{8}$ in. Shear plate	Wood	Full design Value	4	5	7	8	1.0
		Reduced design Value	3	$3\frac{1}{2}$	4	$4\frac{1}{2}$	0.75
	Metal	Full design	3	$3\frac{1}{2}$	4	$4\frac{1}{2}$	1.0

TABLE 4.24 *Continued*

Condition	Factor		
Metal side plates, C_{st}	Species group	P	Q
(for 4 in. shear plate only)	A	1.18	1.00
	B	1.11	1.00
	C	1.05	1.00
	D	1.00	1.00
Group action factor, C_g, and geometry factor, C_Δ			
Multiple connectors			
Split ring or shear plates with wood side members	C_g See NDS Table 7.3.6B		
Shear plate connectors with steel side plates	C_g See NDS Table 7.3.6D		
End and edge distance and spacing	Reduced design value		Full design value
	C_Δ = minimum of C_Δ factor in any direction. Applies to all connectors in a group		$C_\Delta = 1.0$

Moisture content[b]

At fabrication	In service		
Dry	Dry	$C_M = 1.0$	
Partially seasoned	Dry	c	
Wet	Dry	0.80	
Dry or wet	Partially seasoned or wet	0.67	

Notes:
[a]Design capacities for shear plates may not exceed bolt strengths.
    (1) $2\frac{5}{8}$ in. shear plate   2900 lb
    (2) 4 in. shear plate with $\frac{3}{4}$ in. bolt   4400 lb
    (3) 4 in. shear plate with $\frac{7}{8}$ in. bolt   6000 lb
[b]Moisture content limitations apply to a depth of $\frac{3}{4}$ in. below the surface of the wood.
[c]When connectors are installed in wood that is partially seasoned at the time of fabrication but that will dry before full design load is applied, proportional intermediate wet service factors shall be permitted to be used.
Source: National Design Specifications, 1991 edition.

pacity in the row marked "2 faces" using the thickness of the main piece.

(c) Find the capacity of the shear plane as the smaller of the two values (a) or (b).

(d) Follow the procedure above [(a) through (c)] on the other shear plane.

(e) Add the capacity of each shear plane to find the capacity of the total connection.

(f) For multiple members joined on one bolt, the design value can be calculated as the sum of the design values for each shear plane.

8. When more than two connectors are used, the group action factors, C_g, given in NDS Table 7.3.6B or 7.3.6D must be applied. Table 7.3.6B lists the group action factor for 4-in. split rings and 4-in. shear plates with wood side members. This table can also be used conservatively for $2\frac{1}{2}$-in. split rings and $2\frac{5}{8}$-in. shear plates with

wood side plates. Table 7.3.6D lists the group action factor for 4-in. shear plates with steel side plates. This table can also be used conservatively for $2\frac{5}{8}$-in. shear plates with steel side plates. Both tables are derived from equations listed in Chapter 7. These equations can be used to obtain more exact factors for the smaller connectors, if so desired. The equations should also be used for conditions under which the tables are not conservative. These conditions are for spacings between fasteners in a row less than 9 in. or for wood with a modulus of elasticity greater than 1,400,000 psi. These parameters were set so that the designer could use the tables conservatively for most common applications.

As with through bolts, the design values must be modified for specific conditions of moisture content as shown here in Table 4.24. This table also summarizes some of the modifications described above. All modifications are cumulative, except that the maximum design values for shear plates (NDS Table 10.2B, footnote 2) may not be exceeded.

Timber Connectors Used to Carry Moments. The provisions and procedures discussed in Appendix D regarding bolts apply to groups of timber connectors carrying moments. Careful consideration must be paid to maintaining adequate spacing, edge distances, and end distances for this type of connection.

EXAMPLE 4.17—SHEAR PLATES IN A TENSION CONNECTION

Redesign the connection in Example 4.15 using shear plates. The tension connection, shown in Figure 4.31, is made of a 6 × 12 Douglas Fir–Larch (Dense) main member and two 3 × 12 side pieces, also DF-L (D). They carry a tension load of 30,000 lb with a normal duration.

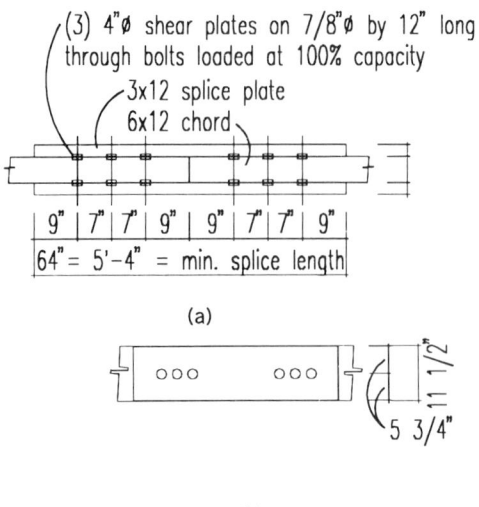

FIGURE 4.31 Shear plates—tension connection.

Start by determining the species group for the wood. Douglas Fir–Larch (Dense) is found in NDS Table 10A to be in group A. Note that Douglas Fir–Larch that is not dense is found under group B. We will use group A for this exercise.

Try a 4-in.-diameter shear plate on either a $\frac{3}{4}$- or $\frac{7}{8}$-in.-diameter bolt. The values shown in NDS Table 10.2B refer to single shear. Since the connection is in double shear, the capacity of the connection will be determined as the sum of the capacities at each shear plane. For each shear plane the capacity based on each piece must be considered and the lesser value used.

Consider first the side pieces. Each side piece has a shear plate on only one face. Find the design value for a piece with the number of faces on the connector equal to 1 and a thickness of $2\frac{1}{2}$ in. For load parallel to grain, $P = 5090^*$ lb. The asterisk indicates that this value might exceed the maximum value for the bolt. For a $\frac{3}{4}$-in.-diameter bolt the maximum bolt value capacity equals 4400 lb; For a $\frac{7}{8}$-in.-diameter bolt the bolt value capacity equals 6000 lb. Select a $\frac{7}{8}$-in.-diameter bolt so that the maximum capacity of the connector can be developed. For the side piece, use $P = 5090$ lb.

Consider next the main piece, which has a shear plate on each face. The design value for a piece with the number of faces on the connector equal to 2 and a thickness of $3\frac{1}{2}$ in. or more shows $P = 5030$ lb. As before, selecting a $\frac{7}{8}$-in.-diameter bolt will maximize the capacity of the connector. The capacity of each shear plane is the lesser of the two values, so $P = 5030$ lb based on the main piece.

For the total connection, with two identical shear planes,

$$P = 2 \times 5030 \text{ lb} = 10{,}060 \text{ lb/shear plate}$$

$$N = \frac{30{,}000 \text{ lb}}{10{,}060 \text{ lb/shear plate}}$$

$$= 2.98 \text{ shear plates}$$

Try 3 shear plates; assume that they will all be in one row.

The modification for number of connectors applies to shear plates, split rings, and lag screws as well as to bolts. Refer to Example 4.15 for the terms A_m and A_s.

$$A_m = 5.5 \text{ in.} \times 11.5 \text{ in.} = 63.25 \text{ in}^2$$

$$A_s = 2 \times 2.5 \text{ in.} \times 11.25 \text{ in.} = 56.25 \text{ in}^2$$

$$\frac{A_m}{A_s} = 1.124$$

Since this is greater than 1, exchange the roles of A_m and A_s. Thus

$$A_m = 56.25 \text{ in}^2 \qquad A_s = 63.25 \text{ in}^2$$

$$\frac{A_m}{A_s} = 0.88$$

In NDS Table 7.3.6B for three fasteners in a row, the group action factors range from $C_g = 0.98$ to 0.95. Take the worst case just to see if the connection is feasible.

$$P' = P \times C_g = 10{,}060 \text{ lb} \times 0.95 = 9557 \text{ lb}$$

$$N = \frac{30{,}000 \text{ lb}}{9557 \text{ lb/shear plate}}$$

$$= 3.14 \text{ shear plates}$$

The stress in each plate is $3.14/3 = 104.6\%$.

This amount of overstress is right at the limit of what would normally be allowed. Because this was the worst case, a slight improvement can be made by calculating the actual C_g using interpolation between 0.95 and 0.98. Doing this, we find that $C_g = 0.973$, $P' = 9792$ lb, $N = 3.06$, and that the stress level is 102%. We can accept this level of stress, so the connection is satisfactory.

Use three 4-in.-diameter shear plates on $\frac{7}{8}$-in.-diameter bolts.

Since each connector is loaded to full capacity, the maximum spacing must be used. For a 4-in. shear plate, loading parallel to grain, from NDS Table 10.3,

End distance, parallel to grain in tension = 7 in.

Spacing between bolts in a row, parallel to grain = 9 in.

The total length of the side plates is twice 9 in. + 7 in. + 7 in. + 9 in., or 64 in. Comparison with the bolted design shows that the shear plate connection will use fewer connectors (three as opposed to six) but will require more room (64 in. as opposed to $42\frac{1}{2}$ in.). This is typical of shear plates and split-ring connectors and is one of the most important factors in their use; although timber connectors provide the highest possible strengths, they require the largest area of wood, in terms of end and edge distances, to do so.

Edge distances are the same top and bottom. For the 4-in.-diameter shear plate, use the tabulated value of $2\frac{3}{4}$ in. The minimum width of either main piece or side piece then is $5\frac{1}{2}$ in. This connection is satisfactory.

EXAMPLE 4.18—SHEAR PLATES IN TENSION: WIND LOAD.

Redesign the connection in Example 4.17 assuming that the 30,000-lb tension load was caused by a combination of wind and dead loads.

In combinations where wind is the load with the shortest duration, the connection design values may be modified by a duration factor, $C_D = 1.6$. Thus, in Example 4.17 the P value for the side piece could be increased to become

$$P' = P \times C_D = 5030 \text{ lb} \times 1.6 = 8048 \text{ lb}$$

which exceeds the bolt capacity of 6000 lb for the $\frac{7}{8}$-in.-diameter bolt. Because the total load is a combination of wind and dead loads, we are allowed a 33% increase in the strength of the bolt itself. This is specified in the AISC *Specifications for Steel Buildings*, Section A5.2 and referred to in the NDS in Section 7.2.3.

$$P' = 6000 \text{ lb} \times 1.33 = 7980 \text{ lb}$$

the bolt strength controls

$$N = \frac{30{,}000 \text{ lb}}{2 \times 7980 \text{ lb/shear plate}}$$

$$= 1.88 \text{ shear plates}$$

Using two connectors the stress in each plate is 1.88/2 = 94%. For two connectors in a row, there is no group reduction, $C_g = 1.0$.

Use two 4-in.-diameter shear plates on $\frac{7}{8}$-inch-diameter bolts.

If the $\frac{3}{4}$-in.-diameter bolt had been selected, the capacity of each shear plane would have been $4400 \times 1.33 = 5852$ lb, and the number of connectors needed would have been three. It is best to stay with the larger bolt diameter.

As an exercise, try using metal side plates. The tabulated design value should be increased by 18% for Douglas Fir–Larch (Dense), a group A wood, according to NDS Table 10.2.4.

EXAMPLE 4.19—SPLIT-RING CONNECTOR AS THE HEEL CONNECTOR.

As our final example using shear plates and split rings, consider the design of the heel connection for the roof structure in Example 4.5. In that example we considered a number of possible solutions that depended on the spacing of the roof rafters. For this example we will take up the last of the options presented—roof rafters at 16 ft on centers.

From Table 4.15 we see that the rafter is a 6×12 Southern Pine No. 1. To use shear plates effectively, let us assume that the ceiling tie will be split into two pieces. Use two 3×10 No. 2. We will design the connections using both split rings and shear plates. Southern Pine is in group B in NDS Table 10A. You could also refer to Table D.1 in the Appendix, which contains the same information.

Try a $2\frac{1}{2}$-in.-diameter split ring on a $\frac{1}{2}$-in.-diameter bolt. The allowable design values for the $2\frac{1}{2}$-in.-thick side pieces are found in NDS Table 10.2A in the row for "one face of member with connector" and a side piece with net thickness greater than 2 in.

$$P = 2730 \text{ lb} \qquad Q = 1940 \text{ lb}$$

These values must be modified by all appropriate factors. In our case, the duration factor due to snow load is 1.15. Our experience with the bolted connectors in Example 4.16 leads us to the conclusion that the design will be controlled by the capacity at the rafter which is loaded at an angle to grain. For an angle of 26.6° the allowable connection capacity is found from Hankinson's formula (or Figure 4.20) to be

$$P_\alpha = 2525 \text{ lb}$$

This is the capacity at each shear plane. The capacity of the total connector, comprising two shear planes is twice that.

$$P_\alpha = 2525 \text{ lb} \times 2 = 5050 \text{ lb}$$

Accounting for the snow-load duration, the design value becomes

$$P'_\alpha = 5050 \text{ lb} \times 1.15 = 5808 \text{ lb}$$

For rafter–ceiling ties spaced 16 ft apart the thrust at this joint equals 988 lb, multiplied by 8 equals 7904 lb. The required number of split-ring connectors then is

$$N = \frac{T}{P'_\alpha} = \frac{7904}{5808} = 1.36$$

Use two $2\frac{1}{2}$-in.-diameter split ring on a $\frac{1}{2}$-in.-diameter bolt.

Each connector will be loaded to 68% of its full allowable capacity. The calculation for the transfer of load from the split ring to the ceiling tie is similar, except that the load is parallel to grain. Therefore, the connection capacity is the tabulated value for *P* multiplied by the duration factor and multiplied by 2 for the two connectors. In this calculation the value for *P* is found opposite the entry for "2 faces of member the connectors on the same bolt." This calculation shows that 1.26 connectors are required, so if 2 are used, they will be stressed to 63% of their allowable value. The requirement at the top chord controls the design.

Table 4.25 summarizes the calculation for the four types and sizes of timber connectors. As can be seen from the table, the connection to the rafter controls the design in all cases. Figure 4.32 shows the layout using four $2\frac{5}{8}$-in.-diameter shear plates. In each option the connectors are loaded to about the same level of stress. Each group of connectors will fit in approximately the same area, so there does not seem to be a preference for any size over the other. The split rings, which do not rely on the bolt to transfer loads, are not limited by the bolt strength and may be somewhat more economical in cases with high levels of stress. Generally, as with most connectors, the best option is to use fewer, larger connectors.

Metal (Proprietary) Connections

A number of companies produce metal structural fasteners for wood connections. Use of these fasteners has increased in the last 10 to 20 years. In areas with large seismic loads, such as portions of California, use of these fasteners is so common that the company designation for the fasteners is included on the plans and is readily understood by designer, building inspector, and contractor alike. Similarly, in southern Florida, the South Florida Building Code recently published a manual referred to as "Deemed-to-Comply," which shows required metal fasteners for one- and two-family dwellings. Because of their common use, all designers working with wood structures should be familiar with the major fasteners used in their area.

TABLE 4.25 COMPARISON OF SPLIT-RING AND SHEAR PLATE CONNECTORS: EXAMPLE 4.19

Connector Type	Member	Number of Faces	Design Value, P	Design Value, Q	Design Value, P, at Angle 26.6°	Design Value × 1.15, Duration Factor	Connector Capacity	Number of Connectors Required	Number of Connectors Used	Percent Stress
2½ in. split ring	Rafter	2	2,730	1,940	2,524	2,903	5,806	1.36	2	68
	Ceiling tie	1	2,730			3,140	6,279	1.26	2	63
	Use								2	68
4 in. dia. split ring	Rafter	2	5,260	3,660	4,837	5,563	11,125	0.71	1	71
	Ceiling tie	1	5,260			6,049	12,098	0.65	1	65
	Use								1	71
2⅝ in. Shear plate	Rafter	2	2,860	1,990	2,630	3,025[a]	5,800	1.36	2	68
	Ceiling tie	1	2,670			3,071[a]	5,800	1.36	2	68
	Use								2	68
4 in. dia. shear plate	Rafter	2	4,320	3,000	3,970	4,566[b]	9,132	0.87	1	87
	(For use with ¾-in. bolt)						8,800	0.90	1	90
	Ceiling tie	1	4,360			5,014[b]	10,028	0.79	1	79
	(For use with ⅞ in bolt)						10,028	0.78	1	79
	Use	(Use with ⅞ in.-diameter bolt)							1	79

[a] Calculated value with snow-load duration exceeds the bolt allowable (2900 lb). Use the lower value for design.
[b] Calculated value with snow-load duration exceeds the allowable (4400 lb) for the ¾ in.-dia. bolt, but not for the ⅞ in.-dia. bolt (6000 lb). The example is continued using both bolt sizes. The larger bolt is selected.

Most of these companies have conducted testing on their fasteners and have had the testing reviewed and accepted by the major code authorities. A listing of approved fasteners may be obtained from the Council of American Building Officials (CABO).[2] Specific fasteners manufactured by the Simpson Strong-Tie Company[3] and TECO Structure Wood Fasteners[4] are used in the examples in the beginning of this chapter and in Chapter 7.

Most lightweight trusses, those made of $2x$ material, are predesigned and manufactured in factories. Typically, these trusses are monoplane trusses in which chords and webs are in the same plane. The connections for these trusses are most often metal plates that have preformed teeth or nails punched into them. The

---

[2] Council of American Building Officials (CABO), 900 Montclair Road, Birmingham, AL 35213.
[3] Simpson Strong-Tie Company, 4637 Chabot Drive, Suite 200, Pleasanton, CA 94588. Simpson connectors are covered by National Evaluation Reports NER-209 and NER-393.
[4] Silver TECO, P.O. Box 203, Colliers Ways, Colliers, WV 26035. TECO Connectors are covered by National Evaluation Report NER-133.

CONNECTION DESIGN   161

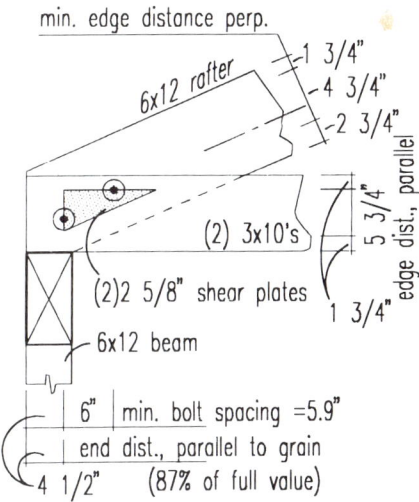

FIGURE 4.32 Heel connection—shear plates.

Truss Plate Institute (TPI) controls the specifications of truss plates and their installation. Plate are galvanized steel of 20 gage (0.036 in. thickness) or more. They can be installed either by a roller press or by a pneumatic press. The design strength of the plates is determined by each plate manufacturer and varies with the type of nail or tooth, their length, and frequency of spacing. Each plate in assigned a strength in terms of pounds per square inch of contact area. The contact area is the amount of wood in contact with the plate after the appropriate end and edge distances have been deducted. An example calculation using truss plates is given in Chapter 5. More information may be obtained from the Truss Plate Institute.[5]

Yield-theory Design

Connection design within the 1991 *National Design Specification* is based on the yield theory of the connection. The yield theory uses equations derived from first

[5] Truss Plate Institute, Inc., 100 West Church Street, Fredrick, MD 21701.

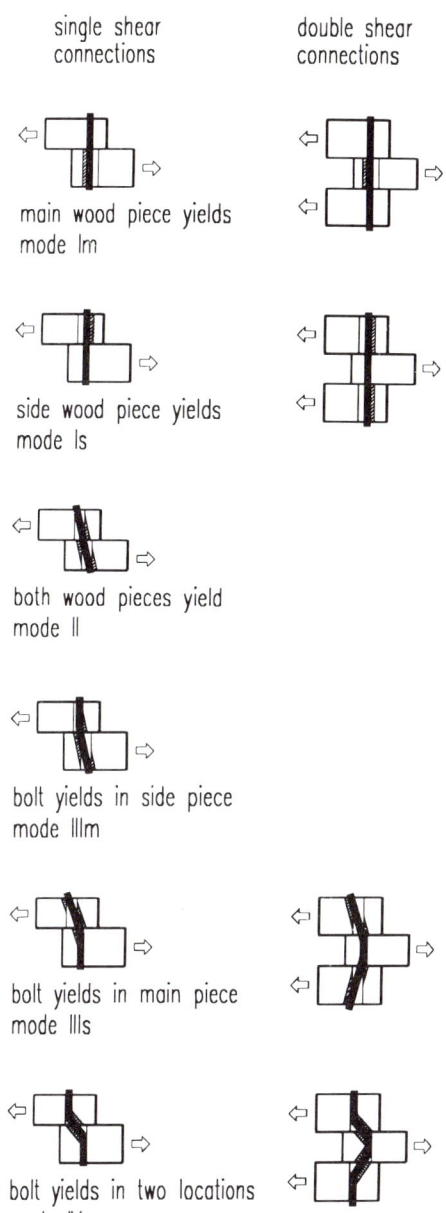

FIGURE 4.33 Yield theory failure modes.

principles to quantifiably explain the yield, or ultimate, strength of a connection. In any connection there are a number of potential modes of failure; for example, the connector may fail in plastic bending at one or more points, either wood piece may

TABLE 4.26 YIELD-THEORY CALCULATIONS

Mode of Failure Number	Description of Failure Geometry	Yield Load, F_u (lb)	Z Value (lb)
I_s	Side piece crushes in uniform bearing	$Z = \dfrac{Dt_s F_{es}}{K_D}$ $= \dfrac{0.162 \times 1.5 \times 4650}{2.2}$ D = nail diameter = 0.162 in. t_s = side member thickness = 1.5 in. F_{es} = side member dowel strength = 4650 psi K_D = factor = 2.2	514
III_m	Connector yields and main piece crushes	$Z = \dfrac{k_1 D p F_{em}}{K_D(1 + 2R_e)}$ $= \dfrac{1.1099 \times 0.162 \times 1.5 \times 4650}{2.2 \times 3}$ $k_1 = 1.1099$ (formula in NDS) p = penetration = 1.5 in. F_{em} = main mem. dowel strength = 4650 psi $R_e = F_{em}/F_{es} = 1.0$	190
III_s	Connector yields and side piece crushes	$Z = \dfrac{k_2 D t_s F_{em}}{K_D(2 + R_e)}$ $= \dfrac{1.1099 \times 0.162 \times 1.5 \times 4650}{2.2 \times 3}$ $k_2 = 1.1099$ (formula in NDS) t_s = side member thickness = 1.5 in. $F_{em} = 4650$ psi	190
IV	Connector yields in both pieces; each crushes	$Z = \dfrac{D^2}{K_D}\sqrt{\dfrac{2F_{em}F_{yb}}{3(1+R_e)}}$ $= \dfrac{0.162^2}{2.2}\sqrt{\dfrac{2 \times 4650 \times 90{,}000}{6}}$ F_{em} = dowel strength = 4650 psi F_{yb} = nail bending yield = 90,000 psi	141

Source: Equations from the *National Design Specification*, 1991 Edition.

fail in crushing under the connector, or the connector, such as a nail or screw, may fail by withdrawal from the holding piece. Yield theory itemizes each potential mode of failure and determines the ultimate load that will produce that failure. For each mode of failure there is an equation that gives the ultimate strength of the connection. Since it stands to reason that the connector will fail whenever any of the modes is reached, the least ultimate load will control a specific connector. The ultimate loads are each a function of the materials in the connector—the yield stress of the connector and the strength of the wood in bearing against the connector, which is referred to as the dowel bearing strength. The ultimate strength is also dependent on the geometry of the connection, such as the number of pieces of wood in a connection, the thickness of each wood piece, and the thickness of steel side pieces if used.

Prior to the 1991 revisions, the basis for wood connection design was a set of empirical relationships derived from test data. These equations are found in the *Wood Engineering Handbook*.[6] The tests used to generate these equations made a series of assumptions about such things as connector strength or thickness of metal side plates or thickness of wood pieces so that a reasonable number of tests could be conducted. Equations, and subsequently tables, based on these tests carried with them the built-in assumptions. With the inclusion of the yield-line approach, the values previously assumed can become variables in the equations. The yield-line approach allows for greater sophistication in the design of connections. Whereas the empirical approach led to some values being much more conservative than others, the yield-line approach will result in most connections having a more consistent factor of safety. In addition, since the yield-line approach is based on formulas, unusual connections with parameters that fall outside the boundaries of the design tables can readily be calculated. Previously, the empirical formulas were available for this purpose, but the designer was not given as much information about the potential modes of failure, nor the probable bounds of any equation, nor the recommended factor of safety to use.

[6] U.S. Forest Products Laboratory, *Wood Engineering Handbook*, Prentice Hall, Englewood Cliffs, NJ, 1972.

The following example is intended to acquaint the designer with the form of the yield-theory equations as well as to compare a design using the equations with a design using the tables. Connection configurations that do not occur within the NDS tables may be calculated using this approach.

EXAMPLE 4.20—YIELD-THEORY DESIGN.

Consider a nail in single shear joining two wood pieces of the same material. Apply the yield-theory equations to determine the allowable load and failure mode.

Consider the failure mechanism of a 16d nail connecting two pieces of Douglas Fir, each 1.5 in. thick. Figure 4.33 shows the possible modes of failure for such a connection: I_s, in which the side piece crushes in uniform bearing; IIIm, in which the connector yields and the main piece crushes; IIIs, in which the connector yields and the side piece crushes; or IV, in which the connector yields and causes crushing in both pieces. The actual mode of failure will depend on the ratio of side to main piece widths and strengths as well as on the strength and thickness of the connector. The yield formulas are given in Table 4.26.

The allowable load that we calculated is the least of all the possible loads, namely 141 lb. The design value for this connection given in the 1991 NDS Table 12.3B is also 141 lb. The strength of the connection is controlled by mode IV, in which the nail forms two plastic hinges, one in each piece of wood near the shear plane. The formulas produce allowable, not ultimate loads. Factors of safety are included in the formulas.

5

TRUSSES

Early people had no trouble building walls for their housing once they began constructing dwellings. The problem of putting on a roof was solved by laying logs down from wall to wall, at least in areas where timbers were available. In other locales or for particular purposes they learned to build arched or dome-shaped stone structures, but the clear openings from wall to wall inside the buildings were not very large. They also learned to put wood rafters up to get a pitched roof of longer span. The open triangle thus formed was not stable since the bottoms of the rafters tended to "kick out" the walls supporting them. These rafters were held in place with a horizontal piece, thus forming a true triangle. Eventually, it was found that triangles could be combined to form an even longer structural unit that was quite stable and self-contained. This structural unit, the truss, was also used as formwork for masonry arches and domes.

The earliest remaining visual record of a wooden truss is a carving in Trajan's Column in Rome (built A.D. 104) of a bridge over the Danube River. Supposedly, this bridge was constructed of about 20 trussed timber arches spanning 100 to 120 ft each. Surely an engineering feat of this magnitude had a long history of precedents.

Palladio (1518–1580), an Italian architect and engineer, apparently did more than any of his predecessors to further development of the truss. His trusses were, of course, of wood. One, a bridge constructed between Trent and Bassano over the Cismone River, is reported to have had a clear span of approximately 100 ft, the largest single-span bridge of any material in its time.

The use of trusses was introduced into England by Inigo Jones (1573–1652) and Christopher Wren (1632–1723). Their sources undoubtedly included Palladio's

Four Books of Architecture (1570) and probably firsthand inspection of the recent Italian works. Most noteworthy, and most novel, of these efforts was the trusses for the Sheldonian Theatre in Oxford, spanning 68 ft, designed in 1663.[1]

Jean-Ulrich Grubenmann, a Swiss carpenter, built a timber truss bridge near Wettingen, Switzerland, in 1778 with a clear span of 390 ft. Able to carry carriage traffic of as much as 25 tons, it was the longest timber-span bridge ever built. It was destroyed in battle by the French only 21 years after being built.

A wrought-iron trussed bridge was built toward the end of the eighteenth century. It was not until about 1840 that a rational method for analyzing trusses was developed. It stands as a testimony to the structural intellect and intuition of these early builders that they could successfully attempt such large spans without the benefit of refined analytic techniques. The graphic technique for analysis that we use now was developed and popularized by Karl Culmann, a professor of structures at the Zurich Polytechnicum in 1866.

Wood bridges made of trusses as well as other types were built in Europe in both the eighteenth and nineteenth centuries, but it was in America where the greatest development took place during this period. Great stands of pine made wood bridges the obvious answer in constructing highways and later, railways to meet the demands of the westward movement of the nineteenth century. The history of the development of the wood bridge is the history of the development of structural engineering in the United States, and these were primarily trusses.

The wood trusses of the nineteenth century used iron bolts and rods for fasteners, although they were dependent primarily on skillful carpentry to obtain the well-fitted joints necessary for the transference of both compression and tension stresses. Modern timber connectors have eliminated the need for the skilled artisans of that period.

Some diagrams of common truss forms are shown in Figure 5.1, and a description of each is given in Table 5.1. The table refers to both lightweight and heavyweight construction. "Lightweight" is used to mean trusses made of nominal 2-in.-thick material. These trusses are often prebuilt by a manufacturer; they can be custom ordered in any shape. They are preengineered for specific conditions and tend to be the most economical solution. Lightweight trusses are typically spaced 16 or 24 in. on center, supporting plywood or similar sheathing, and supported, in turn, by wood or masonry bearing walls. The term "heavyweight trusses" is used to refer to trusses made of much larger material—4×6 to 8×12, used singly or doubly as top and bottom chords. These trusses are efficient only if they are spaced as far apart as possible. The spacing depends on the framing that the truss supports; a typical condition is shown in Figures 5.11 and 5.12, where trusses at 15 ft on center support 2×10 purlins at 24 in. on center. Heavyweight trusses are also used to support single ridge beams or a series of large *purlins* that run parallel to the ridge. The large purlins are ideally located at the *panel points* of the truss, where the top chord is braced by vertical or diagonal webs.

Selection of a type of truss will be influenced by architectural requirements, the shape of the roof, the load to be carried (bridge or building), clearances, and other external factors, but the framing of the truss itself and the design of details are usually left in the hands of the designer, and for purposes of economy they deserve considerable study. Joint details are a major factor in truss cost, and simplicity in

---

[1] David T. Yoemans, *The Trussed Roof: Its History and Development*, Scolar Press, Aldershot, Hants, England, 1992.

TABLE 5.1 TRUSS TYPES

Truss Type	Lightweight Span (ft)	Lightweight Considerations	Heavyweight Span (ft)	Heavyweight Considerations
King post	Less than 12	Used only in small spans, buckling length of top chord is critical; *size:* top chord, 2×6 or 2×8; bottom 2×4 or 2×6; *spacing:* 16 or 24 in. o.c.	18 to 36	Most efficient if loaded at the peak as part of ridge beam and rafter system; *size:* 6×6 and larger; *spacing:* 8 to 16 ft typical.
Queen post	18 to 24	Can carry heavier loads or longer spans than king post	18 to 36	Better than king post for loads distributed on top chord; best if loaded with purlins at panel points; *size and spacing:* same as king post
Fink	24 to 36 (typical), 60 max.	Common configuration of manufactured trusses; top chord 2×4, 2×6, or 2×8 as load increases; bottom chord 2×4 or 2×6 for heavier ceiling load; *spacing:* 16 or 24 in. o.c. *depth:* $\frac{1}{2}$ of span	40 to 60 (typical), 80 max.	Chords 4×6 to 8×12, (typical); *spacing:* 8 to 16 ft o.c.; *depth:* 4 to 10 in. per foot of span; use 6 in. per foot for preliminary design
Howe	24 to 36 (typical), 60 max.	Same uses, sizes, and spacing as Fink; used where heavier ceiling loads are expected	40 to 70 (typical), 90 max.	Same uses, sizes, and spacing as Fink; used where heavier ceiling loads are expected
Fan	24 to 36 (typical), 60 max.	Same uses, sizes, and spacing as Fink; used where heavier top chord loads are expected	40 to 70 (typical), 90 max.	Same uses, sizes, and spacing as Fink; used where "walk-through" attic space is needed
Scissor	18 to 32 (typical), 45 max.	Same uses, sizes, and spacing as Fink except smaller spans; used where vaulted ceilings are desired	30 to 60 (typical), 75 max.	Same uses, sizes, and spacing as Fink except smaller spans; not as efficient in thrust as Fink, so spans are smaller
Monoslope	12 to 24 (typical), 45 max.	Not as efficient as two-slope trusses; for long spans or higher loads, panel points can be added to shorten unbraced length of top chord; *depth:* $\frac{1}{2}$ in. per foot of span	24 to 48 (typical), 60 max.	(See notes for lightweight mono-slope)
Parallel chords Warren Howe Pratt	*Floors:* 18 to 30 (typical), 42 max.; *roofs:* 18 to 45 (typical), 70 max.	Common configuration of manufactured trusses; top chord 2×4, 2×6, or 2×8 as load increases; bottom chord 2×4 up to 2×8 for heavier ceiling load; *spacing:* 16 or 24 in. o.c. *depth:* $\frac{3}{4}$ in. per foot span for floors, 1 in. per foot for roofs	Generally *roofs only:* 30 to 80 (typical), 120 max.	Chords 4×6 to 8×12 (typical); *spacing:* 8 to 16 ft o.c.; *depth:* 1 ft per ft of span; Pratt is slightly more efficient since diagonals are in tension

RESIDENTIAL TRUSS DESIGN   **167**

such details leads to economy in fabrication. For a quick guide to typical truss spans and loads, see *The Architect's Studio Companion*[2] by Edward Allen and Joseph Iano.

As can be seen in the illustration, a truss is actually a long beam with triangular holes punched out of it. Its overall shape can be rectangular, triangular, or arched, depending on requirements. The topmost members constitute the *top chord*, the bottom pieces together make up the *bottom chord*, and the interior struts are called the *web members*. Along its length the truss is divided into several equal segments called *panels*, as in the six-panel Howe truss shown. The point at which web members meet a chord is called a *panel point*. It is assumed for the sake of preliminary analysis that all loads act as concentrated forces on the panel points. Actually, most chords carry uniform loads, and these uniform loads are transferred into the panel points as concentrated forces. The true loading must be accounted for in the final design of these members.

RESIDENTIAL TRUSS DESIGN

EXAMPLE 5.1—TOP CHORD, APPROXIMATE ANALYSIS

Suppose that a house is 30 ft wide, and the gable roof is to have a pitch of 5 on 12. There will be a 325-lb asphalt shingle roof on a 30-lb felt over $\frac{5}{8}$-in. plywood sheathing. The ceiling under the truss will be $\frac{1}{2}$-in. drywall. The trusses are spaced 24 in. on centers. The live load is 20 psf on a horizontal projection.

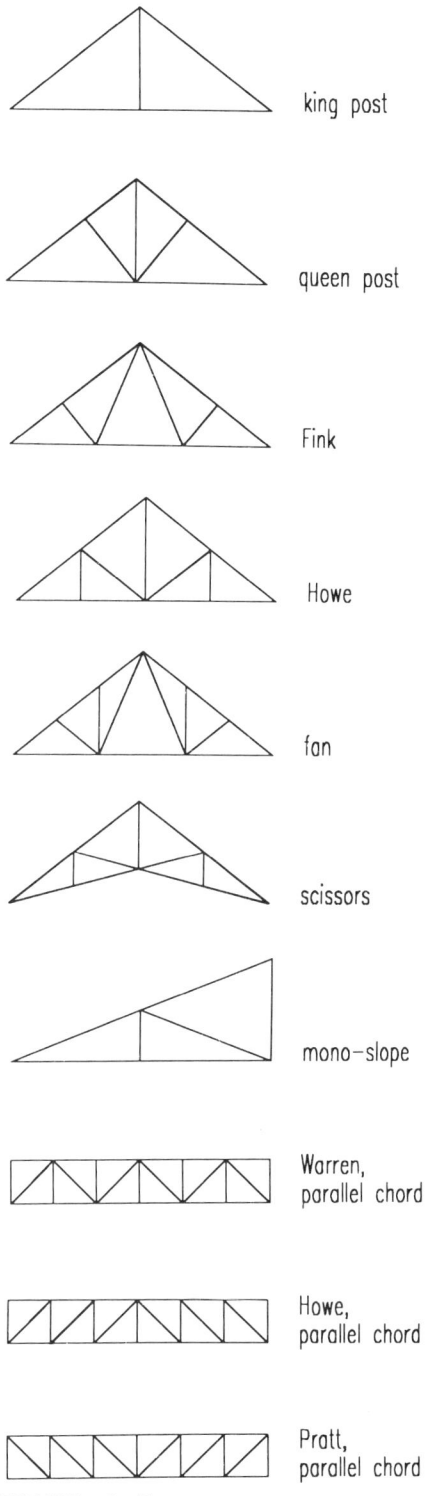

FIGURE 5.1 Truss types.

[2] Edward Allen and Joseph Iano, *The Architect's Studio Companion: Technical Guidelines for Preliminary Design*, Wiley, New York, 1989.

168 TRUSSES

Let us first select our wood. The choice of species will depend largely on the location of the structure and the species most readily available in the area. Structural grades are selected on the basis of stress values. However, certain stress grades are more easily obtained and are therefore more commonly used. In general, the grades with the lower working stresses should be used where possible because they provide the most efficient and economical design. In the case at hand, let us choose Southern Pine No. 2, which has the following allowable stress values:

F_b = 1650 psi, for 2 × 4

1400 psi, for 2 × 6 and larger

F_c = 975 psi, for 2 × 4

1000 psi, for 2 × 6 and larger

F_t = 825 psi, for 2 × 4

625 psi, for 2 × 6 and larger

E = 1,600,000 psi for all sizes

Now we must determine the truss type and loading. A very common type of residential truss is a W truss, a variant of the fan or Fink truss. Figure 5.2 shows the truss layout and loads.

The bottom chord is divided into three equal panels of 10 ft each to make up the 30-ft length. The two top chords are divided into two equal panels. By similar triangles, the length of each top chord is found to be 16.25 ft, and each top chord panel is thus 8.125 ft.

Roofing is measured in squares, and *a square* of shingles means 100 square feet, so that 325-lb shingles weigh 3.25 psf. Similarly, 30-lb felt weighs 0.30 psf. Plywood weighs about 36 lb/ft^3 so that $\frac{5}{8}$-in.-thick

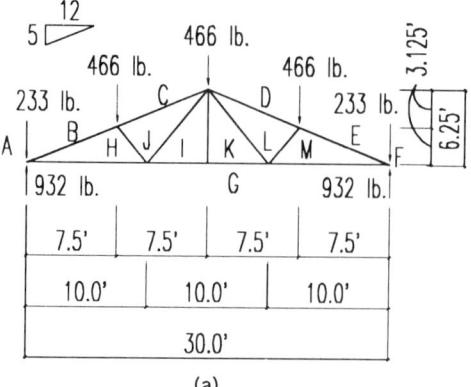

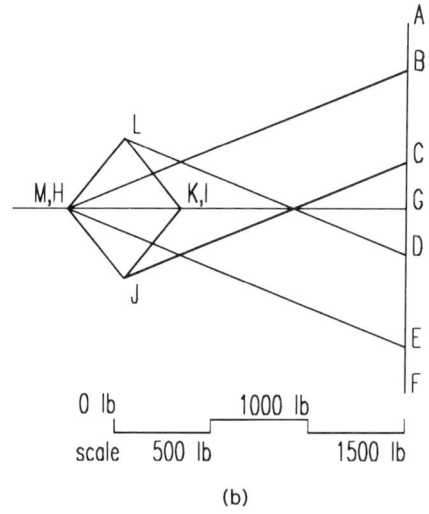

FIGURE 5.2 Residential truss—fan type: (a) space diagram and (b) force diagram.

plywood will weigh about $\frac{5}{8} \times \frac{1}{12} \times 36 =$ 1.88 psf. A pitched wooden truss will weigh

$w = 0.064L$ for a pitched roof truss, or

$w = 0.043L + 1.75$ for a flat-top Pratt truss

where w is the weight in pounds per square foot on the horizontal projection, and L is the span in feet.[3] (*This is a reasonable*

---
[3] W. Fleming Scofield, *Modern Timber Engineering*, 4th Edition. Southern Pine Association, New Orleans, LA, 1954.

estimate for all pitched wood trusses, not just residential.)

Our truss will have a dead load of $0.064 \times 30 = 1.92$ psf. To summarize:

Dead load:	Roofing	3.3 psf
	Felt	0.3 psf
	Sheathing	1.9 psf
	Truss	1.9 psf
	Total	7.4 psf

Since the trusses are 24 in. = 2'-0" on centers, each panel point load will be

8.125 ft × (2 ft × 7.4 psf) = 120 lb

panel point dead load

To be precise, the weight of the drywall ceiling should be taken as acting on the lower panel points. However, it is simpler to add it in with upper chord loads, and this procedure is sufficiently accurate in most cases.

Drywall	3.0 psf
Live load	20.0 psf
Total	23.0 psf

$\frac{1}{4}$ (4 panels) × 30 (ft length)

× 23.0 (psf load) × 2.0 (ft spacing)

= 345 lb   panel point load

Concentrated load on upper panel points:

120 + 345 = 465 lb

Figure 5.2*b* shows a graphic force diagram for the truss as loaded. Table 5.2 summarizes the truss forces for design. Table 5.2 also compares the results from the graphic analysis with those from the mathematical method of joints analysis. If the graphic analysis is done at a sufficient scale and with care, it will produce results with sufficient accuracy for design. The results from the graphic analysis were used in the following design examples.

Now the members can be designed. We first design the top chord, then the bottom chord, and then the web members. The two *top chord* pieces, sometimes called rafters, or truss-rafters, meet at the ridge or crown of the truss. Individually, they meet the bottom chord at the eaves or heels of the truss. For the lengths that we are dealing with, each rafter will be one continuous piece. It is therefore designed for its worst loading condition, an axial compressive force of approximately 1840 lb. In addition, because the load on the roof is actually a uniform load, there will be a bending moment. The dead load is 7.4 psf on the slope, which makes it $7.4 \times \frac{13}{12} = 8.0$ psf on the horizontal pro-

TABLE 5.2 RESIDENTIAL TRUSS: SOLUTIONS

Applied Forces			Bar Forces			
Bar	Force (lb)	Direction	Bar	Diagram Force (lb)	Calculated Force (lb)	Direction
AB	232.5	Down	BH	1840	1814	Compression
BC	465	Down	CJ	1540	1511	Compression
CD	465	Down	DL	1540	1511	Compression
DE	465	Down	EM	1840	1814	Compression
EF	465	Down	HG	1720	1674	Tension
			KG	1115	1116	Tension
			MG	1720	1674	Tension
Reactions			HJ	450	447	Compression
			JK	460	447	Tension
FG	930	Up	KL	450	447	Tension
GA	930	Up	LM	470	447	Compression

jection. Adding in the live load only (the ceiling will act on the bottom chord for bending moment), we get 8.0 + 20.0 = 28.0 psf; and since the trusses are 2 ft on centers,

$$w = 2 \text{ ft} \times 28 \text{ psf} = 56 \text{ lb per linear foot}$$

We use this value in calculating the moment on the chord.

With light loads the top chord may be assumed to be a 2×4 for a first guess. Obviously, we must use the interaction formula. Please note that both the axial stress calculations and the moment calculations are related to buckling and to unbraced lengths. The direction in which a chord buckles, and the unbraced length associated with that buckling, may not always be the same for compression and for bending. We must pay close attention to keep them straight. Now we will start with the compression forces to determine whether the column chord is short, intermediate, or long.

When a 2×4 or smaller truss compression chord has plywood sheathing attached to its narrow face, its buckling stiffness increases by a factor called C_T. The allowable bending strength will be increased by the factor C_T, which is always greater than 1.0. C_T is defined as

$$C_T = 1 + \frac{K_M \times L_e}{K_T \times E}$$

where

$K_M = 2300$ for wood seasoned to 19% MC,

or

 = 1200 for partially seasoned or unseasoned wood

$K_T = 0.59$ for visually graded or MEL lumber, or

 = 0.82 for glulams and MSR lumber

For chords with an effective buckling length greater than 96 in. C_T is taken as the value for a chord having an effective length of only 96 in. In our case we may assume that the plywood will keep the chord from buckling in the lateral direction and that the webs restrain it from buckling up or down. The actual chord length is 8.125 ft or 97.5 in. between lateral supports. Since the chord will buckle in a simple sine curve between these supports, take $K_e = 1.0$ and $L_e = 97.5$ in.

$$C_T = 1 + \frac{2300 \times 96}{0.59 \times 1,600,000}$$

$$= 1.23$$

The top chord must be designed for the combined effects of axial compression and bending. The design formulas and some simplifying assumptions are discussed in Chapter 3. If this material is unfamiliar to the reader, we suggest a review of Chapter 3 before proceeding with the example.

To find the slenderness ratio, L_e/d, it is necessary to notice that the top chord cannot buckle sideways, into the least dimension of the wood, because it is restrained along its entire length by the sheathing. But the sheathing itself is free to move downward if the top chord should sag down, so use $d = 3.5$ in., the depth of the 2×4. Thus for axial compression,

$$\frac{L_e}{d} = \frac{98}{3.5} = 28.0$$

$$F_{cE} = \frac{0.3 E'}{(L_e/d)^2} = \frac{0.3 \times 1,600,000}{(28)^2}$$

$$= 612 \text{ psi}$$

$$F_c^* = F_c \times C_D \times C_M \times C_t$$

$$= 975 \times 1.0 \times 1.0 \times 1.0 = 975 \text{ psi}$$

Note that C_t is the temperature factor, not to be confused with C_T, the truss factor based on plywood stiffness. C_T is used only with bending strength.

$$\frac{F_{cE}}{F_c^*} = \frac{612}{975} = 0.627$$

$$C_P = \frac{1 + (F_{cE}/F_c^*)}{1.6}$$
$$- \sqrt{\frac{[1 + (F_{cE}/F_c^*)]^2}{2.56} - \frac{F_{cE}/F_c^*}{0.8}}$$

$$= \frac{1 + 0.627}{1.6}$$
$$- \sqrt{\frac{[1 + (0.627)]^2}{2.56} - \frac{0.627}{0.8}}$$

$$= 0.517$$
$$F_c' = F_c^* C_P = 975 \text{ psi} \times 0.517 = 504 \text{ psi}$$

Looking at the compression force only, we see that the level of stress is

$$\frac{f_c}{F_c'} = \frac{P/A}{F_c'} = \frac{1840 \text{ lb}/5.25 \text{ in}^2}{504 \text{ psi}} = 0.695$$

This value is less than 1.0. If the chord were loaded so that there was no bending, it would be satisfactory. Since it is loaded in bending as well, we must check the interaction of the two forces. As a quick check, let us find the *approximate* level of stress in the chord. Assume that the maximum moment in the top chord is $M = wL^2/8$, over the web support. This is a *conservative* assumption. Thus

$$M = \frac{56 \times 8.125^2}{8} = 462 \text{ ft-lb}$$

Neglect the moment amplification factor, $(1 - f_c/F_{cE})$. This is a *nonconservative* assumption. The two assumptions offset each other to some extent.

$$\frac{M/S}{F_b' \times C_T} = \frac{462 \times 12/3.06}{1650 \times 1.23} = 0.893$$

Subject to bending only the 2×4 may or may not work, depending on how conservative our assumption of the moment was. Looking at the interaction of the two forces, we see that

$$\left(\frac{f_c}{F_c'}\right)^2 + \frac{f_b}{F_b'(1 - f_c/F_{cE})} \Leftarrow 1.0$$

$$(0.695)^2 + 0.893 = 0.483 + 0.893 = 1.376$$

The simplifications that were made in the calculation were certainly not so gross as to be 40% off. We can conclude only that the 2×4 will not work. We will try a 2×6 instead and use the more exact form of the equations. In using a 2×6 we must remember that the truss factor, C_T, applies only to 2×4's or smaller.

Let us first refine the calculation for the moment.

TPI DESIGN PROCEDURES

As a quick approximation we often assume the moment on the top chord is given by

$$M = \frac{wL^2}{8}$$

In our example this yields a moment of

$$M = \frac{56 \times 8.125^2}{8} = 462 \text{ ft-lb}$$

This assumption is conservative in that it is based on an assumption that the connections at the ends of the chords are pin-ended and will not produce moments. In most cases the chords are continuous over one or two panel points so that negative moments will develop over the supporting webs. Full-scale testing of standard light timber trusses reveals that there is a

TABLE 5.3 TOP CHORD MOMENT FACTORS

	Panel Point Moment		Midpanel Moment	
	Q	L	Q	L
	One panel Not applicable	One panel Not applicable	One panel 0.90	One panel[c] $L_i + S_a$
	Two panels 0.90	Two or three panels[a] Largest of $0.9L_i$, or $\dfrac{L_i + L_a}{2}$, or $0.9L_a$	Two panels[b] $0.58(\cot\theta)^{0.23}$	Two or three panels[c,d] Largest of $0.9(L_i + cS_a)$, or $\dfrac{L_i + L_a}{2} + cS_a$, or $0.9(L_a + cS_a)$
	Three panels 0.85		Three panels[b] $0.53(\cot\theta)^{0.36}$	

[a] If S_t exceeds 24 in., add excess to end (heel) panel L_i or L_a (see Figure 5.3).
[b] $Q = \alpha(\cot\theta)^\beta$ but shall not be less than 0.74; α and β are constants derived from PPSA analysis.
[c] $S_a = S_t - B$ but not less than zero. cS_a shall be added only to the length of the end (heel) panel; [d] $c = 0.5$ for two panels; $c = 0.33$ for three panels; if neither L_i nor L_a are end (heel) panel lengths, then $cS_a = 0$.

Source: *Design Specifications for Metal Plate Connected Wood Trusses*, TPI-85.

substantial reduction in the maximum moment. In 1978 the Truss Plate Institute published their recommended *Design Specifications for Metal Plate Trusses*, revised in 1985 and usually referred to as TPI-85.[4] This procedure has been adopted by all factory-built truss manufacturers and is the basis for their computer-generated design. In general terms this procedure will result in much smaller moments, but as will be seen in the example below, this will not necessarily translate into smaller chord sizes.

TPI-85 recommends that the moments be calculated using a revised length for each section of the chord. The revised length takes into account the stiffness of the connections and the connecting webs or chords. The TPI procedure also uses an effective buckling length which is smaller than the actual length of the chord. The modified panel length is $QL/\cos\theta$, where

[4] *Design Specifications for Metal Plate Connected Wood Trusses*, TPI-85, Truss Plate Institute, 583 D'Onofrio Drive, No. 200, Madison, WI, 53719.

the L refers to the horizontal panel length and Q is found in Tables 5.3 and 5.4. Figure 5.3 defines the terms used in finding the chord lengths. By using the TPI procedure moments will generally be smaller than those calculated using $wL^2/8$.

The recommended lengths, and the modifier, Q, are tabulated here as they appear in the TPI publication. The panel point moment and the midpoint moments are calculated for each panel of the truss using the equation

$$M = \frac{w(QL)^2}{8}$$

The magnitude of the moment derived from this calculation is then used in the interaction formula. It is assumed that the axial force is constant for any panel, so the worst condition always occurs at the larger moment—panel point or midpanel. In the example we calculate the moment for the lower panel, adjacent to the heel. For trusses with two panels the moments in the upper panel will always be the same

TABLE 5.4 BOTTOM CHORD MOMENT FACTORS

Q	L
One panel $\dfrac{}{1.0}$	One panel $\dfrac{}{L_i}$
Two or more panels $\dfrac{}{1.0}$	Two or more panels L is largest of $0.9L_i$, or $\dfrac{L_i + L_a}{2}$, or $0.9L_a$

Source: Design Specifications for Metal Plate Connected Wood Trusses, TPI-85.

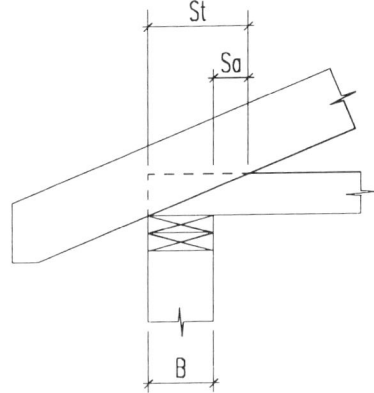

FIGURE 5.3 Truss heel dimensions —TPI formulas.

as in the lower panel. For trusses with more than two unequal panels, you may have to check the moments in each panel. For trusses with gravity loads at each panel point, the axial force in the lowest panel will always be the maximum for the top chord; thus the interaction formula is usually worst for the lowest panel.

EXAMPLE 5.2—TOP CHORD: TPI MOMENTS

As an example, recalculate the moments from Example 5.1 using the TPI procedure.

Refer to Figure 5.3 for the following values:

B = wall thickness = $3\frac{1}{2}$ in.

$S_t = \dfrac{12}{5} \times 3\frac{1}{2}$ in. = 8.4 in.

$S_a = S_t - B$ = 8.4 in. − 3.5 in. = 4.9 in.

  = 0.408 ft

$cS_a = 0.5 \times S_a = 0.5 \times 0.408$ ft = 0.204 ft

$L_2 = 7.5$ ft

$L_1 = 7.5$ ft $- S_t = 7.5 - \dfrac{8.4}{12} = 6.8$ ft

Use the formulas in Table 5.3 to calculate the moment for the first panel, adjacent to the heel.

The *panel point moment* is the negative moment in a chord at the point where the webs support the chord. To find the panel point moment, determine the controlling length to be the largest of

$0.9L_i = 0.9 \times 6.8 = 6.12$ ft

or

$\dfrac{L_i + L_a}{2} = \dfrac{7.5 + 6.8}{2} = 7.15$ ft

or
$$0.9L_a = 0.9 \times 7.5 = 6.75 \text{ ft}$$

Use $L = 7.15$ ft and $Q = 0.9$.

$$M = \frac{w(QL)^2}{8}$$

$$= \frac{56 \times (0.9 \times 7.15)^2}{8} = 290 \text{ ft-lb}$$

The *midpoint moment* is the positive moment near the middle of the chord. To find the midpoint moment, determine L from the largest of

$$0.9(L_i + cS_a) = 0.9 \times (6.8 + 0.204)$$
$$= 6.30 \text{ ft}$$

or

$$\frac{L_i + L_a}{2} + cS_a = \frac{7.5 + 6.8}{2} + 0.204$$
$$= 7.35 \text{ ft}$$

or

$$0.9(L_a + cS_a) = 0.9 \times (7.5 + 0.204)$$
$$= 6.93 \text{ ft}$$

To find Q, use the roof pitch of $\frac{5}{12}$ to calculate

$$\cot \theta = \frac{1}{\tan \theta} = \frac{12}{5} = 2.4$$

$$Q = 0.58(\cot \theta)^{0.23} = 0.58(2.4)^{0.23}$$
$$= 0.709$$

so

$$M = \frac{56 \times (0.709 \times 7.35)^2}{8} = 190 \text{ ft-lb}$$

Use the larger of the two, panel point moment, $M = 290$ ft-lb. Compared with the original estimate of 462 ft-lb, the TPI procedure results in a moment that is about 63% of the previous estimate. Had we designed the top chord using a moment of 290 ft-lb and the moment magnification factor, we would still have decided that a 2×4 top chord was insufficient.

Now analyze the 2×6 chord for a moment of 290 ft-lb and a compression force of 1840 lb. Because the piece is larger than a 2×4, C_T is neglected. For a 2×6, $A = 8.25$ in^2, $S = 7.56$ in^3, $F_c = 1000$ psi, $F_b = 1400$ psi, and $E = 1,600,000$ psi.

For axial compression, the buckling can only occur downward, in the strong direction.

$$\frac{L_e}{d} = \frac{98}{5.5} = 17.8$$

$$F_{cE} = \frac{0.3 E'}{(l_e/d)^2}$$

$$= \frac{0.3 \times 1,600,000}{(17.8)^2} = 1515 \text{ psi}$$

$$F_c^* = F_c \times C_D \times C_M \times C_t$$
$$= 1000 \times 1.0 \times 1.0 \times 1.0 = 1000 \text{ psi}$$

$$\frac{F_{cE}}{F_c^*} = \frac{1515}{1000} = 1.515$$

$$C_P = \frac{1 + (F_{cE}/F_c^*)}{1.6}$$
$$- \sqrt{\frac{[1 + (F_{cE}/F_c^*)]^2}{2.56} - \frac{F_{cE}/F_c^*}{0.8}}$$

$$= \frac{1 + 1.515}{1.6}$$
$$- \sqrt{\frac{[1 + (1.515)]^2}{2.56} - \frac{1.515}{0.8}}$$

$$= 0.812$$

$$F_c' = F_c^* \times C_P$$
$$= 1000 \text{ psi} \times 0.812 = 812 \text{ psi}$$

$$\frac{f_c}{F_c'} = \frac{P/A}{F_c'} = \frac{1840 \text{ lb}/8.25 \text{ in}^2}{812 \text{ psi}}$$

$$= \frac{223 \text{ psi.}}{812 \text{ psi}} = 0.275$$

For bending, $L_e = 0$, and we will take $C_L = 1.0$. $F_b' = 1400$ psi. The interaction equation states that

$$\left(\frac{f_c}{F_c'}\right)^2 + \frac{f_b}{F_b'(1 - f_c/F_{cE})} \leq 1.0$$

$$\left(\frac{P/A}{F_c'}\right)^2 + \frac{M/S}{F_b'(1 - f_c/F_{cE})} \leq 1.0$$

$$\left(\frac{223}{812}\right)^2 + \frac{290 \times 12/7.56}{1400(1 - 223/1515)}$$

$$= (0.275)^2 + \frac{460}{1400 \times 0.853}$$

$$= 0.075 + \frac{460}{1194}$$

$$= 0.075 + 0.385 = 0.46$$

The 2×6 is satisfactory.

Use 2×6 Southern Pine No. 2, top chord.

Using the same revised moment and the complete interaction formula will show that a 2×4 in the top chord will be stressed to approximately twice its allowable stress.

EXAMPLE 5.3—BOTTOM CHORD

The maximum axial thrust in the *bottom chord* is 1720 lb in tension. The drywall ceiling produces a uniform load of 2 (ft spacing) $\times$ 3.0 psf = 6.0 plf. Applying the TPI formulas from Table 5.4 leads to a bending moment of

$$M = \frac{wL^2}{8}$$

$$= \frac{3 \times 10 \times 10}{8} = 37.5 \text{ ft-lb}$$

As often happens in the bottom chord, the TPI moment is the same moment as the standard approximation. This bending moment produces a tension stress on the bottom of the piece, which requires the use of a much simpler interaction formula.

$$\frac{P/A}{F_t} + \frac{M/S}{F_b^*} \leq 1$$

where F_b^* = tabulated value modified by all applicable factors except C_L. Trying a 2×4, with $A = 5.25$ in^2 and $S = 3.06$ in^3 yields

$$\frac{1720/5.25}{825} + \frac{37.5 \times 12/3.06}{1650}$$

$$= 0.397 + 0.089$$

$$0.486 < 1.0 \quad \text{OK}$$

For combined tension and bending a second formula must be checked.

$$\frac{f_b - f_t}{F_b^{**}} \leq 1$$

where F_b^{**} = tabulated value modified by all applicable factors except C_V. Since C_V applies only to glued-laminated members, $F_b^{**} = F_b^*$. It can be seen by inspection that if the first equation is satisfied, this equation will be satisfied unless F_b^{**} is much less than F_b^*. Therefore,

Use 2×4 Southern Pine No. 2, bottom chord.

With such a small sum for the two fractions the piece is obviously overdesigned, and a smaller piece would be theoretically satisfactory. However, to attach the drywall easily, and to have smooth, easily built joints at the panel points, for practicality we will use the 2×4.

EXAMPLE 5.4—WEB MEMBERS

The worst case for web members are HJ and LM, which are 4 ft long and are in compression with an axial load of about 450 lb. There is no uniform load on these members, nor are they laterally braced. Assuming a 2 × 4 size, SP No. 3.

$$\frac{L_e}{d} = \frac{4 \times 12}{1.5} = 32$$

$$F_{cE} = \frac{0.3E}{(L_e/d)^2}$$

$$F_{cE} = \frac{0.3 \times 1,400,000}{32 \times 32} = 410 \text{ psi}$$

$$F_c^* = 975 \text{ psi}$$

$$\frac{F_{cE}}{F_c^*} = \frac{410}{975} = 0.42$$

Find C_P in Table C.4

$$C_P = 0.36$$

$$F_c' = 0.36 \times 975 = 351 \text{ psi}$$

$$A_{\text{req'd}} = \frac{P}{F_c'} = \frac{450}{351} \quad \text{about } 1.28 \text{ in}^2$$

Use 2 × 4 Southern Pine No. 3 webs for practicality.

Note that a 1 × 4 would have $L_e/d =$ (4 × 12)/0.75 = 64 > 50, which is not permitted. Hence a 1 × 4 would be illegal.

Finally, the *long web members* JK and KL are in tension with an axial load of 450 lb.

$$A = \frac{P}{F_t} = \frac{450}{475} < 1 \text{ in}^2$$

use 2 × 4 for practicality

The bottom chord is too long to be made of one piece, so it must be spliced. The ideal location for the splice is at the point of zero moment, the inflection point. The standard rule of thumb, as recommended by the Truss Plate Institute, is to locate splices approximately 1 to 2 ft from the bottom chord panel points (for very long panel lengths a more exact rule is 10 to 20% of the panel length). In Figure 5.4, the inflection points are found using the analogy of a continuous beam. Assuming that the bottom chord is a three-span beam with a continuous load, the inflection point will be near the fifth point of the panels. The actual location compares favorably with the rule. The bottom chord will be in two pieces, one 12 ft long and the other 18 ft long. Thus the splice where the two are joined together will occur 2 ft from a point where web members frame in, as measured toward the center of the truss. All joints will be fabricated using pronged steel plates that are pressed into the pieces of wood in the shop.

Metal Plate Connectors. As mentioned in Chapter 4, most lightweight trusses for residential and light commercial uses are fabricated in factories using teethed metal plate connectors. The connectors are a proprietary item; the strength, size, and type of teeth vary from manufacturer to manufacturer. These plates are usually installed with large rollers or pneumatic presses and their holding strength is critically dependent on correct application. The following example using these plates is included to give some understanding of the principles and to aid the designer or engineer who may have to check shop drawings and calculations from the truss manufacturer.

A similar connection that is easier to install is a nailed-plate connector, which is a metal plate with a matrix of holes. Once the proper-size plate for each joint is selected, the builder aligns it according to the specifications and nails it to the truss members with the correct size and number of nails. Usually special, shorter nails are

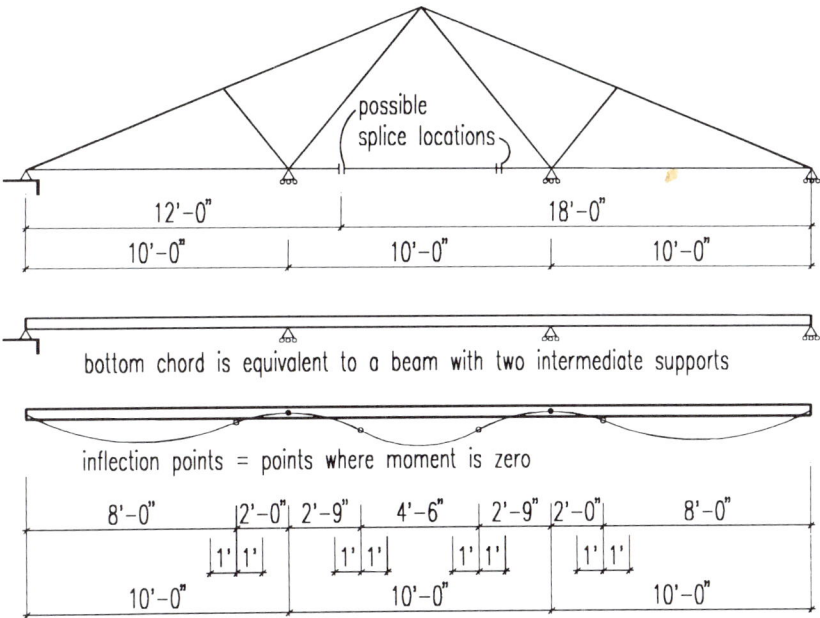

FIGURE 5.4 Bottom chord splice locations.

supplied with the plates. The plates are typically applied to both sides of the truss. In this manner the builder can custom manufacture trusses that are roughly equivalent to those ordered from the larger manufacturers. The procedure for designing the proper plate size and the number of nails is very similar to that shown here for designing deformed-nail plates. Most suppliers of nail-plate connections provide guidelines and design procedures. As an example you may consult *Joints Design Procedure for TECO Nail-on Plate-Connected Trusses*.[5]

EXAMPLE 5.5—TRUSS PLATE CONNECTORS

Consider the truss from the previous examples. Specify the required truss plate sizes for the bottom chord splice and the heel connection (see Figure 5.5).

The strength of truss plates is calculated as pounds per square inch of contact area. This value varies depending on the manufacturer, the type of deformation forming the "nail," their spacing, and length. As a typical value, use 90 psi of plate area and assume that plates will be attached to both sides of the truss. Using the net area approach, the contact area for each piece is the area of overlap between the plate and the wood minus the required edge and end distances. These distances are shown in Figure 5.6 which is taken from TPI-85.

The *tension splice* is placed in the middle panel, where the tension forces are lower; in the example tension in the middle is only 1115 lb, as opposed to 1720 lb at the end panels. Under various load conditions, such as asymmetrical snow, the tension in the middle panel could increase, but we can be assured that it will not exceed 1720 lb. To be conservative we

---

[5] *Joints Design Procedure for TECO Nail-on Plate-Connected Trusses*, 1975, Silver TECO, P.O. Box 203, Colliers Way, Colliers, WV 26035.

will design the splice to carry a force of 1720 lb. To avoid loading the plate in bending, the splice must be located within a foot or two of the panel point.

The contact area must equal

$$A_{contact} = \frac{1720 \text{ lb}}{2 \times 90 \text{ psi}} = 9.55 \text{ in}^2$$

The factor of 2 in the denominator accounts for two plates, one on each side of the truss. The depth of the contact area equals the depth of the chord minus the edge distances of $\frac{1}{4}$ in., top and bottom. For the 2 × 4, the depth of contact area is

$$\text{depth} = 3\frac{1}{2} \text{ in.} - \frac{1}{4} \text{ in.} - \frac{1}{4} \text{ in.} = 3 \text{ in.}$$

The length on each side of the splice equals the required length plus the required end distance of $\frac{1}{2}$ in.

$$\text{length} = \frac{9.55 \text{ in}^2}{3 \text{ in.}} + \frac{1}{2} \text{ in.} = 3.18 + 0.5$$
$$= 3.68 \text{ in.}$$

The total length will be twice this, 7.36 in. Use 8 in. (see Figure 5.7).

Use a 3 in.-wide by 8-in.-long plate on each side at the bottom chord splice.

In a complete design the plates would be checked for tension using their net area.

At the *heel connection* the plate must transfer the horizontal thrust from the bottom chord and the compression from the top chord or rafter. Because of the geometry of this connection, the intersection of these forces lies outside the limits of the plate. The two forces develop a couple. TPI-85 recognizes this couple and establishes a procedure for design using the axial forces by giving a reduction factor for the strength of the connection. The reduction factor also accounts for the necessary end and edge distances, so only the gross area of the plate contact must be determined.

The reduction factor is given as

$$R_F = 0.85 - 0.05(12 \tan \theta - 2.0)$$

but not less than 0.65 or greater than 0.85

For a roof slope of 5 on 12, the equation becomes

$$R_F = 0.85 - 0.05(5 - 2.0) = 0.7$$

The plate capacity at the heel then is 0.7 × 90 psi equals 63 psi.

The thrust is the same as used in the splice connection. The contact area along the bottom chord is

$$A_{contact} = \frac{1720 \text{ lb}}{2 \times 63 \text{ psi}}$$
$$= 13.65 \text{ in}^2 \quad \text{say } 14 \text{ in}^2$$

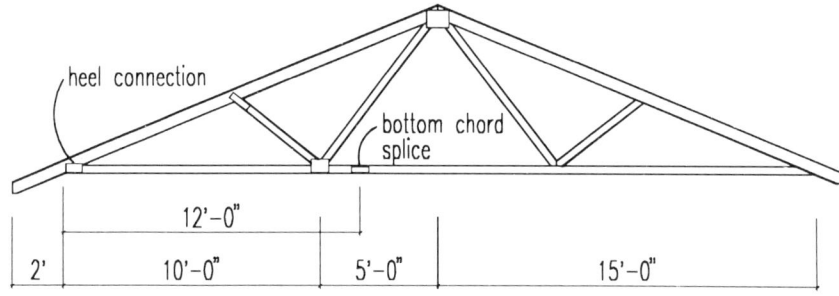

FIGURE 5.5 Residential truss—metal plate connectors.

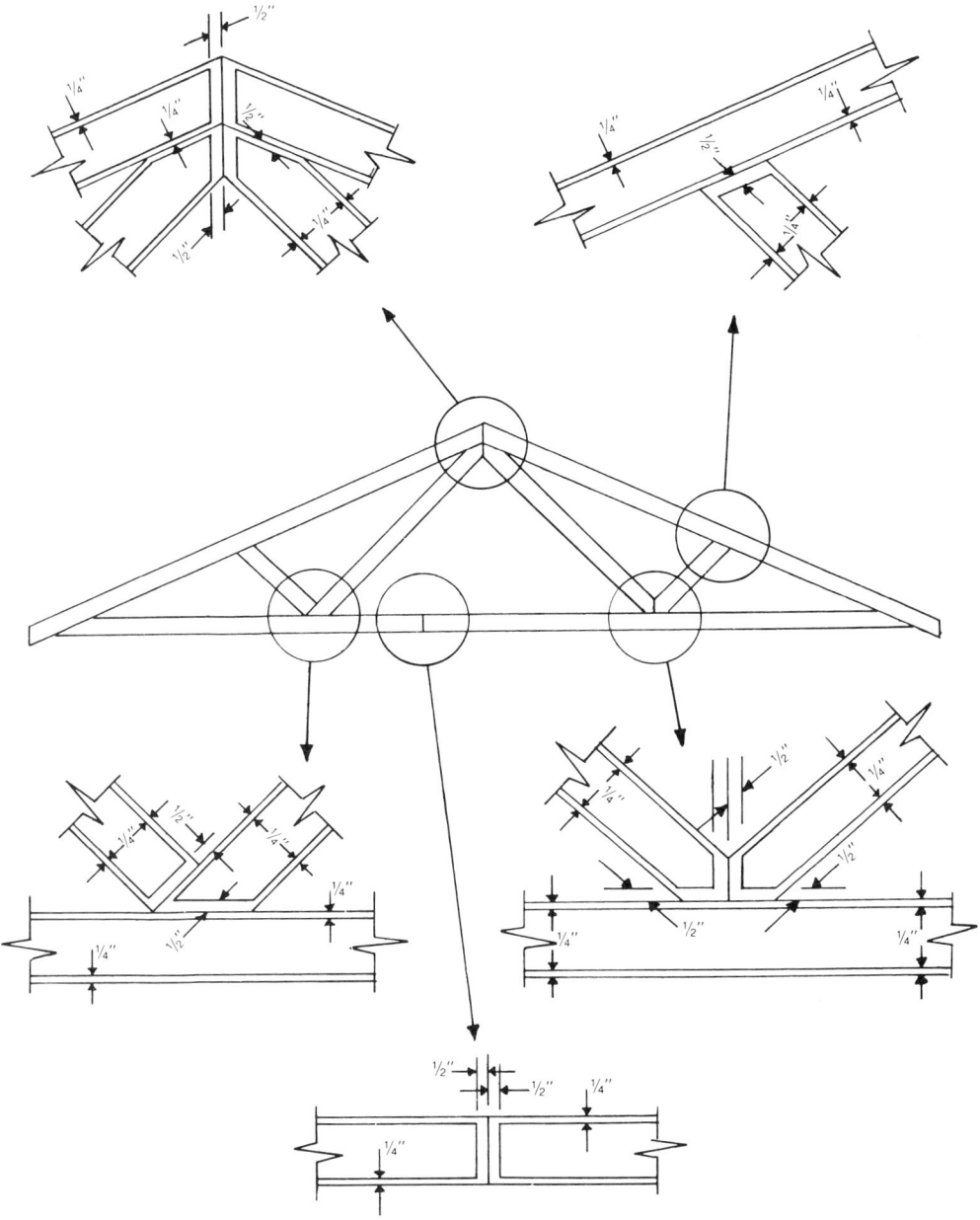

FIGURE 5.6 Metal plate connectors—min. edge distances. Courtesy of Truss Plate Institute

Along the top chord the maximum compression is 1840 lb. If the top chord does not extend beyond the bottom chord, we can assume that there will be some compression transferred through bearing. For joints loaded in compression, as much as 50% of the force can be assumed to be carried directly by wood bearing to wood; however, the plate must develop at least 375 lb (see Figure 5.8).

If the top chord continues beyond the bottom chord to form an eave overhang,

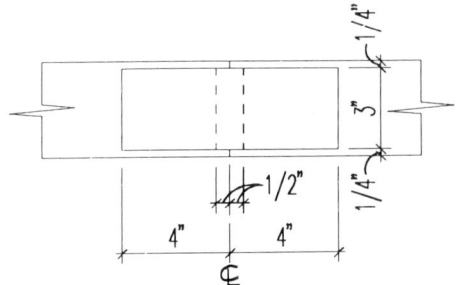

FIGURE 5.7 Bottom chord splice.

the plate in contact with the rafter must be approximately 2×14.6 in^2/3 in., or 9.72 in. The end distances for both chords are incorporated in these values. For the initial trial the total length of the plate needs no more than

$$\text{length} = 9.3 + 9.7 = 19.0 \text{ in.}$$

The best procedure to size any metal plates is to draw the connection at half of full size with the plate superimposed. Use a trial-and-error approach to minimize the plate size while maintaining the required coverage. Figure 5.9 shows the initial step of this process with a plate that is 12 in. long. This results in a plate area of about 20% more than is required; a somewhat

we can assume that there will be very little force transferred through the bearing. (The part of the top chord that extends beyond the bottom chord is called the *rafter tail*.) We will size the top chord connection assuming that the wood-to-wood connection carries no force and that the plate must carry the full force.

$$A_{\text{top chord}} = \frac{1840 \text{ lb}}{2 \times 63 \text{ psi}} = 14.6 \text{ in}^2$$

To size the plate, start with the bottom chord requirements. If the plate covers a rectangular area 3 in. deep, its required length will be $14/3 = 4.66$ in. long. Since the actual coverage area is more like a triangle, assume that the required length must be twice that, namely 9.33 in.

Now check the rafter plate size. To provide an area of 14.6 in^2, the length of

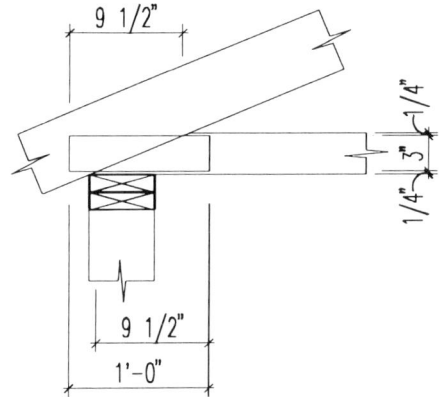

FIGURE 5.9 Detail of heel connection.

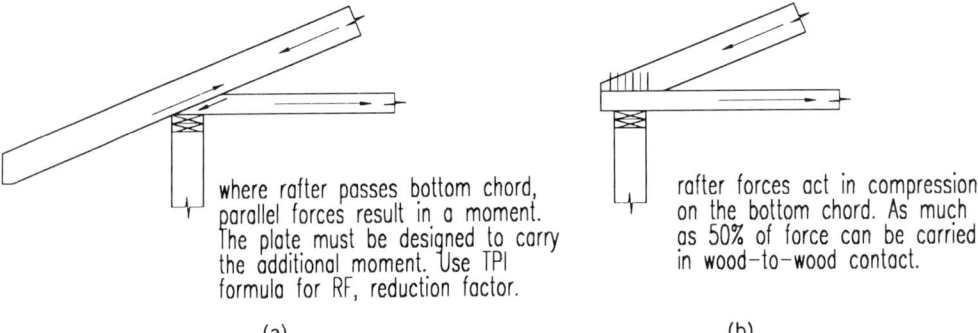

FIGURE 5.8 Comparison of heel geometries: (a) rafter extends over bottom chord and (b) rafter bears on bottom chord.

smaller plate could be specified. It is important in this detail to specify not only the plate length, but also the amount of coverage on the piece. Refer to Figure 5.9 for the acceptable plate.

Use a 3-in.-wide by 12-in.-long truss plate. Maintain 9.5 in. length of plate on the truss top chord.

Truss Bracing Requirements. Trusses must be properly braced to prevent their toppling and to prevent buckling of compression members. The guidelines for bracing presented here are taken from TPI's *Temporary Bracing of Metal Plate Connected Wood Trusses*.[6] Bracing must be considered in three planes to assure that a truss is completely braced: (a) top chord in the plane of the roof; (b) X-bracing of the webs at the ends of the building and at 18- to 20-ft intervals (the X-pattern of the bracing appears in the plane of the webs), and (c) bracing of the bottom chord with diagonals in plan at the ends and at 18- to 20-ft intervals along the length of the building. Longitudinal bracing is also installed running perpendicular to the bottom chords. Usually, two or three sets of longitudinal braces are installed. See Figure 5.10 for each of these types.

There are two conditions for which bracing is needed: permanent bracing and temporary bracing during construction. Compression members are braced using permanent bracing to eliminate buckling under service loads. The determination of permanent bracing is part of the designer's responsibility. The location of the compression member bracing is considered in the truss design when L/d is established. If the ratio L/d is too large, the allowable compression strength of the member might be too small. In general, the L/d ratio should not exceed 50 for any compression member. Specifying additional bracing will lower the L/d ratio. In specifying truss bracing it is important to lay out bracing in more than one plane so that all the compression members are braced to other members or other parts of the structure. In the example the top chord was braced continuously by the roof sheathing, so no further bracing in the plane of the roof is needed. However, bracing should be specified between the trusses tying the compression chord to the tension chord. Webs subject to compression require bracing when they are so long that their L/d ratio exceeds 50. As a practical limit, the maximum L/d ratio should be kept even lower, say 30 or so.

Temporary bracing is installed during erection of the building. The trusses are extremely vulnerable during this time since the sheathing is not applied. The building is often subjected to very heavy loads during erection, due to plywood and roofing material often being loaded on the roof as concentrated loads before it is distributed over the structure. Also, if the roof is on but the walls are not enclosed, the wind uplift forces can be approximately twice the design wind load. If these severe conditions are anticipated, the designer should not rely on standard specifications for temporary bracing but should design for the anticipated loads.

Temporary bracing must be installed to brace the first truss to the ground or some portion of the adjacent structure that has sufficient strength. With the exception of the top chord bracing, the temporary bracing may remain in place as permanent bracing. If possible, the top chord bracing should be installed on the underside of the chords so that it will not interfere with laying the sheathing. Such bracing will normally be left in place. If the top chord bracing needs to be removed to allow installation of the sheathing, bracing should be removed and sheathing should be applied in segments so that a large portion of

---

[6] *Temporary Bracing of Metal Plate Connected Wood Trusses*, DSB 89, Truss Plate Institute, Madison, WI, 1976.

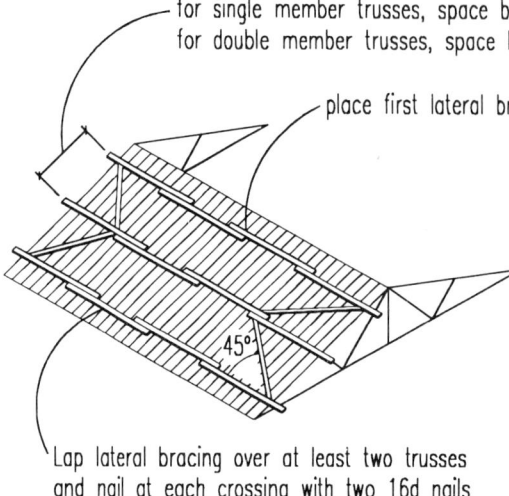

- for single member trusses, space bracing 8' to 10' apart
- for double member trusses, space bracing 10' to 12' apart
- place first lateral brace 6" from ridge line
- Lap lateral bracing over at least two trusses and nail at each crossing with two 16d nails

First truss must be well braced to ground before additional trusses are erected. As additional trusses are erected, they should be braced to the end truss. Temporary bracing is needed in three planes; the plane of the top chord (roof), the plane of the bottom chord (ceiling), and in the plane of the compression webs. Temporary bracing must remain in place until the sheathing is installed. Take care to plumb, align and space trusses properly so that the temporary bracing is not removed during the installation of sheathing. Short spacers between adjacent trusses are not considered bracing. Use 2x3's or larger, bracing should be as long as possible Secure bracing to trusses at all intersections with two 16d nails. If possible the continuous lateral bracing should be located on the underside of the top chord so that trusses will remain braced during the application of the sheathing.

(a)

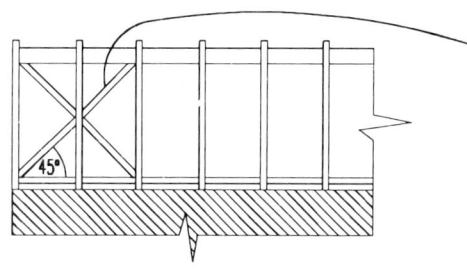

- diagonal bracing – repeat approximately 20' intervals along building length
- spacing: 8' o.c. for floors or 12' to 16' for roofs
- additional diagonals in the plane of the web members will minimize lateral movement

(b)

FIGURE 5.10 Truss bracing requirements: (a) temporary bracing of the top chord, (b) temporary bracing of web member plane, (c) temporary bracing of bottom chord, (d) permanent top chord bracing, and (e) permanent lateral bracing to web members and bottom chords.

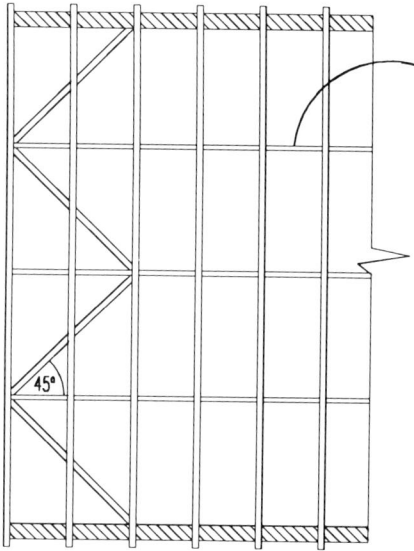

continuous lateral bracing spaced 8' to 10'. locate at or near a panel point

Once temporary bracing in Figures 5.10a through 5.10c are in place, permanent bracing and sheathing can be installed. Where possible leave the temporary bracing to become permanent. Avoid large concentrated loads during sheathing. No more than 8 sheets of plywood should be placed on any pair of trusses and should be located adjacent to supports.

(c)

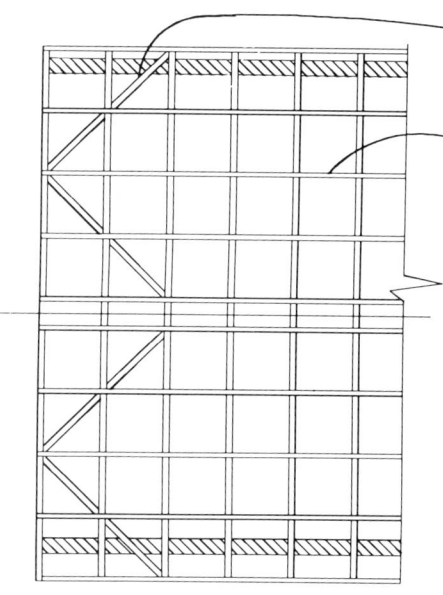

diagonal bracing along the full width of the trusses at each end, nailed to underside of the top chord; repeat at approx. 20' intervals

purlins

A properly designed and installed plywood or OSB floor or roof will develop diaphragm forces to resist lateral movement of the top chord. Additional bracing is usually not required. Metal roofs properly designed and installed with appropriate laps and nails may act as a diaphragm. The selection of these materials is at the discretion of the building designer. If purlins are used, their spacing should not exceed the buckling length of the top chord and diagonal bracing should be applied to the underside of the chords.

Permanent bracing in Figure 5.10e should be installed in all cases

(d)

FIGURE 5.10 (*Continued*)

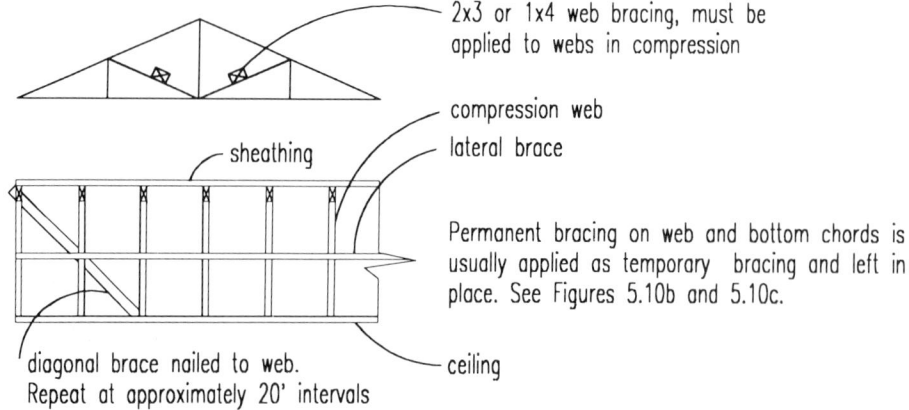

Sources: Temporary Bracing of Metal Plate Connected Wood Trusses, DSB-89; and Light Frame Trusses, CWC Datafile WC-3, Canadian Wood Council

(e)

FIGURE 5.10 (*Continued*)

the trusses will not be left unbraced at any time.

Normally, bracing is made of 2×4's or 2×3's. The bracing itself must have sufficient thickness that the L/d ratio does not exceed 50 for permanent bracing or 75 for temporary bracing. The bracing should be long enough that it is continuous over a number of trusses. Short blocking between trusses will not provide adequate bracing. Bracing should be adequately nailed to all the trusses that it crosses. Usually, two 16d nails are used at each crossing. On temporary bracing double-headed nails are used for easy removal.

HEAVY TRUSSES

EXAMPLE 5.6—HEAVY TRUSS

Let us now design a roof truss with parallel chords having a somewhat longer span of 45 ft. We will design for a snow load of 30 psf and roof construction consisting of built-up roofing on wood sheathing over wood joists.

Actually, a span of 45 ft is not very long, but the procedures we will follow are essentially the same for even longer spans or for a type of truss different from the Howe that will be used here. A Pratt truss would have worked equally well.

As a general rule of thumb, heavy trusses should be spaced apart a distance equal to about one-third of their span. Thus with a 45-ft span we will space them 15 ft on centers. This is a good spacing for the length of the roof joists over the trusses, since $2 \times$ framing is most economical when used at 8- to 16-ft spans. With the load of 30 psf snow on built-up roofing and wood sheathing, select 2×10 joists at 24-in. centers.

Even though we are supposedly using parallel chords, it is necessary to provide drainage for the roof. This can be accomplished by pitching the top chord at least $\frac{1}{4}$ in. per foot, with a maximum of 2 in. per

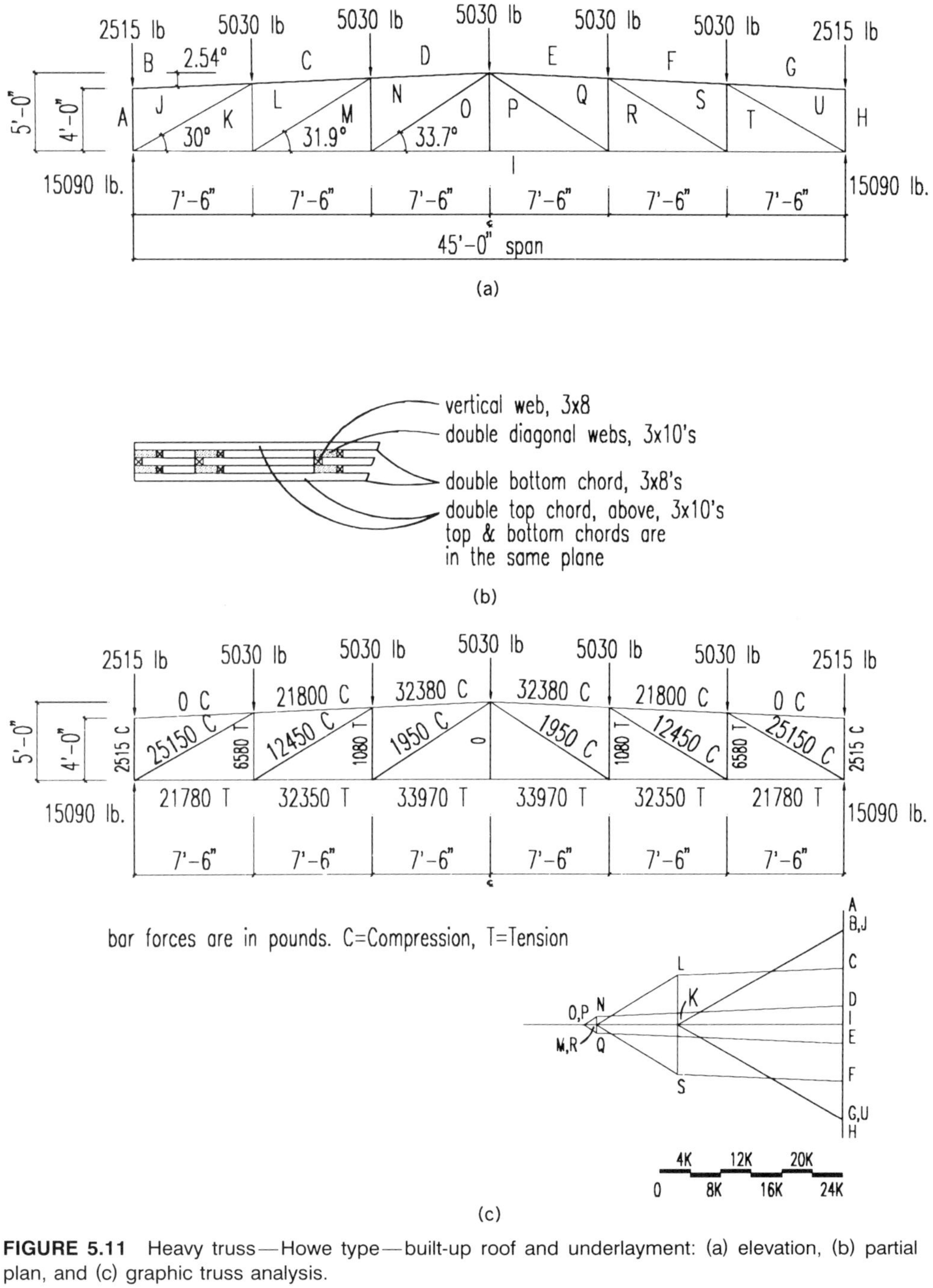

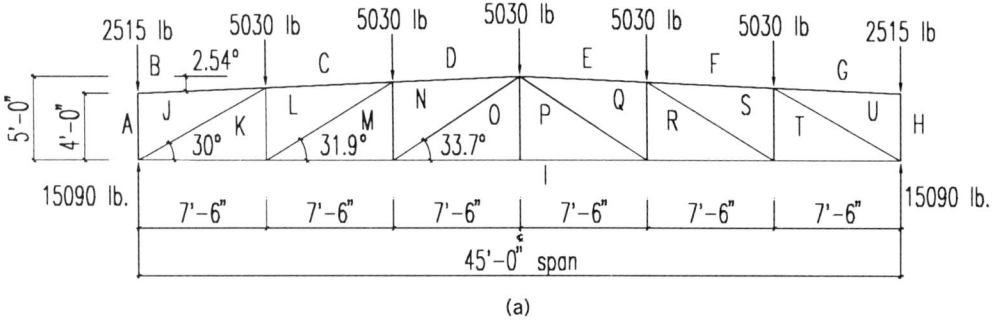

FIGURE 5.11 Heavy truss—Howe type—built-up roof and underlayment: (a) elevation, (b) partial plan, and (c) graphic truss analysis.

foot. The slope can be either one-way, uniformly falling from one side toward the other, or the truss can have its greatest depth at its center, falling off toward either end. In our case, let us put a crown in the center, with a rise of 12 in. from eave to centerline, which comes to slightly more than $\frac{1}{2}$ in. per foot. Figure 5.11 shows the truss layout and the loads.

Having chosen the type of truss to be used, in general the steps in the design procedure are:

1. Determine the truss proportions.
2. Determine the loading.
3. Compute the resulting forces in the members.
4. Select the species and grade of lumber.
5. Determine the size of the truss members.
6. Design the joints, considering first the joints carrying the greatest load. Check the spacing, end distance, and edge distance for each connector. This can be accomplished best by laying out the joint to a scale large enough that angles and distances can be determined with accuracy. Usually, 3 in. = 1'-0" or 6 in. = 1'-0" are the best scales.

TRUSS PROPORTIONS

As stated before, a truss acts like a large beam. In a beam, the applied bending moment is resisted by an internal moment couple. The resisting moment in a truss is created by, usually, a compression force in the top chord coupled with a tension force in the bottom chord. If the distance d between the chords is small, the forces P must be large in order to build up sufficient resisting moment, since $M = Pd$ for a moment couple. Large forces will lead to large members, since the cross-sectional area A of the member is essentially found by using the formula $A = P/F$ or one of its variant forms. On the other hand, if the chords are far apart, the forces will get quite small and the chords will become quite puny, with a resulting wobbly truss. A good balance of depth versus size will produce an economical use of material. The economical ratio of depth of truss to length of span is usually about $\frac{1}{8}$ to $\frac{1}{10}$ for flat trusses and $\frac{1}{6}$ to $\frac{1}{8}$ for bowstring trusses. In our case, let us choose a depth-to-span ratio of $\frac{1}{9}$. A 45-ft span will yield a maximum depth of 5'-0". The maximum depth should occur at the point of maximum moment, the center of the truss. In our analysis we will take this depth to be the distance between the centerlines of the chords, so that the overall depth of the truss will be greater than 5'-0".

Attention must also be given to the number of panels. Several matters of economy are involved here. Increasing the number of panels increases the number of joints, and each joint is expensive to fabricate; but decreasing the number flattens the slope of the diagonal web members and increases their stress, and thus their size and cost. With the number of panels we are using, the slope of the diagonals is about 30° with the horizontal. This is about minimum for good design; the ideal angle will be between 30 and 45° from the horizontal.

LOADING

The estimated dead load weight of a parallel-chord truss is given by

$$w = 0.043L + 1.75 \qquad \text{flat truss weight}$$

where w is the weight in pounds per square foot of horizontal surface, and L is the span in feet. In our case this yields

$$w = (0.043 \times 45) + 1.75$$
$$= 3.7 \text{ psf}$$

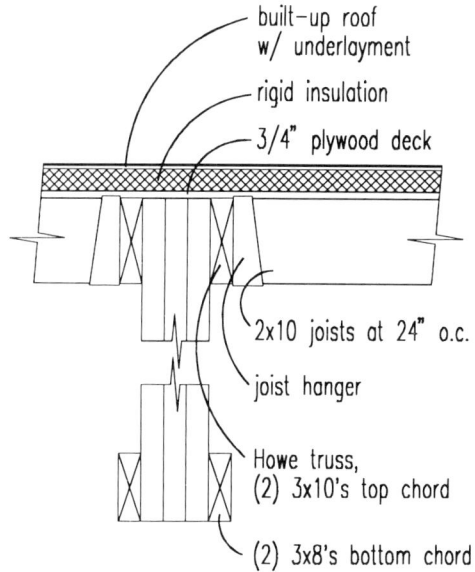

FIGURE 5.12 Cross-section of roof.

The rest of the dead load can be found by looking up the weights of the materials shown in Figure 5.12. The source for weights of materials used here is the American Institute of Timber Construction's *Timber Construction Manual*.[7]

235-lb built-up roof
and 15-lb felt underlayment
$2\frac{1}{2}$-in.-thick rigid insulation
($R = 7.14 \times 2.5$ in. $= 17.75$)
$\frac{3}{4}$-in. plywood
Trusses
2×10 joists at 24 in. o.c.
Allowance for mechanical equipment, sprinklers, lights
Subtotal, dead
Snow
Total, dead + snow

It can be seen that the weight of the truss is only a small part of the total load carried, so that a considerable percentage

[7]American Institute of Timber Construction, *Timber Construction Manual*, Wiley, New York, 1974.

of error in the truss weight will not materially affect the total load on the truss. Following the final design of this truss, the weight of the truss was estimated to be 3.38 psf, a difference of less than 1% in the overall load.

Snow load only will be considered in the design of the truss in this example, except, of course, that the dead-load weight of the structure itself must always be included. With snow load all allowable stresses (except compression perpendicular to grain) may be increased by a duration factor of 15%.

Wind produces a suction force on flat or low-pitched roofs in the amount of 0.7 times the lateral pressure. For a typical design wind speed of 90 mph, the lateral pressure is found from the formula

$P = 0.00256 \times \text{velocity}^2 = 0.00256 \times 90^2$

$= 20.7 \text{ psf}$

The resulting uplift is 0.7×20.7 psf, or 14.5 psf. The standard building codes require that the dead load must equal 1.5 times the uplift load or else mechanical

250 lb/100 ft²	2.5 psf
2 lb/ft³	0.5 psf
3 psf/in. thickness	2.25 psf
3.7 psf	3.7 psf
4.75 lb/lin. ft	2.5 psf
	3.25 psf
	14.7 psf
	30 psf
	44.7 psf

fasteners must be designed to make up the difference. In the example, fasteners will be needed to carry the excess load of

$1.5 \times 14.5 \text{ psf} - 14.7 \text{ psf} = 7 \text{ psf}$

These fasteners will be considered at the end of the design examples.

SPECIES AND GRADE

The span of 45 ft is 50% larger than the residential truss span of 30 ft designed previously, and the spacing of the trusses is $7\frac{1}{2}$ times as much. Hence we can expect much larger loads on the truss and much larger forces in the members. Therefore, let us select a slightly higher grade of lumber, say Select Structural, and assume Douglas Fir–Larch is readily available. This gives

$$F_b = 1.1 \times 1450 = 1595 \text{ psi}$$

$$F_t = 1.1 \times 1000 \text{ psi} = 1100 \text{ psi}$$

$$F_c = 1.0 \times 1700 \text{ psi} = 1700 \text{ psi}$$

$$E = 1,900,000 \text{ psi}$$

all of which assumes that all members will be 2 to 4 in. wide and 10 in. in depth.

Because we are dealing with snow load, these allowable stresses, but not the modulus of elasticity, are increased by 15%.

$$F_b = 1500 \times 1.15 = 1834 \text{ psi}$$

$$F_t = 1100 \times 1.15 = 1265 \text{ psi}$$

$$F_c = 1700 \times 1.15 = 1955 \text{ psi}$$

$$E = 1,900,000 \text{ psi}$$

SIZE OF MEMBERS

The members will be sized using the interaction formula, which combines the effect of axial forces and bending forces. The axial forces are found from the classical graphic analysis, which is presented in Figure 5.11.

TOP CHORD

For the top chord the moments are found by considering the chord as a beam spanning between the web supports. If we assume that the beam is continuous over a series of supports, the maximum moment will be a negative moment over the webs given by the formula

$$\text{bending moment } M = \frac{wL^2}{8}$$

where $w = 44.7$ psf $\times$ 15-ft spacing $= 670.5$ plf.

$$M = \frac{670.5 \times 7.5 \times 7.5}{8} = 4710 \text{ ft-lb}$$

Since the top chord is made of two pieces, it is a spaced column. We need to check the requirements of spaced columns. The total length of the top chord is 22′-6″. Spacing between blocking is 7′-6″. Consider a 3×8 chord to check the slenderness ratios.

$$\frac{L_1}{d_1} = \frac{22.5 \times 12}{2.5} = 108 > 80 \text{ No Good}$$

$$\frac{L_2}{d_2} = \frac{22.5 \times 12}{7.25} = 37.5 < 50. \text{ OK.}$$

$$\frac{L_3}{d_3} = \frac{7.5 \times 12}{3.5} = 25.7 < 40. \text{ OK.}$$

However, since the chord is braced at two foot intervals by the roof purlins, we can take L_1 as 24 in. Assume 3-in. nominal lumber; then $d = 2.5$ in.

$$L_1 = 2 \text{ ft (lateral support from joists)}$$

$$\frac{L_1}{d_1} = \frac{2 \times 12}{2.5} = 9.6 \qquad 9.6 < 11$$

Short, assume that $C_p = 1.0$.

As a trial size, consider two 3 × 8's. Use the simplified version of the interaction formula to test the feasibility of this trial. For two 3 × 8's,

$$A = 2 \times 18.12 = 36.2 \text{ in}^2$$

$$S = 2 \times 21.9 = 43.8 \text{ in}^3$$

$$\frac{P/A}{F_c} + \frac{M/S}{F_b} = \frac{33,950/36.25}{1955}$$

$$+ \frac{4710 \times 12/43.8}{1834}$$

$$= 0.48 + 0.70$$

$$= 1.18 > 1.0$$

The selection is too small; try again with a larger size.

Try two 3 × 10's:

$$A = 2 \times 23.12 = 46.25 \text{ in}^2$$

$$S = 2 \times 35.65 = 71.3 \text{ in}^3$$

Again for a quick check, we will use the simplified interaction formula.

$$\frac{P/A}{F_c} + \frac{M/S}{F_b} = \frac{33,950/46.25}{1955}$$

$$+ \frac{4710 \times 12/71.3}{1834}$$

$$= 0.375 + 0.43$$

$$= 0.805 < 1.0 \quad \text{OK}$$

Since the two 3 × 10's pass the trial, use the more accurate form of the interaction equation with appropriate C_L, C_P, and moment magnification factors. For axial compression, check L/d for the major axis.

$$L = 7'\text{-}6'' = 90 \text{ in.} \quad d = 9.25 \text{ in.}$$

$$\frac{L}{d} = \frac{90}{9.25} = 9.73 < 11$$

$$F_{cE} = \frac{0.3 K_x E}{L/d^2} = \frac{0.3 \times 2.5 \times 1,900,000}{9.73^2}$$

$$= 15051$$

$$\frac{F_{cE}}{F_c^*} = \frac{15051}{1955} = 7.7$$

$$C_P = 0.972$$

$$f_c = \frac{33,950}{46.25} = 734.0$$

$$\frac{f_c}{F_c^* \times C_P} = \frac{734.0}{1955 \times 0.972} = 0.386$$

For bending, L_u, the unrestrained length, is the spacing between the roof rafters, 24 in. $L = 24$ in. Assume that $C_L = 1.0$. Using the interaction formula, then,

$$\left(\frac{P/A}{F_c'}\right)^2 + \frac{M/S}{F_b'(1 - f_c/F_{cE})} \leq 1.0$$

$$\left(\frac{734}{1955 \times 0.921}\right)^2 + \frac{4710 \times 12/71.33}{1834(1 - 734/15051)}$$

$$= (0.386)^2 + \frac{792.7}{1744}$$

$$= 0.149 + 0.454 = 0.603 \quad \text{OK}$$

Use two 3 × 10's for the entire top chord.

BOTTOM CHORD

There is no ceiling, so we may assume that there are no loads supported directly by the bottom chord. Consequently, there is no moment; the chord is loaded in pure tension. The maximum tension is

$$T = 34,000 \text{ lb}$$

$$A_{\text{req'd}} = \frac{T}{F_t}$$

$$= \frac{34,000}{1265} = 26.9 \text{ in}^2$$

A 3×8 has an area of $2.5 \times 7.25 = 18.1 \text{ in}^2$. Two will suffice.

Use two 3×8's for the entire bottom chord.

END DIAGONAL

This member is a spaced column loaded only in compression. For those unfamiliar with spaced column design, refer to Example 3.6. Design all the diagonals for the highest force:

axial compression

$$P = 25,600 \text{ lb}$$

$$L = \sqrt{7.5^2 + 4.33^2} = 8.66 \text{ ft} = 104 \text{ in.}$$

$$d_1 = 2.5 \text{ in.} \quad \text{(assume 3-in. lumber)}$$

$$\frac{L}{d} = \frac{104}{2.5} = 41.6 < 80 \quad \text{OK}$$

To assume end condition a, the centerline of the split ring must be within a distance of $L_1/20$ from the end of the diagonal.

$$\frac{L_1}{20} = \frac{104}{20} = 5.2 \text{ in.}$$

This seems to be manageable. Once the connections are designed and drawn to scale, this assumption can be checked.

$$K_X = 2.5$$

$$F_{cE} = \frac{K_{cE} K_X E}{(L_e/d)^2}$$

$$= \frac{0.3 \times 2.5 \times 1,900,000}{41.6^2} = 823.4 \text{ psi}$$

$$F_c^* = 1955 \text{ psi}$$

$$\frac{F_{cE}}{F_c^*} = \frac{823.4}{1955} = 0.421$$

$$C_P = 0.375$$

$$F_c' = 0.375 \times 1955 = 735 \text{ psi}$$

$$A_{\text{req'd}} = \frac{P}{F_c'} = \frac{25,600}{735} = 34.8 \text{ in}^2$$

Use two 3×8's with an area $= 2 \times 18.1 = 36.2 \text{ in}^2$.

As two final checks for slenderness of the spaced column, check

$$\frac{L_2}{d_2} = \frac{104}{7.25} = 14.3 < 50 \quad \text{OK}$$

$$\frac{L_3}{d_1} < 40 \quad \text{so } L_3 < 40 \times 2.5 \text{ in.}$$

$$= 100 \text{ in.}$$

L_3 is the distance between required between intermediate blocking. Since it is greater than the center-to-center spacing of the end connections, no intermediate blocking is needed.

OTHER WEB MEMBERS

Use 3-in. lumber to avoid filler blocks. Use widths necessary for edge distance and spacing for connectors.

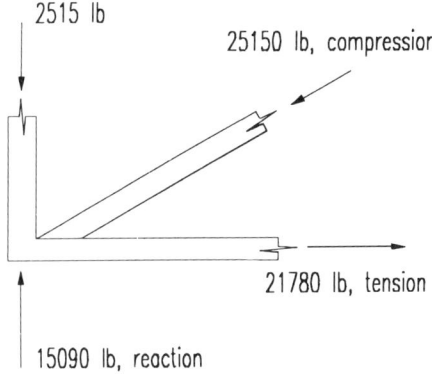

FIGURE 5.13 Heel force diagram—3 × 8 vertical.

DESIGN OF JOINTS

A fuller discussion of connection design is found in Chapter 4. If the designer is unfamiliar with the principles of connection design, referring to the design of split rings may be advisable at this point. In the discussion that follows we concentrate on explaining the most critical aspects of this design, which are the requirements for end and edge distances and for spacing between connectors.

Let us begin the connection design with the hardest connection, that at the heel joint (Figure 5.13). The load from the chord to the diagonal is 21,780 lb at 30° to the grain. We will use 4-in. split rings. Let us specify that the material must be *dense* Douglas Fir–Larch rather than nondense. This is done to improve the capacity of the connections since DF-L (dense) is in group A for split rings rather than group B for DF-L (nondense).

First consider the connection from the bottom chord to the diagonal. This connection carries the tension force of 21,780 lb. We can assume that the capacity of the split ring in the diagonal will be more critical since it is loaded at an angle to grain. The value of one 4-in. connector in two faces of $2\frac{1}{2}$-in. material is 5830 lb parallel to grain, and 4050 lb perpendicular to grain. Since our load is at 30° to the grain, neither parallel nor perpendicular, neither of these two values applies directly. They must be combined using Hankinson's formula. The strength of one 4-in. connector at 30° is determined as

$$P_\alpha = \frac{P \times Q}{P \times \sin^2 \alpha + Q \times \cos^2 \alpha}$$

$$P_{30°} = \frac{5830 \times 4050}{5830 \times \sin^2(30°) + 4050 \times \cos^2(30°)}$$

$$= 5253 \text{ lb}$$

Increase for snow load: $5253 \times 1.15 = 6041$ lb

Number of rings required: $\dfrac{21{,}780}{6041} = 3.6$

If four split rings are used, the level of stress in each will be 90%.

Use four split rings of 4-in. diameter on $\frac{3}{4}$-in.-diameter through bolts.

Since the diagonals are doubled and the chords are doubled, we will use two rings on each side of the truss, for a total of four. See the section drawing in Figure 5.14a.

To complete the design, a large-scale, carefully constructed drawing on the connection is essential. Invariably in the design of split-ring connectors, the critical concern is having large enough timber so that the required spacings and end and edge distances can be met. Figure 5.14b shows the accommodations necessary to meet the spacing requirements. This figure should be referred to for help in understanding the following discussion.

If the split rings were loaded to their maximum capacity, the full required spacings and end and edge distances would be required. For rings loaded less than 100%, a linear interpolation between the full and

192 TRUSSES

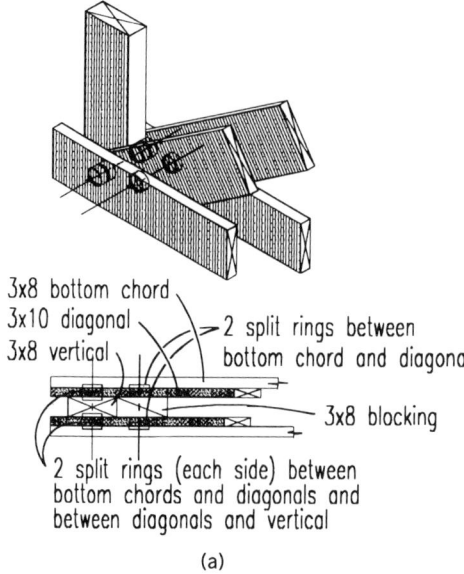

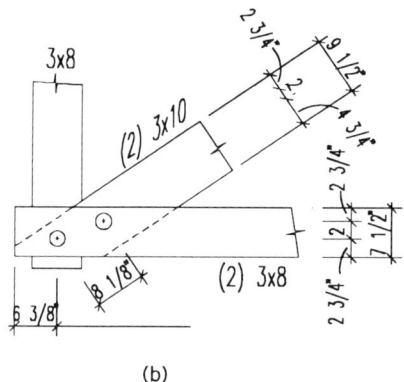

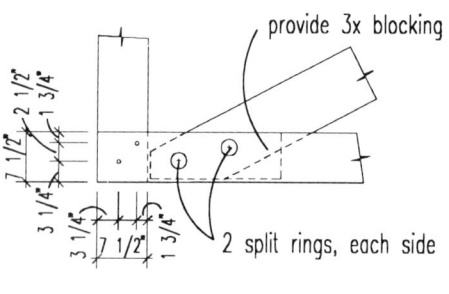

FIGURE 5.14 Heavy truss—heel connection: (a) members and connectors, (b) required dimensions—(2) 1" through bolts, and (c) alternate connection with bolts in vertical.

reduced distances may be taken. Since the spacings are so critical, we will need to take any reduction that is offered.

In the example, each ring will be loaded to a fraction of its allowable load, namely,

$$\frac{3.6}{4} = 90\%$$

Table 10.3 in the *National Design Specification* lists the minimum spacings for full design load and for reduced design load. We will use these distances to establish the connection detail in the following order.

First, lay out the 3×8 bottom chord. It is loaded parallel to grain. The required edge distances are $2\frac{3}{4}$ in. for both top and bottom sides. There are no reductions for reduced load. Lay out these lines on the bottom chord. The resulting area between them, which is 2 in. deep, represents the area in which connectors may be placed. We will assume that one connector will be centered on the top line and one on the bottom, leaving a 2-in. gage between them.

Second, find the required end distance for the bottom chord in tension. The full load distance is 7 in.; the reduced load distance is $3\frac{1}{2}$ in. For 90% of full load the minimum is 6.3 in. Lay out the end distance as $6\frac{3}{8}$ in. The spacing between connectors loaded parallel to the grain is 5 in. for reduced load and 9 in. for full load. For 90% load the required spacing is 8.1 in. This length is laid out as $8\frac{1}{8}$ in., but note that it is measured along the line from center to center of the connectors, not along the horizontal.

Third, place the split rings as shown in the diagram. The gage between them, measured perpendicular to the axis of the diagonal strut, is approximately 2 in. The required edge distance for the diagonal on the unloaded side (top side) is $2\frac{3}{4}$ in. There is no reduction. The required edge distance to the loaded side (bottom) of the diagonal strut is $3\frac{3}{4}$ in. for full load, $2\frac{3}{4}$ in.

for 83% load, and $3\frac{3}{8}$ in. for 90% load. Add these distance $2\frac{3}{4} + 2 + 3\frac{3}{8} = 8\frac{1}{8}$ in. $>$ $7\frac{1}{2}$ in., the width of the 3×8 diagonal. A wider diagonal is required.

Use two 3×10's for diagonal strut to accommodate split rings.

Check the spacing along the diagonal and the end distance as determined in Table 5.5. These are fine, so lay out the diagonal so that the end and edge distances are satisfied.

Finally, lay out the vertical strut so that its centerline coincides with the leftmost split ring. The required distances for the bottom chord, compared with the actual distances provided, are given in Table 5.6.

The vertical strut must transfer 2515 lb in compression. It is connected with two split rings, one on each face. The total capacity of the connectors is

$$P = 2 \times 5830 \times 1.15 = 13{,}409 \text{ lb}$$

The rings are loaded at 19% of their capacity, so reduced end and edge distances will prevail. The minimum end distance required is $3\frac{1}{2}$ in. The bottom of the vertical must extend below the bottom chord to accommodate this requirement.

TABLE 5.5 SPLIT-RING REQUIRED DISTANCES: DIAGONAL WEB

	4-in. Split-Ring Connector, Parallel to Grain Loading		
	Minimum for Reduced Design Value (in.)	Minimum for Full Design Value (in.)	Interpolation for 90% of Full Design Value
Edge distance	2.75 1.00	2.75 1.00	2.75
End distance, compression	3.5 0.625	7 1.00	6.0 (0.90)
Spacing perpendicular to grain	5 1.00	5 1.00	5

Note: Values shown below the required distance are the proportion of full design load allowed with reduced spacings and distances.

TABLE 5.6 SPLIT-RING REQUIRED DISTANCES: BOTTOM CHORD

	4-in. Split-Ring Connector, Parallel to Grain Loading		
	Minimum for Full Design Value	Actual Distances	OK
Edge distance	2.75 1.0	4.625	Yes
End distance, tension	7 1.00	6.375 0.90	Yes
Spacing parallel to grain	9 1.0	8.125 0.90	Yes

Note: Values shown below the required distance are the proportion of full design load allowed with reduced spacings and distances.

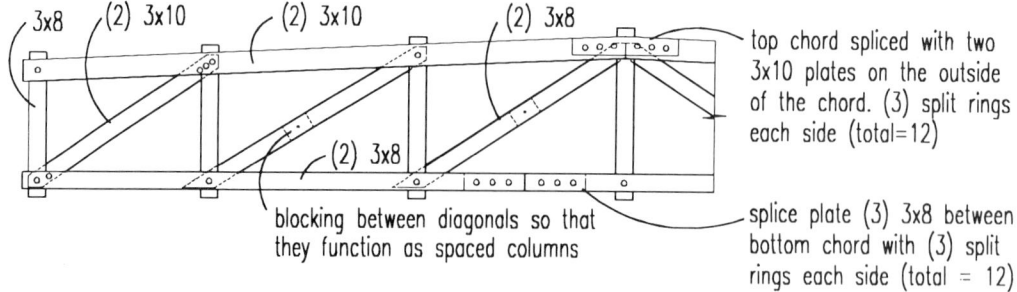

FIGURE 5.15 Heavy truss—final design.

A simpler approach is merely to move the diagonals to the right, avoiding the intersection of the vertical altogether. The forces in the vertical are small enough to be carried by two 1-in.-diameter through bolts. This option, which seems preferable, is shown in Figure 5.14c. In addition to resolving the problem with the vertical strut, this design increases the end distance on the tension chord (always a major concern).

Figure 5.14 summarizes this design. The remaining connectors would be designed in a similar manner. The results of that design are shown in the drawing of the complete truss (Figure 5.15). This truss example is similar to the one in *Modern Timber Engineering* by W. Fleming Scofield.[8] Scofield also shows the results of the design using block and bolt connectors and steel tension rods. Although not commonly practiced now, this construction method may be of interest to architects because of its lighter aesthetic qualities.

[8] W. Fleming Scofield, *Modern Timber Engineering*, 4th Edition, Southern Pine Association, New Orleans, LA, 1954.

6

GLUED-LAMINATED ARCHES

Glued-laminated timbers (glulams), which are used extensively as beams and posts, are also popular in the construction of large arched structures. Some of our larger clear-span structures are made of glue-laminated beams. The largest timber arch yet designed is for an asbestos storage facility at Deception Bay in northern Quebec. Glulam arches 144 ft high span 305 ft from side to side. The arches are built in an I cross section 78 in. deep and are spaced 20 ft apart. The building was completed in 1977. Typical spans of 60 to 80 ft are commonplace for glue-laminated arches. These structures are successful for a variety of reasons.

First, economy is achieved when stresses are axial rather than bending. The arch shape can be selected to eliminate bending stresses in the timber. If the centerline of an arch exactly follows the thrust line developed by the forces, the arch will be in pure compression. No bending will be developed. If the centerline deviates from the thrust line there will be bending and the magnitude of that bending force (moment) will be the deviation (or eccentricity) times the thrust. The thrust line is dependent on the applied loads. Since any structure must be able to withstand a variety of loads from different directions, there will be times when the centerline of the arch will not align with the load thrust line. Under these conditions, arches act in a combination of bending and compression. To the degree that the centerline of the arch can approximate the line of thrust, the magnitude of bending forces will be decreased.

Second, unlike masonry, wood can develop tension and compression. True arch action, such as that seen in masonry arches, develops only compression across the full cross section of the member. Masonry arches, which can sustain very little tension, are restricted in shape: the centerline of the arch must be built so as to closely approximate the thrust line arising from the force. Since wood can carry approximately equal stresses in compression and tension, this restriction does not apply. Wood arches can take a variety of shapes, depending on the function of the space or the desires of the architect.

Wood has other properties that make glulam arch structures feasible. Wood will char in a fire and form an insulating layer that protects the interior of the piece. The larger the piece of wood, the more effective the protection. When exposed to extreme heat, steel will change properties from a stiff, elastic material to a flexible, plastic one. For buildings with public assembly, large pieces of steel must be protected with fireproofing. The aesthetic qualities of wood are therefore used to great advantage in large arch structures by being able to remain exposed.

A commonly used arrangement for wood arches is the three-hinged arch. There are a vast number of possible shapes that can be developed using this principle, as shown in Figure 6.1. One popular form utilizes vertical legs and slanted rafters joined at a central ridge. This profile provides a useful and attractive space for many activities. Adjacent structures can easily be integrated into its rectilinear plan and elevation. It is easily constructed and can readily be sheathed with standard materials. Finally, due to its fabrication as two separate pieces, it can be much more easily shipped than can an equivalent frame or two-hinge arch.

From an engineering point of view, the three-hinge arch is statically determinate. This means that not only is the solution direct, but it is also not greatly dependent

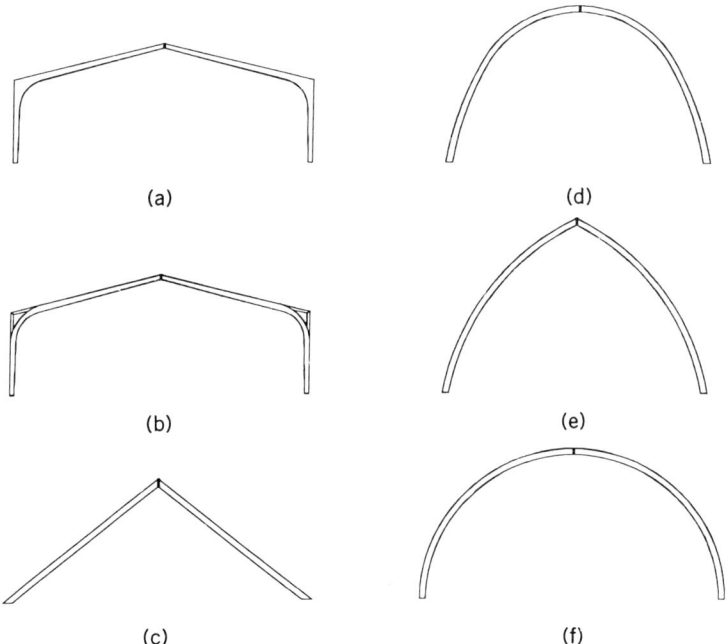

FIGURE 6.1 Timber arch shapes: (a) Tudor, (b) Tudor with outlooker, (c) A-Frame; (d) parabolic, (e) Gothic, and (f) radial.

on assumptions about the shape or stiffness of the arch. With moderate practice a designer can make reasonable estimates of shape, size, and weight that will be commensurate with the final design. Let us proceed to analyze and design a three-hinged arch and use it to explore some of these points.

EXAMPLE 6.1

Consider the three-hinged arch shown in Figure 6.2, which has a span of 64 ft. It has an eave height on the outside of 18 ft and a roof pitch of 3 on 12. Its overall height, or rise, is thus 18 ft plus 8 ft, or 26 ft.

The initial stage of the design is the same as that for a beam: approximate loads and spacings are determined and an estimate of the size and weight of the arch are made. The spacing of the arches is determined mainly by the allowable span of the framing above the arch. We will assume that the designer has chosen an exposed wood purlin system to frame between the arches. Above the purlins will be a tongue-and-groove wood deck that will span between purlins, form the diaphragm, and support the roof insulation and roofing. The dead load of this assembly can be calculated as soon as all of the components are known. Assume that it will be in the range of 15 psf. Assume further that the snow load is 30 psf and the wind load is 20 psf.

The purlin spacing of 8 ft on center is measured along the horizontal. Because of the roof pitch, the spacing when measured in the plane of the roof becomes

$$\text{spacing}_{\text{slant}} = \text{spacing}_{\text{horiz}} \times \frac{1}{\cos \alpha}$$

$$= \frac{8 \text{ ft}}{\cos(14°)} = 8.246 \text{ ft}$$

A review of deck design tables shows that a nominal 2-in.-thick deck can span 8 ft with a load of 45 psf. By inspection this deck is satisfactory for a span of 8'-3".

Let us assume that the arches are to be spaced at 16 ft on center. A purlin design for 6×12 purlins at 8'-3" on center will be satisfactory for Southern Pine, No. 1 grade ($F_b = 1450$ psi and $E = 1,000,000$ psi).

We can now establish dead loads as follows:

Built-up roofing	4.00 psf
2-in.-nominal deck	3.25 psf
6×12 purlins at 8 ft o.c.	
$(36 \text{ pcf})\left(\dfrac{5.5 \text{ in.} \times 11.5 \text{ in.}}{144 \text{ in}^2/\text{ft}^2 \times 8 \text{ ft}}\right)$	
$= 1.98$ psf	2.00 psf
Mechanical work, fans, lights	3.00 psf
Subtotal dead load	12.25 psf

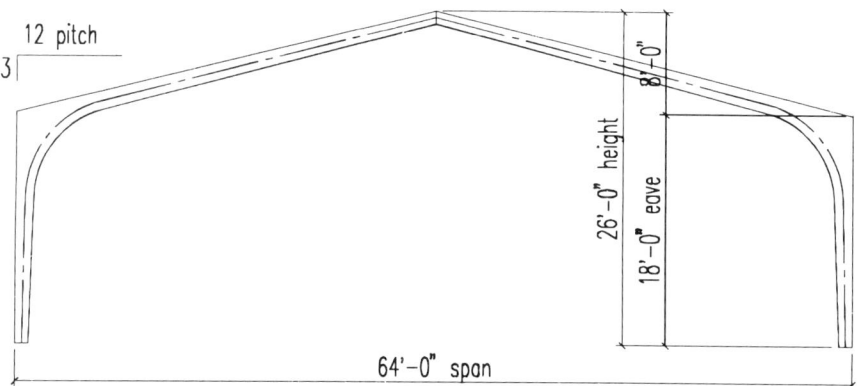

FIGURE 6.2 Three-hinged arch—Example 6.1.

The total dead load, in plf, is

12.25 plf × 16-ft spacing = 196 plf

Estimate that the size of the arch will be approximately 8 in. wide by 24 in. deep, on average. This will yield a self-weight of

$$\frac{8 \text{ in.} \times 24 \text{ in.}}{144 \text{ in}^2/\text{ft}^2} \times (36 \text{ pcf}) \text{ plf} = 48 \text{ plf}$$

The total dead load is 196 plf + 48 plf = 244 plf.

Due to the flatness of the roof there are no snow-load reductions for slope, so the design snow load remains 30 psf. The total snow load is

snow load = 30 psf × 16-ft spacing = 480 plf

We will consider a uniform snow equally placed on both sides of the roof as well as an asymmetric snow load resulting from the snow blowing from one side to the other. The pattern of the asymmetric snow, as specified by code, is one side clear and the other side loaded by 1.25 times the symmetric snow:

asymmetric snow load = 480 plf × 1.25

= 600 plf

The wind load will produce a positive pressure on the windward side and a suction on the leeward wall and across the entire roof. Building codes specify pressure coefficients depending on roof slope and building configuration. A simple approximation is to assume a positive coefficient on the wind ward wall of 0.8 and a negative, suction, coefficient of 0.5 on the leeward wall. For the roof, assume a negative, suction, coefficient of 0.7.

The total wind load from the walls is

20 psf × 1.3 × 16 ft = 416 plf

where the factor 1.3 represents the sum of 0.8 *plus* 0.5. They are added since they operate in the same direction.

The roof wind load is

20 psf × 0.7 × 16 ft

= 224 plf, acting upward

It is very rare that the wind loads will affect the design of a low-pitched roof. The upward force of the wind blowing across the roof will have a tendency to lift up the windward edge of the structure. This effect will probably influence the connections at the base more than it will affect the member design. The dead weight of the structure counteracts the uplift and therefore the combined load is usually relatively small. In this analysis we consider the horizontal component of the wind separately and include the vertical uplift only at the end to check overall uplift of the entire frame.

We can now proceed with the analysis and design of this structure. Normally, we consider three load cases:

Case I: dead + symmetric snow
Case II: dead + asymmetric snow
Case III: dead + wind

Let us start with preliminary design to get some approximate sizes and verify some of our basic assumptions. Because the structure is statically determinate, we can calculate the reactions for each case by referring to the free-body diagrams in Figure 6.3. The vertical and horizontal reactions, found by using the following equations, are summarized in Table 6.1 and in Figure 6.3*b* and *c*.

For a symmetrically applied vertical load, w,

$$V_L = V_R = \frac{wL}{2} \qquad H_L = H_R = \frac{wL^2}{8r}$$

For a uniform vertical load applied to the

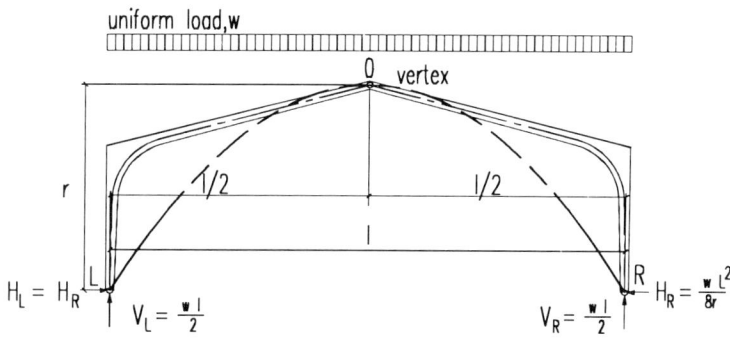

CASE I—Symmetric load

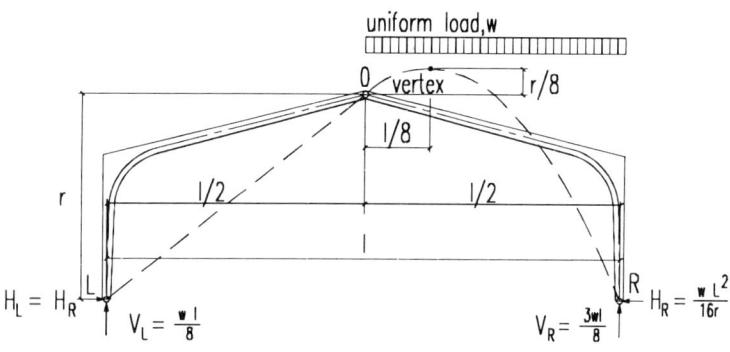

CASE II—Asymmetric load

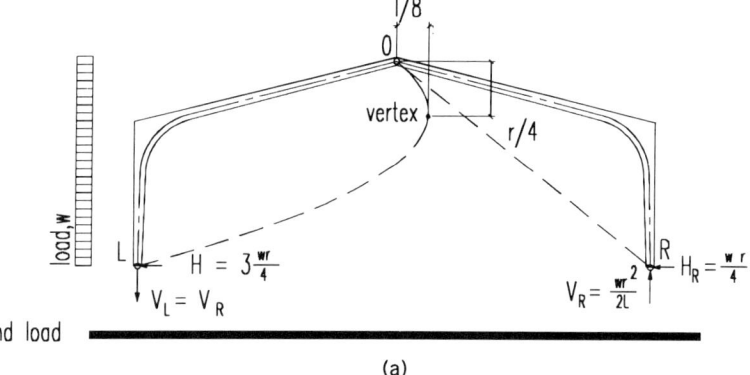

CASE III—Lateral wind load

(a)

FIGURE 6.3 Three-hinged arch—reactions: (a) reaction formulas, (b) reactions under applied loads, and (c) reactions under combined loads.

200 GLUED-LAMINATED ARCHES

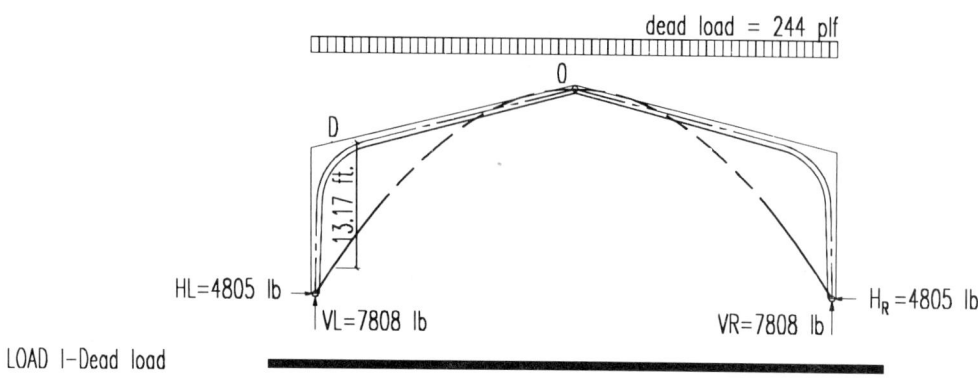

LOAD I—Dead load

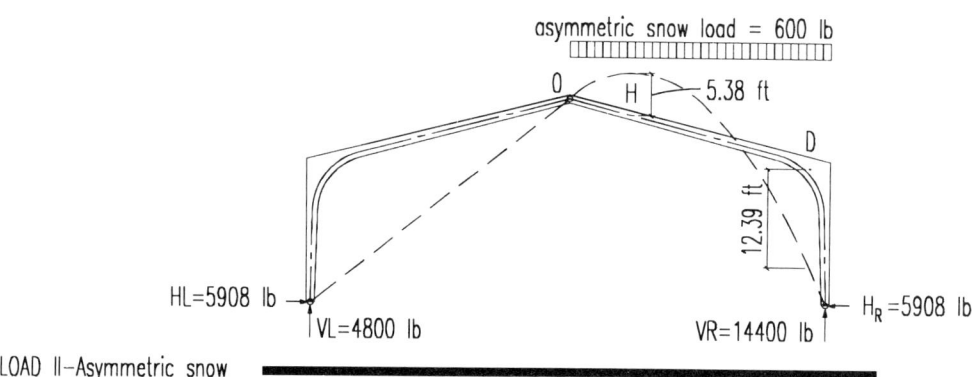

LOAD II—Asymmetric snow

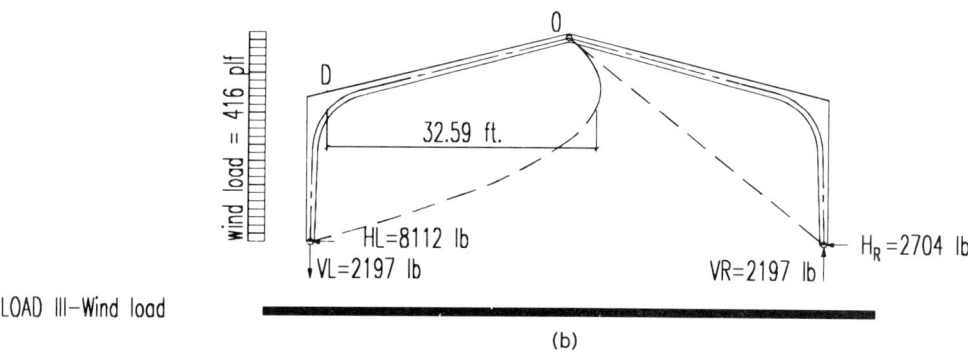

LOAD III—Wind load

(b)

FIGURE 6.3 (*Continued*)

EXAMPLE   **201**

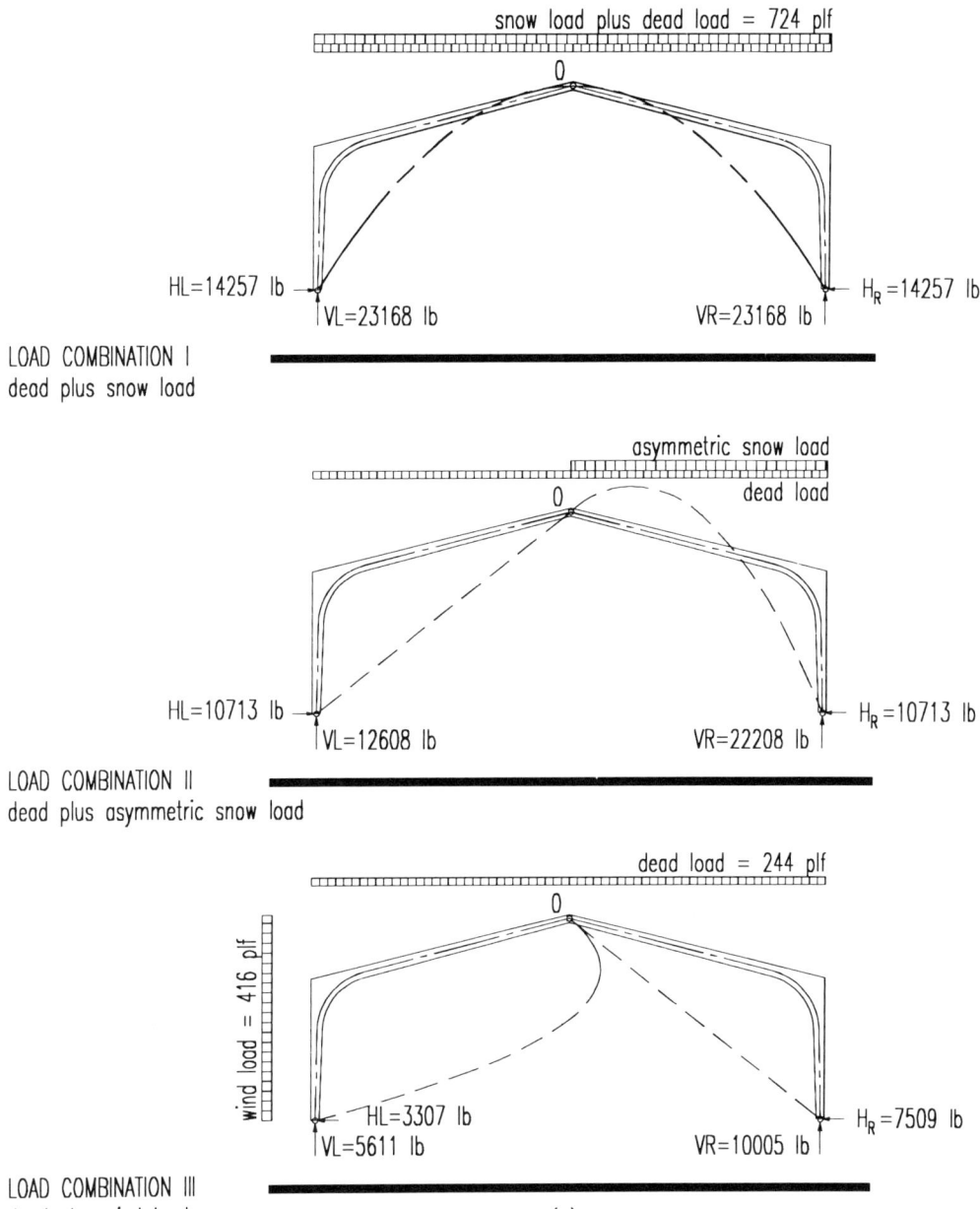

FIGURE 6.3 (*Continued*)

202 GLUED-LAMINATED ARCHES

TABLE 6.1 ARCH EXAMPLE: REACTIONS

Load Case	Applied Load	Vertical Reactions		Horizontal Reactions	
		Left	Right	Left	Right
Dead	244	7,808	7,808	4,805	4,805
Symmetrical snow	480	15,360	15,360	9,452	9,452
Asymmetrical snow	600	4,800	14,400	5,908	5,908
Horizontal wind	416	−2,197	2,197	−8,112	−2,704
Vertical wind	−224	−7,168	−7,168	−4,411	−4,411
Combination I Dead + symmetrical snow		23,168	23,168	14,257	14,257
Combination II Dead + asymmetrical snow		12,608	22,208	10,713	10,713
Combination III Dead + horizontal wind		5,611	10,005	−3,307	7,509
Maximum		23,168	23,168	14,257	14,257
Controlling combination		Dead + symmetrical snow	Dead + symmetrical snow	Dead + symmetrical snow	Dead + symmetrical snow

right half of the arch,

$$V_L = \frac{wL}{8} \qquad H_L = \frac{wL^2}{16r}$$

$$V_R = \frac{3wL}{8} \qquad H_R = \frac{wL^2}{16r}$$

For a lateral load from the left applied over the height, r,

$$V_L = \frac{wr^2}{2L} \text{ acting downward} \qquad H_L = \frac{3wr}{4} \text{ to left}$$

$$V_R = \frac{wr^2}{2L} \text{ acting upward} \qquad H_R = \frac{wr}{4} \text{ to left}$$

PRELIMINARY DESIGN

The largest reactions at each end are due to the symmetrical snow loads plus the dead loads. We can use these to establish the basic dimensions of the arch by sizing the base to carry shear. We will use simple rules of thumb for all other sizes.

Select the glulam species and combination. Choose Southern Pine, 24F-V5, which has the properties noted below. All of the design values are found in NDS Table 5A in the section marked "Bending about $X - X$ Axis" except for compression parallel to grain, F_c which is taken from the section marked "Axially Loaded." This combination was chosen so that the glulam has the same fiber bending strength whether subject to positive or negative bending. One of our checks at the end of the design will be to determine if this precaution is necessary.

F_b (tension on tension face) = 2400 psi
F_b (tension on comp. face) = 2400 psi
F_v (shear) = 200 psi
F_c (compression parallel to grain) = 1700 psi
$F_{c\perp}$ (compression perpendicular to grain) = 750 psi
E (modulus of elasticity) = 1.7×10^6 psi

Size the base for shear, $H_R = 14{,}257$ lb, using the duration factor for snow, 1.15.

$$A_{\text{req'd}} = \frac{3V}{2F_v} = \frac{1.5 \times 14{,}257 \text{ lb}}{200 \text{ psi} \times 1.15} = 93 \text{ in}^2$$

As a rule, assume that the depth at the base is about 1.5 times the width, so

$$A = b \times 1.5b = 93 \text{ in}^2$$
$$b = 7.87 \text{ in.}$$

Refer to Table A.1b for standard sizes of glulams. Use $8\frac{3}{4}$ in. wide. The minimum depth to meet the shear is

$$h = \frac{93 \text{ in}^2}{8.75} = 10.6 \text{ in.}$$

Use the next-highest depth available and select a base depth of 12 in. Glulams are made of $\frac{3}{4}$- or $1\frac{1}{2}$-in.-thick laminations, so the depth will be a multiple of $\frac{3}{4}$ or $1\frac{1}{2}$ in.

The depth of the beam at the ridge can be taken as 1.25 or 1.5 times the width (1.5×8.75 in. = 13.125 in.) We will assume that the depth is equal to 12 in. A common approximate for the depth at the tangent points, where the curve springs from the rafter and column, is 1.5 times the depth at the base. Use 18 in.

The NDS and AITC prescribe minimums for the radius of curvature, R, depending on the thickness of the laminations, t, and the species of the wood. The relationship given by the NDS, which is prescribed by code and cannot be exceeded, is

$\dfrac{t}{R}$ shall not exceed $\dfrac{1}{100}$ for hardwoods and Southern Pine.

$\dfrac{t}{R}$ shall not exceed $\dfrac{1}{125}$ for all other softwoods.

Recommended minimums for the radius of curvature according to the AITC, for a nominal 2-in. lamination (actual thickness is $1\frac{1}{2}$ in.),

- R shall not be less than 18 ft for Southern Pine.
- R shall not be less than 27 ft. 6 in. for any other species.

For a nominal 1-in. lamination (actual thickness, $\frac{3}{4}$ in.),

- R shall not be less than 7 ft. 0 in. for Southern Pine.
- R shall not be less than 9 ft. 4 in. for any other species.

Applying the NDS requirements, and assuming the standard lamination thickness of $\frac{3}{4}$ in., we obtain

$$R = 100 \times 0.75 \text{ in.} = 75 \text{ in.} = 6'\text{-}3''$$

This exceeds the AITC recommendation, so use $R = 84$ in.

We can now construct the drawing of the profile of the arch seen in Figure 6.4. Using this drawing, we derive the coordinates for nine points along the length of the arch. We create an idealized model as a curve going through the centerline of the arch. Another satisfactory approach would be to use the outer face of the arch. In the initial stage of the design the effort to find the centerlines may not be worthwhile since it is possible that the geometry of the arch will change based on the analysis. As with any engineering problem there is an acceptable degree of accuracy in determining the location of the members. The final solution should have an accuracy of 5 to 10%. The dimensions of the members should be at least this accurate.

In the tables below, notice that all coordinates are given in inches, but that moments are given in units of pound-feet.

204 GLUED-LAMINATED ARCHES

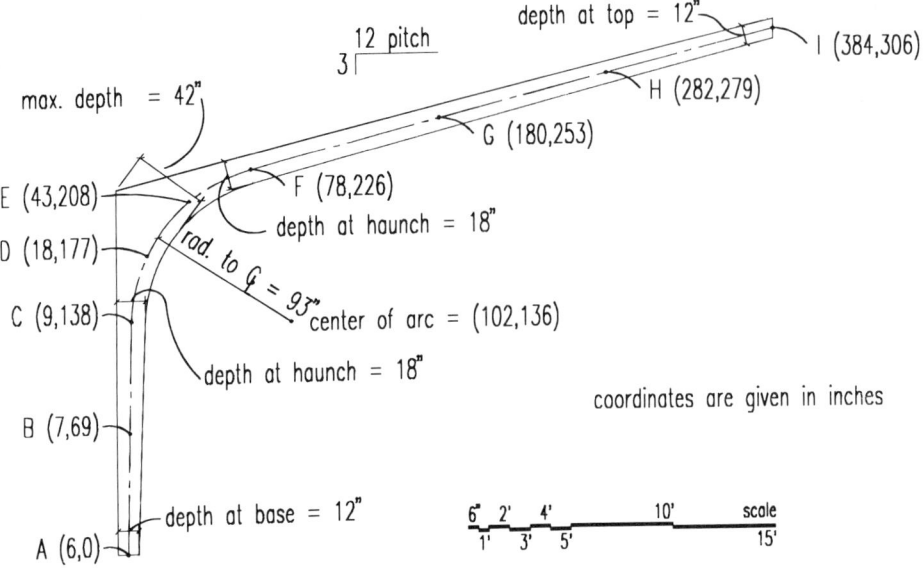

FIGURE 6.4 Three-hinged arch—coordinates.

GRAPHIC ANALYSIS OF THE ARCH

The analysis of the forces in the arch can be done most easily by using the graphic procedure outlined here. To design the arch we must calculate the moment and the axial force at a number of locations along the arch. The analysis for the moment is based on the observation that the moment in the arch at any point is equal to the thrust at that point multiplied by the distance between the arch's centerline and the line of thrust. The line of thrust is called the *equilibrium polygon*.

By way of explanation consider Figure 6.5, which shows the equilibrium polygon for load combination I, dead load plus snow load. The vectorial sum of the reactions at the left end is the force acting on the arch at that point. Proceeding toward the center of the arch, we note that there is an incremental decrease in the vertical component of the force, but the horizontal thrust remains constant. The direction of the force in the arch approaches the horizontal. The dashed line in the figure shows the direction of the force on the arch at any location. The force acts on the arch at an angle that is tangent to the polygon. The magnitude of the force is the vectorial sum of the two components. If the force was applied at the location of the equilibrium polygon, the arch would be in equilibrium due to the force only. Since it is actually applied at a distance away from the equilibrium polygon, a moment is produced to create equilibrium. The magnitude of that moment is the horizontal thrust times the vertical distance between the arch and the equilibrium polygon.

The axial force in the arch can be found as the vectorial sum of the horizontal and vertical forces. The graphic procedure to find the moment is simple and direct.

1. Draw the arch to scale at a sufficient size.
2. Draw over it the equilibrium polygon for each load condition. The equilibrium polygons are drawn to the same scale as the arch using the rules below.

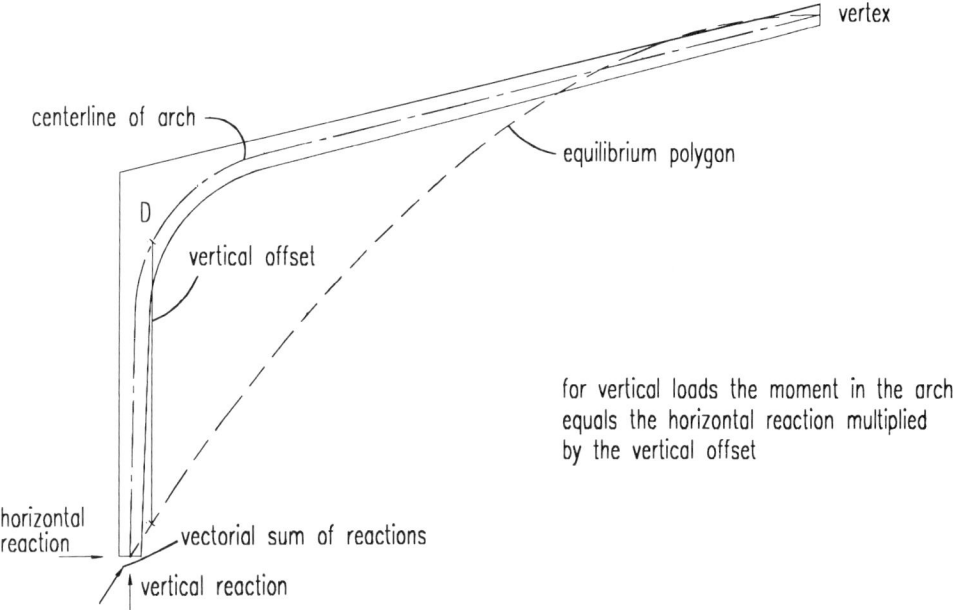

FIGURE 6.5 Equilibrium diagram—load combination I.

3a. For *vertical loads*: measure the vertical distance between the arch centerline and the equilibrium polygon and calculate the moment by

M = horizontal thrust
   × vertical distance between lines

3b. For *horizontal* loads such as wind loads: measure the horizontal distance between the arch centerline and the equilibrium polygon and calculate the moment by

M = vertical reaction
   × horizontal distance between lines

The equilibrium polygon takes a different form for each of the three types of loads that we have considered. These are shown in Figure 6.6. To draw these you will have to calculate a few points along the curve and then sketch in the curve between them. Increasing the number of points will increase the accuracy somewhat, but in reality we are most interested in the moment at critical points: (1) within the curved haunch; (2) at the tangent points, where the straight legs transit into the curve; and (3) near the middle of the rafter, where the positive bending is largest. In the example we have selected nine points to calculate the moment.

In sections where the equilibrium polygon is a parabola, proceed to draw the parabola as follows:

1. Find the vertex of the parabola; call this point (x_v, y_v).
2. For *vertical* loads, at each point along the arch find the *horizontal* distance to the point from the vertex. These points are selected by you. In the example we estimated the two tangent points. This resulted in three sections of the arch: the straight column leg from base to lower tangent, the curved section between tangents, and the straight beam from tangent

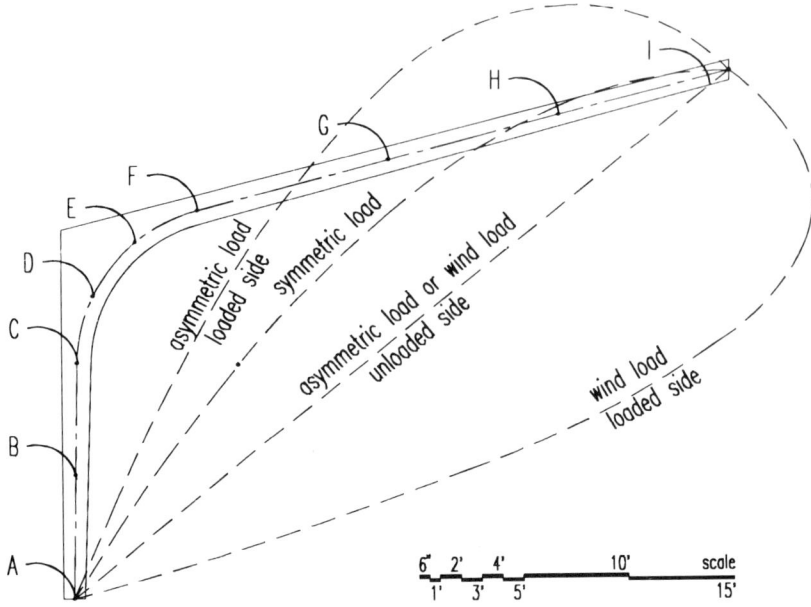

FIGURE 6.6 Equilibrium diagrams—all load conditions.

to ridge. Each section of the arch was divided into equal segments. We designated these points as x_i.

3. Find the *vertical* offset of the parabola by the formula

$$y_i = y_v \left(\frac{x_i}{x_v}\right)^2$$

In the example, refer to Table 6.2. Find the vertical offset of the parabola for point D. The vertex is located at the center of the arch. This point is given as

$$(x_v, y_v) = (384, 306) \text{ in inches,}$$
$$\text{or } (32, 25.5) \text{ in feet}$$

The horizontal distance from the vertex to point D is given in column 3 of the table:

$$x_i = 384 \text{ in.} - 18 \text{ in.} = 366 \text{ in.} = 30.5 \text{ ft}$$

The vertical offset is found from the formula

$$y_i = y_v \left(\frac{x_i}{x_v}\right)^2 = 25.5 \left(\frac{30.5}{31.5}\right)^2 = 23.92 \text{ ft}$$

This value is given in column 5. Please note that all measurements in the example are taken to the centerline of the arch. For this reason the term x_v in the denominator is taken as 31.5 ft, half the distance between the two base point centerlines. Using 32 ft, half the overall length of the arch, would have given sufficient accuracy.

Table 6.2 lists the vertical distance between the arch and the equilibrium polygon in column 6. This is found by measuring the scale drawing (Figure 6.6) or by subtracting column 4 from column 5. Positive values indicate that the arch is above the equilibrium polygon and the resultant moment will be positive (i.e., compression will be at the top and the curved haunch will tend to flatten out). Moments are given in columns 7, 8, and 9 for dead load, symmetrical snow load, and dead load plus snow load. The maximum moment for dead load plus snow load occurs at point D and has a magnitude of

$$M = 187{,}761 \text{ lb-ft (See Table 6.2)}$$

At this point it is advisable to check the depth of the cross section to see if we are in the right range before proceeding with any other calculations. Scale from Figure 6.4 to find the depth of the arch at point D. h scales to be about 42 in. The section modulus

$$S = \frac{bh^2}{6} = \frac{8.75 \times 42^2}{6} = 1143 \text{ in}^3$$

The allowable moment is approximately

$$M_{\text{allowable}} = F_b \times S \times C_f \times C_c$$

where

$$C_f = \text{size factor} = \left(\frac{12}{h}\right)^{1/9}$$

$$= \left(\frac{12}{42}\right)^{1/9} = 0.87$$

$$C_c = \text{curvature factor} = 1 - 2000\left(\frac{t}{R}\right)^2$$

$$= 1 - 2000\left(\frac{0.75}{84}\right)^2 = 0.84$$

$$M_{\text{allowable}} = F_b \times S \times C_f \times C_c$$
$$= (1.15 \times 0.87 \times 0.84 \times 2400 \text{ psi})$$
$$\times 1143 \text{ in}^3 = 2{,}306{,}112 \text{ lb-in.}$$
$$= 192{,}176 \text{ lb-ft, which is greater}$$
$$\text{than } 187{,}761 \text{ lb-ft}$$

The material is stressed to somewhat less than 98% of its allowable. This seems satisfactory; we will have to include the interaction of the compression force but feel that there is sufficient reserve to accommodate this. Further modifications may be made at the end of the problem, if needed, without effecting the geometry of the arch too much.

The next step is to calculate the moments due to the other two load combinations. Table 6.3 through 6.7 show the calculations for these and compare results. Table 6.7 shows that at the critical locations near the haunch, the controlling load case is dead load plus symmetrical snow load. Along the length of the rafter the asymmetrical snow load plus dead load controls; this is reasonable since the snow load on the loaded side is 25% more than the symmetrical load. Also along the rafter, note that load case II produces *negative* moments on the unloaded portion of the arch. A negative moment puts tension into the compression zone. In designing this portion of the arch we must be sure to check the effect of both positive and negative moments.

Table 6.6 shows the moments due to wind forces. The maximum *positive* moment due to wind load plus dead load at point D is 134,884 lb-ft. This value is about 72% of the moment at D due to load case I, dead load plus snow load. To compare the effect of the wind load versus the snow load we must consider the duration factors. Since the wind duration factor is 1.6 and the snow duration factor is 1.15, the effect of the wind load at point D compared to the snow load is

$$0.72 \times \frac{1.15}{1.6} = 0.52$$

The moment due to the wind load combination is about 52% of the moment due to the snow load combination. We can see that the moment due to load case III is so small that we need not consider it further.

Table 6.6 also shows the maximum *negative* moments. At point E that moment is −40,404 lb-ft due to wind load only. The negative moment causes the curvature at the haunch to increase. When the negative wind moment at D is added to the ever-present dead load moment, the dead load more than compensates for the negative moment (column 9 of Table 6.6 reports this as zero). At low locations along the

TABLE 6.2 MOMENTS DUE TO SYMMETRIC SNOW LOADS

Point	Coordinates of Points E(psi) = 1,800,000		Location of Vertex: Vertex at Peak, Point I x = 384 in. y = 306 in.				Horizontal Reaction Due to: Dead 4805	Snow 9452	Dead + Snow 14,257
	x Value (in.)	y Value (in.)	Horizontal from Vertex (ft)	Vertical from Vertex (ft)	Vertical of Parabola (ft)	Vertical Difference (ft)	Moment Due to Dead (lb-ft)	Moment Due to Snow (lb-ft)	Moment Due to Dead + Snow (lb-ft)
A	6	0	31.50	25.50	25.50	0.00	0	0	0
B	7	69	31.39	19.75	25.33	5.58	26,802	52,723	79,526
C	9	138	31.29	14.00	25.16	11.16	53,607	105,451	159,057
D	18	177	30.51	10.75	23.92	13.17	63,280	124,480	187,761
E	43	208	28.46	8.15	20.81	12.66	60,846	119,692	180,539
F	78	226	25.49	6.64	16.70	10.06	48,342	95,095	143,437
G	180	253	17.00	4.43	7.43	3.00	14,415	28,356	42,771
H	282	279	8.50	2.21	1.86	−0.36	−1,714	−3,372	−5,087
I	384	306	0.00	0.00	0.00	0.00	0	0	0
Column	1	2	3	4	5	6	7	8	9

Columns 1 and 2: Measure points from scale drawing.
Column 3: Horizontal distance between vertex and point; subtract $(X_v − X_i)$: $(X_v$ − col. 1).
Column 4: Vertical distance between vertex and point; subtract $(Y_v − Y_i)$: $(Y_v$ − col. 2).
Column 5: Calculate location of parabola: $Y_i = Y_v \times (X_i/X_v)^2$: Plot these points on drawing.
Column 6: Vertical offset between arch and parabola; measure from drawing (or calculate: col. 5-col. 4).
Columns 7, 8, and 9: Calculate moments as horizontal reaction × vertical offset ($R \times$ col. 6).

TABLE 6.3 MOMENTS DUE TO ASYMMETRIC SNOW LOADS: LOADED SIDE[a]

Point	Coordinates of Points		Location of Vertex $L/8$ right of I (in.), x = 289.5 $r/8$ above center (in.), y = 344.25 Loaded Side, Parabola				
	x Value (in.)	y Value (in.)	Horizontal from Vertex (ft)	Vertical from Vertex (ft)	Vertical of Parabola (ft)	Vertical Difference (ft)	Moment on Loaded Side (lb-ft)
A	6	0	23.625	28.688	28.688	0.000	0
B	7	69	23.519	22.938	28.430	5.492	32,448
C	9	138	23.412	17.188	28.173	10.985	64,901
D	18	177	22.633	13.938	26.330	12.392	73,213
E	43	208	20.583	11.338	21.776	10.439	61,671
F	78	226	17.617	9.827	15.951	6.125	36,184
G	180	253	9.125	7.615	4.280	−3.335	−19,702
H	282	279	0.625	5.401	0.020	−5.381	−31,791
I	384	306	−7.875	3.188	3.188	0.000	0
Column	1	2	3	4	5	6	7

[a] Horizontal reaction due to asymmetrical snow, 5908. Column notes are similar to those of Table 6.2.

TABLE 6.4 MOMENTS DUE TO ASYMMETRIC SNOW LOADS: UNLOADED SIDE

Location of Vertex: Vertex at Peak, Point I
$x = 384$ in.
$y = 306$ in.

Point	Unloaded Side, Straight Line				Moment on Unloaded Side (lb-ft)	Compare Loaded with Unloaded Side			
	Horizontal from Vertex (ft)	Vertical from Vertex (ft)	Vertical of Straight Line (ft)	Vertical Difference		Max. Positive Moment, Asym. Snow Only (lb-ft)	Max. Positive Moment, Dead + Asym. Snow	Max. Negative Moment, Asym. Snow Only (lb-ft)	Max. Negative Moment, Dead + Asym. Snow
A	31.500	25.500	25.500	0.000	0	0	0	0	0
B	31.394	19.750	25.414	5.664	33,462	33,462	60,264	0	0
C	31.287	14.000	25.328	11.328	66,924	66,924	120,530	0	0
D	30.508	10.750	24.697	13.947	82,400	82,400	145,681	0	0
E	28.458	8.150	23.038	14.888	87,957	87,957	148,803	0	0
F	25.492	6.639	20.636	13.997	82,694	82,694	131,036	0	0
G	17.000	4.427	13.762	9.335	551,50	55,150	69,565	−19,702	−5,288
H	8.500	2.214	6.881	4.667	27,575	27,575	25,861	−31,791	−33,505
I	0.000	0.000	0.000	0.000	0	0	0	0	0
Column	1	2	3	4	5	6	7	8	9

Columns 1 and 2: Taken from Table 6.2., columns 3 and 4, respectively.
Column 3: Calculate location of funicular (straight line): $Y_i = Y_t \times (X_i/X_t)$; plot these points on drawing.
Column 4: Vertical offset between arch and straight line; measure from drawing (or calculate: col. 3 − col. 2).
Column 5: Calculate moments as horizontal reaction × vertical offset (HR × col. 4).
Columns 6 and 8: Maximum moment from loaded side or from unloaded side.
Columns 7 and 9: Add the moment due to dead load to column 6 or column 8, respectively.

TABLE 6.5 MOMENTS DUE TO WIND LOADS: LOADED SIDE[a]

	Coordinates of Points		Location of Vertex $L/16$ right of I (in.), $x = 431.25$ at $r/4$ below I (in), $y = 229.5$ Loaded Side, Parabola				
Point	x Value (in.)	y Value (in.)	Horizontal from Vertex (ft)	Vertical from Vertex (ft)	Horizontal of Parabola (ft)	Horizontal Difference (ft)	Moment on Loaded Side (lb-ft)
A	6	0	35.438	19.125	35.438	0.000	0
B	7	69	35.331	13.375	17.332	17.999	39,544
C	9	138	35.225	7.625	5.633	29.592	65,013
D	18	177	34.446	4.375	1.854	32.591	71,603
E	43	208	32.396	1.775	0.305	32.091	70,503
F	78	226	29.429	0.264	0.007	29.422	64,641
G	180	253	20.938	−1.948	0.368	20.570	45,192
H	282	279	12.438	−4.161	1.678	10.760	34,639
I	384	306	3.938	−6.375	3.938	0.000	0
Column	1	2	3	4	5	6	7

[a] Vertical reaction due to wind load, 2197.
[b] Since wind load is horizontal, moment is vertical reaction × horiz. offset (VR × col. 6).

rafter leg, between points G and H, there are net negative moment, but their values are much less than the negative moment resulting from the asymmetrical snow, load case II. We can conclude that wind never controls.

These tables were produced using a computerized spreadsheet, and checked with the graphic procedure. In most cases the graphic solution is faster and will produce an acceptable accuracy. The spreadsheet has a disadvantage to the beginner in that it does not reinforce intuition the way the graph does. The spreadsheet can find the maximum moment only if it occurs at one of the selected points. The graphic solution will show the location of the maximum moment if it occurs between selected points. The spreadsheet is advantageous if a large number of load cases need to be considered.

GRAPHIC SOLUTION FOR AXIAL FORCES AND SHEAR

The complete design of the arch must consider shear at each critical cross section and the interaction of axial compression and bending. To include these concerns, we must calculate the axial force and shear at any point on the arch. At any point on the arch the equilibrium force can be found by vectorially summing the vertical and horizontal components. For either load combination I or II, we note that the horizontal thrust is constant throughout the arch. The vertical component at any location can be seen on the free-body diagram (Figure 6.3) to be the reaction minus the applied load between the reaction and the point. Since the applied load is assumed to be uniformly applied, this value can easily be found. Figure 6.7 shows the force diagram for the equilibrium force at a number of selected points, shown simultaneously. This diagram is constructed by laying out the horizontal reaction along the x-axis and the vertical reaction along the y-axis. Along the right edge of the diagram each point is established by measuring down the magnitude of the applied force between the end and the point. These values may easily be calculated first, or the points can be laid out using the proportion of their horizon-

TABLE 6.6 MOMENTS DUE TO WIND LOADS: UNLOADED SIDE

Location of Vertex: Vertex at Peak, Point I
$x = 384$ in.
$y = 306$ in.

Point	Unloaded Side—Straight Line				Compare Loaded with Unloaded Side			
	Horizontal from Vertex (ft)	Vertical from Vertex (ft)	Horizontal of Straight Line (ft)	Horizontal Difference	Moment on Unloaded Side (lb-ft)	Max. Positive Moment, Wind Only (lb-ft)	Max. Positive Moment, Dead + Wind	Max. Negative Moment, Wind Only (lb-ft)	Max. Negative Moment, Dead + Wind
A	31.500	25.500	31.500	0.000	0	0	0	0	0
B	31.394	19.750	24.397	−6.997	−15,371	39,544	66,346	−15,371	0
C	31.287	14.000	17.294	−13.993	−30,743	65,013	118,619	−30,7430	0
D	30.508	10.750	13.279	−17.229	−37,852	71,603	134,884	−37,852	0
E	28.458	8.150	10.068	−18.391	−40,404	70,503	131,349	−40,404	0
F	25.492	6.639	8.201	−17.290	−37,987	64,641	112,983	−37,987	0
G	17.000	4.427	5.469	−11.531	−25,334	45,192	59,607	−25,334	−10,919
H	8.500	2.214	2.734	−5.766	−12,667	23,639	21,925	−12,667	−14,381
I	0.000	0.000	0.000	0.000	0	0	0	0	0
Column	1	2	3	4	5	6	7	8	9

212 GLUED-LAMINATED ARCHES

TABLE 6.7 MAXIMUM MOMENTS AND CONTROLLING LOAD COMBINATION

Point	Symmetrical Snow + Dead		Asymmetrical Snow + Dead		Wind + Dead		Maximum Condition		
	Maximum Positive Moment	Maximum Negative Moment	Maximum Positive Moment	Maximum Negative Moment	Maximum Positive Moment	Maximum Negative Moment	Maximum Positive Moment	Maximum Negative Moment	Controlling Condition
Duration Factor:	1.15		1.15		1.6				
A	0	0	0	0	0	0	0	0	Sym. snow
B	79,526	0	60,264	0	66,346	0	79,526	0	Sym. snow
C	159,057	0	120,530	0	118,619	0	159,057	0	Sym. snow
D	187,761	0	145,681	0	134,884	0	187,761	0	Sym. snow
E	180,539	0	148,803	0	131,349	0	180,539	0	Sym. snow
F	143,437	0	131,036	0	112,983	0	143,437	0	Sym. snow
G	42,771	0	69,565	−5,288	59,607	−10,919	69,565	−10,919	Asym. snow
H	0	−5,087	25,861	−33,505	21,925	−14,381	25,861	−33,505	Asym. snow
I	0	0	0	0	0	0	0	0	Sym. Snow
Column	1	2	3	4	5	6	7	8	9

tal distance from the support applied to the vertical axis.

The angle of the force, shown for example as the force O to D in Figure 6.7, is tangential to the equilibrium polygon and generally not along the axis of the arch. This results in a component that is parallel to the axis of the arch, the axial compression force, and a component that is perpendicular to the axis, the shear force. The ray on the graph, O to D', is laid out parallel to the tangent to the arch at point D. From point D, draw a ray perpendicular to the line $O-D'$. The intersection of this ray and the ray $O-D'$ will determine the location of D'. The length of the line $O-D'$ represents the magnitude of the compression force at D, and the length of the line $D-D'$ represents the magnitude of the shear at point D.

Tables 6.8 and 6.9 list the compression and shear forces for nine points along the arch due to load combinations I and II. These values were actually calculated algebraically using a computerized spreadsheet, but can easily be derived from Figure 6.7.

FINAL DESIGN OF THE ARCH

The design of the arch will now proceed by checking the strength of the cross section at a number of points. Each point needs to be checked for the interaction of the bending and compression forces as well as for shear (Table 6.12). Along the curved section we must also check for radial tension. Each point needs to be considered under the worst-load condition only, but since moments from load combinations I and II are so close, we will check all points for both cases.

The design must satisfy the requirements for interaction of bending and compression. Each cross section must meet the requirement

$$\left(\frac{f_c}{F'_c}\right)^2 + \frac{f_b}{F'_b(1 - f_c/F_{cE})} \leq 1$$

These calculations are explained further in Chapter 3. The lateral stability of the arch must be checked to calculate F'_c and F'_b. Refer to Figure 6.4 to help visualize the possible buckling patterns.

FINAL DESIGN OF THE ARCH   213

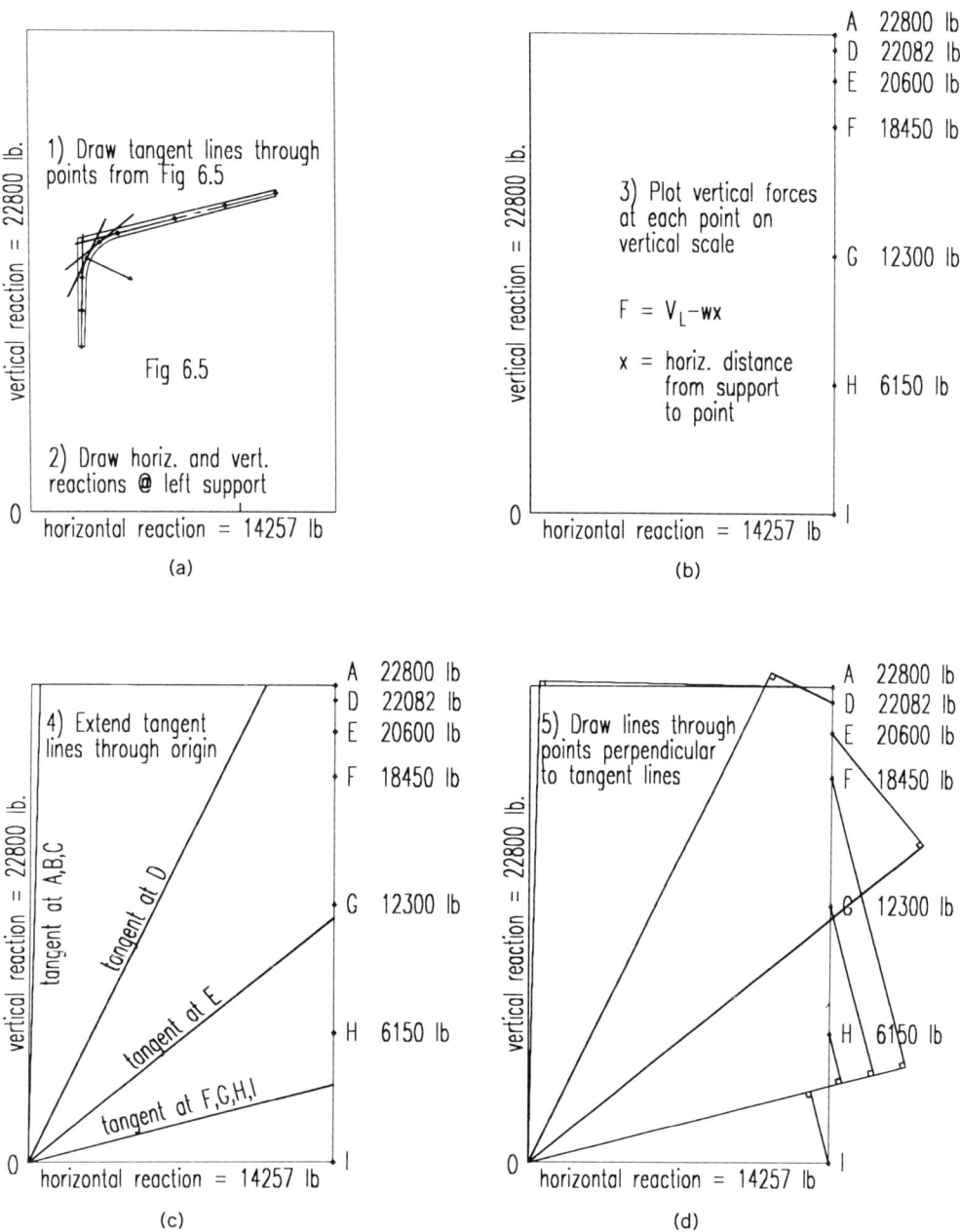

FIGURE 6.7 Force diagram for dead plus symetrical snow load.

214 GLUED-LAMINATED ARCHES

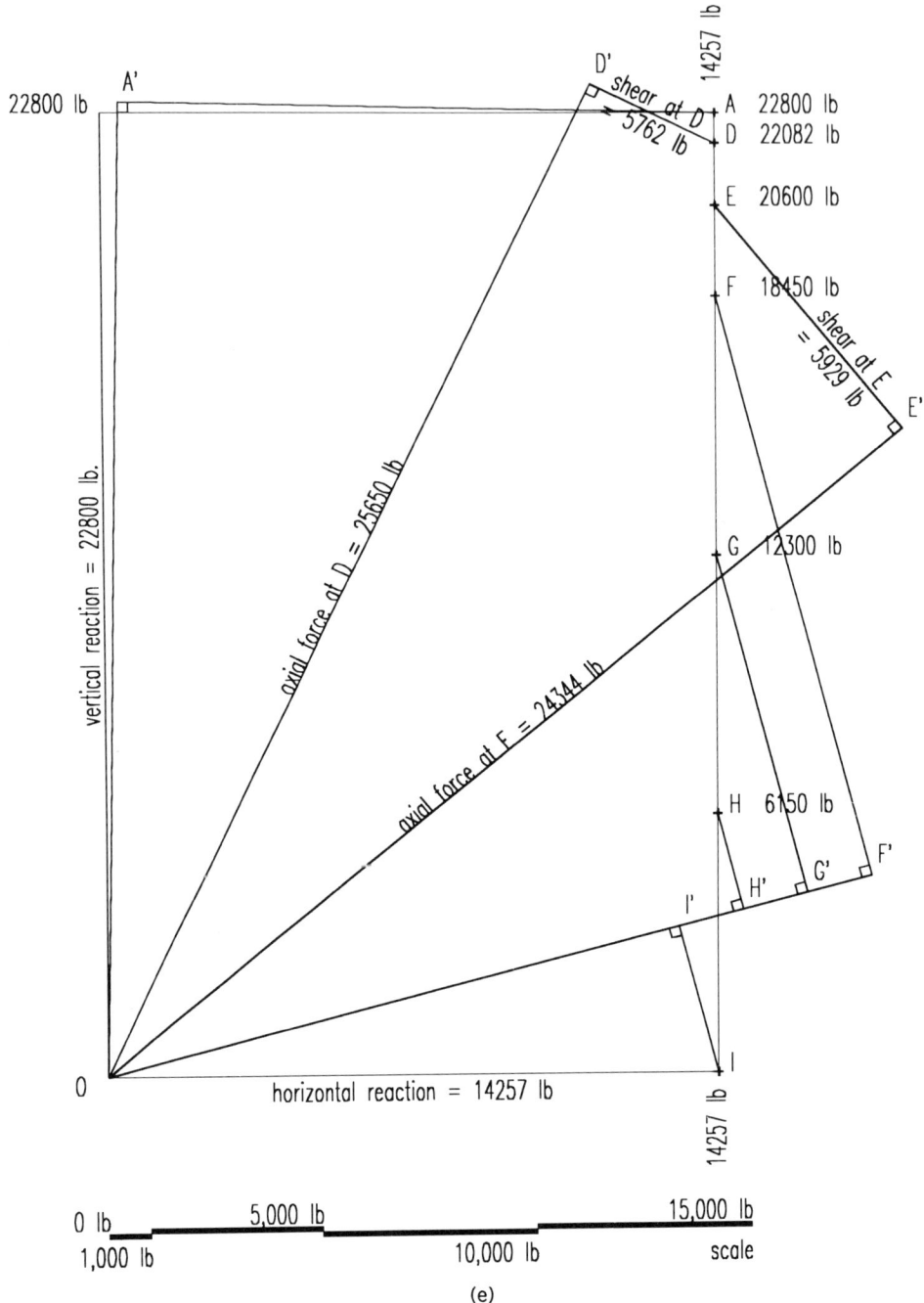

FIGURE 6.7 (*Continued*)

TABLE 6.8 AXIAL FORCES AND SHEAR: SYMMETRIC SNOW LOADS

	Dead Load	Snow Load	Dead + Snow
Vertical load (plf)	244	480	724
Vertical reaction (lb)	7,808	15,360	23,168
Horizontal reaction (lb)	4,805	9,452	14,257

Point	x Value (in.)	y Value (in.)	Tangential Angle at x Due to $d+s$ (deg)	Vertical Force at x Due to $d+s$ (lb)	Vectorial Force at x Due to $d+s$ (lb)	Vectorial Angle at x Due to $d+s$ (deg)	Difference in Angle at x Due to $d+s$ (deg)	Shear Force at x Due to $d+s$ (lb)	Axial Force at x Due to $d+s$ (lb)
A	6	0	88.9	22,806	26,896	58.0	31.0	13,833	23,066
B	7	69	88.9	22,729	26,830	57.9	31.0	13,834	22,989
C	9	138	88.0	22,652	26,765	57.8	30.2	13,470	23,128
D	18	177	69.8	22,088	26,290	57.2	12.7	5,762	25,650
E	43	208	41.6	20,604	25,056	55.3	-13.7	-5,929	24,344
F	78	226	21.1	18,456	23,321	52.3	-31.3	-12,103	19,935
G	180	253	14.6	12,308	18,835	40.8	-26.2	-8,317	16,899
H	282	279	14.6	6,154	15,528	23.3	-8.8	-2,362	15,348
I	384	306	14.6	0	14,257	0.0	14.6	3,593	13,797
Column	1	2	3	4	5	6	7	8	9

Columns 1 through 7: These are *not* needed in the graphic approach, but were used here to derive columns 8 and 9.
Columns 8 and 9: These values may be found through the graphic approach on Figure 6.7.

TABLE 6.9 AXIAL FORCES AND SHEAR: ASYMMETRIC SNOW LOADS

	Dead	Snow	Dead + Snow
Vertical load (plf)	244	600	8844
Vertical reaction (lb)	7,808	14,400	22,208
Horizontal reaction (lb)	4,805	5,908	10,713

Point	x Value (in.)	y Value (in.)	Tangential Angle at x Due to $d+s$ (deg)	Vertical Force at x Due to $d+s$ (lb)	Vectorial Force at x Due to $d+s$ (lb)	Vectorial Angle at x Due to $d+s$ (deg)	Difference in Angle at x Due to $d+s$ (deg)	Shear Force at x Due to $d+s$ (lb)	Axial Force at x Due to $d+s$ (lb)
A	6	0	88.9	21,786	24,278	63.8	25.1	10,308	21,981
B	7	69	88.9	21,696	24,197	63.7	25.2	10,310	21,891
C	9	138	88.0	21,606	24,116	63.6	24.4	9,964	21,962
D	18	177	69.8	20,949	23,529	62.9	6.9	2,828	23,359
E	43	208	41.6	19,219	22,003	60.9	−19.2	−7,248	20,775
F	78	226	21.1	16,715	19,853	57.3	−36.3	−11,751	16,002
G	180	253	14.6	9,548	14,350	41.7	−27.1	−6,540	12,774
H	282	279	14.6	2,374	10,973	12.5	2.1	402	10,966
I	384	306	14.6	−4,800	11,739	−24.1	38.7	7,345	9,159
Column	1	2	3	4	5	6	7	8	9

TABLE 6.10 SECTION PROPERTIES FOR DESIGN (WIDTH = 8.75 IN.) — FIRST ESTIMATE OF DEPTH

Point	x Value (in.)	y Value (in.)	Estimated Depth (in.)	Area (in.^{2})	Section Modulus (in.^{3})	Moment of Inertia
A	6	0	12.0	105	210	1,260
B	7	69	15.0	131	328	2,461
C	9	138	18.0	158	473	4,253
D	18	177	28.0	245	1,143	16,007
E	43	208	42.0	368	2,573	54,023
F	78	226	28.0	245	1,143	16,007
G	180	253	18.0	158	473	4,253
H	282	279	15.0	131	328	2,461
I	384	306	12.0	105	210	1,260
Column	1	2	3	4	5	6

Column 3: Measure depth of arch from scale drawing or interpolate between "known" points such as ends and tangent points.
Columns 4, 5, and 6: Calculate properties for rectangular sections.

First consider the segment of the arch at the roof, referred to here as the *rafter*. The rafter is loaded both in axial compression and in bending. Both forces have a tendency to cause buckling, but since the bracing operates differently in each direction, the buckling effect will be different for axial forces than for bending forces. Axially, the rafter may buckle in either of two directions: to the side or vertically. If it buckles to the side, the purlins act as braces and the unbraced length is the distance between purlins. The dimension of the beam that resists this buckling is the width of the beam. Thus, for sideways buckling,

$$\frac{L}{d} = \frac{8.25 \text{ ft} \times 12 \text{ in./ft}}{8.75 \text{ in.}} = 11.3$$

The other possibility is that the rafter will buckle vertically (in the plane of the arch). In this case the braces will not restrain the buckling. Since the decking runs parallel to the arch, we can assume that it provides no restraint either. The unbraced length is the total length of the rafter, which is 33 ft, measured along the length of the rafter. The depth of the rafter is the critical dimension in this mode of buckling; we will use the estimated depth near the middle, so $d = 15$ in. Therefore,

$$\frac{L}{d} = \frac{33 \text{ ft} \times 12 \text{ in./ft}}{15 \text{ in.}} = 26.4$$

The axially loaded rafter will buckle in the direction of least resistance, so the larger L/d ratio controls. In calculations for the axial stress in the rafter element, use

$$\frac{L}{d} = 26.4$$

For glulams,

$$F_{cE} = \frac{0.418 E}{(L/d)^2} = \frac{0.418 \times 1,700,000}{(26.4)^2}$$

$$= 1019.5 \text{ psi}$$

$$F_c^* = 1.15 \times 1700 \text{ psi} = 1955 \text{ psi}$$

$$\frac{F_{cE}}{F_c^*} = \frac{1019.5}{1955} = 0.521$$

and for glu-lams, c = 0.9

$$C_p = \frac{1 + (F_{cE}/F_c^*)}{2c}$$

$$- \sqrt{\left[\frac{1 + (F_{cE}/F_c^*)}{2c}\right]^2 - \frac{F_{cE}/F_c^*}{c}}$$

$$= \frac{1 + (0.521)}{1.8}$$

$$- \sqrt{\left[\frac{1 + (0.521)}{1.8}\right]^2 - \frac{0.521}{0.9}}$$

$$= 0.477$$

$$F_c' = C_P \times F_c^* = 0.477 \times 1955 \text{ psi}$$

$$= 932 \text{ psi}$$

When loaded vertically the rafter acts like a beam and will buckle laterally (i.e., to the side). Since all of the arches are equally loaded, they all will have the same tendency to buckle. The purlins serve only to brace one arch to the next, so in this type of loading, if one arch were to buckle, they all would deflect laterally. The unbraced length is 33 ft. Applying the design formulas for laterally unbraced beams gives us

$$l_e = 1.54 \times l_u = 1.54 \times 33 \times 12 = 610 \text{ in.}$$

for a simply supported beam with three equal loads at quarter points. Again, take beam dimensions at center of the rafter, so

$$R_b = \sqrt{\frac{l_e \times d}{b^2}} = \sqrt{\frac{610 \times 15}{8.75^2}} = 10.93$$

$$F_{bE} = \frac{0.609 \times E}{(R_b)^2} = \frac{0.609 \times 1,700,000}{(10.93)^2}$$

$$= 8666 \text{ psi}$$

$$F_b^* = C_d \times C_v \times C_C \times F_b$$

C_C is the curvature factor, which will be applied at points C, D, and E only. For the rafter section, now under consideration, use $C_C = 1.0$. C_v is a glu-lam size (or volume) factor.

$$C_v = K_L \times \left(\frac{21}{L}\right)^{1/x} \times \left(\frac{12}{d}\right)^{1/x}$$

$$\times \left(\frac{5.125}{b}\right)^{1/x} \leq 1.0$$

For Southern Pine, $x = 20$; for all other species, $x = 10$.

$$C_v = 1.0 \times \left(\frac{21}{33}\right)^{1/20} \times \left(\frac{12}{15}\right)^{1/20}$$

$$\times \left(\frac{5.125}{8.75}\right)^{1/20} = 0.941$$

$$F_b^* = C_d \times C_v \times C_C \times F_b$$

$$= 1.15 \times 1.0 \times 0.941 \times 2400$$

$$= 2597 \text{ psi}$$

$$\frac{F_{bE}}{F_b^*} = \frac{8666}{2597} = 3.34$$

$$C_L = \frac{1 + (F_{bE}/F_b^*)}{2c}$$

$$- \sqrt{\left[\frac{1 + (F_{bE}/F_b^*)}{2c}\right]^2 - \frac{F_{bE}/F_b^*}{c}}$$

$$= \frac{1 + (3.34)}{1.9}$$

$$- \sqrt{\left[\frac{1 + (3.34)}{1.9}\right]^2 - \frac{3.34}{0.95}}$$

$$= 0.979$$

$$F_b' = C_L \times F_b^* = 0.979 \times 2597 \text{ psi}$$

$$= 2542 \text{ psi}$$

For the vertical portion of the arch, referred to here as the *column*, the same assumptions are made. Thus, for the column, if sideways buckling controls,

$$\frac{L}{d} = \frac{9 \text{ ft} \times 12 \text{ in./ft}}{8.75 \text{ in.}} = 12.34$$

If buckling is in the plane, use the estimated depth near the middle. $d = 15$ in. Therefore,

$$\frac{L}{d} = \frac{18 \text{ ft} \times 12 \text{ in./ft}}{15 \text{ in.}} = 14.4$$

The column will buckle in the plane. Use

$$\frac{L}{d} = 14.4$$
$$F_{cE} = 3427 \text{ psi}$$
$$\frac{F_{cE}}{F_c^*} = \frac{3427}{1955} = 1.75$$
$$C_P = 0.904$$
$$F_c' = 0.904 \times 1955 \text{ psi} = 1767 \text{ psi}$$

Similarly, for consideration of bending, use

$$l_e = 1.63 \times l_u + 3d$$
$$= 1.63 \times 18 \times 12 + 3 \times 15 = 397 \text{ in.}$$

for a simply supported beam with no specified loads.

$$R_b = \sqrt{\frac{397 \times 15}{8.75^2}} = 8.82$$
$$F_{bE} = 9573 \text{ psi}$$
$$C_v = 1.09 \times \left(\frac{21}{18}\right)^{1/20} \times \left(\frac{12}{15}\right)^{1/20}$$
$$\times \left(\frac{5.125}{8.75}\right)^{1/20}$$
$$= 1.06 \quad \text{use } C_v = 1.0$$
$$F_b^* = 1.15 \times 2400 = 2760 \text{ psi}$$
$$\frac{F_{bE}}{F_b^*} = \frac{9573}{2760} = 3.47$$
$$C_L = 0.981$$
$$F_b^* = 0.981 \times 2760 \text{ psi}$$
$$= 2707 \text{ psi}$$

Along the *curve*, we may conservatively assume that its axial stiffness is the same as that of the column section below. We will use

$$F_c' = 1767 \text{ psi}$$

To consider the curve regarding lateral buckling due to bending loads, take the volume factor as equal to 1.0 and calculate the curvature factor as

$$C_c = 1 - 2000\left(\frac{t}{R}\right)^2$$
$$= 1 - 2000\left(\frac{0.75}{84}\right)^2 = 0.84$$
$$F_b^* = C_d \times C_v \times C_C \times F_b$$
$$= 1.15 \times 0.84 \times 1.0 \times 2400$$
$$= 2318 \text{ psi}$$

Since this section is very short and stiff, it is a good estimate to assume that C_L equals 1.0.

$$F_b' = 2318 \text{ psi}$$

These values are taken at the assumed point of buckling for each segment of the arch. Therefore, they are conservative for all other points in each segment. Table 6.11 summarizes the values for F_c' and F_b' calculated earlier.

In the initial trial we see in Tables 6.12 and 6.13 that the original estimate of the depth is not satisfactory. The top of the column and the beginning of the haunch (points C and D) are overstressed in bending by as much as 1.58 times allowable. (*Note:* The tables do not include a column of the total level of stress resulting from the interaction. As can be seen, the ratio of actual to allowable compression stress is so small that it does not affect the outcome of the design.)

All of the stress is in bending; almost none is in compression. This implies that

TABLE 6.11 BUCKLING LENGTHS

	Axial Forces			Bending Forces		
Location	Unbraced Length, l (ft)	F'_c (psi)	F_{cE} (psi)	Unbraced Length, l_u (ft)	Effective Length, l_e (ft)	F'_b (psi)
Rafter	33	932	1019	33	50.83	2,542
Column	18	1767	3427	18	33.08	2,707
Curve	—	1767	3427	—	—	2,318

TABLE 6.12A INTERACTION DESIGN FOR SYMMETRIC SNOW LOADS—FIRST ESTIMATE OF DEPTH

	Axial Stresses					Bending Stresses			
Point	Axial Force at x Due to $d+s$ (lb)	f_c Axial Stress (psi)	F'_c Allowable Axial Stress (psi)	F_{cE} Euler Buckling Stress (psi)	$(f_c/F'_c)^2$	Maximum Positive Moment (lb-ft)	f_b Bending Stress (psi)	Allowable Bending Stress (psi)	f_b/F'_b $1/(1-f_c/F_{cE})$
A	23,066	219.68	1,767	3,427	0.012	0	0.0	2,707	0.000
B	22,989	175.15	1,767	3,427	0.007	79,526	2,908.4	2,707	0.985
C	23,128	146.85	1,767	3,427	0.005	159,057	4,039.6	2,318	1.583
D	25,650	104.70	1,767	3,427	0.003	187,761	1,970.7	2,318	0.763
E	24,344	66.24	1,767	3,427	0.001	180,539	842.2	2,318	0.322
F	19,935	81.37	932	1,019	0.006	143,437	1,505.5	2,542	0.560
G	16,899	107.29	932	1,019	0.010	42,771	1,086.2	2,542	0.415
H	15,348	116.94	932	1,019	0.012	5,087	186.0	2,542	0.072
I	13,797	131.40	932	1,019	0.015	0	0.0	2,542	0.000
Column	1	2	3	4	5	6	7	8	9

Column 1: Taken from Table 6.8, column 9.
Column 2: Axial stress = force/area(col. 1 ÷ col. 4, Table 6.10).
Columns 3 and 4: Calculation shown in text; see NDS, Section 3.7.
Column 5: Ratio of actual to allowable. The sum of cols. 5 and 9 are the level of stress given by the interaction formula.

TABLE 6.12B INTERACTION DESIGN FOR SYMMETRIC SNOW LOADS—REVISED DEPTH FOR BENDING

	Axial Stresses					Bending Stresses			
Point	Axial Force at x Due to $d+s$ (lb)	f_c Axial Stress (psi)	F'_c Allowable Axial Stress (psi)	F_{cE} Euler Buckling Stress (psi)	$(f_c/F'_c)^2$	Maximum Positive Moment (lb-ft)	f_b Bending Stress (psi)	F'_b Allowable Bending Stress (psi)	f_b/F'_b $1/(1-f_c/F_{cE})$
A	23,066	219.68	1,767	3,427	0.012	0	0.0	2,707	0.000
B	22,989	152.31	1,767	3,427	0.006	79,526	2,199.1	2,707	0.739
C	23,128	117.48	1,767	3,427	0.003	159,057	2,585.3	2,318	1.004
D	25,650	118.44	1,767	3,427	0.003	187,761	2,522.2	2,318	0.980
E	24,344	115.92	1,767	3,427	0.003	180,539	2579.1	2,318	1.001
F	19,935	108.49	932	1,019	0.010	143,437	2,676.4	2,542	1.025
G	16,899	107.29	932	1,019	0.010	42,771	1,086.2	2,542	0.415
H	15,348	116.94	932	1,019	0.012	5,087	186.0	2,542	0.072
I	13,797	131.40	932	1,019	0.015	0	0.0	2,542	0.000
Column	1	2	3	4	5	6	7	8	9

Column 1: Taken from Table 6.8, column 9.
Column 2: Axial stress = force/area (column 1 ÷ by col. 5, Table 6.14).
Columns 3 and 4: Calculation shown in text; see NDS section 3.7.
Column 5: Ratio of actual to allowable. The sum of cols. 5 and 9 are the level of stress given by the interaction formula.

FINAL DESIGN OF THE ARCH

TABLE 6.13A INTERACTION DESIGN FOR ASYMMETRIC SNOW LOADS—FIRST ESTIMATE OF DEPTH

	Axial Stresses					Bending Stresses			
Point	Axial Force at x Due to $d+s$ (lb)	f_c Axial Stress (psi)	F'_c Allowable Axial Stress (psi)	F_{cE} Euler Buckling Stress (psi)	$(f_c/F'_c)^2$	Maximum Positive Moment (lb-ft)	f_b Bending Stress (psi)	F'_b Allowable Bending Stress (psi)	$f_b/F'_b \times 1/(1-f_c/F_{cE})$
A	21,980.50	209.34	1,767	3,427	0.011	0	0.0	2,707	0.000
B	21,890.78	166.70	1,767	3,427	0.007	60,264	2,204.0	2,707	0.744
C	21,961.69	139.44	1,767	3,427	0.005	120,530	3,061.1	2,318	1.197
D	23,358.72	95.34	1,767	3,427	0.002	145,681	1,529.0	2,318	0.590
E	20,774.86	56.53	1,767	3,427	0.001	148,803	694.1	2,318	0.265
F	16,002.02	65.31	932	1,019	0.004	131,036	1,375.3	2,542	0.503
G	12,773.69	81.10	932	1,019	0.006	69,565	1,766.7	2,542	0.657
H	10,965.51	83.55	932	1,019	0.006	33,505	1,225.3	2,542	0.457
I	9,157.57	87.21	932	1,019	0.007	0	0.0	2,542	0.000
Column	1	2	3	4	5	6	7	8	9

TABLE 6.13B INTERACTION DESIGN FOR ASYMMETRIC SNOW LOADS—REVISED DEPTH FOR BENDING

	Axial Stresses					Bending Stresses			
Point	Axial Force at x Due to $d+s$ (lb)	f_c Axial Stress (psi)	F'_c Allowable Axial Stress (psi)	F_{cE} Euler Buckling Stress (psi)	$(f_c/F'_c)^2$	Maximum Positive Moment (lb-ft)	f_b Bending Stress (psi)	F'_b Allowable Bending Stress (psi)	$f_b/F'_b \times 1/(1-f_c/F_{cE})$
A	21,980.50	209.34	1,767	3,427	0.011	0	0.0	2,707	0.000
B	21,890.78	145.03	1,767	3,427	0.005	60,264	1,666.5	2,707	0.559
C	21,961.69	111.55	1,767	3,427	0.003	120,530	1,959.1	2,318	0.760
D	23,358.72	107.86	1,767	3,427	0.003	145,681	1,956.9	2,318	0.758
E	20,774.86	98.93	1,767	3,427	0.002	148,803	2,125.8	2,318	0.821
F	16,002.02	87.09	932	1,019	0.007	131,036	2,445.0	2,542	0.915
G	12,773.69	81.10	932	1,019	0.006	69,565	1,766.7	2,542	0.657
H	10,965.51	83.55	932	1,019	0.006	33,505	1,225.3	2,542	0.457
I	9,157.57	87.21	932	1,019	0.007	0	0.0	2,542	0.000
Column	1	2	3	4	5	6	7	8	9

the profile of the arch is not very close to the equilibrium polygon. In most cases the designer does not have a great deal of flexibility to change the shape of the arch. If we were to revise it in this case, we might consider increasing the radius of curvature on the inside face. This would have a number of positive benefits. It would bring the axis of the arch closer to the equilibrium polygon for both snow-load cases, thereby reducing the moment somewhat. It would also add depth to the cross section, although the depth would be added primarily at the haunch, which needs it less than the column does. And it would diminish the curvature reduction factor at the curved section—again a savings that is not necessarily needed in this example.

The simpler solution is to increase the depth of the cross section, primarily along the column face. A set of recommended depths is given on Table 6.14 under column 4. The maximum increase in depth

TABLE 6.14 REQUIRED SECTION PROPERTIES: INTERACTION DESIGN (WIDTH = 8.75 IN.) —REVISED DEPTH FOR BENDING

Point	First Estimate of Depth	Revised Estimated Depth for Dead + Sym. Snow	Revised Estimated Depth for Dead + Asym. Snow	Recom'nded Depth (in.)	Area (in.^{2})	Section Modulus (in.^{3})	Moment of Inertia
A	12.0	12.00	12.00	12.00	105	210	1,260
B	15.0	14.72	12.78	17.25	151	434	3,743
C	18.0	22.47	19.53	22.50	197	738	8,306
D	28.0	24.42	21.47	24.75	217	893	11,055
E	42.0	23.94	21.69	24.00	210	840	10,080
F	28.0	21.00	19.91	21.00	184	643	6.753
G	18.0	11.30	14.44	18.00	158	473	4,253
H	15.0	3.46	9.98	15.00	131	328	2,461
I	12.0	12.00	12.00	12.00	105	210	1,260
Column	1	2	3	4	5	6	7

occurs at point C, which goes from 18 in. to 22.5 in., an increase of 25%. This increase, although substantial, will not have a significant effect on any of the variables used in the design equations. Keeping them as they are is slightly conservative. It will be sufficiently accurate to recalculate the interaction formula and change only the section properties (SAI).

The interaction formulas are recalculated using revised depths. The new depths are shown in Table 6.15, which summarizes three trial designs. The arch has been redesigned to have more of a constant thickness around the curve and to taper linearly from the tangent points to the ends. When the revised values are reused in the interaction formula (Tables 6.12b and 6.13b), we find that the new depth of the arch is satisfactory. The stress ratio at point F is 1.025 in bending and 0.010 in compression. Using the revised depths,

TABLE 6.15 TRIAL BEAM DEPTHS AND INTERACTIONS

Point	Trial Depth		Design for Bending		Design for Deflection	
	Depth	Interaction	Depth	Interaction	Depth	Interaction
A	12.0	0.012	12.00	0.012	12.00	0.012
B	15.0	0.992	17.25	0.745	21.00	0.499
C	18.0	1.588	22.50	1.008	30.00	0.562
D	28.0	0.765	24.75	0.983	30.00	0.665
E	42.0	0.323	24.00	1.005	30.00	0.639
F	28.0	0.565	21.00	1.035	30.00	0.490
G	18.0	0.425	18.00	0.425	24.00	0.233
H	15.0	0.084	15.00	0.084	18.00	0.057
I	12.0	0.015	12.00	0.015	12.00	0.015
Column	1	2	3	4	5	6

RADIAL TENSION 223

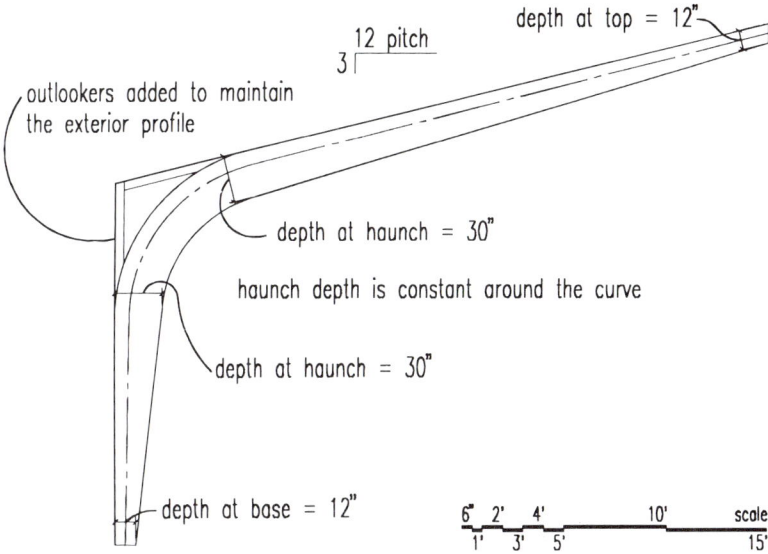

FIGURE 6.8 Three-hinged arch- revised dimensions for deflection.

then, the curved segment of the arch is checked for radial tension and for shear at all points. (See Tables 6.17 and 6.18.)

RADIAL TENSION

When a curved beam has a moment applied to it, radial forces develop. These forces will have the tendency to spread the laminations apart (radial tension) when the moment pushes inward (i.e., when it is positive and produces compression on the outer, convex side). The radial forces will have the effect of compressing the laminations together when the moment pushes outward (i.e., when it is negative and produces compression on the inner, concave face).

Allowable radial tension values for glued-laminated beams are specified by the AITC and the NDS. The allowable radial tension is $f_r = \frac{1}{3} \times f_v$, where f_v is the allowable horizontal shear stress. This equation applies for Southern Pine and California Redwood under any load condition, and for Douglas Fir–Larch under wind or earthquake loadings. For Douglas Fir–Larch under any other loading, the allowable radial tension is $f_r = 15$ psi. The NDS is not explicit, but it is assumed that these values are not to be increased by the duration factor.

The actual radial tension is dependent on the bending moment and is given by the formula

$$f_r = \frac{3M}{2R_c bd}$$

where R_c is the radius of curvature to the centerline of the beam. This formula applies only to beams (or arches) with a constant cross section. For curved beams, which also vary in depth, the formula is

$$f_r = K_r \times \frac{6M}{bd^2}$$

where

$$K_r = A + B\left(\frac{d}{R_m}\right) + C\left(\frac{d}{R_m}\right)^2$$

A, B, and C are constants dependent on the angle β, in degrees, as shown in Table

224 GLUED-LAMINATED ARCHES

TABLE 6.16 COEFFICIENTS FOR RADIAL TENSION

β	A	B	C
(0.0)	(0.0)	(0.2500)	(0.0)
2.5	0.0079	0.1747	0.1284
5.0	0.0174	0.1252	0.1939
7.5	0.0279	0.0937	0.2162
10.0	0.0391	0.0754	0.2119
15.0	0.0629	0.0619	0.1722
20.0	0.0893	0.0608	0.1393
25.0	0.1214	0.0605	0.1238
30.0	0.1649	0.0603	0.1115

Source: Timber Construction Manual, AITC.

6.16 and R_m is the radius of curvature to the centerline of the beam at the midpoint along the length of the curve.

For horizontal beams with an equal slope along the top edges, β is the angle between either top edge and the horizontal. At the haunch of a Tudor arch, such as the one in our design example, the acute angle between the two legs is equal to 2β. To use Table 6.17, find the angle between the two legs and enter the table with $\beta = \frac{1}{2} \times$ angle. In the example the roof pitch is 3 on 12, or 14.0°. Use $\beta = \frac{1}{2} \times 14.0° = 7.0°$. Interpolating from the table above, we find that $A = 0.0258$, $B = 0.1000$, and $C = 0.2117$.

At the center of the curve, midway between points D and E, $R_m = 97$ in. and $d = 25.5$ in.; thus

$$\frac{d}{R_m} = \frac{25.5}{97} = 0.263$$

$$K_r = A + B\left(\frac{d}{R_m}\right) + C\left(\frac{d}{R_m}\right)^2$$

$$= 0.0258 + 0.1(0.263) + 0.2117(0.263)^2$$

$$= 0.0667$$

$$f_r = K_r \times \frac{6M}{bd^2}$$

$$= 0.0667 \times \frac{6M}{8.75 \times 25.5^2}$$

$$= 7.0 \times 10^{-5} \times M$$

where M is in pound-inches and f_r is expressed in psi. In the example the maximum moment at point D is 187,761 lb-ft, resulting from load case I. The resulting radial stress is in tension (the moment flattens the curve and puts compression on the convex side) and the magnitude of the tension stress is

$$f_r = (7.0 \times 10^{-5}) \times M$$
$$= (7.0 \times 10^{-5}) \times 187,761 \times 12$$
$$= 157.7 \text{ psi}$$

Had we calculated the radial tension neglecting the varying cross section, we would have arrived at $f_r = 162$ psi. Tables 6.17 and 6.18 develop the radial tension stress for four points along the curve using the simpler equation.

The radial tension has a tendency to separate the laminations. It essentially puts the wood into shear across the grain. The capacity of the wood to resist this is highly limited. For Southern Pine we are limited to a radial tension stress of one-third the allowable horizontal shear stress, or $\frac{1}{3} \times 200$ psi $= 67$ psi. Since the actual stress exceeds the allowable, the laminations will split apart unless mechanical fasteners are added to carry the tension and hold the laminations together. One common solution is to use a series of lag screws inserted radially into the underside of the curve. The radial tension derives from the axial tension, which extends from the extreme

TABLE 6.17 SHEAR AND RADIAL TENSION: SYMMETRIC SNOW LOADS—REVISED DEPTH FOR BENDING

		For 3/4-in. Lag	For 1-in. Lag
Allowable shear (psi) = 200.0	Root diameter (in.)	0.579	0.780
	Root A	0.263	0.478
Allowable radial tension (psi) = 66.7	Allowable (in.^{2}) T (lb)	6,319	11,468
	Withdrawal	3,744	4,643

			Radial Tension			Shear			
Point	Max. Positive Moment Due to $d+s$ (lb-ft)	Radius of Curvature at Centerline (in.)	Radial Tension Stress Due to $d+s$ (psi)	Required Lag Screw Spacing 3/4-in.-Diameter Screw (in.)	Required Lag Screw Spacing 1-in.-Diameter Screw (in.)	Shear Force at x Due to $d+s$ (lb)	Shear Stress at x Due to $d+s$ (psi)	Recommended Depth to Resist Shear (in.)	Minimum Depth of Cross Section (in.)
A	0	0.00	0.0	None	None	13,833	197.6	10.31	12.00
B	79,526	0.00	0.0	None	None	13,834	137.5	10.31	17.25
C	159,057	95.25	152.7	2.75	3.25	13,470	102.6	10.04	22.50
D	187,761	96.38	161.9	2.50	3.25	5,762	39.9	4.29	24.75
E	180,539	96.00	161.2	2.50	3.25	−5,929	42.4	4.42	24.00
F	143,437	94.50	148.7	2.75	3.50	−12,103	98.8	9.02	21.00
G	42,771	0.00	0.0	None	None	−8,317	79.2	6.20	18.00
H	0	0.00	None	None	−2,362	27.0	1.76	15.00	
I	0	0.00	0.0	None	None	3,593	51.3	2.68	12.00
Column	1	2	3	4	5	6	7	8	9

TABLE 6.18 SHEAR AND RADIAL TENSION: ASYMMETRIC SNOW LOADS—REVISED DEPTH FOR BENDING

	For 3/4-in. Lag	For 1-in.-Lag
Allowable shear (psi) = 200.0		
Root diameter (in.)	0.579	0.780
Root A (in.^{2})	0.263	0.478
Allowable radial tension (psi) = 66.7		
Allowable T (lb)	6,319	11,468
Withdrawal	3,744	4,643

			Radial			Shear			
Point	Max. Positive Moment Due to Dead + Asym. Snow (lb-ft)	Radius of Curvature at Centerline (in.)	Radial Tension Stress, Dead + Asym. Snow (psi)	Required Lag Screw Spacing, 3/4-in.-Diameter Screw (in.)	Required Lag Screw Spacing, 1-in.-Diameter Screw (in.)	Shear Force At x, Dead + Asym. Snow (lb)	Shear Stress at x, Dead + Asym. Snow (psi)	Recommended Depth to Resist Shear (in.)	Minimum Depth of Cross Section (in.)
A	0	0.00	0.0	None	None	10,308	147.3	7.68	12.00
B	60,264	0.00	0.0	None	None	10,310	102.5	7.68	17.25
C	120,530	95.25	115.7	3.50	4.50	9,964	75.9	7.43	22.50
D	145,681	96.38	125.6	3.25	4.00	2,828	19.6	2.11	24.75
E	148,803	96.00	132.9	3.00	3.75	−7,248	51.8	5.40	24.00
F	131,036	94.50	135.8	3.00	3.75	−11,751	95.9	8.76	21.00
G	69,565	0.00	0.0	None	None	−6,540	62.3	4.87	18.00
H	25,861	0.00	0.0	None	None	402	4.6	0.30	15.00
I	0	0.00	0.0	None	None	7,345	104.9	5.47	12.00
Column	1	2	3	4	5	6	7	8	9

fiber to the neutral axis. Therefore, the screws need to extend to the neutral axis as well or at least until the actual tension drops below the allowable. In the example the beam is assumed to be 28 in. deep at D, so we may assume that the neutral axis is approximately 14 in. from the edge. The actual radial stress is more than twice the allowable. Since the tension drops off linearly, the screw will need to extend more than half the distance to the neutral axis, say 9 in.

Consider using a $\frac{3}{4}$-in. diameter by 12-in.-long lag screw. Being loaded in tension, its limiting strength will either be the allowable tension of the bolt itself taken across the root area, or its withdrawal capacity in the wood. A 12-in. screw has a thread length of about $5\frac{1}{2}$ in. and a capacity of 592 lb/in. penetration in Southern Pine. Its withdrawal capacity, adjusting for a snow-load duration, is

$$P_w = 1.15 \times 592 \times 5\tfrac{1}{2} = 3744 \text{ lb}$$

Assuming that the allowable tension stress of the steel is 24,000 psi and that the diameter at the root of the screw is 0.579 in. (approximately $\frac{3}{4}$ the shank diameter) results in an allowable tension on the screw of 6320 lb. The withdrawal controls.

Each lag screw must carry a total tension equal to the radial tension stress times the area between bolts. The area is simply the width of the beam times the spacing (sometimes called the *pitch*) of the screws. Thus the total force per bolt, T, equals

$$T = f_r \times b \times p$$

where p is the spacing between bolts. Solving for p yields

$$p = \frac{T}{f_r \times b} = \frac{3744}{157 \times 8.75}$$
$$= 2.72 \text{ in.} \quad \text{use } 2\tfrac{1}{2} \text{ in.}$$

Tables 6.17 and 6.18 investigate the radial stress along the length of the curve. Because the length of the curve in the example is so short, there is little change in moment and consequently little change in radial tension. The recommended screw spacing is virtually constant throughout the length of the curve.

SHEAR

For a rectangular shape with a constant cross-section the maximum shear stress occurs at the neutral axis and equals

$$f_v = \frac{3V}{2bd}$$

For a tapered beam this relation is true only at the ends of the beam. At other locations along its length the maximum shear stress occurs along the tapered edge. The shear stress is generally less than that given by the formula above.

In the example we are concerned only about shear stress at the ends of the arch pieces. It appears that the increases away from the ends will provide enough strength to carry the shear. Near the ends it is accurate to use the familiar formula for a rectangle; elsewhere, using the formula will be conservative. In Tables 6.17 and 6.18 the shear capacity at all points along the arch are checked for load combinations I and II. The revised depths are satisfactory at all locations for shear. These tables use the depths revised for bending; in the final design, based on deflection, the beam depths are generally greater so the shear capacity will be acceptable for the final design.

COMBINED HORIZONTAL AND VERTICAL WIND LOAD

As mentioned early in the chapter, the total effect of the wind load must be considered. What is of concern is uplift at the

228 GLUED-LAMINATED ARCHES

TABLE 6.19 FASTENER CALCULATIONS

Load Case	Vertical Reactions		Horizontal Reactions	
	Left	Right	Left	Right
Dead	7,808	7,808	4,805	4,805
Horizontal wind	−2,197	2,197	−8,112	2,704
Vertical wind	−7,168	−7,168	−4,411	−4,411
Total wind	−9,365	−4,971	−12,523	−1,707
2/3 × Dead load + wind	−4,160	288	−9,320	1,496
Mechanical fastener?	Yes	No	Yes	No

base of the arch. Table 6.19 shows the vertical and horizontal reactions that were developed previously, as well as the combined effect of the wind on the vertical surface and on the horizontal surfaces. Since the profile of the arch has not changed, these values will remain the same.

The dead load will resist uplift if the reaction due to the dead load is greater than the reaction due to the wind. To assure an extra degree of safety, building codes require that only 2/3 of the dead load be used to resist the wind load. If this is not sufficient, mechanical fasteners must be used that will carry the difference.

DEFLECTION OF THE ARCH

The deflection of the arch can be found fairly easily using the principle of virtual work. An explanation of the procedure is given in the last section of this chapter so that the reader will have a better understanding of where the values come from and also, it is hoped, to whet one's appetite to pursue learning about this topic. At this point, however, it is more important to focus on the relationship of the arch shape and the deflections.

The vertical deflection at the centerline and the horizontal deflection at the haunch

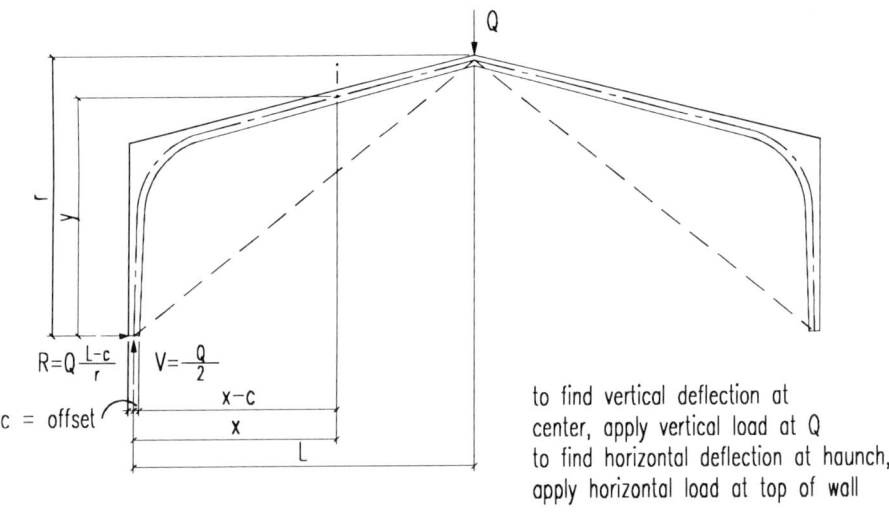

FIGURE 6.9 Equilibrium diagram for virtual load at center.

are calculated using three depths at the curve. The first case uses the preliminary estimates of the arch size and the assumption that the arch is shaped as shown in Figure 6.1a. The second case uses the revised arch dimensions to meet the strength requirements. The third case uses deflection criteria to size the arch. In this case, rather than keeping the profile shown in Figure 6.1a, a rounded profile was used over the haunch. An analysis of the deflection indicated that the arch needs to be deeper at the tangent points where the curved haunch meets the straight legs. A uniform depth of 30 in. is chosen to minimize the deflection. The strength criteria are met with this design.

Tables 6.21 through 6.23 show the design and deflections for the three trial sizes. The results are summarized below.

The deflections are compared with some standard criteria. The maximum vertical deflection for any wood roof without plaster below is usually taken as $L/180$, where L is the length of the spanning member in inches. Load case I, with symmetrical snow and dead loads, will produce the greatest vertical deflection. Based on the revised design with a haunch depth of 24 in., the centerline deflection was 6.02 in. This converts to a deflection of $L/126$—not quite acceptable. The deflection, of course, is related to the moment of inertia, which changes as the depth cubed. A slight increase in depth will produce a significant reduction in deflection. Increasing the depth by about 25%, to 30 in., will reduce the deflection well beyond the deflection limit of $L/180$ to the preferred limit of $L/240$. The centerline deflection with this configuration is shown on Table 6.23 as 2.90 in. or $L/261$.

The maximum horizontal deflection of the wall is usually taken to be $\frac{1}{180}$ of the height, h. In the example the height is measured to the top of the arch at 14'-9". For load case I the arch with a haunch of 24 in. deep gives a horizontal deflection of 1.78 in., equivalent to $h/100$. The arch with a haunch of 30 in. deep gives an even smaller horizontal deflection of 0.87 in., equivalent to $h/204$.

Based on the two deflection criteria for load case I, the revised arch, with a uniform haunch depth of 30 in. will produce satisfactory deflections. We would probably also want to check the horizontal deflection at the eave due to load case III, dead plus wind, as a further check.

ARCH DEFLECTIONS

The deflection of an arch can be found using the principle of virtual work. What follows is a brief discussion of the principle of virtual work and the derivation of the equation that is used to calculate deflections. We feel that with a very little practice, using this approach will not be

TABLE 6.20 TRIAL BEAM DEPTHS AND DEFLECTIONS

	Trial Depth	Design for Bending	Design for Deflection
Depth of haunch	42.00 in.	24.00 in.	30.00 in.
Vertical deflection at centerline	6.35 in.	6.02 in.	2.90 in.
	L/119	L/126	L/261
Horizontal deflection at haunch	2.02 in.	1.78 in.	0.87 in.
	h/88	h/100	h/204

daunting and that the benefit, a better understanding of the behavior of structures, will be worth the effort.

Work is defined as a force moving through a distance and quantitatively is given as

$$\text{work} = Q \times D$$

To understand the principle of virtual work, assume that a structure is subjected to a series of real loads, Q. The loads will cause a series of deformations in the structure at the point of application of the loads, which we will call D. The loads will also result in internal stresses being developed in equilibrium with the external loads; the internal stresses are designated as q. Associated with the internal stresses are internal strains, d. The principle of conservation of work states that the sum of the external work equals the sum of the internal work:

$$\sum QD = \sum qd$$

If a structure is subjected to a series of real loads, it will develop a series of real stresses and strains internally. The real internal work will be the sum of the stresses times the strains, $\sum qd$. If, simultaneously, a structure is subjected to a "virtual" or imaginary load, there will be an external virtual displacement along the line of action of the virtual load. The virtual work done by this force moving through a displacement will equal QD and be equal to the sum of the internal work.

This principle can be applied to statically determinate structures to determine the displacement due to a series of loads by following these steps:

1. Based on the real applied loads, calculate the internal forces within the structure. In the example these will be the moments that we have calculated at various locations along the arch. Designate these as M.
2. Find the displacements resulting from the internal moments, M. Work done by a force is the force times a distance; work done by a moment is the moment times a rotation. Since we are calculating work done by a moment, the displacements that we need will be in the form of rotations. The rotation, or change in slope, of an element is the integral of the moment divided by EI.

$$\theta = \int \frac{M}{EI} dx$$

This term can be approximated by the sum

$$\theta = \sum \frac{M}{EI} \Delta L$$

where ΔL is the length of each segment in the arch.

3. Select the location and direction of the displacement that is to be found.
4. Apply a virtual force, $Q = 1$, at the location selected, with its line of action in the direction of the displacement.
5. Draw a free-body diagram and write the equation for internal moments, m, that result from the applied virtual force.
6. Write the work equation, $Q \times D = m \times \theta$:

$$Q \times D = \sum \frac{M}{EI} m \Delta L$$

7. $\sum (M/EI) \Delta L$ is constant for any set of real applied loads. The work equation can be rewritten to solve for D by dividing both sides by Q. Q is the virtual force and can be taken as any value. We will set $Q = 1$, so that it will drop out of the equation. We are left with

TABLE 6.21   DEFLECTIONS DUE TO DEAD PLUS SYMMETRICAL SNOW LOADS: HAUNCH DEPTH = 42 IN.; EAVE HEIGHT = 177 IN.

Points	x Value (in.)	y Value (in.)	Delta L (ft)	Moment Due to Dead + Snow (lb-ft)	Deflection at Centerline		Deflection at Haunch	
					Moment, m, Due to Virtual Force, Q (lb-ft)	Sum $(Mm/EI) \times$ Delta L	Moment, m, Due to Virtual Force, Q (lb-ft)	Sum $(Mm/EI) \times$ Delta L
A	6	0	0.00	0	0.000	0.00	0.000	0.00
B	7	69	5.75	−79,526	−6.997	1,300.26	−2.424	450.49
C	9	138	5.75	−159,057	−13.993	3,009.97	−4.848	1,042.84
D	18	177	3.34	−187,761	−17.229	675.41	−6.218	243.76
E	43	208	3.31	−180,539	−18.391	203.49	−4.714	52.16
F	78	226	3.33	−143,437	−17.290	515.83	−3.840	114.57
G	180	253	8.78	−42,771	−11.531	1,017.72	−2.561	226.01
H	282	279	8.78	5,087	−5.766	−104.67	−1.280	−23.25
I	384	306	8.78	0	0.000	0.00	0.000	0.00
						6,618.01		2,106.59
					Deflection at Centerline		Deflection at Haunch	
Delta = Sum $(Mm$ Delta $L/I) \times 1{,}728/E =$					Delta (in.) = 6.35 Delta/L = 1/119		Delta (in.) = 2.02 Delta/h = 1/88	
Column	1	2	3	4	5	6	7	8

TABLE 6.22 DEFLECTIONS DUE TO DEAD PLUS SYMMETRICAL SNOW LOADS: HAUNCH DEPTH = 24 IN.; EAVE HEIGHT = 177 IN.

Points	x Value (in.)	y Value (in.)	Delta L (ft)	Moment Due to Dead + Snow (lb-ft)	Deflection at Centerline		Deflection at Haunch	
					Moment, m, Due to Virtual Force, Q (lb-ft)	Sum $(Mm/EI) \times$ Delta L	Moment, m, Due to Virtual Force, Q (lb-ft)	Sum $(Mm/EI) \times$ Delta L
A	6	0	0.00	0	0.000	0.00	0.000	0.00
B	7	69	5.75	−79,526	−6.997	752.47	−2.424	260.70
C	9	138	5.75	−159,057	−13.993	1,541.10	−4.848	533.93
D	18	177	3.34	−187,761	−17.229	977.95	−6.218	352.95
E	43	208	3.31	−180,539	−18.391	1,090.59	−4.714	279.56
F	78	226	3.33	−143,437	−17.290	1,100.53	−3.840	244.44
G	180	253	8.78	−42,771	−11.531	900.41	−2.561	199.96
H	282	279	8.78	5,087	−5.766	−90.42	−1.280	−20.08
I	384	306	8.78	0	0.000	0.00	0.000	0.00
						6,272.63		1,851.46

Delta = Sum(Mm Delta L/I) × 1,728/E =

	Deflection at Centerline	Deflection at Haunch
	Delta (in.) = 6.02 Delta/L = 1/126	Delta (in.) = 1.78 Delta/h = 1/100

| Column | 1 | 2 | 3 | 4 | 5 | 6 | 7 | 8 |

TABLE 6.23 DEFLECTIONS DUE TO DEAD PLUS SYMMETRICAL SNOW LOADS: HAUNCH DEPTH = 30 IN.; EAVE HEIGHT = 177 IN.

Points	x Value (in.)	y Value (in.)	Delta L (ft)	Moment Due to Dead + Snow (lb-ft)	Deflection at Centerline		Deflection at Haunch	
					Moment, m, Due to Virtual Force, Q (lb-ft)	Sum $(Mm/EI) \times$ Delta L	Moment, m, Due to Virtual Force, Q (lb-ft)	Sum $(Mm/EI) \times$ Delta L
A	6	0	0.00	0	0.000	0.00	0.000	0.00
B	7	69	5.75	−79,526	−6.997	473.86	−2.424	164.17
C	9	138	5.75	−159,057	−13.993	650.15	−4.848	225.25
D	18	177	3.34	−187,761	−17.229	549.13	−6.218	198.19
E	43	208	3.31	−180,539	−18.391	558.38	−4.714	143.13
F	78	226	3.33	−143,437	−17.290	419.39	−3.840	93.15
G	180	253	8.78	−42,771	−11.531	429.35	−2.561	95.35
H	282	279	8.78	5,087	−5.766	−60.58	−1.280	−13.45
I	384	306	8.78	0	0.000	0.00	0.000	0.00
						3,019.69		905.79

Delta = Sum(Mm Delta L/I) × 1,728/E =

	Deflection at Centerline	Deflection at Haunch
	Delta (in.) = 2.90	Delta (in.) = 0.87
	Delta/L = 1/261	Delta/h = 1/204

| Column | 1 | 2 | 3 | 4 | 5 | 6 | 7 | 8 |

$$D = \sum \frac{M}{EI} m \, \Delta L$$

The great advantage of this approach is that we have already done the hardest part, finding the moments caused by the external forces. Refer to Table 6.21, which calculates the downward deflection of the arch at the center due to load case I, dead plus symmetrical snow loads.

Refer to Figure 6.9, which shows the application of the virtual load, $Q = 1$ lb, at the center. In the free-body diagram the reaction at the base must equilibrate the moment caused by Q.

$$R \times r = Q \times L, \quad \text{so } R = Q \times \frac{L}{r} = \frac{L}{r} \quad \text{(lb)}$$

since $Q = 1$ lb

If a free-body diagram is cut at any point along the arch, such as point i, there will be a moment in the arch, m, needed to balance the reactions. Taking the sum of the moments about point i yields the relationships

$$Ry - Q(x - c) + m = 0$$

or

$$m = Q(x - c) - Ry = (x - c) - \frac{L}{r} y$$

where c is the offset of the support, usually the base depth divided by 2.

In the example, find the moment at point D due to the virtual force Q.

$$R = \frac{L}{r} \quad \text{(lb)} = \frac{32 \text{ ft}}{26 \text{ ft}} \quad \text{(lb)} = 1.231 \text{ lb}$$

At point D, $x_i = 18$ in. $= 1.5$ ft and $y_i = 177$ in. $= 14.75$ ft, and $c = 6$ in. $= 0.5$ ft.

$$m = (x_i - c) - \frac{L}{r} \times y_i$$
$$= (1.5 - 0.5) - 1.231 \times 14.75$$
$$= -17.213 \text{ lb-ft}$$

The term ΔL is the length of each segment, so at point i,

$$\Delta L = \sqrt{(x_i - x_{i-1})^2 + (y_i - y_{i-1})^2}$$

Once these terms have been derived, the mathematics readily lend themselves to a tabular format. In Table 6.21 the downward deflection at the centerline is calculated as 6.35 in. due to dead load plus symmetrical snow load. Divide this deflection by the length to get a relative magnitude for the deflection of

$$\frac{\Delta}{L} = \frac{6.35 \text{ in.}}{(2 \times 378 \text{ in.})} = \frac{1}{119}$$

or

$$\Delta = \frac{L}{119}$$

As with beams, a deflection in the range of $L/180$ under dead plus snow loads is acceptable. The deflection shown on Table 6.21 is slightly higher than the recommended ratio. In Table 6.23 the depth of the haunch is increased to 30 in. and an acceptable deflection is found. Tables 6.21 through 6.23 also show the horizontal deflection at the eave. This deflection is a result of the thrust effect of the arch.

7

PLYWOOD: Diaphragms, Shearwalls and Stressed-Skin Panels

Plywood is a manufactured wood-based product that was developed as a replacement for board sheathing. Plywood is a flat panel made from a series of thin layers that are glued together to form a single large board which replaces a number of smaller boards. It minimizes the amount of labor to install the sheathing and results in a more uniform and more unified construction. During the original development, the designers of plywood attempted to make the best use of certain properties of wood and to overcome the adverse effect of other properties. Wood does not have the same strength properties in all directions; the strength parallel to the grain of the fibers is much greater than that perpendicular to fibers. To create a more homogeneous product, the layers of plywood, called the *plies*, are alternated so that the grain of the wood in one layer is at right angles to the grain in the adjacent layer. Furthermore, many defects, such as holes or knots, which diminish the strength of the wood, may be cut out and patched. The number and size of the knots that are retained determine the grade of the plywood.

Plywood is manufactured by peeling a thin strip of wood, called *veneer*, from a log. Short sections of logs, usually $8\frac{1}{2}$ ft long, are placed in a lathe, called a peeling mill, in which a long blade peels the veneer from the log. The veneer can range in thickness from $\frac{1}{16}$ to $\frac{1}{4}$ in. The veneers are then cut to length and dried to a very low moisture content of about 2%. They are then visually graded for defects and repaired as necessary to meet the grade desired. Following this, glue is applied and the veneers are laid together to form panels. The glued panel is then subjected to heat and pressure so that the glue will set properly. Depending on the use, some panels will be sanded after this process; sanding is usually done on panels that will be exposed to view in their final use or panels that will require painting.

Panels are made of an odd number of plies or laminations. Since the grain of adjacent layers alternate, in a panel with an odd number of layers, both of the face plies will be in the same orientation. Typically, the grain of the face plies runs parallel to the long direction of the panel.

Each panel fabricated in a mill that is a member of the American Plywood Association (APA) will be grade-stamped with information about the plywood material, thickness, strength, and possible end uses. Examples of grade stamps are shown in Figure 7.1 and are explained in the section on plywood specifications. The fabrication of grade-stamped plywood panels conforms to the U.S. Product Standard PS 1-83.

In recent years a number of similar products have been developed which are referred to generically as *particleboard*, or by the APA as *nonveneer boards*. These boards are composed of small pieces of wood that have been fabricated into boards by coating the particles with a binder and then forming them into mats. The mats are pressed, and in some cases heated, until the binder sets.

Particleboard classifications account for two aspects of the board: (1) the particle size and composition, and (2) the type of glue or binder that holds the particles together. The most commonly used boards are waferboard (1-MW or 2-MW), flakeboard (classified as 2-MF), or oriented-strand board (OSB). A *wafer* is defined as a particle of length $1\frac{1}{4}$ in. aligned with the grain of the wood. A *strand* is a particle of uniform thickness, of predetermined length. The length must be at least three times the width of the particle. *Flakes* can be of any predetermined size and thick-

TYPICAL APA REGISTERED TRADEMARKS

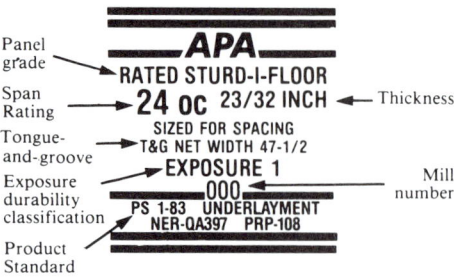

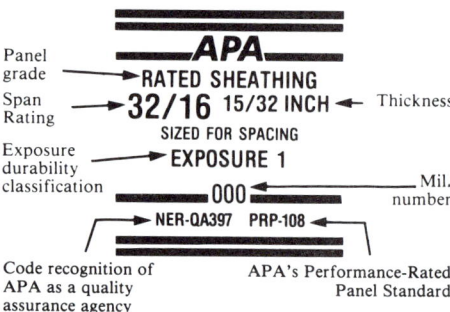

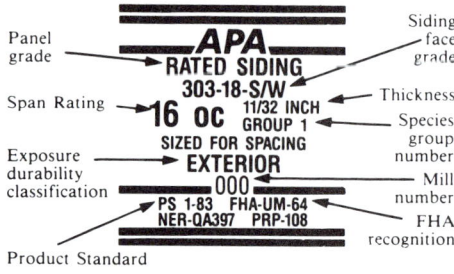

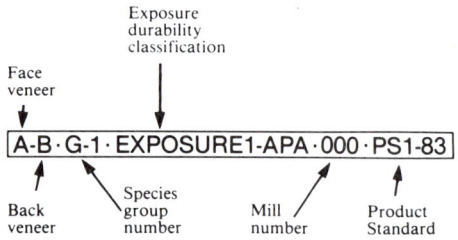

FIGURE 7.1 Typical APA registered trademarks.

ness. They are of uniform thickness and the grain runs in the plane of the flake. The numbers in the classifications refer to the glue types. These classifications are used throughout the building codes to reference particular types of particleboard.

The strength of the panels is more dependent on the binders than on the wood species or flake size. Two types of binders are used; particleboard is classified by the two binders as type 1 when a urea-formaldehyde resin is used, or as type 2 when the binder is phenol-formaldehyde. In all strength measurements the type 2 panel is equal to or stronger than the type 1 panel. The fabrication of particleboard at participating plants is monitored and graded by the American Plywood Association according to an American National Standards Institute Standard, ANSI A208.1-1979. Because particleboard is an engineered wood product it can be designed to meet any reasonable standard. To simplify construction specifications and to keep the market as large as possible, particleboard have been designed to match the strengths of common plywood. Throughout the discussion in this chapter, particleboard can be interchanged with plywood. Where it is advisable in some instances to use one instead of the other, the advantages and disadvantages are discussed. Where information is available, allowable loads are provided for both plywood and particleboard.

Another type of board used extensively in construction is fiberboard. Unlike particleboard, which is made of fairly large, discrete particles, fiberboard is made of microscopic cellulose fibers that are interlocked to form boards. Fiberboard derives its strength from the interfelting of the fibers rather than from glues or binders. Fiberboard, which is sometimes called nail-base, has less strength and nail-holding capability than particleboard. It is sometimes used as a shearwall sheathing material in situations where there are very light loads, but it is never recommended in a horizontal diaphragm. Fiberboard that is formed under heat and pressure is referred to as *hardboard* and is used frequently as the underlayment for floors. The shear capabilities of hardboard are superior to those of standard fiberboard but are only about 50% of the capacity of the same thickness of plywood or particleboard.

DESCRIPTION OF THE BUILDING AS A WHOLE

Plywood and other structural-use panels are the most common structural sheathing materials used in wood construction. Because of the strength properties of plywood and because of its location in a building, it is called upon to perform a large variety of tasks, some structural and some nonstructural. The proper specification for plywood in a building, then, needs to address its ability to perform each specific task.

Plywood is installed primarily in floors, roofs, and walls as a sheet (see Figure 7.2). In these locations it must carry loads ap-

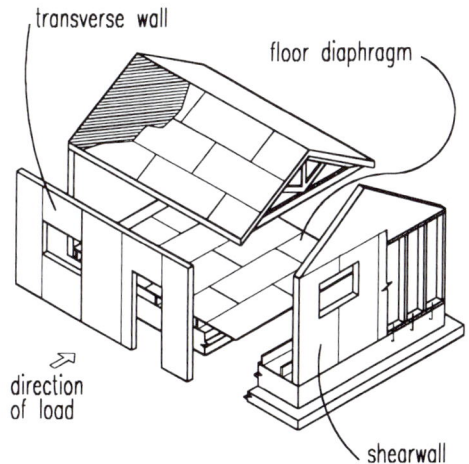

FIGURE 7.2 Schematic building diagram.

plied perpendicularly to its surface. For example, as a floor, it must carry the weight of the building occupants and furniture (live loads), as well as other portions of the building that rest on the plywood (dead load). Gravity causes these loads to act downward. In a similar manner, plywood on a wall must carry wind loads which are horizontal and are applied perpendicularly, or transverse, to the surface of the plywood. The walls are in turn supported by the floor, by the roof, and by walls parallel to the direction of the wind. The walls parallel to the wind, called *shearwalls* or *vertical diaphragms*, will also carry the wind load, but they do it by developing their shear strength. The shear strength is also referred to as the *racking strength* or *racking resistance* (see Figure 7.3).

The roof and floors also participate in carrying the lateral loads as diaphragms. Under a wind load a transversely loaded wall works like a vertical beam, transferring loads up to the roof and down to the floor below. The floor and roof transfer the load horizontally to the adjacent walls (the shearwalls, which are parallel to the wind direction). To do this the floor and roof operate as large, deep beams and develop shear and bending. These elements are referred to as *horizontal diaphragms*. Proper design of the roof and floor structure then must account for transverse gravity loads, as well as the shear and bending loads developed to resist lateral loads.

Plywood is used in a number of other applications on the modern construction site. In addition to its use as a sheathing material, plywood is used extensively in the construction of concrete formwork, in structural components such as box beams (see Chapter 3) or stress-skin panels (later in this chapter), as truss gusset plates (see Chapter 6), as a part of treated wood foundation walls, and as spacers in headers.

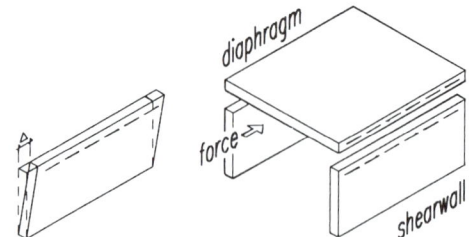

FIGURE 7.3 Racking action of shearwall.

SPECIFICATIONS

There are a number of grade and rating systems that are applied to plywood and nonveneer products. Each system is intended to be used for a particular type of application. To help avoid confusion the various systems and their uses are discussed here.

Perhaps the most commonly used and easiest to understand system is the *span rating system* or *Panel Identification Index*. The span rating system refers to the ability of the panel to carry transverse loads. It is used to specify sheathing that will be used primarily on horizontal planes (roofs and floors) and to identify allowable spans between supports. The span rating system was developed so that plywood for standard conditions could be specified accurately without requiring calculations. The system contains two numbers, such as 32/16. The first number refers to the maximum allowable span for the plywood when used on a roof (in this case 32 in.); the second number refers to the maximum allowable floor span (16 in.). The Panel Identification Index is included on the grade stamp for all rated panels (see Figure 7.1). Table 7.1 correlates nominal panel thickness with span rating. For each span rating a number of possible thicknesses are shown, the most common thickness for each is the least thick, which is shown in boldface type.

TABLE 7.1 NOMINAL PANEL THICKNESS (INCHES) CORRELATED TO SPAN RATING[a]

Span Rating	$\frac{3}{8}$	$\frac{7}{16}$	$\frac{15}{32}$	$\frac{1}{2}$	$\frac{19}{32}$	$\frac{5}{8}$	$\frac{23}{32}$	$\frac{3}{4}$	$\frac{7}{8}$	1	$1\frac{1}{8}$
APA-Rated Sheating											
$\frac{24}{0}$	**0.375**	0.437	0.469	0.500							
$\frac{24}{16}$		**0.437**	0.469	0.500							
$\frac{32}{16}$			**0.469**	0.500	0.594	0.625					
$\frac{40}{20}$					**0.594**	0.625	0.719	0.750			
$\frac{48}{24}$							**0.719**	0.750	0.875		
APA-Rated Sturd-I-Floor											
16 o.c.					**0.594**	0.625					
20 o.c.					**0.594**	0.625					
24 o.c.							**0.719**	0.750			
32 o.c.									**0.875**	1.000	
48 o.c.											**1.125**

[a] Nominal thickness is given in decimal inches for each span rating. The predominant thickness for each span rating is given in boldface type. Neither $\frac{1}{2}$-in. nor $\frac{3}{4}$-in.-thick panels are used predominantly; the next smaller panel, $\frac{15}{32}$ in. and $\frac{23}{32}$ in., respectively, are more common. The table applies to three-, four-, and five-ply plywood and to OSB particleboard.

Source: APA Technical Note N375A, September 1991.

Table 7.2 shows the allowable live loads for plywood used as roof decking, and Table 7.3 shows the allowable loads for floors. Both of these tables can be used only when the face grain of the plywood spans perpendicularly across the framing members. This condition is diagrammed in Figure 7.4a. The fourth column in each of these tables is labeled "unsupported edge maximum length (in.)," which refers to the maximum distance that the unsupported edge can span. The unsupported edge is the edge perpendicular to the framing. To meet the requirements, the unsupported edge must be supported by blocking, tongue-and-groove edges, or plywood clips (see Figure 7.4 for these details). Usually, the most expensive option is blocking, which will not be used unless it is also needed as part of the diaphragm system or to meet bracing requirements in Figure 2.8. The most common solution is to use tongue-and-groove (T&G) plywood; the initial material cost will be more than when clips are used, but the long-term results will be better.

In Table 7.2, for roof loads, the minimum load that is listed for each panel identification index is 30 psf. This load corresponds to the rated span. For a roof live load of 30 psf, the span rating will produce a safe design. For roof live loads greater than 30 psf, all panels will have to be supported at intervals less than the listed rating. For floors, shown in Table 7.3, the span rating is based on floor live loads of 100 psf or a concentrated load of 2000 lb. For residential applications the live load is usually 40 psf, and for business application the live load is 50 psf. Usually, the highest live loads that are used for wood floors is 125 psf for light storage. Because the concentrated load usually controls the design, the use of the span rating is therefore safe for almost all floor applications and is very conservative for

240 PLYWOOD

TABLE 7.2 ALLOWABLE UNIFORM LIVE LOADS FOR ROOF DECKS
Long dimension perpendicular to supports[e]

Panel Span Rating	Minimum Panel Thickness (in.)	Maximum Span (in.) With Edge Support[a]	Maximum Span (in.) Without Edge Support	Allowable Live Loads[b] (psf) for Spacing of Supports Center to Center (in.): 12	16	20	24	32	40	48	60
APA-Rated Sheathing (Plywood, OSB and Com-ply Panels)[c]											
12/0	5/16	12	12	30							
16/0	5/16	16	16	70	30						
20/0	5/16	20	20	120	50	30					
24/0	3/8	24	20, 24[d]	190	100	60	30				
24/16	7/16	24	24	190	100	65	40				
32/16	15/32	32	28	325	180	120	70	30			
40/20	19/32	40	32		305	205	130	60	30		
48/24	23/32	48	36			280	175	95	45	35	
APA-Rated Sturd-I-Floor (Plywood, OSB and Com-ply Panels)[f]											
16 o.c.	19/32	24	24	185	100	65	40				
20 o.c.	19/32	32	32	270	150	100	60	30			
24 o.c.	23/32	40	36		240	160	100	50	30	25	
32 o.c.	7/8	48	48			295	185	100	60	40	
48 o.c.	1-3/32	60	48				290	160	100	65	40

[a] Tongue-and-groove edges, panel clips (one midway between each support, except two equally spaced between supports 48 in. o.c.), lumber blocking, or other.
[b] 10 psf dead load assumed.
[c] Includes APA-rated sheathing/ceiling deck.
[d] Maximum unsupported length 20 in. for 3/8-in. plywood and 24-in. for 15/32-in. and 1/2-in. plywood.
[e] Applies to panels 24 in. or wider.
[f] Also applies to c-c pluged grade plywood.
Source: APA, *Design/Construction Guide,* "Residential & Commercial," Table 21, April 1993.

TABLE 7.3 ALLOWABLE UNIFORM LIVE LOADS FOR FLOOR DECKS
Long dimension perpendicular to supports

Panel Span Rating	Minimum Panel Thickness (in.)	Maximum Span (in.)	Allowable Live Loads (psf)[a] for, Spacing of Supports Center to Center (in.): 12	16	20	24	32	40	48
APA-Rated Sturd-I-Floor (Plywood, OSB and Com-ply Panels)									
16 o.c.	19/32	16	185	100					
20 o.c.	19/32	20	270	150	100				
24 o.c.	23/32	24	430	240	160	100			
32 o.c.	7/8	32		430	295	185	100		
48 o.c.	1-3/32	48			460	290	160	100	55

[a] 10-psf dead load assumed.
Live-load deflection limit is $L/360$.
Source: APA, *Design/Construction Guide,* "Residential & Commercial," Table 6, April 1993.

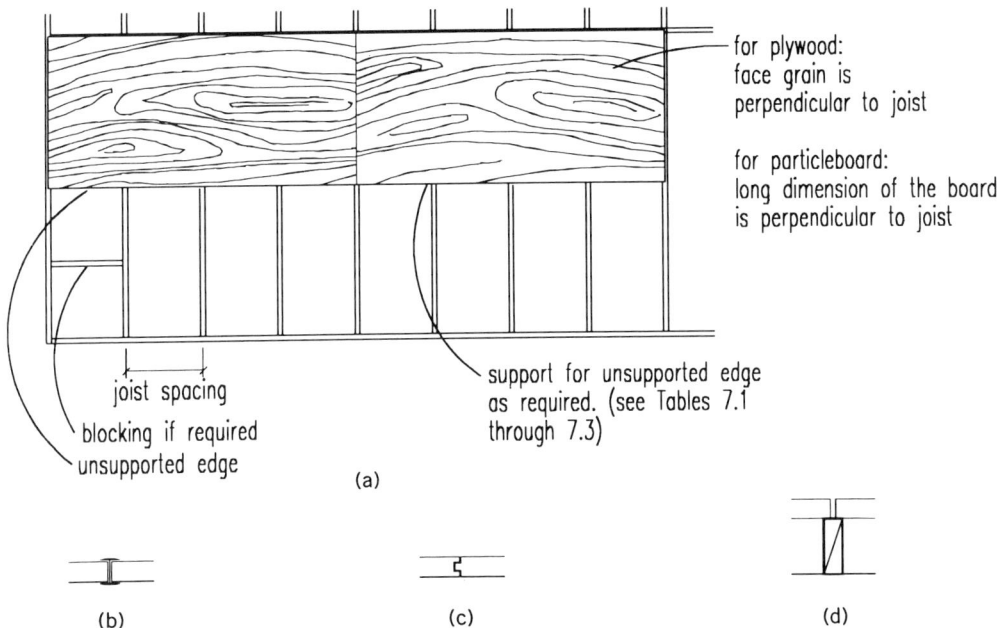

FIGURE 7.4 (a) Face orientation to supporting framing, (b) plywood clips, (c) tongue and groove edge, and (d) edges supported by solid wood blocking.

TABLE 7.4 ALLOWABLE UNIFORM LIVE LOADS FOR HEAVY-DUTY FLOOR DECKS[a]

Uniform Live Load (psf)	Center-to-Center Spacing (in.) (Use nominal 2-in.-wide framing at supports unless noted)					
	12[b]	16[b]	20[b]	24[b]	32	48[c]
50	$\frac{32}{16}$, 16 o.c.	$\frac{32}{16}$, 16 o.c.	$\frac{40}{20}$, 20 o.c.	$\frac{48}{24}$, 24 o.c.	48 o.c.	48 o.c.
100	$\frac{32}{16}$, 16 o.c.	$\frac{32}{16}$, 16 o.c.	$\frac{40}{20}$, 20 o.c.	$\frac{48}{24}$, 24 o.c.	48 o.c.	$1\frac{1}{2}^d$
150	$\frac{32}{16}$, 16 o.c.	$\frac{32}{16}$, 16 o.c.	$\frac{40}{20}$, 20 o.c.	$\frac{48}{24}$, 48 o.c.	48 o.c.	$1\frac{3}{4}^e$, 2^d
200	$\frac{32}{16}$, 16 o.c.	$\frac{40}{20}$, 20 o.c.	$\frac{48}{24}$, 24 o.c.	48 o.c.	$1\frac{1}{8}^e$, $1\frac{3}{8}^d$	2^e, $2\frac{1}{2}^d$
250	$\frac{32}{16}$, 16 o.c.	$\frac{40}{20}$, 24 o.c.	$\frac{48}{24}$, 48 o.c.	48 o.c.	$1\frac{3}{8}^e$, $1\frac{1}{2}^d$	$2\frac{1}{4}^e$
300	$\frac{32}{16}$, 16 o.c.	$\frac{48}{24}$, 24 o.c.	48 o.c.	48 o.c.	$1\frac{1}{2}^e$, $1\frac{5}{8}^d$	$2\frac{1}{4}^e$
350	$\frac{40}{20}$, 20 o.c.	$\frac{48}{24}$, 48 o.c.	48 o.c.	$1\frac{1}{8}^e$, $1\frac{3}{8}^d$	$1\frac{1}{2}^e$, 2^d	
400	$\frac{40}{20}$, 20 o.c.	48 o.c.	48 o.c.	$1\frac{1}{4}^e$, $1\frac{3}{8}^d$	$1\frac{5}{8}^e$, 2^d	
450	$\frac{40}{20}$, 24 o.c.	48 o.c.	48 o.c.	$1\frac{3}{8}^e$, $1\frac{1}{2}^d$	2^e, $2\frac{1}{4}^d$	
500	$\frac{48}{24}$, 24 o.c.	48 o.c.	48 o.c.	$1\frac{1}{2}^d$	2^e, $2\frac{1}{4}^d$	

[a] Plywood continuous over two or more supports with face grain across supports. Use tongue-and-groove edges, or provide structural blocking at panel edges, or install a separate underlayment.
[b] A-C group 1 sanded plywood panels may be substituted for span-rated Sturd-I-Floor panels ($\frac{1}{2}$ in. for 16 o.c.; $\frac{5}{8}$ in. for 20 o.c.; $\frac{3}{4}$ in. for 24 o.c.).
[c] Nominal 4-in.-wide supports.
[d] Group 1 face and back, any species inner plies, sanded or unsanded, single layer.
[e] All Group 1 or Structural I plywood, sanded or unsanded, single layer.
Source: APA, *Design/Construction Guide,* "Residential & Commercial," Table 11, April 1993.

most. For roofs it is not overly conservative. In Tables 7.1 through 7.3 the designation "16 o.c." refers to a panel rating, not the actual joist spacing.

For floors with live loads exceeding 100 psf, APA recommends the plywood sizes and joist spacings shown in Table 7.3. These values are based on an allowable deflection of $L/360$. If the higher required deflection limit of $L/480$ is needed, read the required plywood thickness from the next-higher joist spacing. As an example, if the deflection is limited to $L/480$ (0.05 in.), the required live load is 100 psf, and the actual joist spacing is 24 in., select the plywood thickness from the column for 32-in. support spacing (i.e., $\frac{7}{8}$ or $1\frac{3}{32}$ in. thick).

The span rating system is a performance-based system: panels of different thicknesses can have the same span rating. Plywood made of better-quality wood will carry a higher rating than the equivalent thickness made from wood of lesser quality (see Table 7.1). For applications other than those where gravity loads are carried, the span rating system is not useful. For diaphragm design and shearwall design the specific engineering qualities of the panel must be considered. Therefore, plywood is graded in a manner similar to dimensional lumber. The complete identification of a piece of plywood is based on four primary factors: the species of wood used in the outer veneers, the plywood grade, the grade of the veneers, and the type of glue.

Species of Wood

Wood for plywood is divided into five groups based on the strength and density of the wood. Group 1 has the greatest strength and is used where strength is most needed. Because the stresses in the plywood are greatest at the extreme fibers, plywood is often made with a better group of wood on the outer veneers and with the interior made from weaker groups. If a species group number is specified, it will refer to the outer veneer only. Table 7.5 lists the species by groups.

The structural grade of plywood is based on the species of wood used on the outer veneers. Structural I is limited to plywood with group 1 species only. This grade is commonly available and should be used when a higher strength of plywood is needed. Structural II is limited to plywood with species groups 1, 2, or 3 on the outer veneers. This is not as readily available as Structural I. In the section on diaphragm design it will be seen that the increase in strength between Structural I and Structural II is not very great. If the additional strength is not needed, "Structural II or better" may be specified so that whichever is available may be used. To avoid confusion, it is recommended that on a single job site, only one grade specification be used.

Veneer Grades

Veneer grades are visually assigned to the veneer before it is made into a panel. The grade is based on the type, size, frequency, and quality of defects. The grades are designated as N, A, B, C, C-plugged, and D. The highest grade is N, which is intended to be left visible and natural (unpainted). This grade has no knots and is virtually free of all other defects. Grade A is intended to be painted; it is free of knots and has limited splits, checks, or other open defects. Grade B permits knots up to 1 in. wide provided that they are sound (not rotten) and tight. Grade B veneer is often used on the back side of panels with an A veneer or in concrete formwork where Exterior B-B panels are needed because of their exposure to wet concrete. Grades C and D have larger allowable knots and other defects and larger allowable repairs. These veneer grades are used extensively in construction, the most common designations being Exterior C-C

TABLE 7.5 PLYWOOD SPECIES BY GROUPS

Group 1	Group 2	Group 3	Group 4	Group 5
Beech, American		Alder, Red	Aspen, Bigtooth Quaking	Basswood
Birch, Sweet Yellow		Birch, Paper		
	Cedar, Port Orford Cypress	Cedar, Alaskan	Cedar, Incense, Western Red	
			Cottonwood, Eastern Black (Western Poplar)	Poplar, Balsam
Douglas Fir 1[a]	Douglas Fir 2			
	Fir, Balsam California Red Grand Noble Pacific Silver White	Fir, Subalpine		
Larch, Western	Hemlock, Western	Hemlock, Eastern		
Maple, Sugar	Maple, Black	Maple, Bigleaf		
Pine, Caribbean Ocote	Pine, Pond Red Virginia West White	Pine, Jack Lodgepole Ponderosa Spruce	Pine, Eastern White Sugar	
Pine, Southern Loblolly Longleaf Shortleaf Slash				
		Redwood		
	Spruce, Black Red Sitka Sweetgum Tamarack Yellow-Poplar	Spruce, Engelmann White		
Apitong	Luaun Almon Bagtikan Mayapis Red Tangile White		Cativo	
Kapur	Mengkulang			
Keruing	Meranti, Red[b]			
Tanoak	Mersawa			

[a] Douglas Fir from trees grown in the states of Washington, Oregon, California, Idaho, Montana, Wyoming, and the Canadian Provinces of Alberta and British Colombia shall be classed as Douglas Fir No. 1. Douglas Fir from trees grown in the states of Nevada, Utah, Colorado, Arizona, and New Mexico shall be classed as Douglas Fir No. 2.

[b] Red Meranti shall be limited to species having a specific gravity of 0.41 or more based on green volume and oven dry weight.

where weather exposure is likely or C-D with the C-grade face on the exterior and protected soon after it is applied.

As is the case with visually graded lumber, knotholes usually establish the grade. Knotholes or any other defects that interrupt the wood fibers will diminish the strength of the wood, so the larger the knot or the more of them there are, the lower the wood strength. Knotholes can be repaired with tight-fitting "boat" patches (football-shaped patches) so that the strength of the surface can be maintained. In this manner a C veneer can be upgraded to an A, and a D veneer can be upgraded to a B—for strength but not for appearance.

Exposure

Plywood is graded for different types of exposure to weather or changes in the moisture content of the plywood.

Exterior-grade plywood is plywood that has been developed to withstand exposure to weather. To be marked as "exterior plywood," two conditions must be met: an exterior glue must be used and the veneer grade on both sides must be C or better. Plywood that is made with exterior glue and a D veneer cannot be marked "Exterior." "Exterior" glue is a water-insoluble glue that will continue to provide the necessary bonding strength when subjected to moisture or repeated cycles of high and low moisture content. Interior glue is not insoluble and must be used only where the plywood is well protected from the weather and from cyclical changes in moisture content. The restriction on the veneer grade for exterior plywood reduces the size of knotholes so that water cannot easily penetrate between the plies.

Exterior plywood may be left completely unprotected, but in severe climates it can rot like any other wood product. Simple precautions should always be followed with exterior plywood. Protect it if possible with paint, sealer, varnish, or the like. Water entering the end grain will cause the worst damage. Detail the plywood so that end grain is lapped with another material. Detail the plywood so that it is never dead flat; instead, provide sufficient slope so that the water may run off. If possible, avoid situations where the plywood is in contact with concrete or the ground. If this is not possible, specify pressure-treated plywood.

Interior-grade plywood can have any veneer grade and either kind of glue. Interior plywood is to be used only in locations that are well protected during use and during construction, such as interior walls. There are three strength specifications for interior plywood made with interior glue: Underlayment, C-D Plugged, and C-D. In areas with high humidity or in portions of a building such as a crawl space where high humidity is likely, an interior plywood with an intermediate glue (Interior IMG) should be specified. The intermediate glue includes pentachlorophenol, which inhibits the growth of molds and bacteria. An interior plywood made with exterior glue is also available and is referred to as "C-D Structural." For this panel, the exterior glue is specified for its greater strength and not for its weather resistance. Plywood designated Structural I or Structural II uses exterior glue. It is possible that this is the most commonly used plywood for construction, particularly in areas with large lateral forces, such as California and the southeast Atlantic states.

Plywood Grades

Plywood is grade stamped with a grade that incorporates all or most of the points raised above. There are two types of grades based on the end use of the plywood. Appearance grades are concerned mainly with the surface of the plywood; engineering grades are concerned more with the strength of the plywood than with appearance. The lowest stamp in Figure 7.1 shows a typical appearance grade stamp; the top three stamps show typical engineering grade stamps.

Allowable Strengths

For many common applications, the designer can readily design and specify plywood by using the span rating tables (Tables 7.2 and 7.3) and by using the diaphragm or shearwall design tables (Tables G.1 through G.10). There are other uses of plywood where the material strength must be known to proceed with the design. These applications include plywood boxbeams and headers using plywood plates (covered in Chapter 3) and stress-skin panels, which will be discussed at the end of this chapter. The allowable strengths of plywood are based on the species of the wood in the face plies and the grade of the plywood. Table 7.6 lists the allowable stresses of plywood as developed by the APA.

Specific values for the strength of particleboard are not published in the same format as those for plywood. The American Plywood Association publishes the design strengths of both plywood and particleboard in APA Technical Note N375A.[1] Strengths are shown for panels by referring to the Panel Identification Index. Included in the tables are multipliers for various panel types, including OSB. For panels with the same PII, the strength of OSB is equivalent to five-ply plywood in most directions; OSB has 10% less strength than five-ply in bending stiffness and in shear in the plane (rolling shear). Both of these values are critical in the design of stress-skin or sandwich panels used for floor or roof loads.

[1] American Plywood Association, Technical Note N375A, *Design Capacities of APA Performance Rated Structural-Use Panels*, September 1991, APA, P.O. Box 11700, Tacoma, WA 98411-0700.

Gravity Load Path

The first function of the roof or floor diaphragm is to receive loads applied perpendicularly to its surface and transfer them horizontally as a beam to its supports. Because the thickness of the diaphragm panel is rather thin, the span between supports, such as joists or rafters, must be very short. Tables 7.2 through 7.4 show the allowable spacing for various plywood grades and thicknesses. In the design of a plywood roof and diaphragm, one of the first steps is to select the thickness of the plywood to resist gravity loads. Refer to Example 7.1 for the selection of roof sheathing to carry gravity loads.

Lateral Load Path

In addition to carrying gravity loads, all buildings must withstand applied lateral loads and carry them from their points of application to resolution in the foundation and ultimately into the earth. Lateral loads can arise from a variety of forces, such as static water pressure during flooding, dynamic water surging, or even explosive blasts. For the majority of structures, though, the lateral loads most commonly considered in design are wind and seismic loads.

For both types of loads the lateral load path that transmits the forces to resolution are almost the same. At this point we discuss their similar effects and consider the differences later. In almost all wood structures the lateral load path consists of four major elements: horizontal diaphragms, vertical shearwalls, foundations, and the connections between each component. Figure 7.2 shows these elements diagrammatically.

The diaphragm is the horizontal element that carries the lateral force due to wind loads. It receives the force from the walls that are perpendicular to the wind load and transfers them to the walls (the

TABLE 7.6 ALLOWABLE STRESSES FOR PLYWOOD (PSI)

Type of Stress	Species Group of Face Ply	Grade Stress Level[a]				
		S-1		S-2		S-3
		Wet	Dry	Wet	Dry	Dry Only
Extreme fiber stress in bending, F_b	1	1,430	2,000	1,190	1,650	1,650
Tension in plane of plies, F_t	2, 3	980	1,400	820	1,200	1,200
Face grain parallel or perpendicular to span (at 45° to face grain use $\frac{1}{6}$ ft) F_t	4	940	1,330	780	1,110	1,110
Compression in plane of plies, F_c	1	970	1,640	900	1,540	1,540
	2	730	1,200	680	1,100	1,100
Parallel or perpendicular to face grain	3	610	1,060	580	990	990
(at 45° to face grain use $\frac{1}{3} F_c$)	4	610	1,000	580	950	950
Shear through the thickness,[b] F_v	1	155	190	155	190	160
Parallel or Perpendicular to face grain	2, 3	120	140	120	140	120
(at 45° to face grain use $2 F_v$)	4	110	130	110	130	115
Rolling shear (in the plane of plies), F_s	Marine and Structural I	63	75	63	75	—
Parallel or perpendicular to face grain (at 45° to face grain, use $1\frac{1}{3}) F_s$	All others[c]	44	53	44	53	48
Shear modulus (or modulus of rigidity), G	1	70,000	90,000	70,000	90,000	82,000
	2	60,000	75,000	60,000	75,000	68,000
Shear in plane perpendicular to plies (through the thickness)	3	50,000	60,000	50,000	60,000	55,000
At 45° to face grain use 4G	4	45,000	50,000	45,000	50,000	45,000
Bearing (on face), F_c	1	210	340	210	240	240
	2, 3	135	210	135	210	210
Perpendicular to plane of plies	4	105	160	105	160	160
Modulus of elasticity in bending	1	1,500,000	1,800,000	1,500,000	1,800,000	1,800,000
in plane of plies, E	2	1,300,000	1,500,000	1,300,000	1,500,000	1,500,000
Face grain parallel or	3	1,100,000	1,200,000	1,100,000	1,200,000	1,200,000
perpendicular to span	4	900,000	1,000,000	900,000	1,000,000	1,000,000

[a] See "Guide to Use of Allowable Stress and Section Properties Tables" in APA *Plywood Design Specifications*. To qualify for stress level S-1, glue lines must be exterior and only veneer grades N, A, and C (natural, not repaired) are allowed in either face or back. For stress level S-2, glue lines must be exterior and only veneer grades B, C-Plugged, and D are allowed in either face or back. Stress level S-3 includes all panels with interior or intermediate (IMG) glue lines.
[b] Shear-through-the-thickness stresses for Marine and Special Exterior grades may be increased 33%. See Section 3.8.1 of APA PDS for conditions under which stresses for other grades may be increased.
[c] Reduce stresses 25% for three-layer (four- or five-ply) panels over $\frac{5}{8}$ in. thick. Such layups are possible under PS 1-83 for APA rated sheathing, APA rated Sturd-I-Floor, Underlayment, C-C Plugged, and C-D Plugged grades over $\frac{5}{8}$ in. through $\frac{3}{4}$ in. thick.

Source: APA *Plywood Design Specifications* (PDS), August 1986.

If the diaphragm is properly designed and detailed, the transverse wall will act as a short vertical beam distributing loads primarily into the roof diaphragm and the foundation. The load will be in proportion to the stiffness of the wall in each direction. If the wall is 2 times as long as it is high (2h = l), then the roof will receive 16 times as much load as will the sidewallls. The transverse wall moment in the transverse wall will be 1/4 as much with the diaphragm as it would be without.

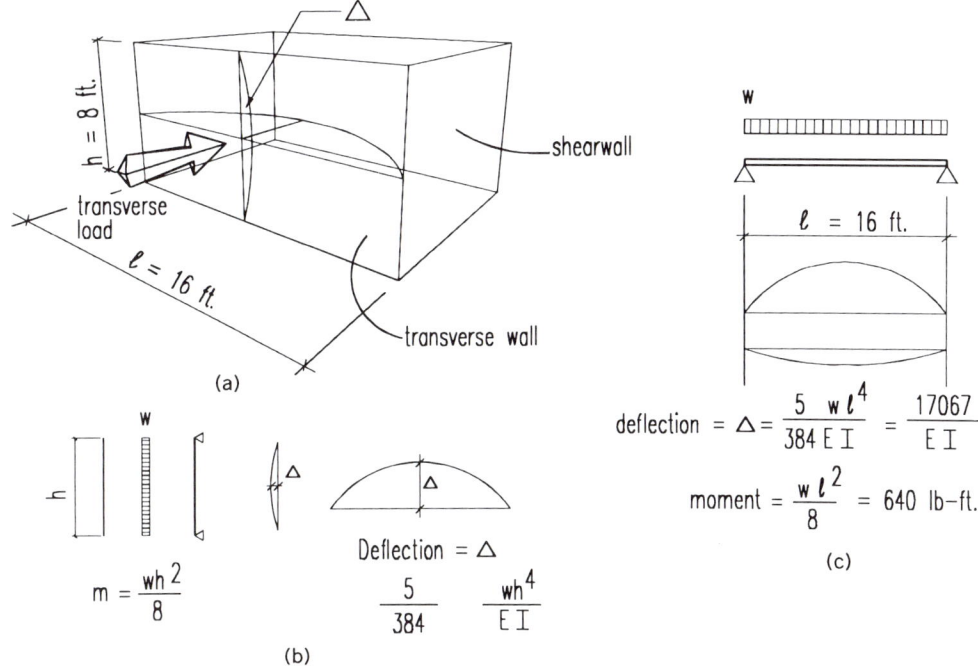

FIGURE 7.5 Effect of inadequate diaphragm.

shearwalls) that are parallel to the wind load. Under seismic forces a diaphragm operates in basically the same way except that as a major mass element of the building it generates the seismic force as well as collects it from the walls that frame into it. It is fairly easy to imagine what will happen if the shearwall is too small; it will either fail or deflect excessively, and ultimately the building will collapse.

What happens if there is no diaphragm or if it is underdesigned? Figure 7.5 shows an idealized example of what happens under wind loads; when the diaphragm is adequate, the transverse wall acts as a vertical beam spanning between two diaphragms. When the diaphragm fails or deflects too much, the transverse wall must act as a beam, transferring its load horizontally to the shearwall supports. Since typically the horizontal span is much greater than the vertical span, the shear and moment forces in the transverse wall will increase dramatically and can easily lead to failure.

DIAPHRAGMS

In its most general form, a diaphragm is a thin structural element loaded in its own plane. It develops shear and moment forces to carry the load to the supports. In rectilinear box-type structures, the diaphragm is the horizontal element that receives the lateral force from the walls perpendicular to the wind load and transfers them to the shearwalls parallel to the wind load.

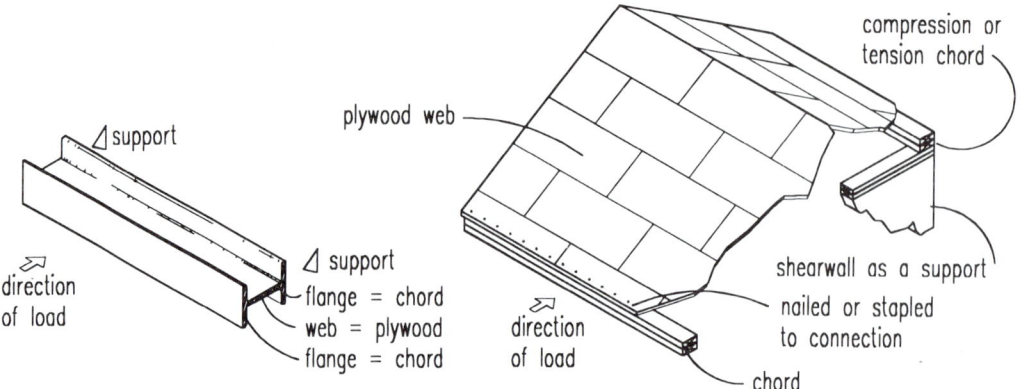

FIGURE 7.6 Analogy of diaphragm as an I-beam.

The action of the diaphragm can be thought of as being analogous to that of an I-beam. For each part of an I-beam there is a corresponding part in a plywood diaphragm, as demonstrated in Figure 7.6. In the diaphragm the plywood sheathing performs the same function as the beam web—it carries shear forces and ties the two chords into a unified beam. The chords of the diaphragm are elements that run along the outer edge of the diaphragm and are perpendicular to direction of the load. The chords develop opposing forces of compression and tension and allow the diaphragm to operate like a flat, deep beam, carrying moment. The most common approach is to use the top plate of the wall as the diaphragm chord, provided it is big enough. The chords along the top plate of the shearwalls are called the struts; these struts are parallel to the direction of the wind. Their major role is to transfer shear across openings in the shearwall below. The design of a diaphragm must consider that the wind can be applied in two orthogonal directions. What was a diaphragm chord under one direction of load becomes a strut under loads in the perpendicular direction. The last set of components in the diaphragm are the connections. Their function depends on their location. Nails, or staples, connect the diaphragm skin to the chords and also to the top plate of the shear walls. These members along the diaphragm edge and parallel to the direction of the load are called struts. The connections at the diaphragm edges and along panel joints are one of the most critical steps in the design of the diaphragm.

Ratios

Diaphragm dimension ratios, length over width, are given in the *Uniform Building Code*, Table 25-I, reproduced here as Table 7.7. Many of the design assumptions concerning plywood diaphragms are valid only within the ratio of length to width listed. In particular, the ratios were established to contain the deflection of the diaphragm within acceptable limits. Diaphragms should not exceed the ratios given in the table. If the building configuration cannot be altered by the designer, the offending diaphragm can be segmented into shorter diaphragms that will maintain the ratio. Figure 7.7 defines the terms in the ratio.

The capacities of the most commonly specified diaphragms have been determined by testing and calculations and are published by the American Plywood Association or other interested manufacturers. Most of these tables have been adopted directly or by reference by the major building codes. The most commonly used tables are incorporated in Appendix G for reference.

DIAPHRAGMS

TABLE 7.7 MAXIMUM DIAPHRAGM DIMENSION RATIOS

Material	Horizontal Diaphragms: Maximum Span/Width Ratios	Vertical Diaphragms (Shearwalls): Maximum Height Width Ratios
Diagonal sheathing, conventional[a]	3 : 1	2 : 1
Diagonal sheathing, special[b]	4 : 1	$3\frac{1}{2}$: 1
Plywood and particleboard, blocked and nailed all edges	4 : 1	$3\frac{1}{2}$: 1
Plywood and particleboard, blocking omitted at intermediate joints	4 : 1	2 : 1

[a]Conventional diagonal sheathing refers to one layer, nailed at each joist and at ends.
[b]Special diagonal sheathing refers to two layers of sheathing.
Source: UBC Table 25-I.

In Examples 7.1 through 7.4 the design of a simple rectangular building is explored. These examples lead the designer through wind load and seismic load derivations and the design of diaphragms and shearwalls. Examples 7.5 through 7.7 investigate the design of a smaller, but more complex structure, the woodframe house shown in Figures 7.25 through 7.27. In these examples we concentrate on preliminary design decisions, such as type and direction of framing, location of shearwalls, extent of diaphragms, and identification of problem areas. In Examples 7.8 through 7.11 we investigate further the residence design for the resolution and design of some problem areas encountered in the preliminary analysis.

EXAMPLE 7.1—DIAPHRAGM DESIGN

Select the plywood sheathing and connections for the roof diaphragm for the structure shown in Figure 7.8a. The initial step is to check the diaphragm ratio. The ratio is taken as the long side over the short side. This value is compared with the limits given in Table 7.7.

$$\text{Diaphragm ratio} = \frac{\text{long}}{\text{short}} = \frac{120 \text{ ft}}{45 \text{ ft}}$$

$$= 2.66 < 4.0 \quad \text{OK}$$

The design will start with an investigation of the gravity loads to determine the

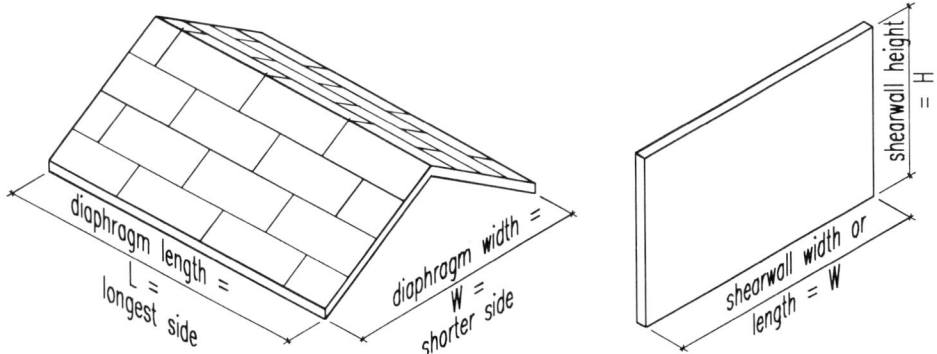

FIGURE 7.7 Diaphragm dimension ratios.

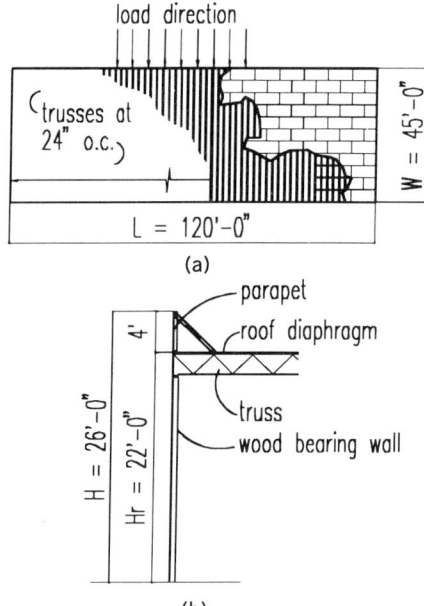

FIGURE 7.8a and 7.8b Warehouse building—roof plan and section.

grade and thickness of the plywood. Once the plywood has been selected, the lateral loads are calculated and the diaphragm action considered. The diaphragm design will check the plywood thickness, determine the required nailing, and size the diaphragm chords. Chord splices are considered and the estimated deflection of the diaphragm checked. These steps constitute the complete design of this diaphragm.

GRAVITY LOADS

The building is a simple manufacturing facility which requires that the clear span of 45 ft be maintained with no columns. A wooden roof truss that spans the 45 ft has been designed for a spacing of 2 ft on center using a snow load of 30 psf and a dead load of 15 psf. Table 7.2 shows allowable live loads for various thicknesses, specifications, and spans of plywood. Referring to this table for plywood for a 2-ft spacing of trusses indicates that $\frac{15}{32}$- or $\frac{1}{2}$-inch plywood, rated 24/0, is the least thick plywood acceptable and that it can carry a live load of 50 psf. As a trial,

select $\frac{15}{32}$-in. plywood, rated 24/0.

LATERAL LOADS

The roof diaphragm is to be designed to withstand wind loads of 70 mph. Figure 7.8b shows a section through a portion of the building which indicates that the roof plane is at an elevation of 22 ft above grade and that there is a parapet which extends 4 ft above the roof. Let us assume that the wall will be a wood stud wall that will act as a vertical beam to transfer the lateral loads into the foundation and into the roof diaphragm. Thus the roof diaphragm will receive load from the upper 11 ft of wall plus the load from the parapet wall of 4 ft. The total load to the diaphragm is given by a wind pressure of 13 psf with configuration coefficients $+0.8$ on the windward side and -0.5 on the leeward side.

Net wind pressure = 13 psf × (0.8 + 0.5)
  = 16.9 psf
Net wind load = 16.9 psf × (11 ft + 4 ft)
  = 253.5 plf   say 255 plf

The wind force must be considered acting in any direction. For any building composed of basically rectangular elements, considering the wind parallel to each side of the building in turn will be sufficient. By inspection it can be seen that the worst condition will be for wind perpendicular to the long side. This condition results in the maximum wind force and the minimum width of diaphragm to develop resistance.

The diaphragm acts like a deep horizontal beam, receiving a uniform load of $w = 255$ plf and transferring it to the supports, which are the side shearwalls. The force at each reaction, or shearwall, is

$$V = \frac{w \times L}{2} = \frac{255 \text{ plf} \times 120 \text{ ft}}{2} = 15{,}300 \text{ lb}$$

The diaphragm transfers this shear force into either the shearwall below it or the collector strut. The shear transfer is made by the connections (nails or staples) between the plywood diaphragm and the material below it. The capacity of the diaphragm is measured in terms of the connectors as pounds per linear foot. It is assumed that the shear flow is uniform along the length of the diaphragm, so the shear flow, v, is given as

$$v = \frac{V}{B} = \frac{15{,}300 \text{ lb}}{45 \text{ ft}} = 340 \text{ plf}$$

Let us look at Table G.1 in the Appendix. It is divided into three main sections. The first five columns describe the diaphragm: plywood type and thickness, nail size, penetration, and width of framing. The next four columns give the allowable shear for blocked diaphragms in plf based on the nail spacing, and the final two columns on the right give the allowable shear for unblocked diaphragms. Because blocking is so labor intensive, it is to be avoided whenever possible.

Assume that $\frac{15}{32}$-in. Structural II grade plywood is to be used. Assume also that 8d nails are to be used. Under this set of specifications there are four design values for unblocked diaphragms that may be considered. These depend on the load case (see the top of the right-hand two columns) and the size of the framing (2 or 3 in.; see the fifth column from the left). Of the four values the largest is 265 plf, which is not acceptable. Therefore, a blocked diaphragm is required.

To select the appropriate value from the table, the load direction must be considered. At the base of the table are six drawings that indicate the relation of the direction of the load to the direction of the continuous edges of the plywood. When unblocked diaphragms are used, the case is also based on the direction of the framing with respect to the continuous panel edge. For blocked diaphragms, either the framing or the blocking can be considered continuous, so the case is independent of the direction of framing. Compare these diagrams with Figure 7.8a, which shows the assumed plywood layout pattern. In the figure the continuous edges of the plywood are the 8-ft-long edges, and these are laid out so that they are perpendicular to both the load and the framing. This pattern corresponds to case 1 in Table G.1.

For $\frac{15}{32}$-in. Structural II plywood and 8d nails under case 1, the allowable shear is 360 plf when the nail spacing is 4 in. on center at the diaphragm boundaries. In Table G.1, immediately below the specification for 4-in. boundary nailing is an entry of 6-in. panel edge nailing. Figure 7.9 identifies the location of each type of nailing. Footnote 1 of Table G.1 requires that wherever the plywood crosses support members, it be nailed at 12 in. on center for roofs and 10 in. for floors. This nailing is referred to as *field nailing*. The complete specification for nailing this panel is

8d nails at 4 in. o.c. at diaphragm boundaries,
at 6 in. o.c. at all panel edges, and
at 12 in. o.c. in the field

The shortened specification, which is generally understood by building inspectors and builders is

8d nails at 4"/6"/12"

The complete specification gives plywood and nail information and is shown as

$\frac{15}{32}$-in. Structural II plywood (24/0), blocked, with 8d nails at 4"/6"/12".

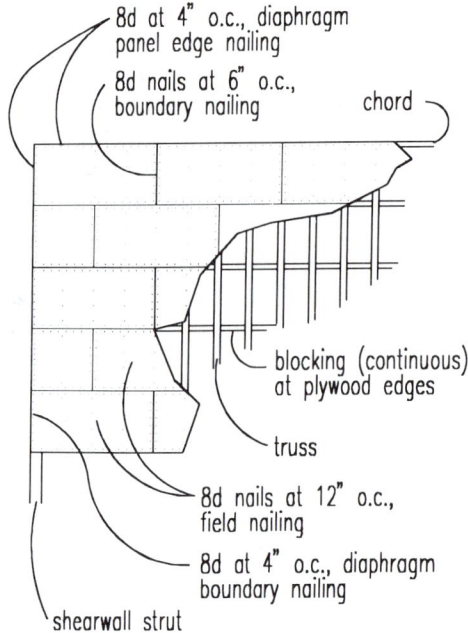

FIGURE 7.9 Diaphragm nailing pattern.

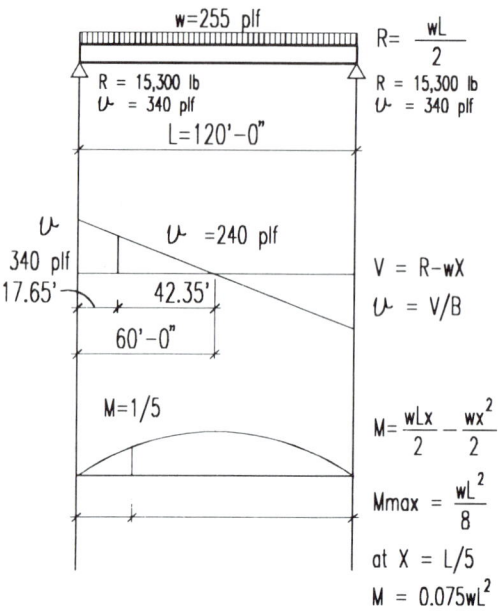

FIGURE 7.10 Diaphragm—shear and moment diagram.

In our initial investigation of the diaphragm capacity, it was noted that an unblocked diaphragm has the capacity to carry 240 plf. In a diaphragm, just as in a uniformly loaded simply supported beam, the shear decreases linearly from a maximum at the ends to zero at the center of the span (see Figure 7.10).

At the point that the shear equals 240 plf, the blocking will no longer be needed. This point is measured from the center of the span as

$$x = 60 \text{ ft} \times \frac{240 \text{ plf}}{340 \text{ plf}} = 42.35 \text{ ft}$$

Measured from the ends, the blocked portion will extend 60 ft − 42.35 ft = 17.65 ft. In even increments of 4 ft, the blocked portion will extend 20 ft. The middle 80 ft of the diaphragm will be unblocked and follow the specification, shown also in Figure 7.11,

$\frac{15}{32}$-in. Structural II plywood (24/0), unblocked, with 8d nails at 6″/6″/12″.

For large diaphragms it is economical, and common practice, to reduce the nail spacing and eliminate blocking wherever possible.

As explained earlier, horizontal diaphragms are composed of three elements: the sheathing, the connections, and the chords. The chords must be designed to be able to carry tension and compression forces that compose the moment in the beam action. As with a simply supported beam,

$$M = \frac{wL^2}{8} = \frac{255 \text{ plf} \times (120 \text{ ft})^2}{8}$$

$$= 459{,}000 \text{ lb-ft}$$

The chord forces, T or C, are

$$T \text{ or } C = \frac{M}{B} \quad \text{where } B \text{ equals the distance between chords}$$

$$= \frac{M}{B} = \frac{459{,}000 \text{ lb-ft}}{45 \text{ ft}} = 10{,}200 \text{ lb}$$

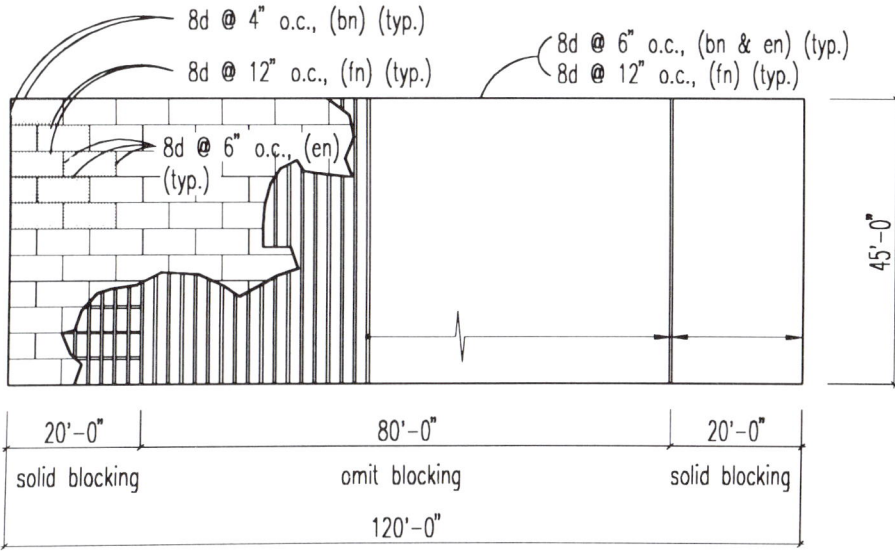

FIGURE 7.11 Diaphragm—blocking and nailing pattern.

Typically, the top plate of the wall is used as the chord. It is made of the same material as the framing members of the wall and is attached via the roof framing to the underside of the plywood. Let us assume that the chords will be made of 2 × Douglas Fir–Larch and that they will be connected as shown in Figure 7.12. The chords are sized to meet the required tension and compression forces. Assume that the wall will use 2 × 6 studs and that the top plate will be 2 × 6's, Douglas Fir–Larch No. 1.

$$F'_t = F_t \times C_F \times C_D = 675 \text{ psi} \times 1.3 \times 1.6$$
$$= 1404 \text{ psi}$$
$$F'_c = F_c \times C_F \times C_D = 1450 \text{ psi} \times 1.1 \times 1.6$$
$$= 2552 \text{ psi}$$

We can assume that the wall sheathing and studs will brace the plate in its weak direction and that the diaphragm itself will brace the chord in the stronger direction. No reduction for unbraced length is necessary, so the full F'_c can be used. Tension will control.

$$A_{\text{req'd}} = \frac{T}{F'_t} = \frac{10{,}200 \text{ lb}}{1404 \text{ psi}} = 7.3 \text{ in.}^2$$

One 2 × 6 provides 8.25 in^2, but two 2 × 6's are necessary so that the short lengths, say 16 ft, may be spliced. The splice has to be designed to carry the full load through the connection. For a diaphragm with this length, the loads are fairly large and the typical connection using nails will result in too many nails. Instead, consider using through bolts with steel side plates. Each bolt will pass through 3 in. of wood; to use the maximum strength available from this connection, it will be best to splice both 2 × 6's at the same point rather than the typical arrangement of alternating splices. Select $\frac{3}{4}$-in. bolts, loaded in double shear. Each bolt has a capacity of

$$Z' = Z \times C_D = 3150 \text{ lb} \times 1.6 = 5040 \text{ lb/bolt}$$
$$N = \frac{T}{Z} = \frac{10{,}200 \text{ lb}}{5040 \text{ lb/bolt}} = 2.02 \text{ bolts}$$

Let us use three $\frac{3}{4}$-in. bolts each side of the splice.

Figure 7.12a and b compare the ramifications of using two $\frac{3}{4}$-in. diameter bolts or three $\frac{1}{2}$-in. diameter bolts. If three bolts

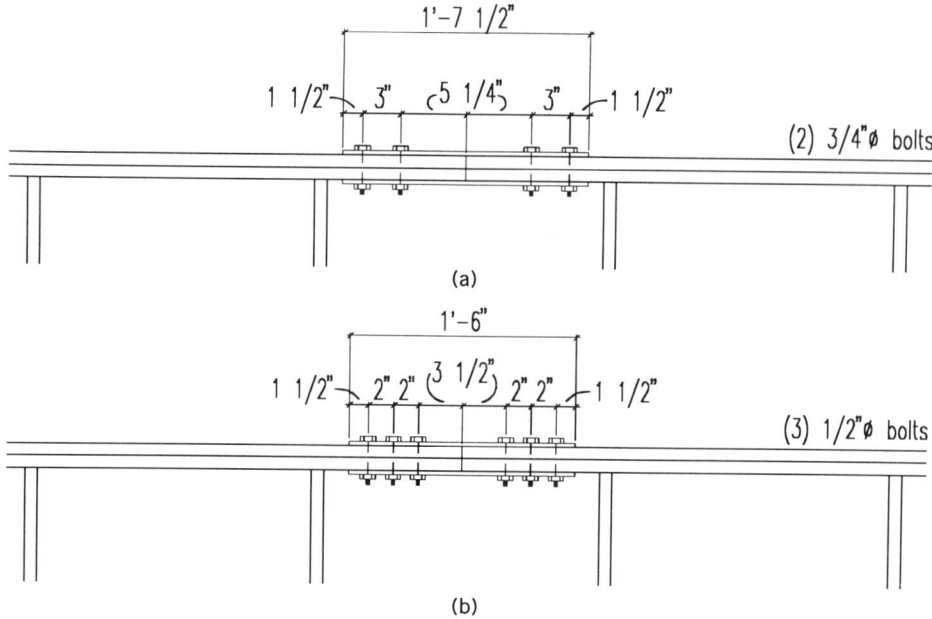

FIGURE 7.12a and 7.12b Comparison of chord bolt patterns.

are used, the required spacing and end distance are based on the bolt diameter of $\frac{1}{2}$-in. The reduced spacing allows three bolts to be used in the same length of splice as two bolts.

In normal practice we would specify two bolts, each of which would be stressed to 101% of its allowable strength. The 1% is well within our range of confidence of about 3 to 5%. For the purpose of a being conservative we have consistently rounded all values up to the next integer; hence here we specify three bolts. As designers see more of their work under construction, they will gain the confidence to allow for that 3 to 5% overstress. However, in no cases do we recommend exceeding this limit.

The connection requirements may be reduced toward the ends of the diaphragm in proportion with the moment. Since the moment curve is parabolic, the reduction is not as great as it is with the shear. This reduction is usually not considered for short to moderate-length diaphragms (say less than 40 to 50 ft). By calculating the reduction in moment, we can discover that at the fifth points, the moment will equal two-thirds of the maximum moment. In the example, one bolt can be eliminated at connections within the first 24 ft of the diaphragm. It is felt that a note to this effect will be more confusing than it is worth and could possibly lead to leaving out bolts in the wrong location.

As a final step, quickly check the effect of the wind in the other direction, perpendicular to the short side of the building.

$$V = \frac{w \times B}{2} = \frac{255 \text{ plf} \times 45 \text{ ft}}{2} = 5738 \text{ lb}$$

$$v = \frac{V}{L} = \frac{5738 \text{ lb}}{120 \text{ ft}} = 48 \text{ plf}$$

This load case corresponds to case 3 at the bottom of Table G.1. Using the least nailing pattern already designed (unblocked, with 8d nails at 6″/6″/12″) provides an allowable shear of 180 plf, which is more than sufficient.

The previously designed chords may be used, or smaller chords may be considered. Usually, the depth of the chords must be consistent on all sides, and the only savings that can be found will be in reducing the number of chord elements. In this case two elements is the practical minimum. Often, the chord must also act as a drag strut, and its size is controlled by those forces. The chord splice can be reduced. It is left as an exercise to the reader to determine the type and size of connection needed in the splices along the short direction (lap 2×6's; use three 16d nails at each side of the lap).

Diaphragms that meet the ratios in Table 7.7 will generally meet the deflection criteria of $L/180$. The deflection is calculated only for very large diaphragms or for cases when more restrictive deflection criteria are needed. Deflection calculations are not included in this book.

DETAILS

Figure 7.13 shows details of the connections for this diaphragm and the connections to a plywood-sheathed shearwall, which will be designed in Example 7.5. Figure 7.14 shows alternate details for connecting this diaphragm to a shearwall. In Figure 7.14 the wall continues past the diaphragm rather than the diaphragm bearing directly on the wall. In this situation the diaphragm bears on a ledger that is lag-bolted to the wall. This connection requires the addition of a metal twist strap to connect the diaphragm to the wall. Without the strap, diaphragm forces perpendicular to the wall will be transferred through the ledger. The force in the ledger forms a couple since the point of application, at the diaphragm, is separated from

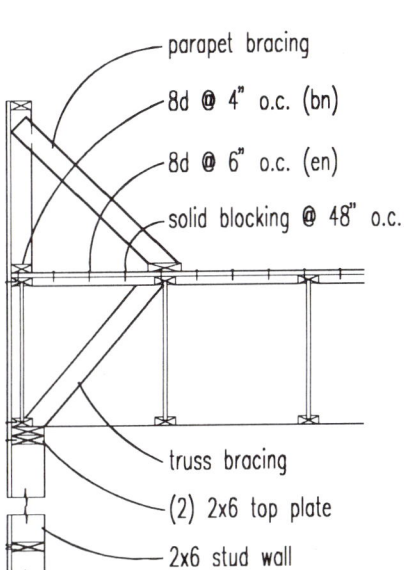

FIGURE 7.13 Connection of diaphragm to shearwall—wall is continuous.

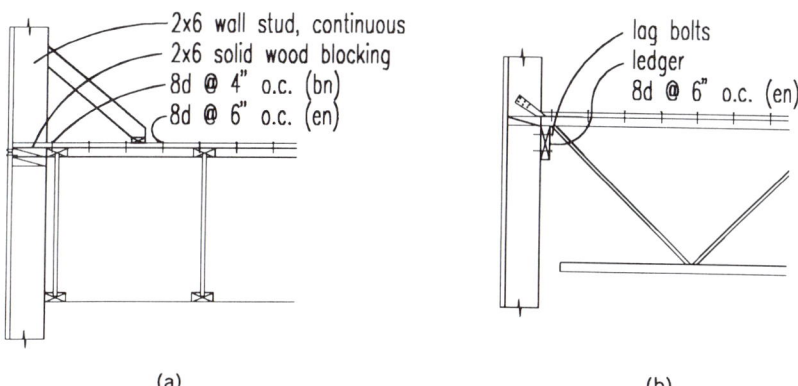

(a)   (b)

FIGURE 7.14 Connection of diaphragm to shearwall—wall is continuous.

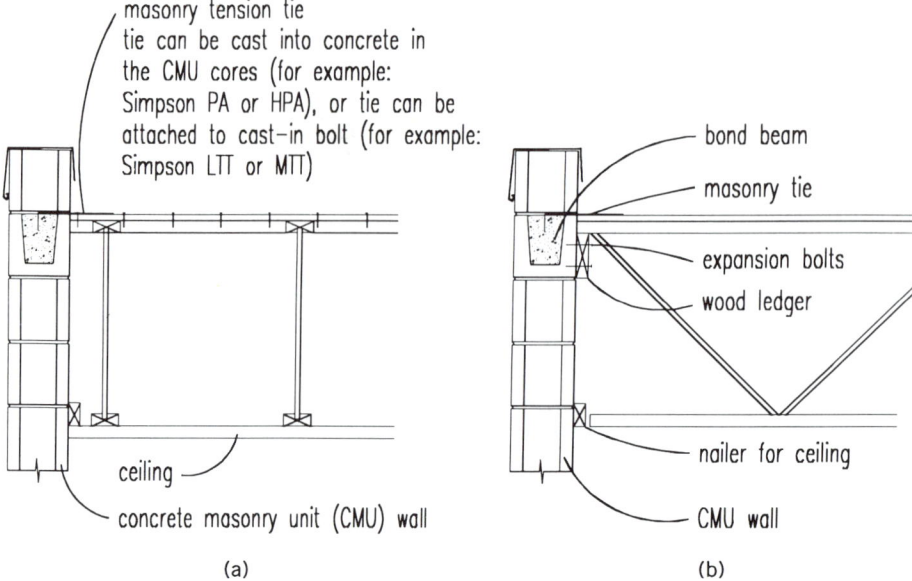

FIGURE 7.15 Connection of diaphragm to shearwall—masonry wall.

the point of transfer, at the lag bolts. This moment forms cross-grain bending and cross-grain shear in the ledger. Experience has shown that wood is very weak in this direction, and therefore major building codes prohibit using wood in cross-grain loading conditions. The twist strap will transfer the lateral forces directly into the wall, thereby eliminating cross-grain forces in the ledger. The ledger is still needed to carry the vertical roof loads to the wall.

Figure 7.15 shows the connection of the diaphragm to a masonry wall. This is a very common construction type and a common detail. Notice again the use of the ledger for bearing and a metal tension tie to eliminate cross-grain forces in the ledger.

EXAMPLE 7.2—DIAPHRAGM WITH OPENING.

The diaphragm from Example 7.1 is to be redesigned to accomodate a large opening in the roof. The location and size of the opening are shown in Figure 7.16*a*

Start with the wind load perpendicular to the long direction. The calculations for reactions and shears are the same as before. The shear diagram in Figure 7.16*b* shows the total shear in pounds at the edges of the opening for winds perpendicular to the long side. The free-body diagrams in Figure 7.16*d* show the shear in plf along the edges of the two 10-ft sections of diaphragm. The magnitudes of shear at these two locations are

$$v_b = \frac{V}{B} = \frac{8160 \text{ lb}}{2 \times 10 \text{ ft}} = 408 \text{ plf}$$

$$v_c = \frac{V}{B} = \frac{6120 \text{ lb}}{2 \times 10 \text{ ft}} = 306 \text{ plf}$$

For panels this small it is better to design the entire panel to meet the required strength than to specify different nailing along line c. To reach these values, the small diaphragms will have to be blocked and 8d nails will need to be spaced $2\frac{1}{2}$ in. o.c. along the boundaries, resulting in an allowable shear of 530 plf. The roof framing must be at least a nominal 3 in.

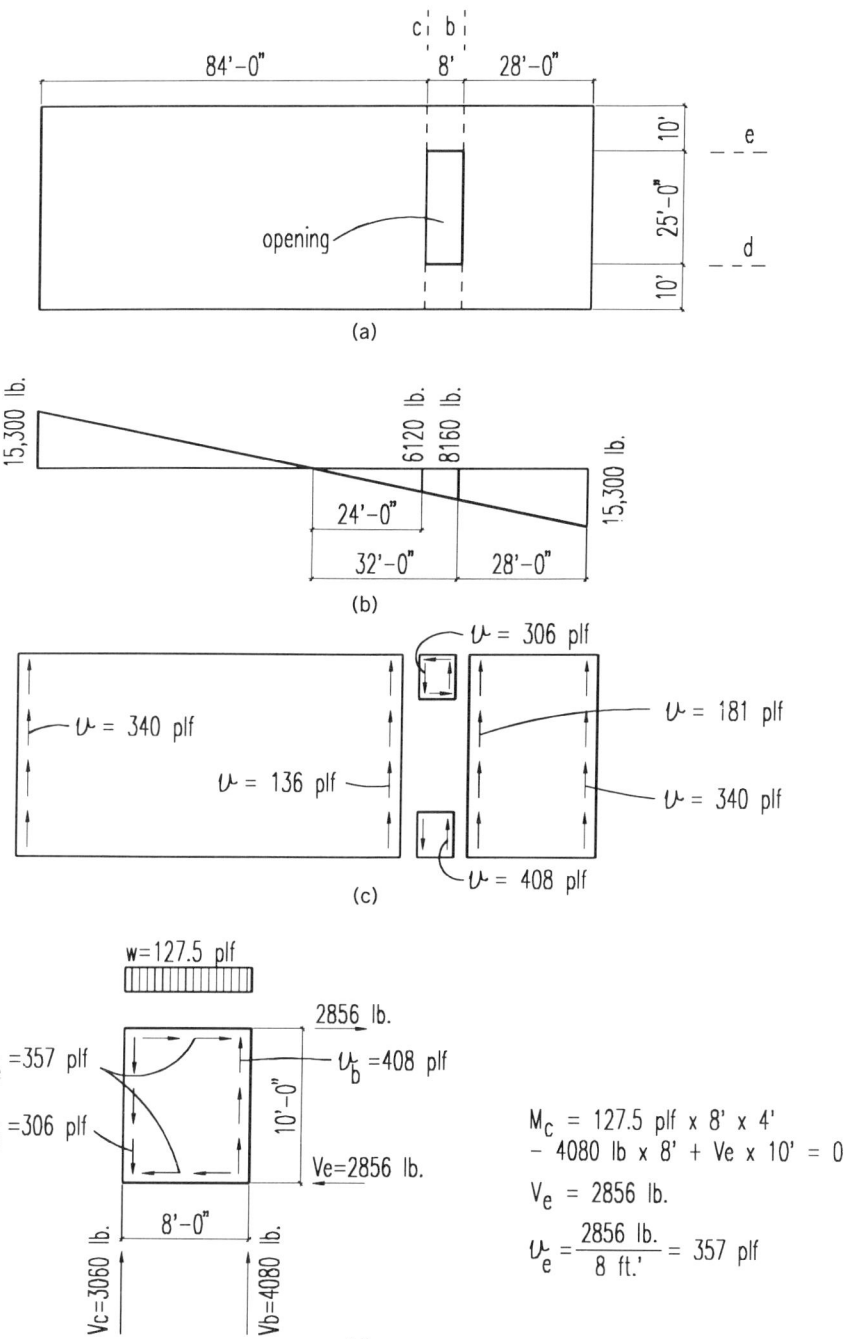

FIGURE 7.16 Opening in diaphragm: (a) plan, (b) shear diagram, (c) component shears, and (d) shear on 8 × 10 panel.

wide. Each of these sections will have to be framed separately anyway and bear on the trusses along lines b and c. This can be done with 3 × material, or as is more commonly done, by doubling 2 × material.

Figure 7.16d also shows the shear of 357 plf along the 8-ft edges perpendicular to the wind. This shear is developed to balance the moment that results from twist caused by the unequal shears at lines b and c. The shear of 357 plf is acting at 90° to the framing, so the operable load case is case 3. The allowable shear applies to load case 3. The new specification is good in both directions.

For the 8 × 10 ft panels, use double 2 × 10 framing with solid blocking. Use 8d nails at $2\frac{1}{2}$ in. o.c. staggered at all boundary edges, 8d nails at 4 in. o.c. all panel edges, and 8d nails at 12 in. o.c. in the field.

or

For the 8 × 10 ft panels, use double 2 × 10 framing with solid blocking, 8d at $2\frac{1}{2}$ in. o.c. staggered (bn) / 4" (en) / 12" (fn).

The effect of the two small panels between b and c is to maintain the integrity of the diaphragm. All other design aspects, such as chords and chord splicing, will be unchanged. The deflection can be assumed to be much greater, but we will not attempt to quantify it here.

If seismic conditions were a concern here, we would have to check for the code provisions for horizontal irregularities (see Table 7.9). In this case the hole in the diaphragm is less than 50% of the gross diaphragm area, so the design would proceed without change. If the opening area exceeded 50% of the gross, then for the design of the connectors and struts the 33% increase in strength due to the duration factor would not be used. All allowable diaphragm shears in the tables include the duration factor. To disregard the duration factor, divide the tabulated values by 1.33.

EXAMPLE 7.3—FLEXIBLE DIAPHRAGM DESIGN

Consider now the same building (45 ft × 120 ft) with an additional shear wall added as shown in Figure 7.17a. The shearwall divides the diaphragm into two sections, 76 ft long and 44 ft long. A wood diaphragm is a flexible diaphragm. It is best conceived of as a series of simply supported beams rather than a continuous beam. Therefore, each section of the diaphragm can be treated separately, and the force on each wall can be assumed to be proportional to the tributary area of the load.

Refer to Figure 7.17b, which shows the shear forces for both sections of the diaphragm. For the section on the left the maximum shear equals

$$V = \frac{w \times L}{2} = \frac{255 \text{ plf} \times 76 \text{ ft}}{2} = 9690 \text{ lb}$$

$$v = \frac{V}{B} = \frac{9690 \text{ lb}}{45 \text{ ft}} = 215.3 \text{ plf}$$

For the section on the right the maximum shear equals

$$V = \frac{w \times L}{2} = \frac{255 \text{ plf} \times 44 \text{ ft}}{2} = 5610 \text{ lb}$$

$$v = \frac{V}{B} = \frac{5610 \text{ lb}}{45 \text{ ft}} = 124.6 \text{ plf}$$

The design will use the same plywood specification and nail size as before.

$\frac{15}{32}$-in. Structural II plywood (24 / 0) with 8d nails.

For the left side, with $v = 215.3$ plf, and using case 1, Table G.1 shows that an

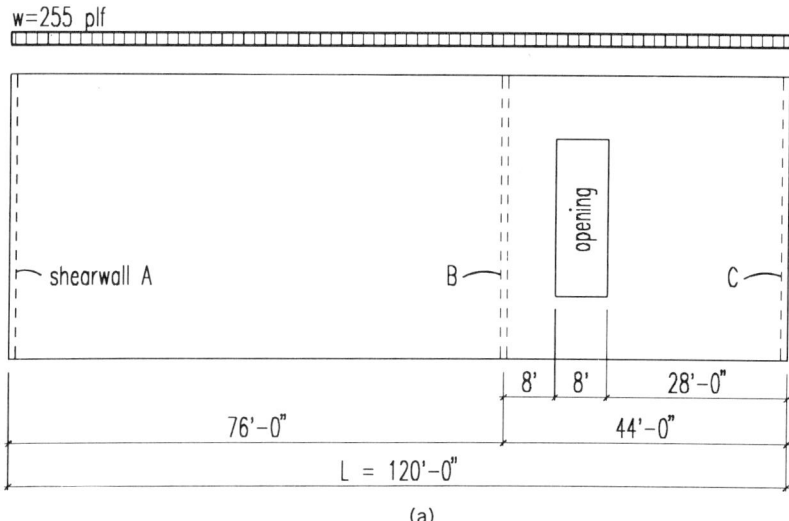
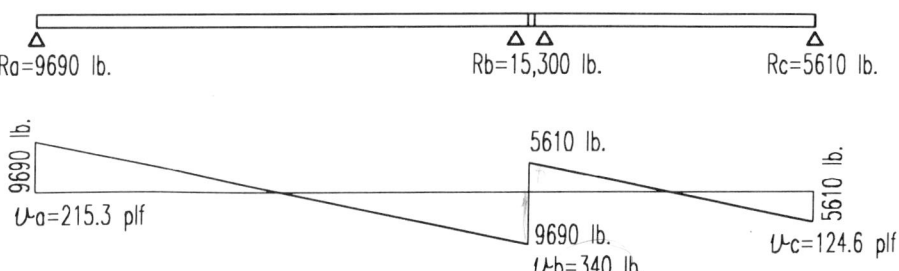
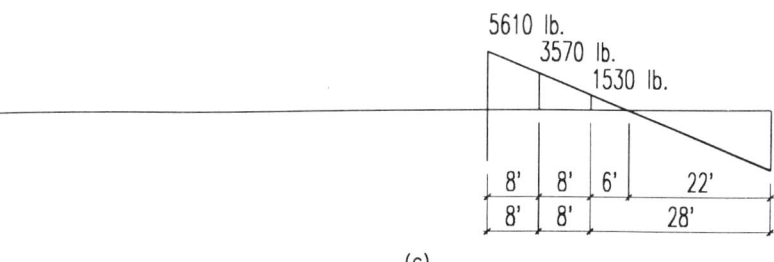

FIGURE 7.17 Diaphragm with interior shearwall: (a) diaphragm plan, (b) shear diagram, and (c) shear diagram.

unblocked diaphragm with 8d nails at 6 in. o.c. at supported edges will carry 240 plf and therefore be satisfactory. By inspection, for the right side with $v = 124.6$ plf, the same specification will work. Specify

$\frac{15}{32}$-in, Structural II plywood (24 / 0), unblocked, with 8d nails at 6 in. o.c. edges and 12 in. o.c. in the field.

Figure 7.17c also shows the amount of shear at either side of the opening, which is reduced with the presence of an interior shearwall. The shear flow for these, based on $B = 2 \times 10$ ft $= 20$ ft is V at left $= 255$ plf $\times 14$ ft $= 3570$ lb, so

$$v_{\text{left}} = \frac{V}{B} = \frac{3570 \text{ lb}}{20 \text{ ft}} = 178.5 \text{ plf}$$

V at right $= 255$ plf $\times 6$ ft $= 1530$ lb. So

$$v_{\text{left}} = \frac{V}{B} = \frac{1530 \text{ lb}}{20 \text{ ft}} = 76.5 \text{ plf}$$

Both of these are less than 240 plf, so the minimum specification for the diaphragm is acceptable.

Design the chords for the largest moment, which is in the left section.

$$M_{\text{max}} = \frac{w \times L^2}{8} = \frac{255 \times 76^2}{8}$$

$$= 184{,}110 \text{ lb-ft}$$

$$C \text{ or } T = \frac{M}{B} = \frac{184{,}110 \text{ lb-ft}}{45 \text{ ft}} = 4091 \text{ lb}$$

In Example 7.2 the chord was designed for a larger load using two 2×6's. These will be used again in this example. It was determined that one through bolt $\frac{3}{4}$ in. in diameter will carry 5040 lb. It is ill advised to use one bolt in most connections, so specify

two 2×6 top plates with two $\frac{3}{4}$-in.-diameter bolts each side of splice.

Rigid Versus Flexible Diaphragms

Until now the design has concentrated on the diaphragm with no discussion of the shearwalls. In the next section we deal extensively with shearwalls. As an introduction we discuss the relationship between the diaphragm and the shearwalls in Example 7.3. Figure 7.17a shows a plan view of the three short shearwalls, marked A, B, and C. At wall A, the total diaphragm shear, 9690 lb, must be transferred into the shearwall. Wall C on the right side must carry a shear of 5610 lb. Wall B must carry the shear from both the left section and the right section of the diaphragm, a total of 9690 lb + 5610 lb = 15,300 lb. Refer to Example 7.1 to see that with no interior shearwall at B, the shear on walls A and C was 15,300 lb. The addition of the interior wall B has reduced the loads on walls A and C considerably, but there still is a wall that must carry the maximum load of 15,300 lb.

This effect stems from the assumption that wood diaphragms are flexible. If we had assumed that the diaphragm was rigid, it would act as a continuous beam and the distribution of forces into the walls would be an indeterminate solution. Figure 7.18 shows the forces at each wall based on the later assumption.

It is often suggested that the designer calculate the forces for both assumptions and use the larger values for each component. In the example the diaphragm boundary at C would be designed for 5610 lb (flexible assumption). The boundary to the right of B would be designed for 8774 lb (rigid assumption) rather than 5610 lb (flexible assumption), an increase of 56%.

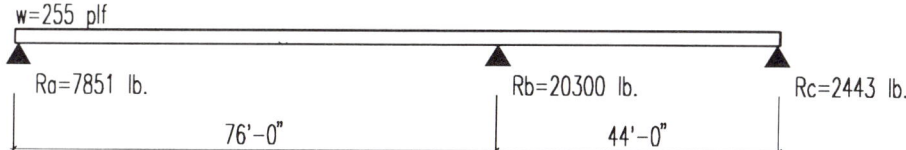

rigid diaphragm acts like a beam continuous over an internal support. find the reaction at b using compatability of deformations

(a)

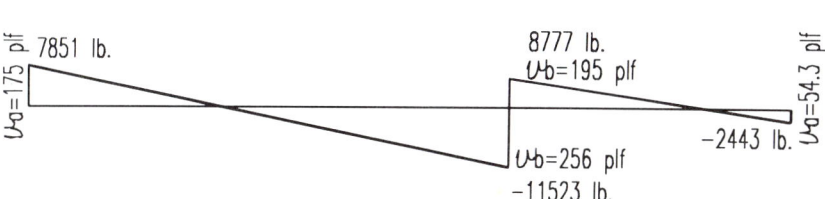

the maximum shear is about 20% greater with the rigid diaphragm than with the flexible diaphragm

(b)

FIGURE 7.18 Shear distribution for rigid diaphragm: (a) rigid diaphram as continuous beam and (b) shear diagram.

Full-scale tests, in particular those conducted by Tarpy, Thomas, and Soltis in 1984, indicate that the actual response of wood diaphragms is closer to that of the flexible diaphragm. The use of the rigid assumption, in this case, would be overly conservative. All values throughout the design, such as forces at the edge of the openings, chord forces, and deflections, would have to be developed for each assumption. The final design would, in fact, be based in part on the flexible assumption (chords and deflections) and in part on the rigid assumption (shears). Although the research is not conclusive, we recommend that unless the building code specifically requires both analyses, the designer use only the flexible diaphragm assumption for wood diaphragms.

Lumber Diaphragms

As late as 1950 a significant number of buildings were framed with lumber subfloors rather than plywood or particleboard sheathing. This procedure, in which lumber boards, usually 1 or 2-in. nominal thickness, are laid directly on and nailed directly to the joists, is still used where the underside of the board serves as the ceiling to the space below. Since it is also encountered in rehabbing or remodeling work, it is worthwhile to consider the effect of lumber flooring and roofing as diaphragms. In Figure 7.43 a portion of a transversely sheathed lumber floor with diaphragm forces is shown. *Transversely sheathed* means that the boards are perpendicular to the joists. From the figure we can see that the shear resistance is developed by the paired nails that develop a couple. The magnitude of the couple is the lateral resistance of the nail times the distance between them. Each pair of nails, which occur at every intersection with a joist, contributes the same amount of moment; therefore, closer joist spacing will provide more connections and a greater moment capacity. The effect of these couples is additive, so that each board contributes the same amount regardless of its location. The allowable shear in a transversely sheathed lumber diaphragm is given by the equation

$$v_s = m \times \frac{b}{j}$$

where m is the nail couple $= P_L \times$ nail spacing, b is the board width, j is the joist spacing, and P_L is the lateral strength of the nail.

Table G.6 shows the capacity of lumber-sheathed diaphragms. It can be seen that the transversely sheathed diaphragms carry a substantial amount of load and are adequate for many conditions involving small diaphragms and light loads. However, they provide far less shear than do plywood diaphragms, while consuming more material and labor.

If the boards were wider, more nails could be used and there would be an increase in the capacity of the outer pairs of nails. Although this effect is not actually multiplicative, it can be seen that it is greater than the effect of a series of narrow boards. This is why plywood sheathing develops a greater capacity in diaphragms. An alternative method is to unite the individual boards into a unit by gluing them together. For this to be effective, the board edges must be tongue-and-groove to provide a sufficient area for the glue. If the boards are laid out in a diagonal pattern, their diaphragm shear capacity is greatly increased, as demonstrated in Table G.6.

SHEARWALLS

Shearwalls are vertical elements in the lateral load path that receive loads from diaphragms or transverse walls and conduct them down to other elements and eventually to the foundation. In wood-frame construction, shearwalls are the most common vertical element. They are most often wood-framed walls sheathed with plywood or some other sheathing, or they may be

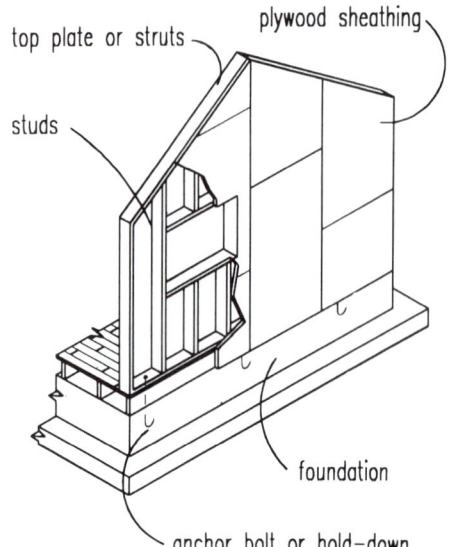

FIGURE 7.19 Shearwall components.

masonry. Other shear-resisting elements, such as frames, are common in steel or concrete construction, but are normally used only in heavy timber or timber-frame construction. For the purpose of this book we confine our discussion to shearwalls.

The shearwall is made up of four components: the framing, such as 2×4 studs; the sheathing; the connections between frame and sheathing; and the connections between the wall and other elements, such as diaphragms or foundations (see Figure 7.19). The complete design of a shearwall must account for all these elements.

There is an additional element, the collector strut or drag strut, that is used in shearwalls with openings or in discontinuous shearwalls in a single plane. The purpose of the drag strut is to collect the forces from the diaphragm and deposit them into the adjacent portions of the shearwall. The drag strut is loaded along its axis so that it operates in tension or compression. Frequently, the drag strut is the top plate of the wall, but occasionally it will be a part of the header. When it is, the header will have to be designed for the interaction of axial forces and bending. Example 7.4 shows the design of a shearwall and its components.

Shearwall Capacities

Tables G.7 through G.10 show the shear capacity of shearwalls commonly used in construction. Definitions for particleboard, fiberboard, and hardboard, used in these tables, were given earlier in this chapter. Comparing materials in these tables shows that plywood and particleboard are virtually interchangeable, whereas all of the materials shown on Table G.10 are less effective. In comparing nails and staples for plywood shearwalls, we see that with the proper selection of staples the same strength as produced by nails can be gained with staples. Now that both nails and staples are driven with pneumatic guns, the choice between them must be based on economics and availability.

Table G.10 lists the shear capacity for a number of materials that are used to sheath interior or exterior walls. The values given in these tables are drawn from a number of sources, some of which will be recognized by local building officials and some of which may not. In general, most of these applications will prove satisfactory for small to moderate loads. By using these values carefully, the designer will be able to quantify and make use of the inherent strength available in these assemblies and may avoid calling out plywood shearwalls in cases where they are not needed. Special notice should be given in this table to let-in corner bracing. The let-in brace, traditionally a wood 1 × 4, or more recently proprietary bent metal straps, are intended only to brace and plumb the walls during construction. Let-in bracing should not be relied on to carry any of the lateral loads applied to the permanent structure. Other sheathing systems, found in these tables, must be installed to carry these loads.

All of the values listed in the tables, unless noted otherwise, are for sheathing applied to one side of studs only. If sheathing is applied to both sides of the studs, the strength may increase, but special precautions must be taken. The shear capacity of an assembly is governed by any of three factors: the strength of the sheathing material, the nailing capacity, and the strength of the stud frame. If we assumed that the only factor that applied was the strength of the panel, we could safely assume that doubling the number of panels would result in twice the strength. However, in the tests that are the basis of these values, the connections controlled most frequently. Therefore, the following considerations must be taken in determining the capacity of a two-sided shearwall:

1. The strength of two sheathing materials may be added only if the sheathing on both sides is the same material and the same thickness. In addition, the same connections, size, and spacing are used on both sides. (See number 4, below, for dissimilar materials.)

2. If sheathing is applied to both sides, assume that the allowable shear strength is equal to the strength listed for the sheathing with a connection spacing of one-half the value used. For example, if a layer of $\frac{3}{8}$ in. Structural II plywood with 6d nails at 6 in. o.c. (200 plf) is used on two sides, the acceptable shear strength is taken from the column for 6d nails at 3 in. o.c. (390 plf) (see Table G.7).

3. To use the value described above, all of the footnotes and precautions for a closer spacing must be followed. Specifically, see footnote 4 in Table G.7, which requires that "where plywood is applied on both faces of a wall and nail spacing is less than 6 in. on center both on either side, panel joints shall be offset to fall on

264 PLYWOOD

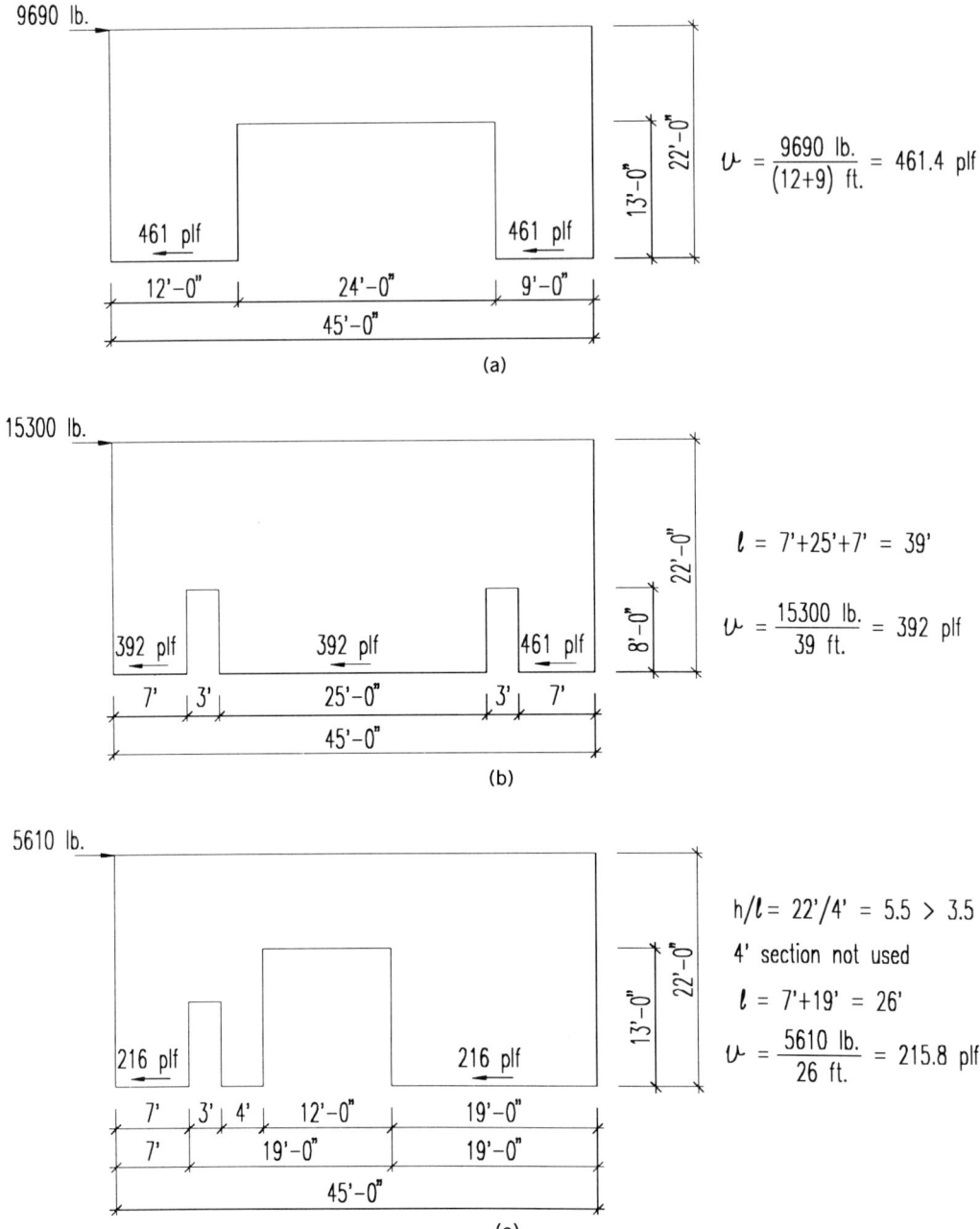

FIGURE 7.20 Shearwall elevations.

different framing members or framing shall be 3-in. nominal or thicker and nails on each side shall be staggered."

4. In shearwalls with dissimilar materials on either side (or with different segments made of dissimilar materials) only the strength of the stronger assembly may be counted. The strengths of dissimilar materials may not be added. If the two materials have greatly different stiffnesses, they will deflect much differently under the same load. The stiffer element will carry the greater load and must be designed to carry the full load.

The most efficient shearwall usually tends to be the longest. Therefore, it is generally better to sheath a wall continuously in one material than to use a strong material such as plywood at the corners and a weaker material for the remainder. The longer wall will better avoid problems associated with shearwall height-to-width ratios, design of collectors, and diaphragm chords.

EXAMPLE 7.4—SHEARWALL DESIGN

Consider the building described in Example 7.3 and Figure 7.17. Figure 7.20 shows the three shearwalls in the short direction and the openings in them. Let us design one of these shearwalls to resist the lateral force applied to it.

SHEARWALL A

Shearwall A has a large opening in it that serves as a loading door into the warehouse. The height of the door is given in Figure 7.20a as 13 ft and the width is given as 24 ft. Short shearwalls at either end measure 12 and 9 ft in length. Each shearwall is 22 ft high.

The first step in the design is to design the wall for gravity loads. Since this wall is not a bearing wall we can skip this step. The next consideration is to design the wall to carry the lateral loads applied transversely against its surface. The design for both of these considerations can usually be accomplished by referring to the *Uniform Building Code* Table 25-R-3, which is based on standard practice. If we were to use the values in the table to estimate allowable wind loads, we will find that the table is based on relatively low wind loads of about 45 mph. Therefore any wall designed through calculations will need to use much larger sections than those shown here.

In UBC Table 25-R-3 the tallest nonbearing wall listed is 20 ft of unsupported height using 2×6's at 24 in. o.c. Although no grade is given, we may construe from the note that a grade higher than "Utility" must be used. At the least we should specify "Stud" grade, or in this case "Construction" grade. Because the height in the problem is close to the listed height, we can feel comfortable in using 2×6's at a reduced spacing, say 16 in. o.c.

Since the allowable height is probably based on allowable relative deflections, and since relative deflections vary as the length cubed, we can calculate an acceptable spacing as

$$\text{spacing} = \left(\frac{L_{\text{actual}}}{L_{\text{listed}}}\right)^3 \times S_{\text{listed}}$$

$$= \left(\frac{20}{22}\right)^3 \times 24\,\text{in.} = 18\,\text{in.} > 16\,\text{in.}$$

Specify studs as 2×6 at 16 in. o.c. Douglas Fir–Larch, Construction grade.

The second step in the design is to check the allowable height-to-length ratio of the shearwalls given in Table 7.7. The table shows that the maximum height-to-length ratio for an unblocked plywood diaphragm is 2 to 1. The 12-ft.-long wall will meet this requirement (1.83 to 1), but the 9-ft.-long section will not (22/9 equals 2.44 to 1). The 9-ft-long section does meet the requirement for a blocked plywood shearwall ($3\frac{1}{2}$ to 1). At the least the 9-ft-long portion of the wall will need to be blocked. For convenience in construction it is recommended that the entire length of the wall be blocked.

The total shear to be carried is 9690 lb. This amount is to be distributed to the two sections of the wall. We assume that they will each carry the same force per linear foot. So

$$v = \frac{V}{B} = \frac{9690 \text{ lb}}{(12+9) \text{ ft}} = 461.4 \text{ plf}$$

Table G.7 lists the allowable shear for plywood shearwalls. Enter the table horizontally at Structural II plywood and vertically under "plywood applied directly to framing." Look for shear values in the range of 461 plf. It seems that $\frac{15}{32}$ in.-thick plywood with 10d nails at 4 in. o.c. will give an allowable shear of 460 plf. Alternatively, $\frac{15}{32}$-in.-thick plywood with 8d nails at 3 in. o.c. will give an allowable shear of 490 plf. Normally, we would prefer to use the larger nail (10d) with the larger spacing, assuming that this would be less likely to chew up the framing than would smaller nails. In this case the shearwall would be slightly overstressed (less than 1%) and the designer would have to determine the level of confidence in the figures: wind load, heights, and so on. To be conservative we will specify

$\frac{15}{32}$-in.-thick Structural II plywood, blocked with 8d nails at 3 in. o.c. at all plywood edges and 12 in. o.c. in the field.

Now we design the connectors at the base of the wall to transfer the shear into the foundation. The typical specification for this base connection is $\frac{1}{2}$-in.-diameter anchor bolts at 4'-0" o.c. for seismic zones 3 and 4, or $\frac{1}{2}$-in. anchor bolts at 6'-0" o.c. for all other locations. Table 4.19 lists the shear capacity of $\frac{1}{2}$-in. anchor bolts in concrete as 2000 lb for a bolt with 4-in. minimum embedment. However, the strength of the wood in the connection governs, as seen in the bolt values tabulated in NDS Table 8.2A. In wood-to-concrete connections, look up the value for a main member twice as thick as the actual wood side member. Thus for $t_m = 3$ in. and $t_s = 1\frac{1}{2 \text{ in.}}$,

$$Z = 610 \text{ lb/bolt} \times 1.33 = 811 \text{ lb/bolt}$$

The factor 1.33 is the duration factor for wind and seismic accepted by the *Uniform Building Code*. BOCA and SBCCI use a factor of 1.6 for wind. On the longer section of wall the total shear force at the base is

$$V = v \times L = 461.4 \text{ plf} \times 12 \text{ ft} = 5537 \text{ lb}$$

$$N = \text{number of bolts} = \frac{5537 \text{ lb}}{811 \text{ lb/bolt}}$$

$$= 6.8 \text{ bolts}$$

A minimum of seven bolts will be needed to carry this load.

Anchor bolts should not be closer than 1 ft from the end of any piece of the sill, so the spacing that is needed is 1'-8" between bolts. For the smaller wall, use five bolts, spaced at 1'-9". As a trial, use

seven $\frac{1}{2}$-in.-diameter anchor bolts at the 12-ft section of shearwall A.

After checking the wall for overturning, we may revise the number of bolts.

The next step is to make sure that the load is actually transferred into the shearwall. In the diaphragm design the connections were designed between the edge of the diaphragm and the top plate of the wall. If the wall were continuous with no opening, this would satisfy the requirement. However, at the opening it is necessary to assure that the forces above the opening are carried into the lengths of wall that are actually carrying the load to the foundation. The top plate of wall above the opening will act as a collector strut, receiving forces from the diaphragm and depositing them into the adjacent sections of the shearwall. Refer to Figure 7.21b. In this figure the force from the diaphragm is shown as a constant 215.3 plf (9690 lb/45 ft) acting to the left. The force developed at the top of the shearwall is shown as 461.4 plf (9690 lb/21 ft). The 461.4 plf is developed to resist the 215.3 plf so it is acting in the opposite direction, toward the right. The net force on any section of the wall is the sum of these two forces. Over the opening the net force amounts to 215.3 plf acting to the left. Over the active sections of the shearwall the net force is 246.1 plf acting to the right (215.3 plf minus 461.4 plf). The collector strut acts in tension or compression to collect the force and then distribute it into the wall. Assuming that the wind force is applied from the left the section of the collector strut over the wall will be in compression. The lower diagram shows the compression force as it builds up in the collector. The magnitude of the force is the rate of force times the length; thus at the left of the opening the compression in the collector is

$$C = 246.1 \text{ plf} \times 12 \text{ ft} = 2953 \text{ lb.}$$

Along the length of the opening the compression drops off and becomes tension; the magnitude of tension at the right of the opening is

$$T = 2953 \text{ lb.} - (215.3 \text{ plf} \times 24 \text{ ft})$$
$$= -2215 \text{ lb.}$$

Since the wind can come from either direction, the collector needs to be designed for these values as either compression or tension. As with the diaphragm chords, design for compression will result in the size of the strut and design for tension will result in the connection design at the splice.

The collector must be designed to carry either tension or compression with a maximum magnitude of 2953 lb. The collector can occur vertically at any height in the wall, but it must be located so that there can be a transfer of forces into the two active sections of shearwalls. It is best to use the top plate of the wall as the collector. Previously, we designed the top plate to act as the diaphragm chord and found that a single 2×6 was satisfactory to carry an axial tension of 10,200 lb. Therefore, let us use standard top plate configuration of (two) 2×6's. To assure that the collector operates as required, we must be sure that it can transfer a 2953-lb force into the shearwall. The standard connection between top plates is 16d nails at 16 in. o.c. If we look up the capacity of a 16d common wire nail in single shear holding two Douglas Fir–Larch $2 \times$'s together, we will find the tabulated value of 141 lb/nail on NDS Table 12.3B. The capacity of 16d nails at 16 in. o.c. with the wind load duration factor becomes

$$P = 141 \text{ lb/nail} \times 1.6 \times \frac{12 \text{ in./ft}}{16 \text{ in./nail}}$$
$$= 169.2 \text{ lb/ft}$$

The length of the splice then must become

$$L_{\text{splice}} = \frac{2953 \text{ lb}}{169.2 \text{ lb/ft}}$$
$$= 17.45 \text{ ft for a single row of nails}$$

This is far from practical. Let us consider using two rows of 16d nails and assume that the splice length, the amount of

268 PLYWOOD

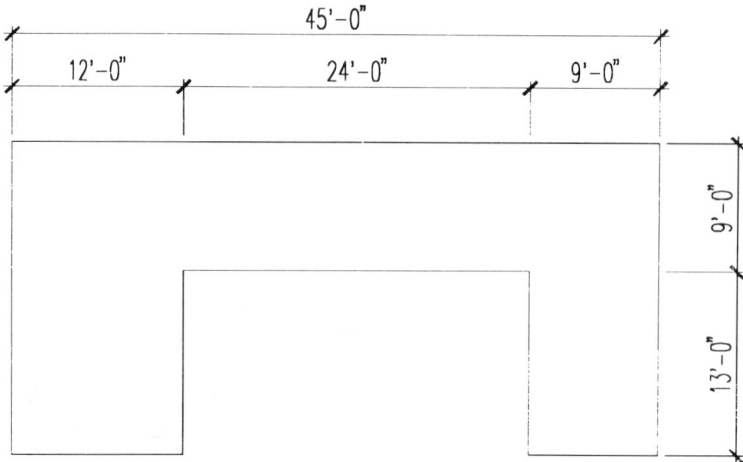

7.21a elevation of shearwall

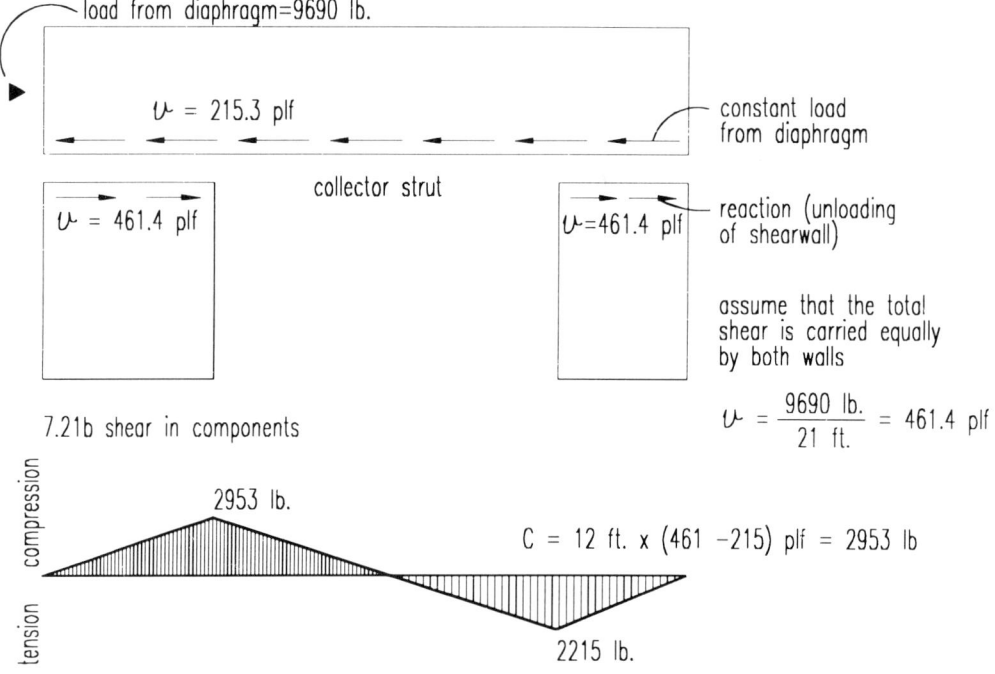

7.21c collector strut forces

FIGURE 7.21 Shearwall A—compression and tension in collector strut: (a) elevation of shearwall, (b) shear in components, and (c) collector strut forces.

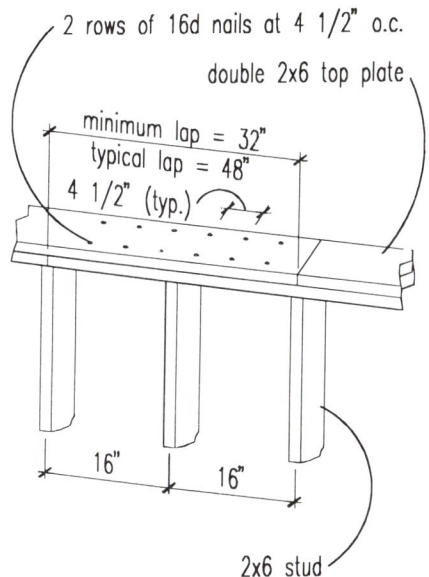

FIGURE 7.22 Shearwall—collector strut splice.

overlap of the two plates is 32 in. (two stud spaces). What is the required nail spacing?

The capacity of each nail is

$$P = 141 \text{ lb/nail} \times 1.6 = 225.6 \text{ lb/ft}$$

Each row of nails must carry half of 2953 lb. or 1476.5 lb. The necessary number of nails in each row is

$$N = \frac{1476.5 \text{ lb}}{225.6 \text{ lb/ft}} = 6.5 \text{ nails}$$

use 7 nails per row

The required spacing is 32 in./7 = 4.5 in., which seems practical. Specify the collector and splice as shown in Figure 7.22 as

(two) 2 × 6 Douglas Fir–Larch No. 1, top chord. Splice with two rows of 16d common nails at $4\frac{1}{2}$ in. o.c. staggered. Splice over lap to be 32 in. long.

The walls that are parallel to the wind direction are the shearwalls and they contain the collector struts; the perpendicular walls contain the diaphragm chords. In both locations we have opted to use the top plates to carry these forces. When the wind changes direction the role of the collectors and chords switch. Because their roles change, we will not have a situation where the top plate will be fully loaded as a collector at the same time that it is fully loaded as diaphragm chord, so we will not have to consider these forces as additive.

The final consideration for this design is for the overturning of the various sections of the walls. In part the height-to-length ratio is intended to limit the possibility of overturning, but this aspect should always be checked. In this example the short sections of shearwall are very close to the height-to-length limit, so it seems that overturning is a distinct possibility. The forces that resist overturning are the dead weight of the wall and any dead loads that it carries. In this case the shearwalls are not load bearing, so it seems that there will be very little resistance to overturning and that the walls will have to be connected to the foundation so that the overturning forces can be carried by the hold-down connectors.

Figure 7.23 shows the smaller 9-ft-long wall and the free-body diagram of the forces acting on it. The dead load is taken for the wall itself and consists of

Wall sheathing ($\frac{1}{2}$ in. plywood)	= 4.5 psf
2 × 6 stud at 16 in. o.c.	= 2.0 psf
Total	= 6.5 psf
$W = 22 \text{ ft} \times 9 \text{ ft} \times 6.5 \text{ psf}$	= 1287 lb

Taking the moments about the right bottom corner of the wall, the resisting moment due to the dead weight is the weight times half the length of the wall

$$\text{RM} = W \times \frac{L}{2} = 1287 \text{ lb} \times \frac{9 \text{ ft}}{2} = 5792 \text{ lb-ft}$$

The total lateral load on the wall is the

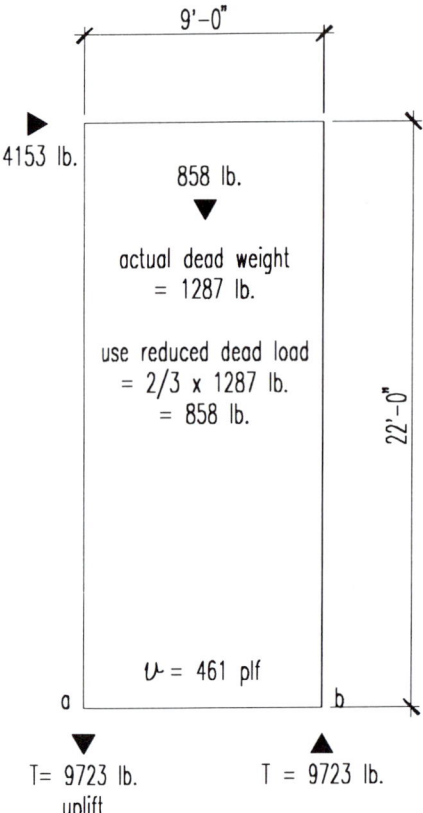

FIGURE 7.23 Nine-foot-wide shearwall and free body diagram.

shear at the base, in units of plf, times the length of the wall.

$$V = v \times L = 461.4 \text{ plf} \times 9 \text{ ft} = 4153 \text{ lb}$$

The lateral force is applied at the top of the wall, so its overturning moment is

$$\text{OTM} = V \times h = 4153 \text{ lb} \times 22 \text{ ft}$$
$$= 91{,}366 \text{ lb-ft}$$

The overturning moment is over 15 times larger than the resisting moment. Clearly, the wall will tip over unless it is fastened to the foundation.

Because it is possible that the assumed dead loads may not be in place for the entire life of the building, or that portions of them may fall off in severe winds or earthquakes, the codes seek to assure a greater degree of safety against overturning than for other loads. The codes require that connectors be designed to resist the overturning if

$$\text{OTM is greater than } \tfrac{2}{3} \times \text{RM}$$

The connectors must be designed to resist the difference between these two forces:

$$\text{mechanical moment} = \text{OTM} - \tfrac{2}{3} \times \text{RM}$$

In this example, the connectors must carry a moment of

$$\text{mechanical moment} = 91{,}366 - \tfrac{2}{3} \times 5792$$
$$= 87{,}505 \text{ lb-ft}$$

Assuming that the connector is located about 8 in. from the edge of the panel, the moment arm for the connector is 8'− 4". The uplift force on the connector is

$$T = \frac{87{,}505 \text{ lb-ft}}{8.33 \text{ ft}} = 10{,}500 \text{ lb}$$

Reference to Table 4.27 will show that a Simpson HD-9 (or equal) hold-down will be needed. The hold-down must be bolted with three 1-in.-diameter bolts into a stud with at least a $3\tfrac{1}{2}$-in. thickness. Since the wall is nominally 6 in. thick, we will specify a 4 × 6 corner stud. To assure adequate anchorage into the foundation, use a $1\tfrac{1}{8}$-in.-diameter anchor bolt with 15 in. of embedment. Figure 7.24 shows the detail for the end of the panel. Because the load could come from either side, the hold-downs must be placed on both ends of the 9-ft section. The specification for the

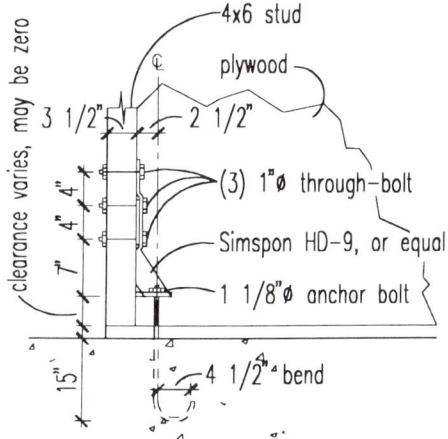

FIGURE 7.24 Nine-foot-wide shearwall hold-down.

hold-down is:

Simpson HD-9 (or equal) hold-down with three 1-in.-diameter bolts into 4 × 6 corner stud and $1\frac{1}{8}$ in.-diameter anchor bolt with 15-in. embedment.

At the 12-ft section the resisting force will grow as the length of the wall grows, but so will the total shear force. As a quick check assume that the mechanical moment is the same and find the uplift force based on the longer moment arm to the hold-down.

$$T = \frac{87{,}505 \text{ lb-ft}}{11.33 \text{ ft}} = 7721 \text{ lb}$$

A smaller hold-down may be used here. Maintaining the 4 × 6 corner stud will allow us to select a Simpson HD-7 (or equal). In regions of the country where builders are quite familiar with the use of hold-downs, a small saving can be found in specifying different hold-downs for various locations; otherwise, with the slight difference between the HD-7 and the HD-9, the designer may feel more comfortable in specifying the larger one in all four locations.

In the anchorage design we had originally specified seven $\frac{1}{2}$-in.-diameter anchor bolts to transfer 5537 lb into the foundation. By the same logic as was used before, we can account for the capacity of the $1\frac{1}{8}$-in. bolts as 1.33 times 1590 lb equals 2115 lb, where 1590 is the largest capacity listed for 1-in.-diameter bolts. The force to be carried by additional anchor bolts is 5537 minus 2115 for each hold-down, or 1308 lb. Two $\frac{1}{2}$-in.-diameter anchor bolts with a capacity of 811 lb each will suffice. Therefore, the original specification for anchor bolts can be changed from seven bolts to

$1\frac{1}{8}$-in.-diameter anchor bolt each side and two $\frac{1}{2}$-in.-diameter anchor bolts at the 12-ft section of shearwall A.

As a final step the designer may consider the shearwall deflection. This could be critical for shearwalls that are very near the maximum allowable ratio and which are highly loaded, or for special situations in which a great extent of glass is used. Normally, the deflection is not considered. It is the intent of the shearwall ratio requirements that any wall which meets the ratio will develop deflections which are within normal limits.

RESIDENTIAL DESIGN EXAMPLE

This example problem was chosen in part to represent a typical suburban house but also to draw attention to some of the common problems that have been observed in houses of this nature when subjected to large lateral loads such as earthquakes. This house can be viewed as being in two parts. The western section with the bedrooms is a very straightforward "ranch" house with moderate spans, uninterrupted diaphragms, and adequate opportunity for developing shearwalls. The eastern portion of the house, containing the garage and the two-story room with cathedral ceiling, has a series of problems that will require attention. Also, the connections between the two portions poses some problems that will require careful engineering and details.

EXAMPLE 7.5—PRELIMINARY DESIGN OF STRUCTURE

The design of any structure must be responsive to both gravity and lateral loads. In the case of wood-frame buildings, particularly platform stud framing, the elements used to carry one set of loads (such as gravity) are also used to carry the other set (lateral). The two systems are inextricably mixed and the location and sizing of any element must be considered for both sets of loads.

Figures 7.25 through 7.27 show plans and elevations of a typical wood-studded two-story ranch house. In the example we determine a possible framing plan and diaphragm design for the roof and the second floor. We will select certain shearwalls to be designed, identify some problem areas, and propose some details that will accommodate the problems. The design will be carried out using the provisions of the *Uniform Building Code* for seismic zone 4 and for a wind speed of 80 mph.

To determine the framing plans for each level, we will need to consider the diaphragm design and the interaction of diaphragms and framing. Before we can clearly identify the diaphragms, though, we must first identify all potential shearwalls. These consist of all exterior walls, shown in Figures 7.28 and 7.29, as well as the interior load-bearing E-W wall along the centerline of the 28-ft-wide section (SW2), the wall separating the garage from the great room (SW4) and the wall joining both sections of the house (SW7). At this

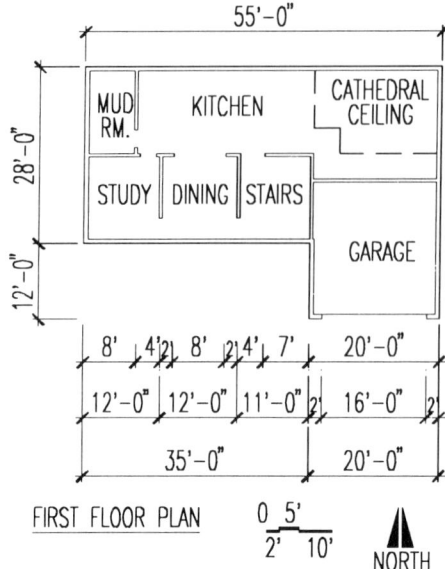

FIGURE 7.25 Wood frame residence—first floor plan.

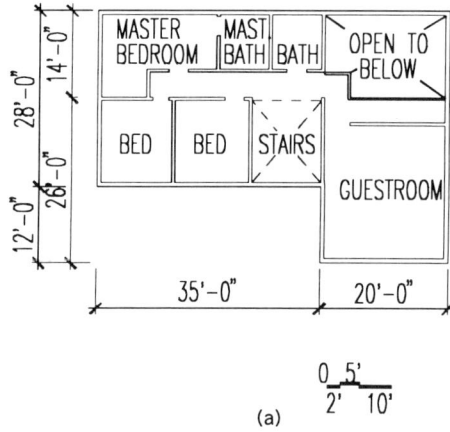

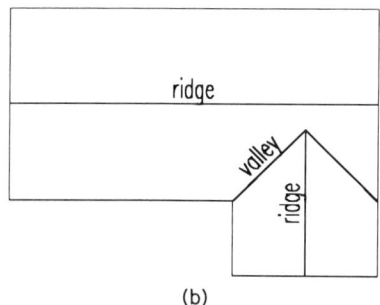

FIGURE 7.26 Wood frame residence second floor and roof plan.

FIGURE 7.27 Wood frame residence—elevation.

point we may note problems, most noticeably the large opening for the garage door in SF5, which must be resolved later. Another question that must be resolved will be whether SW2, the interior E-W shearwall, will be needed. There are a number of framing problems that will have to be resolved for this wall to transfer shear effectively from the roof through the basement to the foundation.

Having identified possible shearwalls, our next step will be to indicate the extent of the diaphragms which span from shearwall to shearwall. The diaphragms must be bounded on each edge by a shearwall or a collector. The collector will carry the shear from the diaphragm into an adjacent shearwall; for this transfer of forces to be successful, the collector must be in line with the shearwall as seen in plan and be at or near the top of the shearwall as seen in section. The size of the resulting diaphragms must be within the limits of the diaphragm ratio given in Table 7.7. The ideal proportion of a diaphragm is 1:1. In the example we have identified SW 2 as a potential shearwall, but ignored it in identifying diaphragm D1. In the design phase we will need to compare both of these options to determine the most feasible.

The east portion of the house poses the most problems. Notice that we have identified SW4 as a shearwall. The two resulting diaphrgms, D2 and D3, are both relatively small, 20 by 20 ft. Had SW4 been omitted, the diaphragm for the east portion of the house would have been 20 by 40 ft. Although this diaphragm is easily within the required ratio of 3.5 to 1.0, we can look ahead to problems at the garage opening. By using SW4 the forces at the front of the garage will be greatly reduced and the construction of the garage door frame will be simplified. Also considered in this decision is the roof and floor framing. The most direct way to frame is to install 28-ft-wide trusses over the complete 55-ft length of the house, leaving the garage to be framed with its own 20-ft trusses in the front. To achieve the roof profile as shown in Figure 7.28, a triangular portion is doubly framed—special modified trusses, or 2 × 6 rafters, are installed over the 28-ft trusses and the plywood diaphragm. This construction will result in the roof diaphragm having a

274 PLYWOOD

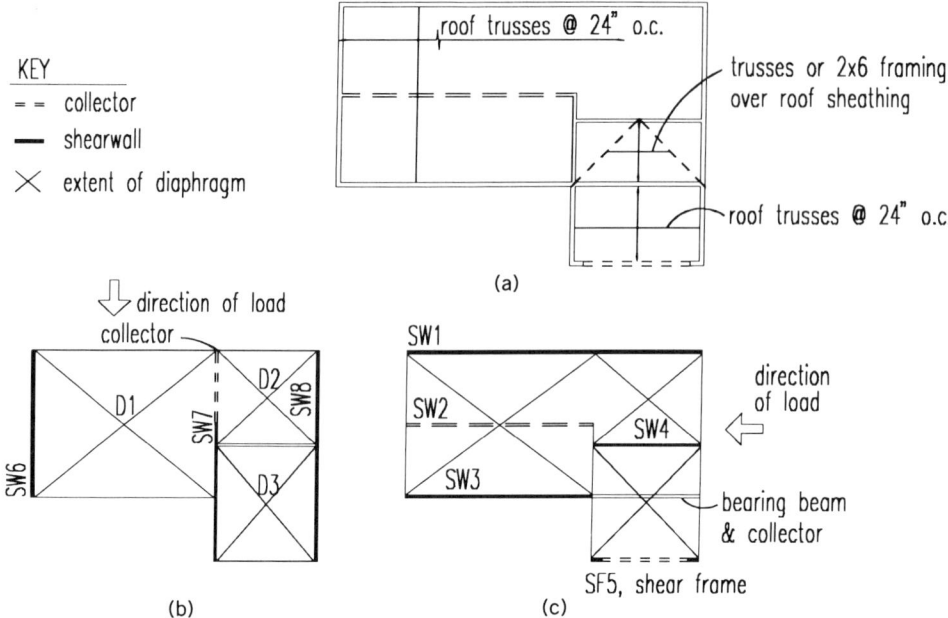

FIGURE 7.28 Wood frame residence—roof framing plan and shearwall locations: (a) roof framing plan, (b) N-S shearwalls and diaphragms, and (c) E-W shearwalls and diaphragms.

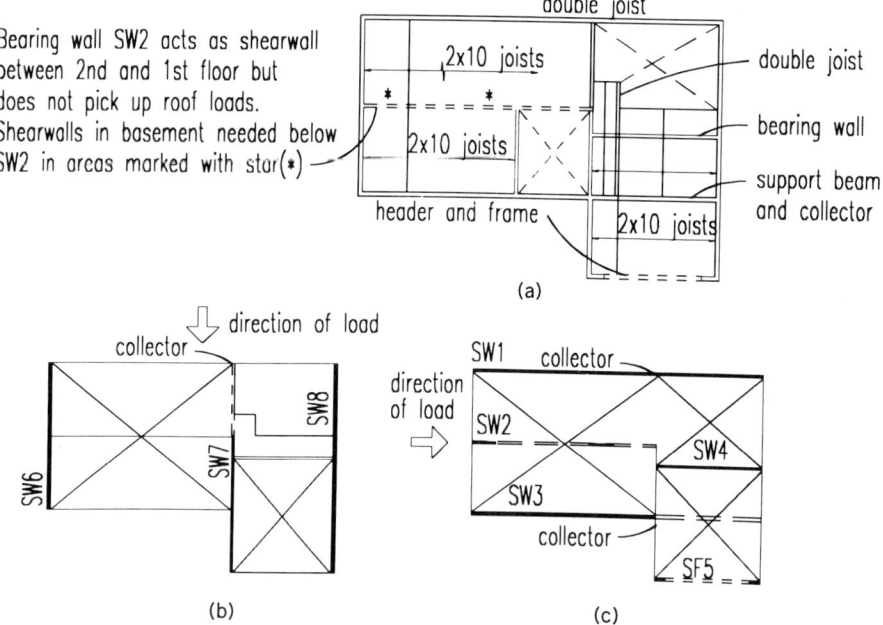

FIGURE 7.29 Wood frame residence—second floor plan and shearrwall locations: (a) second floor framing plan, (b) N-S shearwalls and diaphragms, and (c) E-W shearwalls and diaphragms.

different configuration from the floor diaphragm below. In the roof D1 extends the full 28 ft by 55 feet and is supported at the garage by a collector that is connected to the top plate of shearwall SW3. In the floor D1 is only 35 ft long and the diaphragm over the garage spans 20 ft by 20 ft, being supported in the rear by SW4. This implies that between the roof and the second floor SW4 does not pick up any shear load and does not function as a shearwall. It must be detailed so that there will be a minimum of shear transferred.

All of the considerations above were made in concert with concerns for the gravity load framing. The choice to frame the full 55-ft roof necessitated a bearing beam in line with the front wall of the house. This beam can also be used as the collector for the roof diaphragm. Framing the floor of the guest room over the garage presents a number of problems. The 20-ft span is further than we can comfortably span with 2×10's. If columns are to be avoided in the garage, then either a different material, such as truss joists, must be used, or intermediate beams must be located to carry the floor joists. Since SW4 is desirable as a shearwall, it is sensible to use it as a bearing wall as well. By adding

WIND LOADS
USE UBC

70 MPH, EXPOSURE B
$P = q \, C_e = 13 \, psf \times 0.8 = 10.4 \, psf$

roof slope is 6:12 $\theta = 26.5°$
USE METHOD 1 for pressure coefficients C_q

FOR ROOF USE VERTICAL PROJECTED HEIGHT, 7'
THE HORIZONTAL PROJECTIONS OF ROOF FORCES CANCEL EACH OTHER

FOR WIND IN NORTH-SOUTH DIRECTION
(NEGLECT AREA OF GARAGE DORMER)
FOR EACH LEVEL ASSUME LOAD AT DIAPHRAGM COMES FROM ½ HEIGHT ABOVE PLUS ½ HEIGHT BELOW

DIAPHRAGM	HEIGHT ABOVE	PRESSURE ABOVE	HEIGHT BELOW	PRESSURE BELOW	TOTAL
ROOF	7'	0	4.5	13.5	60.75 plf
2ND FLOOR	4.5	13.5	4.5	13.5	121.5 plf

TOTAL FORCE (N-S) $F = (60.75 + 121.5) \, plf \times 55' = \underline{10,025 \, lbs}$

FIGURE 7.30

276 PLYWOOD

WIND, CONTINUED

<u>EAST-WEST</u>

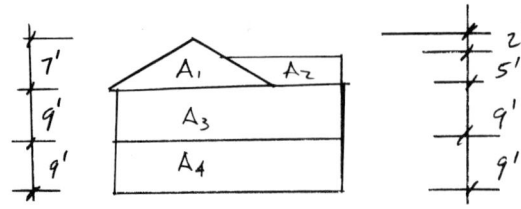

SECTION	AREA	COEFFS	PRESSURE $(10.4 \times C_q)$	FORCE = AREA × PRESSURE
A_1	$\frac{28 \times 7}{2} = 98^{\square}$	(0.8+0.5)	13.5 psf	1323 lbs
A_2	~5'×12' = 60$^{\square}$	(0.7+.7)	0	0
A_3	40×9 = 360	(0.8+0.5)	13.5	4860 lbs
A_4	40×4.5 = 180	(0.8+0.5)	13.5	2430 lbs
			TOTAL FORCE =	8613 lbs

DISTRIBUTION TO DIAPHRAGMS

DIAPHRAM	FORCE ABOVE	FORCE BELOW	FORCE TOTAL	$w = \frac{FORCE}{LENGTH}$ (PLF)
ROOF	1323	$\frac{4860}{2}$	3753 lbs	$\frac{3753}{40}$ = 94 plf
2ND FLOOR	$\frac{4860}{2}$	$\frac{4860}{2}$	4860	$\frac{4860}{40}$ = 121.5 plf

<u>UPLIFT</u>

CONSIDER EACH SECTION SEPARATELY. LOOK AT 28 WIDE

w_U = 0.7 × 10.4 psf = 7.28 psf

UPLIFT FOR 1 FOOT STRIP = 7.28 psf × 28' = 204 plf

OVERTURNING MOMENT FOR 1 FOOT STRIP

OTM = 204 plf × $\frac{28'}{2}$ = 2856 lb-ft/ft

FIGURE 7.30 (*Continued*)

a beam to align with SW3, the span of the joists is reduced to 12 ft and 8 ft. The rear joists will rest on the bearing wall (SW4) and cantilever 4 ft to form the walkway overlooking the great room. Figure 7.30 demonstrates the wind and seismic load-calculations for the residence.

Seismic Loads

One definition of a force is mass multiplied by an acceleration. For gravity loads the acceleration of gravity is used to determine the weight of an object. For seismic loads we assume that the earthquake imparts a horizontal force that is proportional to the weight of the building. For relatively small buildings, less than six stories high, we can approximate the dynamic effect of the seismic load with a static load. Since in the United States the most frequent seismic events occur in California, the *Uniform Building Code* (UBC) is probably the most advanced seismic code in force in the United States. All problems in this book that deal with seismic loads will refer to the UBC.

Wood-framed structures, particularly plywood-sheathed stud bearing-wall structures, inherently perform very well under seismic loads. Understanding the advantages of wood structures will aid a designer in accentuating the benefits of wood construction and avoiding the potential problems. A brief description of seismic design is given here with the emphasis on the reaction of wood structures. The reader may need to review seismic forces and design in greater detail than is offered here.

Two of the inherent advantages of typical wood-frame construction are demon-

WIND, CONTINUED

FOR 110 MPH, EXPOSURE C

$q_s = 31\,psf \qquad C_e = 1.3 \qquad p = 40.3\,psf$

RATIO $\dfrac{110\,MPH}{70\,MPH}$ IS $\dfrac{40.3}{10.4} = 3.875$

MULTIPLY ALL VALUES FOR 70 MPH BY 3.875

	70 MPH EXP B		110 MPH EXP C	
	N-S	E-W	N-S	E-W
TOTAL SHEAR	10025 LBS	8613	38850 LBS	33375
ROOF DIAPH.	60.75 plf	94 plf	235 plf	364 plf
2ND FLOOR	121.5 plf	121.5 plf	471 plf	471 plf
UPLIFT	204 plf		791 plf	
UPLIFT OTM	2856 #-ft/ft		11067 #-ft/ft	

FIGURE 7.30 (*Continued*)

SEISMIC

USE 1989 UBC, COMPARE ZONES 2 AND 4

$$V = \frac{ZIC W}{R_W} \quad \text{(BASE SHEAR)}$$

FOR ZONE 2, $Z = 0.2$
FOR ZONE 4, $Z = 0.4$

$R_W = 8.0$ (WOOD SHEARWALL)
$I = 1.0$ (LEAST IMPORTANCE)

$T = \text{period} = C_t (h_n)^{3/4}$

$C_t = 0.02$ for WOOD STRUCTURES
$h_n = 25'$

$T = 0.02 (25)^{3/4} = 0.02 \times 11.18 = 0.22 \text{ seconds}$

S = SOIL COEFFICIENT, ASSUME $S = 1.5$ FOR MODERATE QUALITY OF SOIL

$$C = \frac{1.25 S}{(T)^{2/3}} \text{ but} \leq 2.75$$

$$C = \frac{1.25 \times 1.5}{(.22)^{2/3}} = \frac{1.875}{0.368} = 5.09 \quad \text{use } C = 2.75$$

	ZONE 2	ZONE 4
$V = \frac{ZIC W}{R_W}$	$\frac{(0.2)(1.0)(2.75)}{8} W$	$\frac{(0.4)(1.0)(2.75)}{8} W$
BASE SHEAR	$0.06875 \, W$	$0.1375 \, W$

FIGURE 7.30 (*Continued*)

strated in the UBC equation used to calculate the static equivalent force from an earthquake. The UBC uses the equation.

$$V = \frac{ZIC}{R_W} W$$

where V is the base shear, Z refers to the zone as detailed in the UBC map, I is an importance factor based on the building occupancy, C is a factor related to the interaction of site conditions with the building configuration, $1/R_W$ corresponds to the building configuration, and W is the weight of the building, including permanent items attached to it.

Wood-frame construction is almost always the least-weight construction. Reducing the weight of a building reduces the seismic shear. A typical wood-frame structure has a floor weight in the range of 10 to 20 psf depending on length of spans, type of ceiling and amount of mechanical equipment or lighting. The average weight of a gyp-board wall is 15 to 20 psf, depending on type and thickness of sidings. So the total weight of wood structures averages between 30 to 50 psf; most wood structures are at the low end of this range. A concrete structure, designed to carry the same live load of, say, 50 psf may have a floor weight ranging from 50 to 125 psf. Its

SEISMIC, CONT.

WEIGHT OF BUILDING

ROOF ASSEMBLY

ASPHALT SHINGLE (235)	2.35 psf
15# FELT	.15 psf
1/2" plywood	1.5 psf
2X 6 T&B TRUSS @ 24" °/c assume 4 plf × 12/24	2.0 psf
R-30 BATT INSULATION	3.0 psf
1/2" GYPSUM CEILING	1.5 psf
	10.5 psf

2ND & 1ST FLOOR

CARPET & PAD	0.5 psf
3/4" plywood	2.25 psf
2×10 @ 16" JOIST	2.0 psf
1/2" GYPSUM CEILING	1.5 psf
	6.25 psf

INT. WALLS

1/2" GYPSUM BOARD	1.5 psf
2×4 @ 16" °/c STUDS	1.0 psf
1/2" GYPSUM BOARD	1.5 psf
	4.0 psf

EXT. WALLS

7/8" STUCCO	10.0 psf
3/8" plywood	1.2 psf
2×4 STUDS @ 16" °/c	1.0 psf
R-13 BATT INSULATION	1.5 psf
1/2" GYPSUM BOARD	1.5 psf
	15.2 psf

PLAN AREAS & WTS

		AREA	PSF	WEIGHT (LBS)
ROOF	$(28 \times 55 + 12 \times 20) \, 1/\cos\theta =$	1990	10.5	20900
2ND FLOOR	$1780 - (11 \times 14) - (14 \times 20) =$	1350	6.25	8450
1ST FLOOR	1780 =	1780	6.25	11125

FIGURE 7.30 (*Continued*)

total building weight will range from 70 to 150 psf. The weight of a steel structure will be intermediate between these two types.

The term $1/R_W$ responds to the type of construction and the building configuration. Wood-framed systems are typically boxlike in plan. Table 7.8 lists the R_W values for four types of wood construction. Stud-wall wood-frame structures, referred to as "bearing wall system—light-framed walls with plywood shear panels," have an R_w of 8. If the plywood walls are not load bearing and are attached to a wooden structural frame that carries the gravity loads, the system is referred to as "building frame system—light-framed walls with

SEISMIC, CONT.

WALL DISTRIBUTION:

ASSUME ALL 2ND FLOOR INTERIOR WALLS PRODUCE A FORCE ON 2ND FLOOR DIAPHRAGM

ASSUME ALL 1ST FLOOR INTERIOR WALLS PRODUCE A FORCE ON THE 1ST FLOOR DIAPHRAGM

ASSUME EXTERIOR WALLS DISTRIBUTE LOADS BY RATIO OF ½ ABOVE AND ½ BELOW

WALL WEIGHTS (ALL WALLS ARE 9' HIGH)

LOCATION	INTERIOR LENGTH	WT (PSF)	WEIGHT	EXTERIOR WALLS LENGTH	WT (PSF) (85%)	WEIGHT
2ND FLOOR	~135 FT	4.0	4860 lbs	190	15.2	22,100 LBS
1ST FLOOR	~80 FT	4.0	2880 lbs	174	15.2	20,250

DISTRIBUTION OF WEIGHTS TO LEVELS

	INT	EXT	ROOF OR FLOOR	TOTAL	
ROOF 2ND WALL	—	11050	20900	31950	@ ROOF
2ND WALL 2ND FLOOR 1ST WALL	4860 —	11050 10125	8450	34485	@ 2ND FLOOR
1ST WALL 1ST FLOOR	2880	10125	11125	24130	@ 1ST FLOOR

90565 LBS

BASE SHEAR

ZONE 2	ZONE 4
.06875 W	.1375 W
6226 LBS	12453 LBS

FIGURE 7.30 (*Continued*)

plywood shear panels" and has an R_W of 9. For typical wood-frame construction a value of R_W equals 8 is appropriate.

The dynamic analysis of a building under a seismic force demonstrates that an important feature of any building is to remain ductile and to deflect enough so that the energy of the earthquake can be dissipated in the connections. Under this consideration a structure that has a large number of small connections will perform better than a structure with fewer connections. In stud-framing there is a large number of connection. As mentioned in

SEISMIC, CONT.

DISTRIBUTION OF SHEAR FORCES TO DIAPHRAGMS

FORCE TO ROOF $F_t = 0.07 \, T \, V$

(FOR ZONE 4) $= 0.07 \times .22 \times 12453 \, LBS$

$= 192 \, LBS$

F_t MAY BE 0 IF $T \leq .7$ SECONDS

F_t IS FORCE AT TOP, IN THIS CASE, RIDGE OF ROOF
$F_t = 0$. FORCE AT ROOF DIAPHRAGM WILL BE CONCENTRATED AT $h_x = 18'$

$F_x = \dfrac{(V - F_t) \, w_x \, h_x}{\Sigma \, w_x \, h_x}$

ELEMENT	w_x	h_x	$w_x h_x$ K-FT	$\dfrac{w_x h_x}{\Sigma}$	ZONE 2 V=6226 LB F_x	ZONE 4 V=12453 LBS F_x
ROOF	31950	18	575.1	.632	3936	7873
2ND FLOOR	34485	9	310.3	.341	2124	4248
1ST FLOOR	24130	1	24.1	.026	165	330
$\Sigma W =$	90565		909.6 K·FT	1.00 ✓	6225 ✓	12450 LBS ✓

FORCE w (plf) AT EACH DIAPHRAGM

ELEMENT	N-S DIRECTION		E-W DIRECTION	
	ZONE 2	ZONE 4	ZONE 2	ZONE 4
BASE SHEAR	6226 LB	12450	6226	12450
ROOF DIA.	71.6 plf	143 plf	98.5	197 plf
2ND FLOOR	38.6 plf	77.2 plf	53 plf	106 plf

FIGURE 7.30 (*Continued*)

Chapter 4, nails are designed with a large factor of safety, in the range 4 to 5. This factor will ensure that in a well-detailed stud building most of the connections will stay within the elastic range. If this is the case, the building will not only remain safe and standing, it will also be able to withstand a series of seismic events near the size of the design load. These factors are reflected in the $1/R_W$ term. A wood-frame structure is optimal where it forms a closed rectangle, where there are a large number of connections, and where the shearwalls and diaphragms cover the largest areas possible.

Table 7.9 lists a variety of irregular configurations. All of the irregularities can be thought of as deviations from a simple box; the closer the form of the building is to a box, with continuous diaphragms and

COMPONENT	WIND				SEISMIC				CONTROLLING CASE			
	70 MPH EXP B		110 MPH EXP C		ZONE 2		ZONE 4		70 MPH vs ZONE 2		110 MPH vs ZONE 4	
	N-S	E-W	N-S	E-W	N-S	E-W	N-S	E-W	N-S	E-W	N-S	E-W
TOTAL SHEAR	10025 LBS	8613	38850	33375	6226	6226	12450	12450	10025 WIND	8613 WIND	38850 WIND	33375 WIND
ROOF DIAPHRAGM	60.7 plf	94 plf	235	364	71.6	98.5	143	197	71.6 SEISMIC	98.5 SEISMIC	235 WIND	364 WIND
2ND FLOOR	121.5 plf	121.5 plf	471	471	38.6	77.2	53	106	121.5 WIND	121.5 WIND	471 WIND	471 WIND
UPLIFT	204 plf		791						204 WIND		791 WIND	
OTM	2856 #-ft/ft		11067						2856 #-ft/ft		11067 #-ft/ft	

FIGURE 7.30 (Continued)

balanced pairs of shearwalls, the better it will perform under an earthquake. For each of these irregularities special considerations or allowable heights are listed in the *Uniform Building Code*. Any structure that is under 65 ft in height, which includes virtually all wood-frame buildings, can be analyzed using the static equation. The irregularities are usually permissible for such structures, although they may require special design or may limit the height in some instances. The designer should become familiar with these irregularities since they will inevitably be the location of failure or damage during an earthquake. Wherever possible, they should be avoided;

TABLE 7.8 R_w **FOR WOOD STRUCTURAL SYSTEMS**

Basic Structural System	Lateral Load-Resisting System	R_w	H^a
Bearing wall	Light-framed walls with plywood shear panels, three stories or less	8	65
	Braced heavy timber frames where bracing carries gravity loads	4	65
	All other light-framed walls	6	65
Building frame system	Light-framed walls with plywood shear panels, three stories or less	9	65
	Braced heavy timber frames where bracing carries gravity loads	8	65
	All other light-framed walls	7	65

Note: Bearing Wall System (BWS) is a structural system without a complete load-carrying space frame. Bearing walls or bracing systems provide support for all or most gravity loads. Resistance to lateral loads is provided by shear walls or building frame. Typical stud-wall residential wood construction is an example of BWS.

Braced Frame System (BFS) is an essentially complete frame which provides support for gravity loads. Resistance to lateral loads provided by shear walls or braced frames. A building in which steel or wood frames (as shown in Figures 7.34 and 7.35) are used extensively are braced frames. This type of wood building is used primarily in light commercial and warehouse applications.

Concentrically braced from (CBF) is a braced frame in which the members are subjected primarily to axial forces.

[a]H is the maximum allowable height (feet) in seismic zones 3 and 4.
Source: UBC Table 23-O

TABLE 7.9   IRREGULARITIES FOR WOOD STRUCTURAL SYSTEMS

Irregularity Type	Limitations	Solutions	Detail
Vertical Irregularities			
Soft story: any story with lateral *stiffness* less than 70% of adjacent story (for walls of the same height and material, the stiffness is proportional to the length of wall raised to the fourth power)	(UBC requires dynamic calculation for structure over 65 ft.) This is common in split-levels or other houses with a story over the garage. Occurs anywhere with a large opening.	Severe problem; avoid if possible. Use wood frame or steel moment frame at large openings.	See steel frame (Figure 7.35) or wood frame (Figure 7.33)
Mass distribution irregularity. any story that weighs more than 150% of an adjacent story (neglect roofs lighter than floors below)	(UBC requires dynamic calculation for structure over 65 ft.) This is less common in wood construction and usually will not pose a problem unless the heavier floor is above the lighter one.	Usually easy to avoid. Where an intermediate floor is irregular, shear walls above and below must be designed to carry the shear.	
Vertical geometry irregularity. occurs when the shearwall system in one floor is 130% longer in plan than that of an adjacent floor	(UBC requires dynamic calculation for structure over 65 ft.) This is common in split levels or other houses with a smaller second story.	Requires collectors and straps in the diaphragm to transfer forces into the full length of the lower wall.	See collector strut details (Figure 7.22 and 7.37)
In-plane discontinuity of shear wall: shear wall is discontinuous so that the openings are longer than the length of the adjacent wall	(UBC requires that overturning be checked.) This is a common situation and has a number of common solutions.	Drag struts over the openings; limit height-to-length ratios of shearwalls, check for overturning, hold-downs if needed.	See Figures 7.23 and 7.24
Weak story: any story with lateral *strength* less than 80% of adjacent story. (for walls of the same material, the strength is directly proportional to the length of wall, and independent of the height)	(Structures in which the weak story has less than 65% of the strength of the adjacent story, may not exceed two stories or 30 ft in height.) This is similar to the soft story condition, but can be remedied more easily.	Avoid if possible. Increase plywood thickness or number of nails if possible. For severe cases, use wood frame or, for extreme conditions, use steel moment frame.	See steel or wood frames (Figures 7.33 and 7.35)
Plan Irregularities			
Torsional irregularity	(Not considered for flexible diaphragms.) Most wood diaphragms are flexible.	See discussion of flexible vs. rigid diaphragms and torsion of flexible diaphragms.	See garage opening (Example 7.11 and Figure 7.33)

TABLE 7.9 *Continued)*

Irregularity Type	Limitations	Solutions	Detail

Plan Irregularities

Reentrant corners: plans where the projections in both directions beyond the edge of the corner exceed 15% of the horizontal dimension	In seismic zones 3 and 4, UBC requires that collectors and connections shall be designed without the 33% strength increase normally used in seismic design. Design shall consider independent movement of both parts: movement in the same direction and in opposing directions.	Separate the elements. Determine expected deformations and provide flexible joint that can bridge the gap. Not often done on small buildings. Tie the two elements together. Provide shearwalls or collectors.	See Figure 7.37.
Diaphragm discontinuity: abrupt changes in stiffness, such as openings greater than 50% of the gross area, or where the stiffness is less than 50% of the adjacent story	In seismic zones 3 and 4, UBC requires that collectors and connections shall be designed without the 33% strength increase normally used in seismic design.	See Example 7.2.	See Figure 7.16, and Example 7.2
Out-of-plane discontinuity of shear wall	Where columns support the upper wall, the UBC requires they must be designed to carry the gravity load plus a portion of the lateral load. In seismic zones 3 and 4, collectors and connections shall be designed without the 33% strength increase normally used in seismic design. Typically shear walls which are within 2 ft of each other may be considered as acting along the same line.		
Nonparallel systems: the gravity load-carry system is not parallel with the lateral load-resisting system.	System must be designed to carry a portion of the loads arising from two orthogonal directions. Since the gravity system and the lateral system are usually identical in wood frame construction, this condition is not often encountered.		

Source: Derived from UBC Table 23-0.

where they cannot be avoided, the designer should recognize that special attention, additional calculations, and careful detailing will be needed at the location of these irregularities. The most commonly encountered irregularities in woodframing, and possible remedies for them, are listed in Table 7.9. In Examples 7.8 through 7.11 we look at the effects of some irregularities and show the design for the connections.

The example problem was chosen in part to represent a typical suburban house but also to draw attention to some of the common problems that have been observed in houses after earthquakes. This house has a number of problems; careful engineering, specification of details, and construction quality control are needed to assure that the problems are resolved.

1. Garage Door Opening. The garage door opening is placed in the wall so that the adjacent shearwalls are only 2 ft long. (This wall is designated SF5 on Figure 7.29.) The height-to-length ratio is 4.5, which exceeds the recommended ratio of 3.5. In addition, the second-story living space above the garage contributes a large mass and seismic force to the wall. Since this shearwall is 12 ft in front of the front wall of the house (SW3), none of the load can be transferred to the front wall. Two alternate details for the shearwall surrounding the garage door are shown in Figures 7.33 and 7.35, and the calculations for the wood-frame opening are given in Figure 7.34.

2. Diaphragm Discontinuity. The second-floor diaphragm is broken into two separate elements by the openings at the stairs and the cathedral ceiling. The shearwall between the two sections (SW7) is long enough to carry the force, but collectors will have to be used between the kitchen and the living room to transfer the forces into the wall. Example 7.10 develops the design of these collectors.

3. Connections at Shearwall SW2. The long wall running east-to-west in the 28-ft-wide portion of the house (SW2) aligns with the walls below. This is an advantage because the shear forces from the roof could work their way directly down these walls. However, for these walls to act as shearwalls, there must be shear-resisting elements below them at the basement level. Often, steel lally columns are used which have no shear capacity. The columns should be replaced with walls in the basement. The best location for these are shown on Figure 7.29 as (1) in line with the wall between study and mudroom, and (2) below the wall separating stairs and kitchen. In addition, for the second-story part of SW2 to function as a shearwall, there needs to be a connection between the wall and the roof diaphragm. The plywood sheathing on the roof constitutes the diaphragm, but it is almost 6 ft above the top of the wall, which ends at the level of the exterior wall plates. A shearwall would have to be built in the attic space with sufficient sheathing and connections to transfer shear from the roof diaphragm to the top of SW2. This would be very difficult to do with a trussed roof construction, not worth the effort, and probably not necessary. Consequently, only the first floor wall acts as a shear-wall carrying second floor loads to the basement.

4. Discontinuous Shearwalls. The center E-W wall (SW2) on the first floor has a number of large openings. The short 2-ft-long sections should not be considered capable of carrying any shear. Of the 35-ft length of the wall, the shearwalls make up an 8-ft section on the west and a 7-ft section on the east, with a 20-ft opening in the middle. A collector strut would be needed along the 20-ft length. This concern, combined with the problems listed above, might persuade us to ignore SW2 as a shearwall. In reality, we must depend on it as a load-bearing wall to carry the gravity loads from the second floor. Therefore, we will keep it as a shearwall carrying second-floor seismic and wind forces but not carrying roof seismic or wind forces. Were we to analyze this wall, we would find that because of the low level of forces, the problems can be resolved, without extreme difficulty.

5. Reentrant Corner. The corner between the entrance and the garage constitutes a reentrant corner. Special connections will be needed at both the roof diaphragm and the second-floor diaphragm to tie these two elements together. Details of the typical connections for this location are shown in Figures 7.38 and 7.39.

EXAMPLE 7.6—SEISMIC LOADS ANALYSIS

The residential building in Example 7.5 now is analyzed for seismic loads. The forces arising from an earthquake and the distribution of those forces to the diaphragms are shown for two earthquake zones: Zone 4, which is the maximum level in the United States and occurs primarily in coastal California and the Sierras; and Zone 2, which covers the majority of the western states, most of New England, and the southern Appalachians (see Figure 7.31b). Calculations for the seismic load on the entire building are given in Figure 7.30.

EXAMPLE 7.7—WIND LOAD ANALYSIS

The plans and elevations of a typical wood-frame house is shown in Figures 7.25 through 7.27. In Figure 7.30, wind loads in each direction on the house are calculated and distributed to the floors and roof diaphragms. Two wind conditions are calculated: 70 mph, exposure B, which is the typical design load for the majority of the United State (see Fig. 7.31a); and 110 mph, exposure C, which is a moderately high hurricane wind, used along portions of the gulf coast and the Atlantic coast. The forces produced by these winds are then compared with the forces caused by an earthquake.

Comparison of Wind and Seismic Loads

The standard approach in design of small wood structures is to calculate the base shear from the seismic load and to compare it with the total shear, in each direction, due to the wind load. Whichever load is larger will control the design and is used; the other is neglected. While this approach is usually adequate for the performance of the overall building, certain precautions should be considered.

Wind blowing over low-pitched roofs causes uplift and overturning forces. The uplift coefficients increase, sometimes by a factor of 2 or more, at overhangs, ridge, edges, and corners. Wind uplift needs to be considered and, if it is a factor, connectors—called hurricane clips—may be needed (see Example 4.1).

Wind uplift forces must be transferred down into the foundation or at least beyond a level where they are countered by the dead load. This requires a continuity of the load path. In many cases this continuity can be made through the exterior sheathing; in many cases it requires that metal connectors be used to connect all lines of principal framing. See Figure 4.6 for an example.

Both wind loads and seismic loads require continuity of lateral load paths. In the case of seismic loads, this condition becomes more critical. Discontinuous diaphragms must be bridged together using collector struts and straps. The bridging must be able to operate in compression and tension, since the direction of the load is never known. Figures 7.38 and 7.39 show sample details that may be needed in these areas. In many cases the connection re-

TABLE 7.10 COMPARISON OF WIND (70 MPH) AND SEISMIC FORCES

	Wind: 70 mph, Exposure B		Seismic: Zone 4		Maximum	
	N-S	E-W	N-S	E-W	N-S	E-W
Total shear (lb)	10,025	8,613	12,450	12,450	12,450 (seismic)	12,450 (seismic)
Roof diaphragm (plf)	60.75	94	143	197	143 (seismic)	197 (seismic)
Second-floor diaphragm (plf)	121.5	121.5	77.2	106	121.5 (wind)	121.5 (wind)

quirements will need to be calculated for specific cases.

Under wind loads, large, light elements such as high parapet or privacy walls will act like sails. Under seismic loads, because of their lightness, they may not be affected. On the other hand, under seismic loads, massive elements such as chimneys may experience very large seismic forces whereas due to their relatively small surface, they will not experience large wind forces. Figure 7.40 shows typical connec-

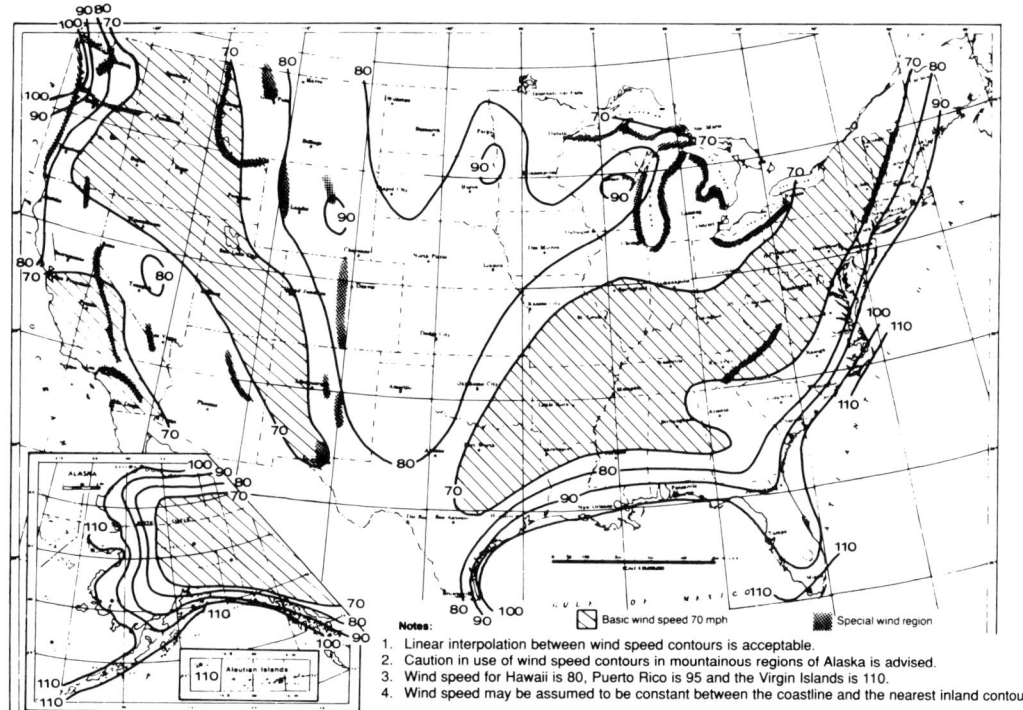

FIGURE 7.31 (a) minimum basic wind speeds in miles per hour. (Reprinted Courtesy of ICBO)

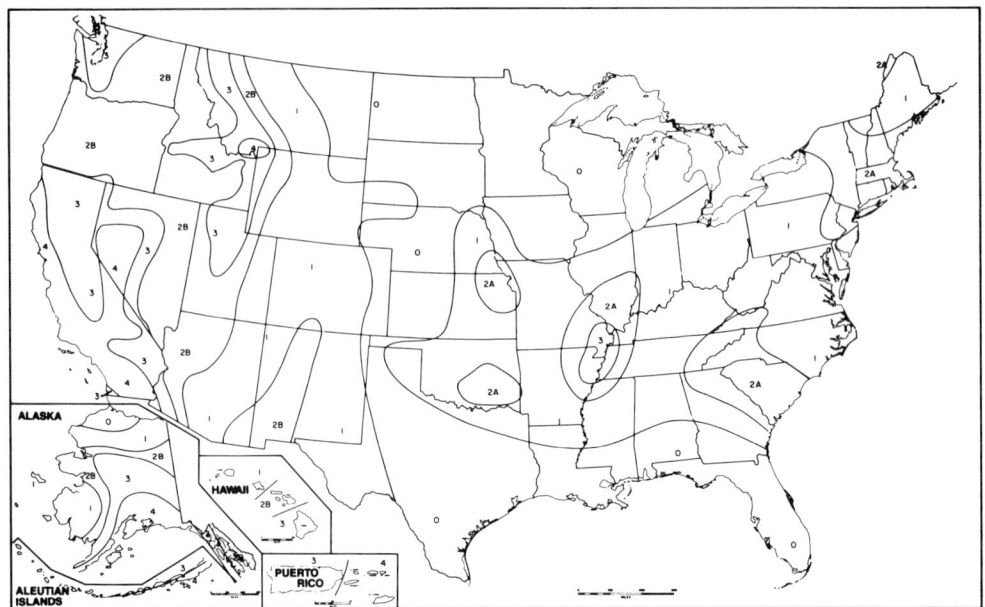

FIGURE 7.31 (b) seismic zone map of United States. (Reprinted Courtesy of ICBO)

tion details for masonry chimneys in high seismic zones.

Comparison of Forces in Examples 7.6 and 7.7. In example 7.7 the effect of a 70-mph wind (13 psf) and a 110-mph wind (40 psf) were developed for a typical residence. In example 7.6 the static equivalent seismic load was calculated for the same building based on seismic zones 2 and 4. These forces are compared in Table 7.10.

If we consider first the conditions on the west coast, we should compare the greatest seismic intensity (zone 4) with a moderate wind intensity (70-mph wind). Compare Figures 7.31a and 7.31b. The comparison shows that the maximum base shear comes from seismic forces but that the greatest shear at the second floor arises from the wind loads. The distribution formula for seismic loads places a greater portion of the load higher up on the building. The distribution of the wind loads places a relatively lower force at the roof level, due primarily to the canceling of the wind load on either side of a low-pitched roof. Also note that the seismic shear is the same in both directions, whereas the wind shear is greater when it blows parallel to the shorter dimension of the building. In scismic zone 3 the base shear would be three-fourths of the zone 4 shear (9340 lb). Had this been the case, wind would control the design in the north-south direction and seismic in the east-west direction. This is a common situation. It must be remembered that both the seismic and wind forces need to be calculated to determine which controls.

Although we tend to associate earthquakes with California, there are some important exceptions to this. Charleston, South Carolina suffered what was probably the second worst earthquake in the United States on record in 1886, estimated at an intensity of about 10 MMI (Modified Mercali Intensity). As a result this area is in seismic zone 3. It is also in the 100- to 110-mph wind zone. The Boston area had

TABLE 7.11 COMPARISON OF WIND (110 MPH) AND SEISMIC FORCES

	Wind: 110 mph, Exposure C		Seismic Zone 4		Maximum	
	N-S	E-W	N-S	E-W	N-S	E-W
Total shear (lb)	38,850	33,375	12,450	12,450	38,850 (wind)	33,375 (wind)
Roof diaphragm (plf)	235	364	143	197	235 (wind)	364 (wind)
Second-floor diaphragm (plf)	471	471	77.2	106	471 (wind)	471 (wind)

earthquakes estimated at 8 MMI in 1755 and again in 1817. It is in seismic zone 3 and a 100-mph wind zone. Table 7.11 compares the hurricane force wind (110 mph) with the zone 4 earthquake.

For this situation the wind forces far exceed those due to earthquakes. In part the wind forces are actually larger, but the building codes also use a lower seismic force than may actually occur. In the example the zone 4 earthquake force equals 13.75% of the acceleration of gravity. Actual ground accelerations in large earthquakes have often been measured at 1.0 times the acceleration of gravity and sometimes slightly higher. Because most buildings designed under the code have performed well in large earthquakes, the building codes attempt to model the response of the building rather than the ground, and the factors employed by the code are deemed accurate and conservative for buildings.

The conclusions to be drawn from this are important to bear in mind. On any project both wind and seismic forces can play a role in the design. Each type of force causes different problems. The proper configuration for a building to minimize seismic problems will not be the proper configuration when considering high wind loads. In each case special attention to details is required. The proper connection and details for each type of load are different.

Duration Factors

With wood construction the duration factor recommended by the NDS for seismic design is 1.6. This value corresponds to a load duration of 10 minutes. The actual duration of an earthquake is usually 30 seconds to 4 minutes, with the major shock occurring in the first few seconds. The duration factor for loads of that duration are closer to 2.0. Because of wood's time-dependent qualities and the provision in the code, wood construction has an additional "hidden" capacity to carry seismic loads. This capacity, in part, allows the underestimation of the acceleration and force of earthquakes on wood structures. The NDS also recommends a duration factor of 1.6 for wind loads. This factor, a revision to the 1.33 factor used until 1991, reflects the fact that the wind force considers the maximum force to occur during gusts. The 1.33 factor correlates to a 1-day duration, which is commensurate with a "steady-state" wind, not wind gusts. In the late 1980s both the BOCA *National Building Code* and the Southern (SBCCI) *Standard Building Code* adopted wind loads based on ANSI A58.1, the American National Standards Institute standard for loads on buildings.[2] The ANSI standard

[2] ANSI A58.1/ASCE Standard 7-88, *Minimum Design Loads for Buildings and Other Structures*, American Society of Civil Engineers, New York, 1988.

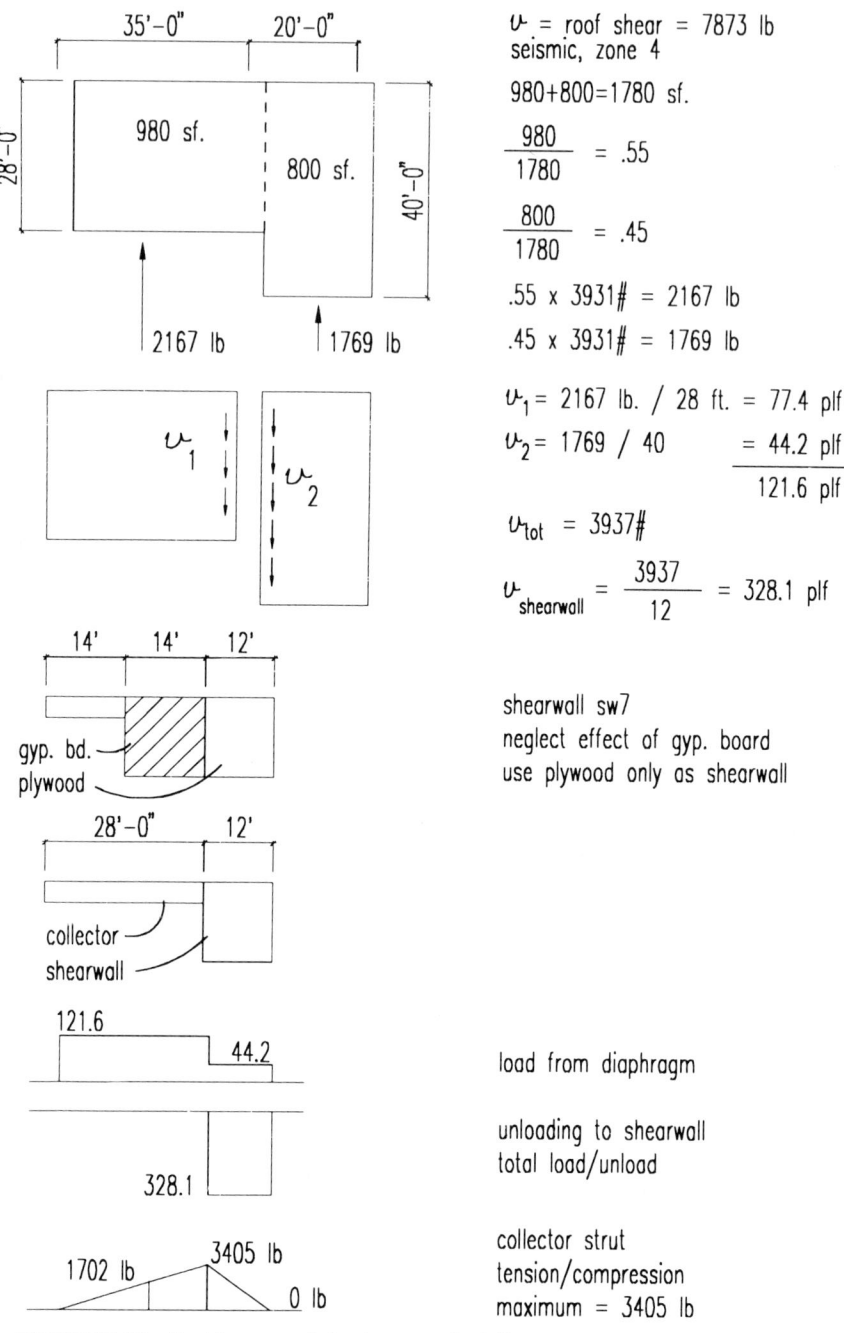

FIGURE 7.32 Residence roof diaphragm calculations.

uses gust modifiers and both BOCA and SBCCI recognize the use of the 1.6 duration factor for wind forces figured using their codes. Whereas the ICBO *Uniform Building Code* uses gust factors, their wind computations are not ANSI-based, and ICBO requires that a duration factor of 1.33 be used with winds calculated from the *Uniform Building Code.* Does this sound confusing? Before starting a project it is always good practice to check with the local building code department to clarify a number of important design conditions. We recommend that you consult with them concerning the appropriate duration factor for wind and seismic.

EXAMPLE 7.8—TWO-STORY SHEARWALL DESIGN

Design shearwalls for the house in Figure 7.25. Use the zone 4 seismic load for the design. Consider the force in the N-S direction. The main portion of the roof acts as a one diaphragm, 28 ft wide by 55 ft long. This configuration is well within the diaphragm ratio of 3.5 to 1. The smaller dormer portion will function as its own diaphragm. Because the lengths of the shearwall are the same as the lengths of the diaphragm edges above them, they will all have very similar loads in terms of pounds per foot. Choose the wall with the most openings for the design, say the west elevation.

The total shear on the roof is 7874 lb. This is the seismic load, so it operates in either the E-W direction or the N-S direction. Rather than recalculating an exact analysis of the tributary area for the west wall (SW6), assume that half the total load will be on that wall. This is slightly conservative because it places some of the garage dormer on the west wall. The west elevation shows that at the second-floor level the windows take up about 8 ft along the wall; the available length of shearwall will be 28 ft minus 8 ft, or 20 ft. The height-to-width ratio for plywood shearwalls is 2.5 to 1.0. For a wall height of 8 ft, a wall shorter than 8 over 2.5, or 3.2 ft, will fail to meet the requirement and may not be included in the length of actual shearwall. If there were any sections of wall that are less than 3.2 ft long, we would include them in the sum of doors and windows.

$$v = \frac{V}{b} = \frac{7874 \text{ lb}/2}{20 \text{ ft}} = 197 \text{ plf}$$

Refer to Table G.7 for plywood shearwall capacities. Assume that either $\frac{3}{8}$- or $\frac{1}{2}$-in.-thick plywood will be used. The $\frac{3}{8}$-in.-thick plywood will accept an allowable shear of 200 plf when nailed with 6d nails at 6 in o.c. This would be satisfactory, but it will require that the full length of the wall be sheathed in plywood.

$$\text{Required shearwall length} = \frac{7874 \text{ lb}/2}{200 \text{ plf}}$$
$$= 19.68 \text{ ft}$$

As an alternative, determine the required plywood specification if only 4-ft widths of plywood are used only at the two outside corners.

$$v = \frac{V}{b} = \frac{7874 \text{ lb}/2}{8 \text{ ft}} = 492 \text{ plf}$$

The $\frac{3}{8}$-in.-thick plywood is satisfactory if the nails are spaced 2 in. o.c. This spacing requires nominal 3-in.-thick studs to avoid splitting. This option is not economical. The $\frac{3}{8}$-in. plywood will also work with 8d nails at 3 in. o.c. The tabulated shear for this condition is 410 plf, but footnote 3 allows that value to be increased 20% when the plywood is nailed directly to the framing.

Allowable $v = 410 \text{ plf} \times 1.20 = 492 \text{ plf}$ OK

The $\frac{1}{2}$-in. plywood can also be used with 8d nails at 3 in. o.c. and will be the best option. If the full extent of the wall is not sheathed in plywood, then the $\frac{1}{2}$-in. plywood is preferable since other sheathing materials, such as fiberboard are typically available in $\frac{1}{2}$-in. thickness.

Referring to the elevation, the windows and doors at the first-floor level consume about the same 8 ft of wall leaving the length of available shearwall as 20 ft also. To consider how much of the wall needs to be sheathed in plywood, the additional load from the second floor must be added. The second-floor level contributes an additional shear of 4248 lb. Consider first sheathing only the corners.

$$v = \frac{V}{b} = \frac{4248 \text{ lb}/2}{8 \text{ ft}} + 492 = 265.5 + 492$$
$$= 757.5 \text{ plf}$$

This value is off the chart, so a longer extent of shearwall is needed. Assume that we will use $\frac{1}{2}$-in. Structural II plywood with 8d nails at 3 in. o.c. The allowable shear is 490 plf, and the required length of shearwall at the first floor level is

$$l = \left[\frac{(7874 + 4248)\text{lb}}{2} \times \frac{1}{490 \text{ plf}} \right]$$
$$= 12.36 \text{ ft}$$

Typically, it will be most efficient to sheath the entire length of the first floor wall, 20 ft, with

$\frac{1}{2}$-in. Structural II plywood with 8d nails at 3 in. o.c.

EXAMPLE 7.9 — INTERIOR SHEARWALL DESIGN

Design an interior shearwall for the house in Example 7.8. Consider SW7, the wall on the west side of the cathedral ceiling between the kitchen and the living room. This wall is 26 ft long (see Figure 7.32). Twelve feet are exterior and 14 ft have both sides as interior walls. Because the wall is not as long as the diaphragm edge, a collector strut will be needed to transfer the force from the southern 14 ft of diaphragm. The collector will be analyzed in Example 7.10.

Design the wall for seismic loads arising in zone 4. The roof lateral force is 7874 lb. We will assume that the roof diaphragm is flexible, therefore half of the 7874 lb will be transferred to wall SW7. The second-floor seismic load is 4248 lb, of which half falls on wall SW7. In situations such as this where there are large openings in the floor diaphragm, a careful analysis of the distribution of loads might be required. In this case the analysis results in half of the floor seismic load going into SW7, as we might intuitively expect.

Considering the exterior portion of the wall only, the shear from the roof is

$$v = \frac{V}{b} = \frac{7874 \text{ lb}/2}{12 \text{ ft}} = 328 \text{ plf}$$

The shear at the first floor, including the roof and second-floor shear combined, is

$$v = \frac{V}{b} = \frac{(7874 \text{ lb} + 4248 \text{ lb})/2}{12 \text{ ft}} = 505 \text{ plf}$$

According to Table G.7 we must specify $\frac{15}{32}$-in.-thick plywood nominally called $\frac{1}{2}$ inch. If we use Structural II, 10d nails at 3 in. o.c. are needed to develop 600 plf shear. The footnote attached to the 10d nails requires that the framing behind adjoining panel edges must be a nominal 3 in. thick. We could use this panel and require double 2 × 4 studs every 4 ft. The studs would be nailed to each other with 16d nails at 6 in. on center.

Alternatively, we could use Structural I plywood with 10d nails at 4 in. on center. This requires no additional framing and is preferred. Usually, Structural I plywood is as available as Structural II. Consider now only the interior portion of the wall, the 14 ft that is covered with gypsum drywall on each side. Consider using the gypsum wallboard as a shear panel. Table G.10 shows the allowable shear for a variety of materials other than plywood. Normal construction will call for $\frac{1}{2}$-in. gypsum wallboard (not sheathing) without blocking. This specification can carry 100 plf for gypsum board on one side. When board is used on both sides, the value can be increased, provided that we understand the controlling condition. Table G.10 shows that $\frac{1}{2}$-in. gypsum nailed at 7 in. o.c. carries 100 plf and when nailed at 4 in. o.c.. it carries 125 plf. Although the nail spacing is almost doubled, the assembly strength increases only 25%. We can surmise from this that the critical element is not the nailing or the framing, but the gypsum board itself. Testing and common experience indicate that the most common form of failure occurs when the heads of the nails pull through the gypsum. Consequently, the effect on the framing of increased nail spacing, with nails coming from both sides, does not seem to be an issue. We assume that the framing can carry the highest value on the table (250 plf), so the doubling is permissible. The wall with $\frac{1}{2}$-in. gypsum unblocked wallboard can carry 200 plf; the allowable force on the full length of wall is 200 plf multiplied by 14 ft, or 2800 lb.

Unfortunately, this is far less than the 6061 lb that is needed. The assembly will not meet the required load. Furthermore, note 1 requires that in seismic zones 3 and 4, the allowable value must be reduced by 50% because the brittle nature of gypsum will not withstand the dynamic load of an earthquake. This factor will cancel the increase due to using two sides. Plywood must be used.

Shearwalls with Dissimilar Materials

The plan of the house shown in Figure 7.25 shows that the wall between the garage and the stair continues from the outside of the house to the inside. Along the west face of the wall the outer portion will most likely be sheathed with plywood, while the inner portion will be covered with gypsum board. The east face, which faces the garage, will need to be covered with $\frac{1}{2}$-in. type-X gypsum wallboard to protect the structure from fire. In calculating the shear capacity of this wall, the strengths of dissimilar materials may not be added. The allowable shear strength can be taken as the plywood strength only. Since the plywood and the gypsum board have greatly different stiffnesses, they will deflect much differently under the same load. The actual distribution of forces will be indeterminate; the stiffer element will carry the greater load. Analysis of this condition shows that most of the load is carried by the plywood. If it was assumed that the gypsum carried a substantial amount of the load, say half, it would fail and transfer all the load to the plywood, which would be designed to carry the other half of the load rather than the full load. The plywood wall would fail as well, leaving no shearwall in this vital location.

EXAMPLE 7.10—COLLECTOR STRUT FOR WALL SW7

Since the size of the diaphragms on each side of wall SW7 is different, the wall is not equally loaded from both sides. The west diaphragm applies loads along 28 ft of the collector–shearwall assembly while the east diaphragm, over the garage, applies its load over the full 40-ft length. To account for these differences the total seismic load at the roof is proportioned to the areas of each diaphragm. Calculations for this are given in Figure 7.32.

The results of the calculations shows that the shear flow from the west diaphragm (D1) is 77.4 plf. The shear flow

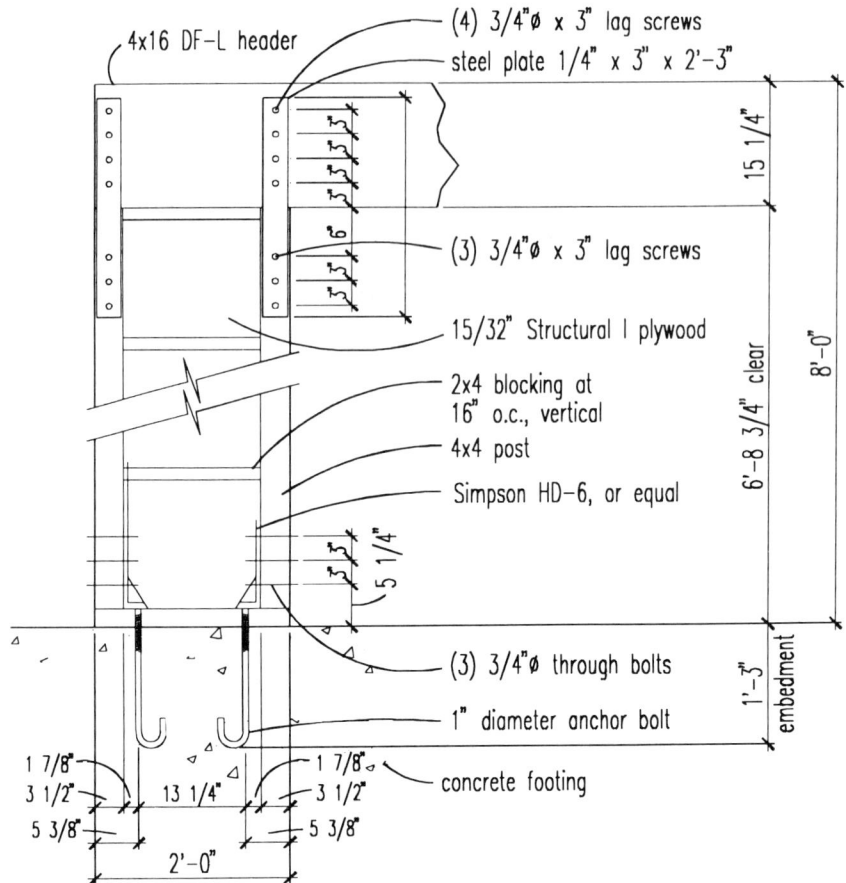

FIGURE 7.33 Special wood moment—resisting frame.

from the east diaphragm is 44.2 plf. Along the same lines as the derivation in Example 7.9, the shear carried by the 12-ft length of shearwall is 3937 lb divided by 12 ft or 328 plf. Figure 7.32b shows the forces that are applied to the wall and collector at the roof, and Figure 7.32c shows the accumulated tension or compression in the collector at the roof plate level. At the intersection of the wall and the collector the maximum compression or tension is 1702 lb. The force continues to increase for another 14 ft along the wall to the maximum value of 3405 lb. Should the collector be designed for 1702 lb or 3405 lb? Only the final 12 ft of wall, the plywood-covered exterior wall, functions as shearwall. The collector must be able to carry the shear over the middle 12 ft of drywall; from the point of view of the shearwall this portion of the wall contributes no strength and must actually be protected. Design the collector for 3405 lb.

The analysis and design for the collector at the second-floor level is identical. Rather than recalculating it, multiply all of the values above by the ratio of floor forces over roof forces, 4248 divided by 7874, or 54%. The total shear on the lower wall will be half of 4248 plus half of 7874 or 6061 lbs, but the collector will only carry the second floor load.

If the wall is well nailed to the roof truss at this location, the second floor wall will act as a shearwall and receive half of the roof shear, 3937 lb. The typical detail at this location, calling for two 16d toenails at 16 in. o.c. between truss and wall plate, will be able to carry about 200 plf. This load is carried over the full length of the wall and collector, namely 40 ft. The total force that can be transferred through that connection is 8000 lb. This example indicates that for normal residential construction, the recommended nailing schedule is adequate.

Wood Moment Frames

In cases where extremely short shearwalls cannot be avoided, the height-to-length ratios will be exceeded. The solution to this is to design a moment frame that works as a unit to resist overturning. It has become common practice in southern California to use steel moment frames around large openings, even in one- or two-story residences. Figure 7.35 shows the details of a typical steel frame used in residential buildings.

An alternative procedure is to use a wooden frame that is able to carry the additional moment due to the lateral load by increasing the size of the header beam and providing additional connections between the wall and the beam. Figure 7.33 shows the overall configuration for a frame

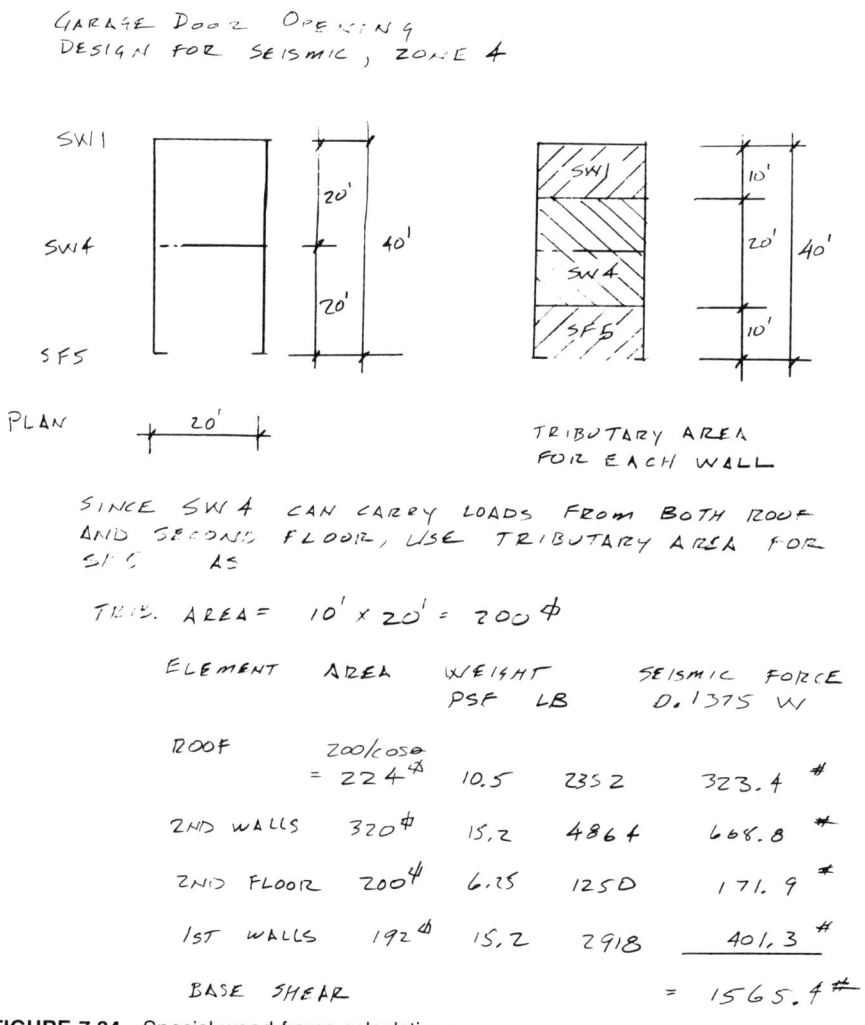

FIGURE 7.34 Special wood frame calculations.

296 PLYWOOD

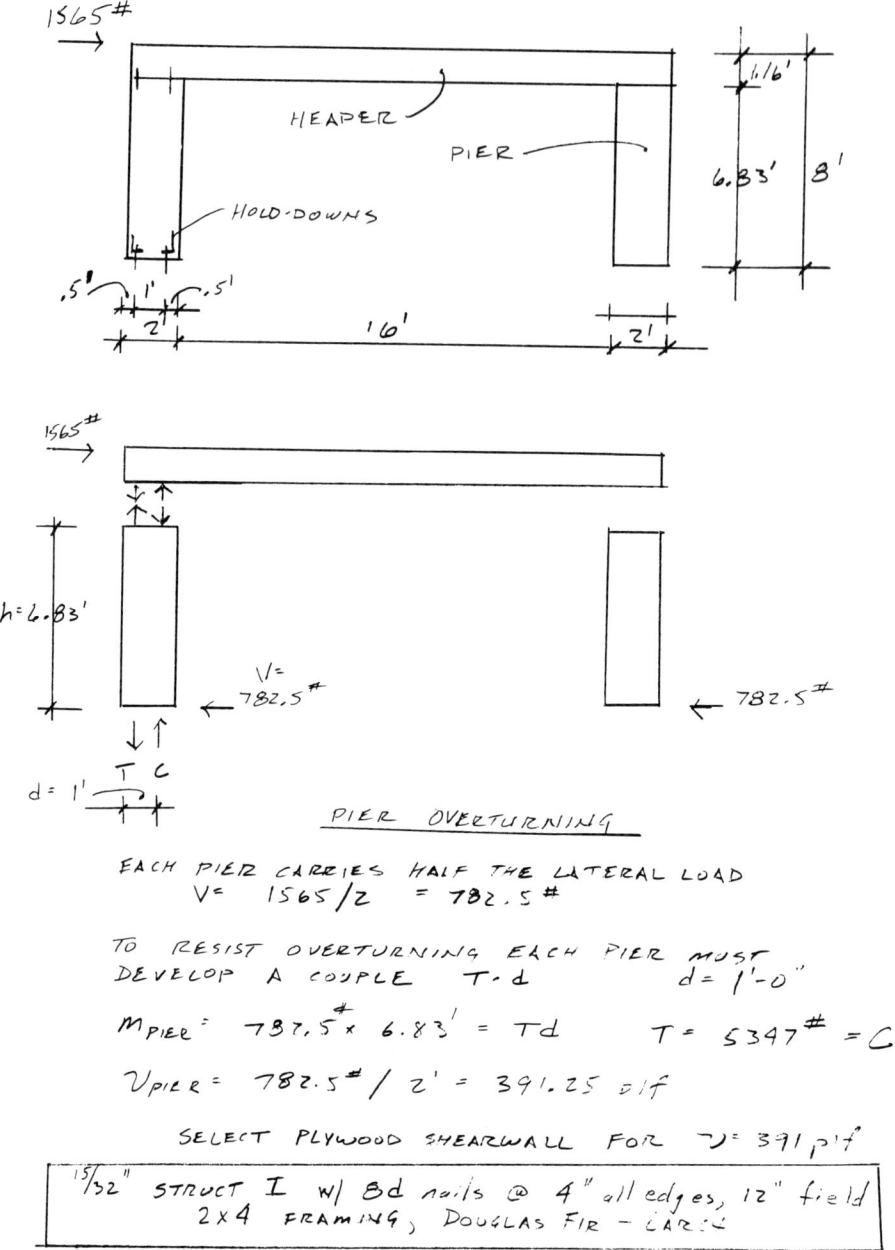

PIER OVERTURNING

EACH PIER CARRIES HALF THE LATERAL LOAD
$V = 1565/2 = 782.5^{\#}$

TO RESIST OVERTURNING EACH PIER MUST
DEVELOP A COUPLE $T \cdot d$ $d = 1'-0"$

$M_{PIER} = 782.5^{\#} \times 6.83' = Td$ $T = 5347^{\#} = C$

$\nu_{PIER} = 782.5^{\#} / 2' = 391.25 \, plf$

SELECT PLYWOOD SHEARWALL FOR $\nu = 391 \, plf$

$15/32"$ STRUCT I w/ 8d nails @ 4" all edges, 12" field
2×4 FRAMING, DOUGLAS FIR - LARCH

FIGURE 7.34 (*Continued*)

FRAME OVERTURNING

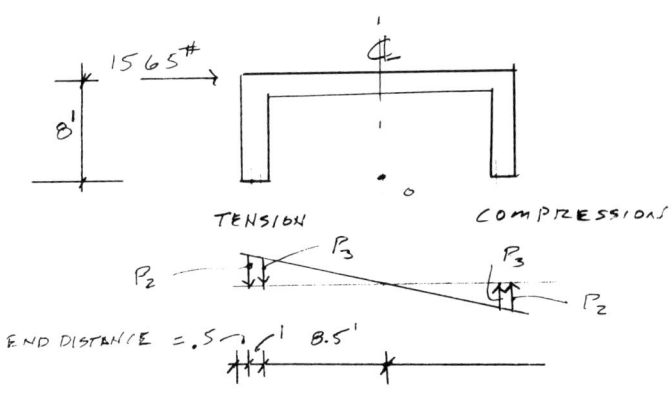

FOR FRAME TO ACT AS A UNIT IT MUST DEVELOP FORCES P_2 & P_3 AT EACH PIER. ASSUME COMPATIBLE DEFORMATIONS, THEN P_2 AND P_3 ARE LINEARLY RELATED

$$\frac{P_2}{9.5} = \frac{P_3}{8.5} \qquad P_3 = \frac{8.5}{9.5} P_2 = 0.8947 P_2$$

$\Sigma M_O = 0$
OTM $= 1565^\# \times 8' = 12520^{\#-1}$
RM $= 2 \cdot (P_2 \times 9.5' + P_3 \times 8.5') =$ OTM $= 12520^{\#-1}$

SO $2(P_2 \times 9.5 + P_2 \times 8.5 \times .8947) = 12520^{\#-1}$

$P_2 = 366^\#$ $\qquad P_3 = 327.5^\#$

ACTUAL FORCES ON PIER ARE SUM OF PIER EFFECT AND FRAME EFFECT

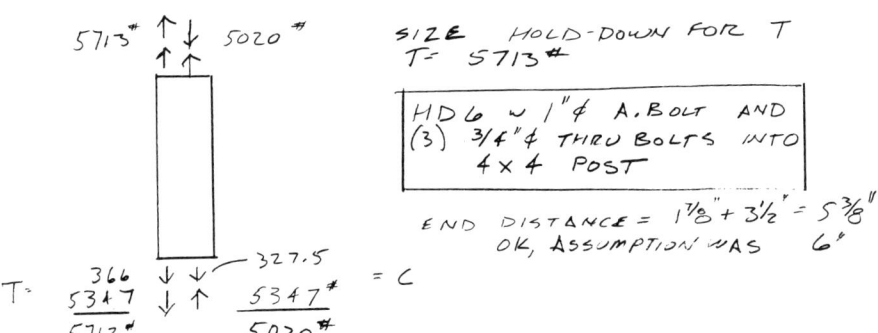

SIZE HOLD-DOWN FOR T
T = $5713^\#$

HD6 w/ 1"ɸ A.BOLT AND
(3) 3/4"ɸ THRU BOLTS INTO
4 × 4 POST

END DISTANCE = $1\frac{7}{8}" + 3\frac{1}{2}" = 5\frac{3}{8}"$
OK, ASSUMPTION WAS 6"

FIGURE 7.34 (*Continued*)

HEADER DESIGN

HEADER WILL HAVE MOMENT FROM FRAME RACKING
AND FROM GRAVITY LOADS

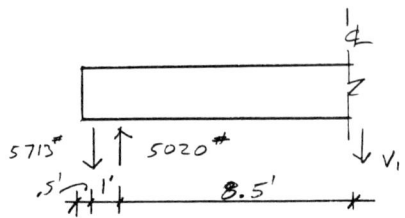

$V_1 = 5713 - 5020 = 693^\#$

NEGLECT V OVER SUPPORT

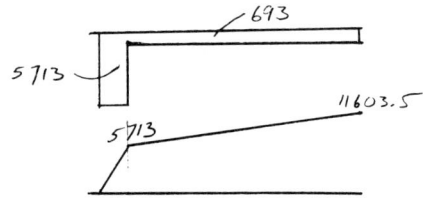

M_{max}, RACKING @ ℄
$M = 5713 \times 1' + 693^\# \times 8.5'$
$= 11603.5^{\#-'}$
$= 139242^{\#-''}$

FOR DEAD LOADS, ASSUME TRIB. WIDTH IS $12'/2 = 6'$
$w = (40\,psf\,live + 10\,psf\,dead) \times 6' = 300\,plf$

$M_{max} @ \, ℄ = \frac{wL^2}{8} = \frac{300 \times 17^2}{8} = 10837.5^{\#-'} = 130050^{\#-''}$

$M_{TOT} = 139242^{\#-''} + 130050^{\#-''} = 269,292\,lb\text{-}in$

FOR DFL Select Structural, assume 4×14
$F_b = 1450\,psi \times 1.0 \times 1.6\,(DURATION) = 2320\,psi$

$S_{reqd} = \frac{M}{F_b} = \frac{269292}{2320} = 116\,in^3$

USE 4×16 $S = 135.7\,in^3$ 85% stress OK

DESIGN STRAPS TO CONNECT HEADER TO PIER
$T = 5713^\#$
try 1/4" plate w/ 3/4"⌀ lag screw

$Z_\perp = 1.6 \times 1060 = 1696$

$N = 5713/1696 = 3.36$ use (4) 3/4" lag screws
end spacing $= 4d = 3"$ use 3"
spacing $= 2.5d = 1.875$ use 2" $3 + 3 \times 2 = 9"$ OK

FIGURE 7.34 (*Continued*)

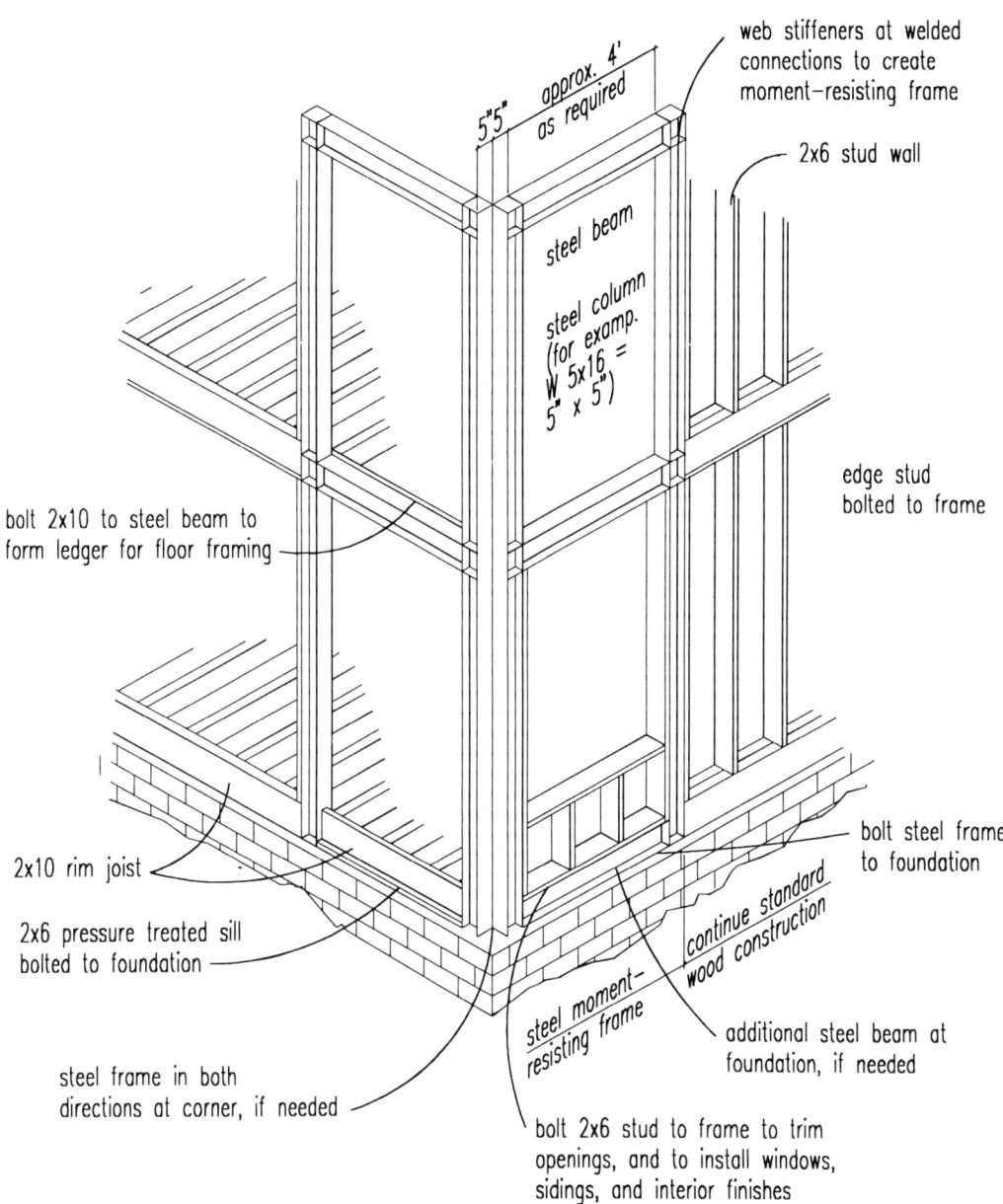

FIGURE 7.35 Steel moment—resisting frame.

of this nature and the details at the base and at the beam.

EXAMPLE 7.11—SHEARWALL DESIGN WITH GARAGE DOOR OPENING

In our preliminary analysis we identified the wall with the garage door, SF5, as a potential problem since the short sections of wall flanking the garage door fail to meet the required shearwall ratio. In Figures 7.25 and 7.27 the plan and elevation of the garage and its dimensions are given. The calculations are shown in Figure 7.34, and the final design in FIgure 7.33.

This procedure was developed by Ralph W. Goers & Associates and is described in a pamphlet published for the Applied Technology Council by the U.S. Department of Housing and Urban Development.[3] In this report the author states: "The design of rigid frames in wood is at best questionable.... The essential function of [this] detail is to inhibit excessive rotation of the second-floor diaphragm.... In order not to place too much reliance on the frame, [we recommend that] twice the seismic load be taken by the rear wall."

An alternative procedure is to apply plywood sheathing to both sides of the wall. To do this, the capacity can be determined using the guidelines given previously. The deflection of the panel is controlled mostly by plywood stiffness and nail slip. Doubling the plywood will reduce the deflection to about half. The height-to-width ratios are required, in part, to control deflections. The extension of this logic, then, is to allow less restrictive ratios for plywood sheathing applied to both sides of the wall. In our practice we have found that plywood sheathing on both sides is satisfactory for shearwalls with a height-to-width ratio of 4 to 1.

[3] Ralph W. Goers & Associates, *A Methodology for Seismic Design and Construction of Single-Family Dwellings*, U.S. Department of Housing and Urban Development, Washington, DC, 1976.

Torsion of Flexible Diaphragms

Figure 7.36 shows a situation of a diaphragm with unbalanced shearwalls. On one side there is a large opening in the shearwall that is not balanced by a similar opening on the other side. The solid wall

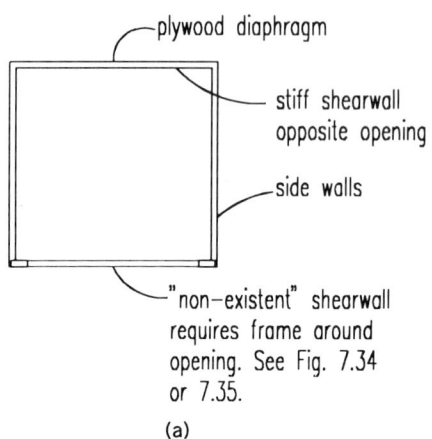

(a)

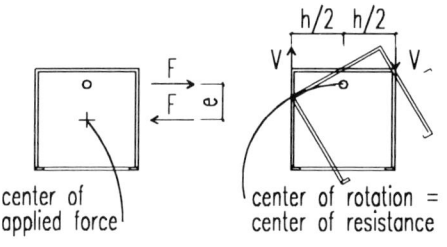

e = torsional eccentricity

$F_e = V \times h$

$V = F \times e/h$

if opening is not reinforced with a moment-resisting frame, sidewalls must develop the shear V to resist rotation

large deformations will occur at the edges of the opening, possibly causing glass, drywall or siding to break

(b)

FIGURE 7.36 Torsion of diaphragm: (a) plan of a diaphragm and (b) torsion without frame at opening.

will have considerably more stiffness than the wall with the opening. Since the resistance is not symmetrical, there is a point nearer the stiff wall that will be the center of rotation. The load can be assumed to act as if it were applied at the geometric center of the diaphragm. The distance between the center of the load application and the center of resistance is the torsional eccentricity. The magnitude of the torsion is found as the product of the applied force times the torsional eccentricity. The larger the eccentricity, the greater the torsional effect.

In systems with rigid diaphragms (such as concrete slabs) a great deal of the torsion can be transferred into the walls perpendicular to the load. These walls will develop equal and opposite shear forces to resist the moment induced by the torsion. The designer could calculate these forces simply by equating their moment resistance with the applied torsion.

In wood construction this approach is not a very accurate portrayal of actual conditions. The diaphragm is flexible and does not transfer forces in the same manner; and in the analysis of many houses that collapsed during the San Fernando earthquake of 1971, it was found that the walls perpendicular to the load twisted out of plane and were unable to carry the additional loads.

The recommendation for this condition is to assume that a flexible diaphragm will transfer half of its load to each shearwall. In the case of wood shearwalls, the smaller shearwall will have to be designed and

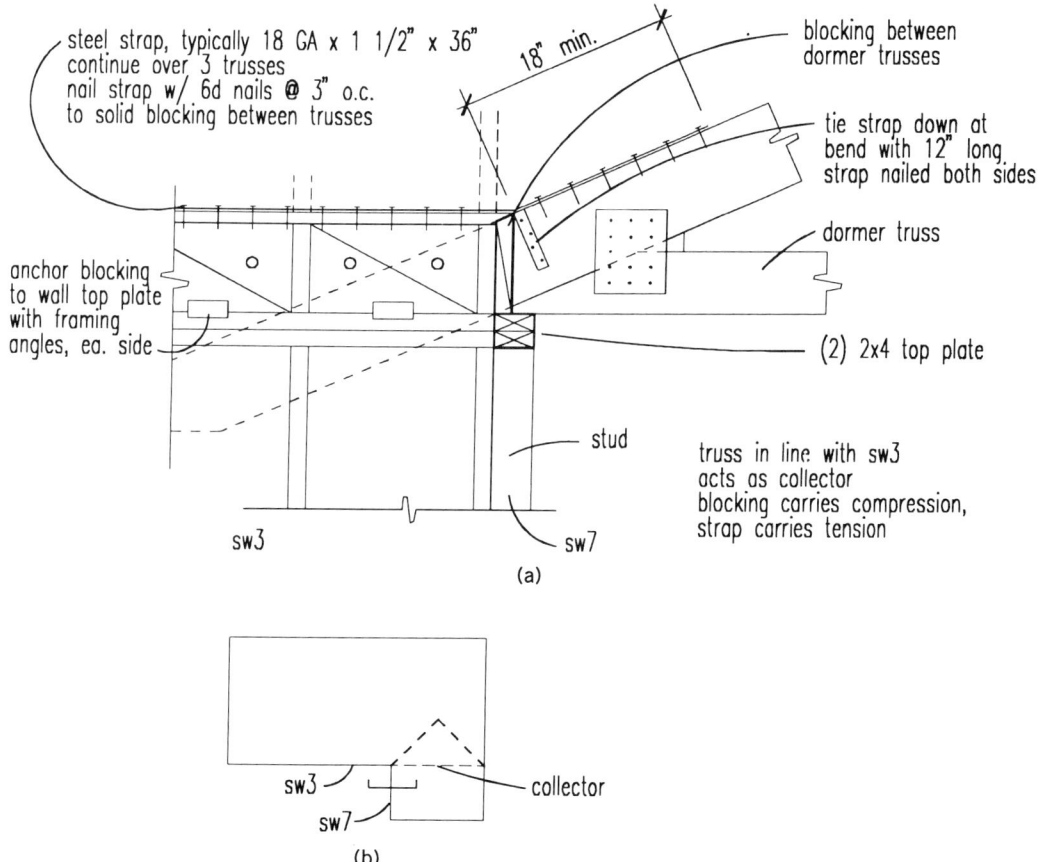

FIGURE 7.37 Collector strap for roofs at dormer levels: (a) collector strap detail and (b) roof plan.

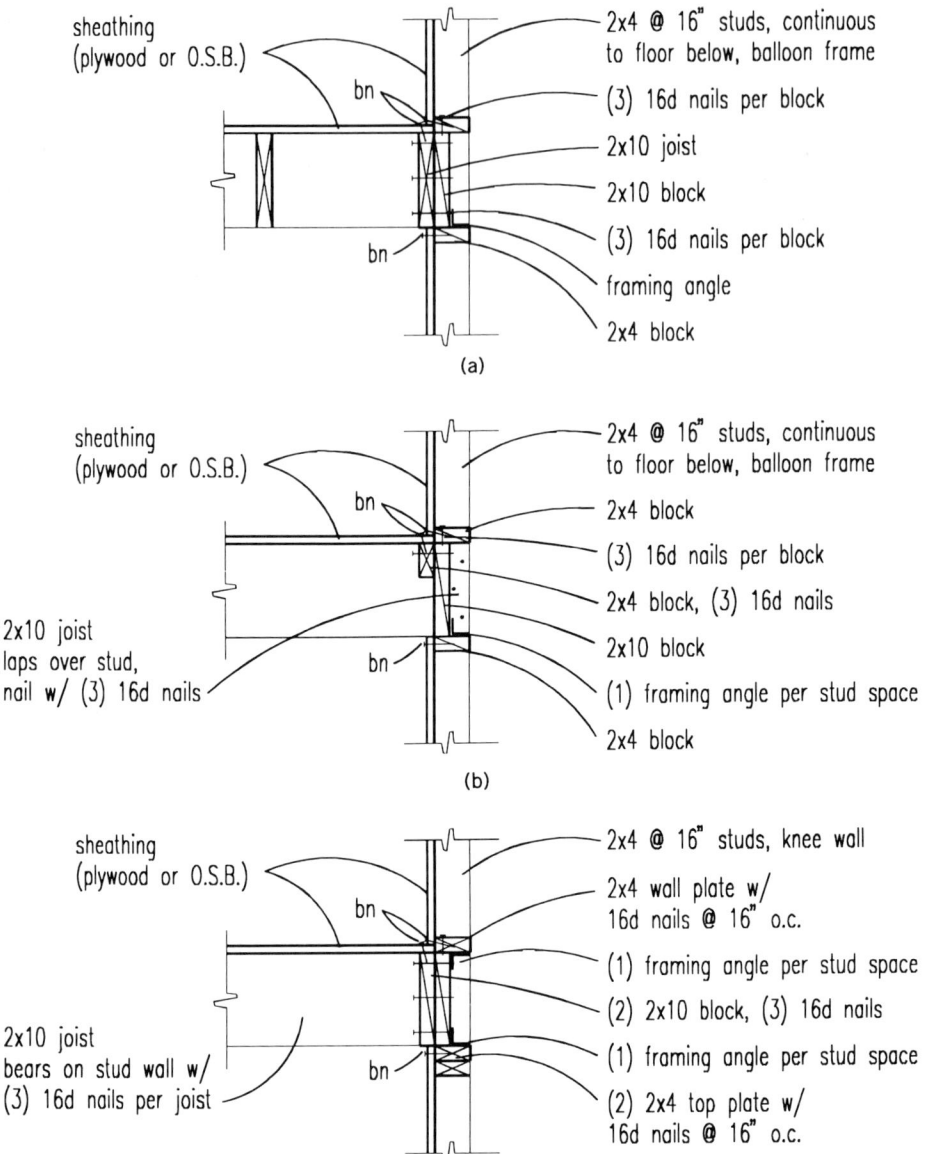

FIGURE 7.38 Connection at split level floor or roof: (a) joists parallel to wall, (b) joist perpendicular to wall, and (c) knee wall.

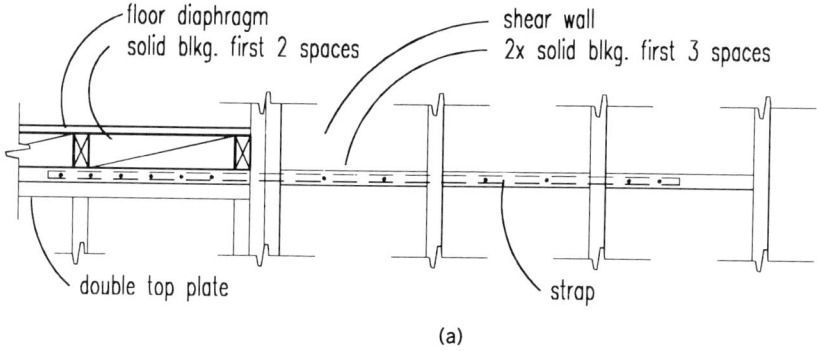

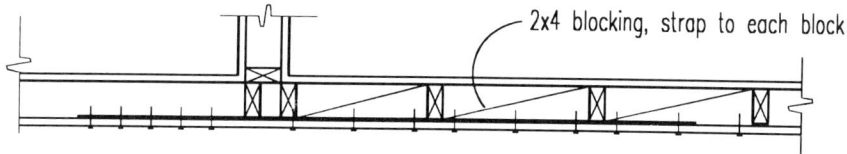

fasten finish material to blocking with same nailing as required at edges of sheetsh

(b)

FIGURE 7.39 Connection for diaphragm at different level: (a) elevation—connection where diaphragms are not at the same level and (b) plan view.

detailed to carry the load assigned to it. In the case of shearwalls made of other materials, usually masonry or concrete, the analysis can proceed based on the assumption that torsion will operate and that perpendicular walls can carry the shear. The calculations for this are rather complex and must assure that the perpendicular walls will have sufficient stiffness to withstand twisting out of plane, that the diaphragm can withstand the compression forces due to torsion without buckling, and that the connections between diaphragm and all shearwalls will resist out-of-plane twisting and buckling of all elements.

Stressed-Skin Panels and Sandwich Panels

A recent innovation in the use of plywood has been the development of stressed-skin panel and sandwich panels. Although similar conceptually and in appearance, there are some notable differences between the two. *Stressed-skin panels* consist of panels with plywood (or particleboard) on one or both faces. The skins are supported by a series of lumber or plywood stringers. The plywood skins are stressed in compression, or in tension if a bottom skin is used. Because of the high stresses in the assembly, the skins must be glued to the stringers and the plywood skins must be connected with splice plates. The purpose of the splice plates is to maintain continuity of the compression or tension in the skins. Figure 7.41, taken from the APA PDS Supplement 3, shows examples of one- and two-sided, as well as T-flange stressed-skin panels. In general, stressed-skin panels are engineered for a particular application and are not marketed as an "off-the-shelf" product. The design of stressed-skin pan-

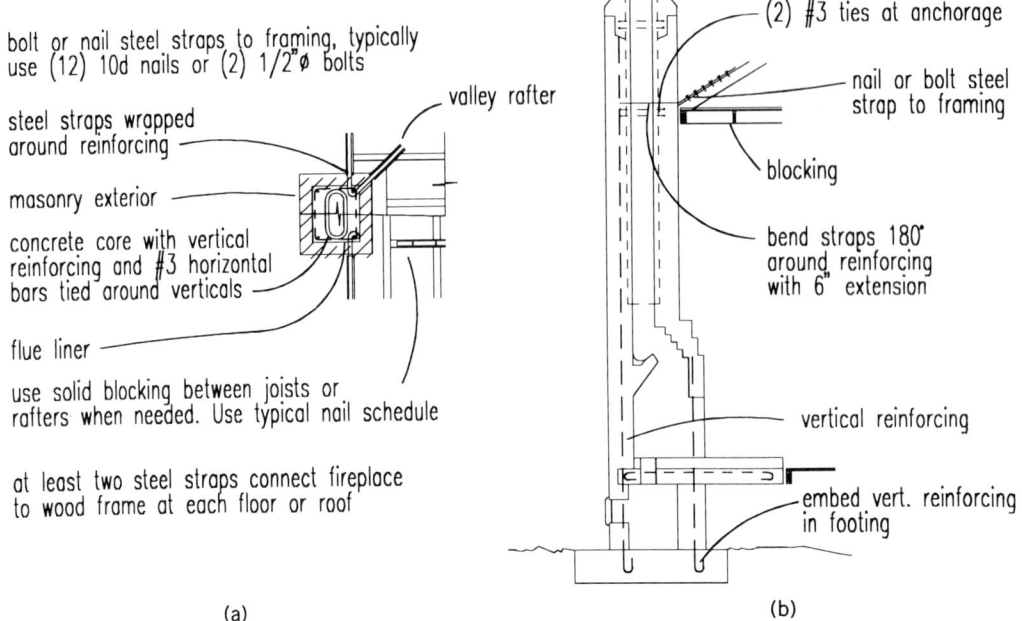

FIGURE 7.40 Fireplace details: (a) plan and (b) section.

els is outlined in the APA PDS, Supplement 3.[4]

Sandwich panels are structural assemblies in which a lightweight core is joined between two layers of plywood or particleboard. Figure 7.42 shows the construction of a typical sandwich panel. The lightweight core is usually an insulating foam. The energy benefits deriving from the foam explains one of the advantages and interests in this assembly. These walls or roofs can develop heat-resisting R-values as much as twice that of batt-filled stud construction. The skins in the sandwich panel are not as highly stressed as those of the stressed-skin panel. In addition, the deflection of the sandwich panel is dependent on the bulk modulus of the foam. In many cases the allowable loads of the sandwich panel are limited by deflection criteria.

Sandwich panels are manufactured by a number of producers. More information on the availability and performance of sandwich panels can be obtained from the Structural Insulated Panel Association (SIPA).[5] The design of sandwich panels is outlined in the APA PDS Supplement 4.[6]

Tables G.11 and G.12 give allowable transverse loads for four sizes of stressed-skin panels and four sizes of sandwich panels acting as roof panels. In both cases the panel is placed so that the long dimension of the particleboard aligns with the long dimension of the panel. This dimen-

---

[4] American Plywood Association, *Plywood Design Specification*, Supplement 3 "Design and Fabrication of Plywood Stressed-Skin Panels," August 1990, APA, P.O. Box 11700, Tacoma WA 98411-0700.

[5] Structural Insulated Panel Association (SIPA), 1511 K St., NW, Suite 600, Washington, DC 20005.

[6] American Plywood Association, *Plywood Design Specification*, Supplement 4, "Design and Fabrication of Plywood Sandwich Panels," March 1990, APA, P.O. Box 11700, Tacoma WA 98411-0700.

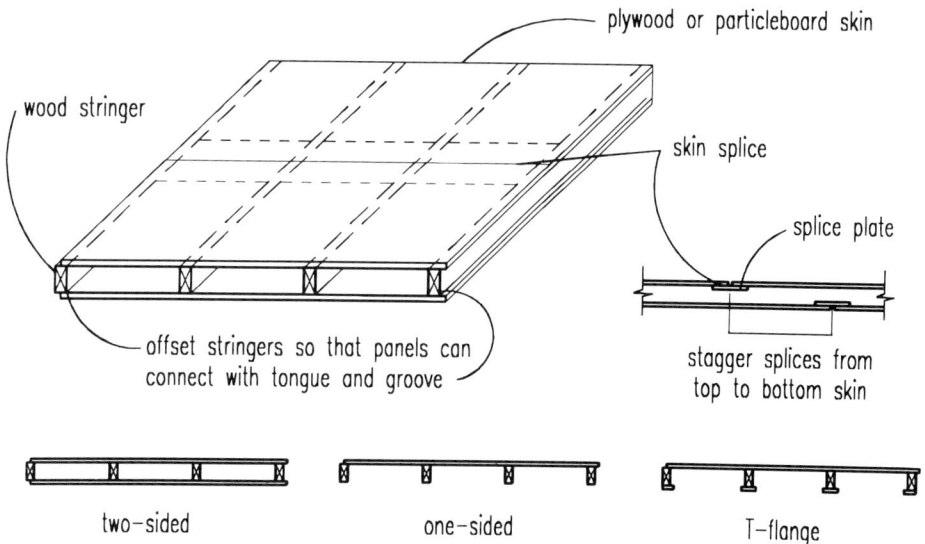

FIGURE 7.41 Stress-skin panel construction.

sion is indicated in the tables as panel length. The tables show the allowable imposed live load. In all cases the maximum load is determined by the deflection. The allowable deflection limit was taken as $L/180$, which is appropriate for most roof conditions. For situations in which the deflection limit is more restrictive, divide the tabulated values by 1.33 for $L/240$ or 2.0 for $L/360$. The dead load is assumed to be 10 psf. Where deflection controls, no duration factor adjustments can be taken, so the values in this table are appropriate for snow load, roof live loads, or wind loads. Other parameters, such as particleboard strengths, thicknesses, and stiffness, as well as foam core or framing properties, are given in the heading of the table. These tables apply to panels made with these properties only and cannot be extrapolated for other materials. Stressed-skin and sandwich panels may also be used as wall

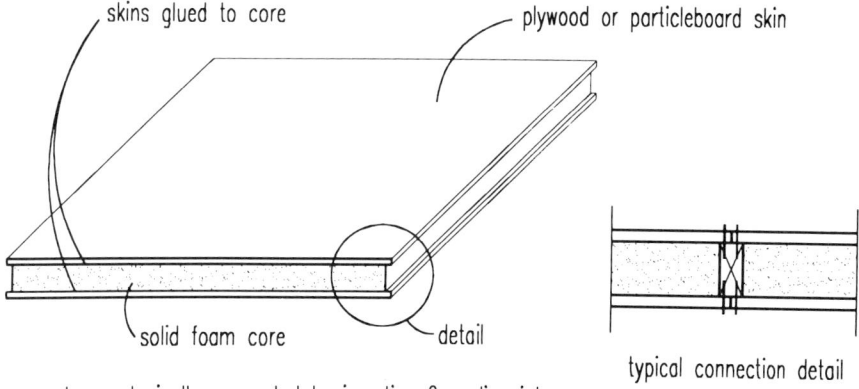

FIGURE 7.42 Sandwich panel construction.

306 PLYWOOD

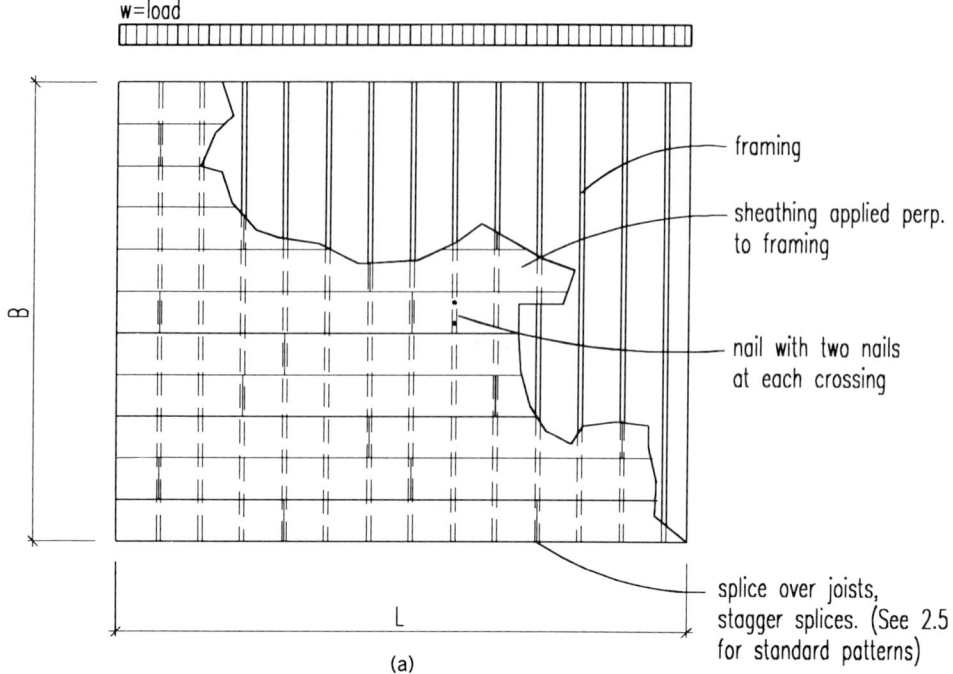

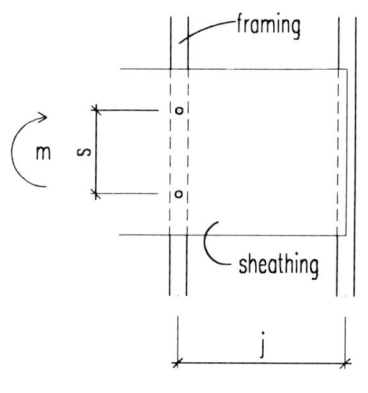

moment developed by the nail couple
$m = s \times P_{lateral\ of\ nail}$
m must equal applied moment

$$m = \frac{\upsilon\, j\, B}{NB}$$

$$\upsilon = \frac{wL}{2} \text{ so}$$

$$m = \frac{w\, L\, j\, B}{2NB}$$

where w = applied lateral load (plf)
       L = diaphragm length
       j = joist spacing
       B = diaphragm width
and   NB = number of sheathing
            boards along the width B

FIGURE 7.43 Transversely sheathed lumber diaphragm: (a) transversely sheathed diaphragm and (b) moment developed by the nail couple.

panels. Where such panels are load-bearing and receive axial loads, their capacity to carry transverse loads is greatly diminished. The values in the tables apply to wall panels carrying wind loads only where gravity loads are carried by an independent frame and the wall panels are not load bearing.

APPENDIX A

LUMBER PROPERTIES

TABLE A.1a PROPERTIES OF SECTIONS: SAWN LUMBER

Nominal Size (in.)	Surfaced Size for Design (in.)	Area, A (in.^{2})	Section Modulus, S (in.^{3})	Moment of Inertia, I (in.^{4})	Board Feet per Lineal Foot of Piece
Joists and Beams					
2 × 2	1.5 × 1.5	2.25	0.56	0.42	0.33
2 × 3	1.5 × 2.5	3.75	1.56	1.95	0.50
2 × 4	1.5 × 3.5	5.25	3.06	5.36	0.67
2 × 6	1.5 × 5.5	8.25	7.56	20.8	1.00
2 × 8	1.5 × 7.25	10.9	13.1	47.6	1.33
2 × 10	1.5 × 9.25	13.9	21.4	98.9	1.67
2 × 12	1.5 × 11.25	16.9	31.6	178.0	2.00
2 × 14	1.5 × 13.25	19.9	43.9	290.8	2.33
3 × 3	2.5 × 2.5	6.25	2.60	3.26	0.75
3 × 4	2.5 × 3.5	8.75	5.10	8.93	1.00
3 × 6	2.5 × 5.5	13.8	12.6	34.7	1.50
3 × 8	2.5 × 7.25	18.1	21.9	79.4	2.00
3 × 10	2.5 × 9.25	23.1	35.7	164.9	2.50
3 × 12	2.5 × 11.25	28.1	52.7	296.6	3.00
3 × 14	2.5 × 13.25	33.1	73.2	484.6	3.50
3 × 16	2.5 × 15.25	38.1	96.9	738.9	4.00
4 × 4	3.5 × 3.5	12.3	7.15	12.5	1.33
4 × 6	3.5 × 5.5	19.3	17.7	48.5	2.00
4 × 8	3.5 × 7.25	25.4	30.7	111.2	2.67
4 × 10	3.5 × 9.25	32.4	49.9	230.8	3.33
4 × 12	3.5 × 11.25	39.4	73.8	415	4.00
4 × 14	3.5 × 13.25	46.4	102	678	4.67
4 × 16	3.5 × 15.25	53.4	136	1034	5.33
6 × 6	5.5 × 5.5	30.3	27.7	76.3	3.00
6 × 8	5.5 × 7.5	41.3	51.6	193	4.00
6 × 10	5.5 × 9.5	52.3	82.7	393	5.00

TABLE A.1a *Continued*

Nominal Size (in.)	Surfaced Size for Design (in.)	Area, A (in.^{2})	Section Modulus, S (in.^{3})	Moment of Inertia, I (in.^{4})	Board Feet per Lineal Foot of Piece
6 × 12	5.5 × 11.5	63.3	121	697	6.00
6 × 14	5.5 × 13.5	74.3	167	1,128	7.00
6 × 16	5.5 × 15.5	85.3	220	1,707	8.00
6 × 18	5.5 × 17.5	96.3	281	2,456	9.00
6 × 20	5.5 × 19.5	107	349	3,398	10.00
8 × 8	7.5 × 7.5	56.3	70.3	264	5.33
8 × 10	7.5 × 9.5	71.3	113	536	6.67
8 × 12	7.5 × 11.5	86.3	165	951	8.00
8 × 14	7.5 × 13.5	101	228	1,538	9.33
8 × 16	7.5 × 15.5	116	300	2,327	10.67
8 × 18	7.5 × 17.5	131	383	3,350	12.00
8 × 20	7.5 × 19.5	146	475	4,634	13.33
8 × 22	7.5 × 21.5	161	578	6,211	14.67
10 × 10	9.5 × 9.5	90.3	143	679	8.33
10 × 12	9.5 × 11.5	109	209	1,204	10.00
10 × 14	9.5 × 13.5	128	289	1,948	11.67
10 × 16	9.5 × 15.5	147	380	2,948	13.33
10 × 18	9.5 × 17.5	166	485	4,243	15.00
10 × 20	9.5 × 19.5	185	602	5,870	16.67
10 × 22	9.5 × 21.5	204	732	7,868	18.33
12 × 12	11.5 × 11.5	132	253	1,458	12.00
12 × 14	11.5 × 13.5	155	349	2,358	14.00
12 × 16	11.5 × 15.5	178	460	3,569	16.00
12 × 18	11.5 × 17.5	201	587	5,136	18.00
12 × 20	11.5 × 19.5	224	729	7,106	20.00
12 × 22	11.5 × 21.5	247	886	9,524	22.00
12 × 24	11.5 × 23.5	270	1058	12,437	24.00
Planks					
3 × 2	2.5 × 1.5	3.75	0.94	0.70	0.50
4 × 2	3.5 × 1.5	5.25	1.31	0.98	0.67
6 × 2	5.5 × 1.5	8.25	2.06	1.55	1.00
8 × 2	7.25 × 1.5	10.9	2.72	2.04	1.33
10 × 2	9.25 × 1.5	13.9	3.47	2.60	1.67
12 × 2	11.25 × 1.5	16.9	4.22	3.16	2.00
4 × 3	3.5 × 2.5	8.75	3.65	4.56	1.00
6 × 3	5.5 × 2.5	13.8	5.73	7.16	1.50
8 × 3	7.25 × 2.5	18.1	7.55	9.44	2.00
10 × 3	9.25 × 2.5	23.1	9.64	12.0	2.50
12 × 3	11.25 × 2.5	28.1	11.7	14.6	3.00
14 × 3	13.25 × 2.5	33.1	13.8	17.3	3.50
16 × 3	15.25 × 2.5	38.1	15.9	19.9	4.00
6 × 4	5.5 × 3.5	19.3	11.2	19.7	2.00
8 × 4	7.25 × 3.5	25.4	14.8	25.9	2.67
10 × 4	9.25 × 3.5	32.4	18.9	33.0	3.33
12 × 4	11.25 × 3.5	39.4	23.0	40.2	4.00
14 × 4	13.25 × 3.5	46.4	27.1	47.3	4.67
16 × 4	15.25 × 3.5	53.4	31.1	54.5	5.33
Decking					
2	12 × 1.5	18.0	4.50	3.38	2.00
3	12 × 2.5	30.0	12.5	15.6	3.00
4	12 × 3.5	40.0	24.5	42.9	4.00

TABLE A.1b SECTION PROPERTIES: GLUED-LAMINATED TIMBER[a]

Number of Laminations	Depth (in.)	Form Factor, C_F	Area, A (in.^{2})	Section Modulus, S	Moment of Inertia, I (in.^{4})
$3\frac{1}{8}$-in. Width					
2	3.00	1.00	9.4	4.7	7.0
3	4.50	1.00	14.1	10.5	23.7
4	6.00	1.00	18.8	18.8	56.3
5	7.50	1.00	23.4	29.3	110
6	9.00	1.00	28.1	42.2	190
7	10.50	1.00	32.8	57.4	302
8	12.00	1.00	37.5	75.0	450
9	13.50	0.99	42.2	94.9	641
10	15.00	0.98	46.9	117	879
11	16.50	0.97	51.6	142	1,170
12	18.00	0.96	56.3	169	1,519
13	19.50	0.95	60.9	198	1,931
14	21.00	0.94	65.6	230	2,412
15	22.50	0.93	70.3	264	2,966
16	24.00	0.93	75.0	300	3,600
$5\frac{1}{8}$-in. Width					
3	4.50	1.00	23.1	17.3	38.9
4	6.00	1.00	30.8	30.8	92.3
5	7.50	1.00	38.4	48.0	180
6	9.00	1.00	46.1	69.2	311
7	10.50	1.00	53.8	94.2	494
8	12.00	1.00	61.5	123	738
9	13.50	0.99	69.2	156	1,051
10	15.00	0.98	76.9	192	1,441
11	16.50	0.97	84.6	233	1,919
12	18.00	0.96	92.3	277	2,491
13	19.50	0.95	99.9	325	3,167
14	21.00	0.94	108	377	3,955
15	22.50	0.93	115	432	4,865
16	24.00	0.93	123	492	5,904
17	25.50	0.92	131	555	7,082
18	27.00	0.91	138	623	8,406
19	28.50	0.91	146	694	9,887
20	30.00	0.90	154	769	11,530
21	31.50	0.90	161	848	13,350
22	33.00	0.89	169	930	15,350
23	34.50	0.89	177	1,017	17,540
24	36.00	0.88	185	1,107	19,930
$6\frac{3}{4}$-in. Width					
4	6.00	1.00	40.5	40.5	122
5	7.50	1.00	50.6	63.3	237
6	9.00	1.00	60.8	91.1	410
7	10.50	1.00	70.9	124	651
8	12.00	1.00	81.0	162	972
9	13.50	0.99	91.1	205	1,384
10	15.00	0.98	101	253	1,898

TABLE A.1b *Continued*

Number of Laminations	Depth (in.)	Form Factor, C_F	Area, A (in.^{2})	Section Modulus, S	Moment of Inertia, I (in.^{4})
$6\frac{3}{4}$-in. Width					
11	16.50	0.97	111	306	2,527
12	18.00	0.96	122	365	3,281
13	19.50	0.95	132	428	4,171
14	21.00	0.94	142	496	5,209
15	22.50	0.93	152	570	6,407
16	24.00	0.93	162	648	7,776
17	25.50	0.92	172	732	9,327
18	27.00	0.91	182	820	11,070
19	28.50	0.91	192	914	13,020
20	30.00	0.90	203	1,013	15,190
21	31.50	0.90	213	1,116	17,580
22	33.00	0.89	223	1,225	20,210
23	34.50	0.89	233	1,339	23,100
24	36.00	0.88	243	1,458	26,240
25	37.50	0.88	253	1,582	29,660
26	39.00	0.88	263	1,711	33,370
27	40.50	0.87	273	1,845	37,370
28	42.00	0.87	284	1,985	41,670
29	43.50	0.87	294	2,129	46,300
30	45.00	0.86	304	2,278	51,260
31	46.50	0.86	314	2,433	56,560
32	48.00	0.86	324	2,592	62,210

[a] Dimensions shown for typical sizes available using Western Wood species. Southern Pine species typically use $1\frac{3}{8}''$ laminations and widths of $3''$, $5''$, and $6\frac{3}{4}''$ widths. For properties using Southern Pine species, see NDS Supplement, Table 1C.

Source: Courtesy of American Forest and Paper Association, Washington, D.C.

TABLE A.2 BASE DESIGN VALUES FOR VISUALLY GRADED DIMENSION LUMBER: DOUGLAS FIR-LARCH[a,b]

Species and Commercial Grade	Size Classification	Design Values in Pounds per Square Inch (psi)						
		Bending F_b	Tension Parallel to Grain F_t	Shear Parallel to Grain F_v	Compression Perpendicular to Grain $F_{c\perp}$	Compression Parallel to Grain F_c	Modulus of Elasticity E	Grading Rules Agency
Douglas Fir-Larch								
Select Structural	2-4 in. thick	1,450	1,000	95	625	1,700	1,900,000	WCLIB WWPA
No. 1 and Better	2 in. and wider	1,150	775	95	625	1,500	1,800,000	
No. 1		1,000	675	95	625	1,450	1,700,000	
No. 2		875	575	95	625	1,300	1,600,000	
No. 3		500	325	95	625	750	1,400,000	
Stud		675	450	95	625	825	1,400,000	
Construction	2-4 in. thick	1,000	650	95	625	1,600	1,500,000	
Standard		550	375	95	625	1,350	1,400,000	
Utility	2-4 in. wide	275	175	95	625	875	1,300,000	

[a]Lumber dimensions: Tabulated design values are applicable to lumber that will be used under dry conditions, such as in most covered structures. For 2–4 in. thick lumber, the *dry* dressed sizes shall be used (see Table A.1a) regardless of the moisture content at the time of manufacture or use. In calculating design values, the natural gain in strength and stiffness that occurs as lumber dries has been taken into consideration, as well as the reduction in size that occurs when unseasoned lumber shrinks. The gain in load-carrying capacity due to increased strength and stiffness resulting from drying more than offsets the design effect of size reduction due to shrinkage.

[b]Stress-rated boards: Stress-rated boards of nominal 1 in., $1\frac{1}{4}$ in. and $1\frac{1}{2}$ in. thickness, 2 in. and wider, of most species, are permitted the design values shown for Select Structural, No. 1 and Better, No. 1, No. 2, No. 3, Stud, Construction, Standard, Utility, Clear Heart Structural, and Clear Structural grades, as shown in the 2–4 in. thick categories herein, when graded in accordance with the stress-rated board provisions in the applicable grading rules. Information on additional values may be available from the respective grading agency.

Source: NDS Supplement, Table 4A. Courtesy of American Forest and Paper Association, Washington, D.C.

TABLE A.3   SIZE FACTORS[a], C_F

		Size Factors, C_F			
		F_b			
		Thickness			
Grades	Width	2″ & 3″	4″	F_t	F_c
	2, 3, and 4 in.	1.5	1.5	1.5	1.15
Select	5 in.	1.4	1.4	1.4	1.1
Structural,	6 in.	1.3	1.3	1.3	1.1
No. 1 and Better	8 in.	1.2	1.3	1.2	1.05
No. 1,	10 in.	1.1	1.2	1.1	1.0
No. 2,	12 in.	1.0	1.1	1.0	1.0
No. 3	14 in. and wider	0.9	1.0	0.9	0.9
Stud	2, 3, and 4 in.	1.1	1.1	1.1	1.05
	5 and 6 in.	1.0	1.0	1.0	1.0
Construction and Standard	2, 3, and 4 in.	1.0	1.0	1.0	1.0
Utility	4 in.	1.0	1.0	1.0	1.0
	2 and 3 in.	0.4	—	0.4	0.6

[a]Tabulated bending, tension, and compression parallel to grain design values for Douglas Fir-Larch dimension lumber 2–4 in. thick (Table A.2) shall be multiplied by the size factors shown above. The results for selected sizes and grades of Douglas Fir-Larch are given in Table A.8. For Southern Pine, appropriate size adjustment factors have already been incorporated in the tabulated design values shown in Table A.9. For sizes and grades other than those shown in Table A.2, refer to NDS Supplement, Tables 4A and 4B. Courtesy of American Forest and Paper Association, Washington, D.C.

TABLE A.4   REPETITIVE MEMBER FACTOR, C_r

Repetitive Member Factor, C_r: Bending design values, F_b, for dimension lumber 2–4 in. thick shall be multiplied by the repetitive member factor, $C_r = 1.15$, when such members are used as joists, truss chords, rafters, studs, planks, decking, or similar members that are in contact or spaced not more than 24 in. on centers, are not less than three in number, and are joined by floor, roof, or other load-distributing elements adequate to support the design load.

Source: NDS Supplement, Tables 4A and 4B. Courtesy of American Forest and Paper Association, Washington, D.C.

TABLE A.5   FLAT USE FACTOR[a], C_{fu}

Flat Use Factor, C_{fu}

	Thickness	
Width	2 and 3 in.	4 in.
2 and 3 in.	1.0	—
4 in.	1.1	1.0
5 in.	1.1	1.05
6 in.	1.15	1.05
8 in.	1.15	1.05
10 in. and wider	1.2	1.1

[a]Bending design values adjusted by size factors are based on edgewise use (load applied to narrow face). When dimension lumber is used flatwise (load applied to wide face), the bending design value, F_b, shall also be multiplied by the flat use factors shown above.

Source: NDS Supplement, Tables 4A and 4B. Courtesy of American Forest and Paper Association, Washington, D.C.

TABLE A.6 WET SERVICE FACTOR[a,b], C_M

	F_b	F_t	F_v	$F_{c\perp}$	F_c	E
Dimensional lumber (2–4 in. thick)	0.85[c]	1.00	0.97	0.67	0.8[d]	0.9
Timbers (5 × 5 in. or larger)	1.00	1.00	1.00	0.67	0.91	1.00

[a] When dimension lumber or timbers are used where moisture content will exceed 19% for an extended time period, design values shall be multiplied by the appropriate wet service factors from the table above.
[b] For Southern Pine and Mixed Southern Pine Timbers, and for Southern Pine and Mixed Southern Pine dimensional lumber marked Dense Structural 86, Dense Structural 72, and Dense Structural 65, use tabulated values without further adjustment for wet service use.
[c] When $(F_b)(C_F) \leq 1150$ psi, $C_M = 1.0$.
[d] When $F_c \leq 750$ psi, $C_M = 1.0$.

Source: NDS Supplement, Tables 4A–4D. Courtesy of American Forest and Paper Association, Washington, D.C.

TABLE A.7 SHEAR STRESS FACTOR, C_H, FOR DIMENSION LUMBER[a]

Shear Stress Factor, C_H, for Dimensional Lumber

Length of split on wide face of 2 in. (nominal) lumber	C_H	Length of split on wide face of 3 in. (nominal) and thicker lumber	C_H	Size of shake[b] in 2 in. (nominal) and thicker number	C_H
No split	2.00	No split	2.00	No shake	2.00
1/2 × wide face	1.67	1/2 × narrow face	1.67	1/6 × narrow face	1.67
3/4 × wide face	1.50	3/4 × narrow face	1.50	1/4 × narrow face	1.50
1 × wide face	1.33	1 × narrow face	1.33	1/3 × narrow face	1.33
1-1/2 × wide face or more	1.00	1-1/2 × narrow face or more	1.00	1/2 × narrow face or more	1.00

[a] Tabulated shear design values parallel to grain have been reduced to allow for the occurrence of splits, checks, and shakes. Tabulated shear design values parallel to grain, F_v, shall be permitted to be multiplied by the shear stress factors specified in the table above when length of split or size of check or shake is known and no increase in them is anticipated. When shear stress factors are used for Southern Pine and Mixed Southern Pine, a tabulated design value of $F_v = 90$ psi shall be assigned for all grades of Southern Pine and Mixed Southern Pine dimension lumber. Shear stress factors shall be permitted to be linearly interpolated.
[b] Shake is measured at the end between lines enclosing the shake and perpendicular to the loaded face.

Source: NDS Supplement, Tables 4A and 4B. Courtesy of American Forest and Paper Association, Washington, D.C.

TABLE A.8 DESIGN VALUES FOR VISUALLY GRADED DIMENSION LUMBER: DOUGLAS FIR-LARCH[a,b]

Species and Commercial Grade	Size Classification	Design Values in Pounds per Square Inch (psi)						Grading Rules Agency
		Bending F_b	Tension Parallel to Grain F_t	Shear Parallel to Grain F_v	Compression Perpendicular to Grain $F_{c\perp}$	Compression Parallel to Grain F_c	Modulus of Elasticity E	
Douglas Fir-Larch								
Stud	2–4 in. thick	742	495	95	625	866	1,400,000	WCLIB WWPA
Construction	2–4 in. wide	1,000	650	95	625	1,600	1,500,000	
Standard		550	375	95	625	1,350	1,400,000	
Utility		275	175	95	625	875	1,300,000	
Stud	2–4 in. thick	675	450	95	625	825	1,400,000	
Construction	6 in. wide	1,000	650	95	625	1,600	1,500,000	
Standard		550	375	95	625	1,350	1,400,000	
Utility		275	175	95	625	875	1,300,000	
Select Structural	2–3 in. thick	1,740	1,200	95	625	1,785	1,900,000	
No. 1 and Better	8 in. wide	1,380	930	95	625	1,575	1,800,000	
No. 1		1,200	810	95	625	1,522	1,700,000	
No. 2		1,050	690	95	625	1,365	1,600,000	
No. 3		600	390	95	625	787	1,400,000	
Select Structural	2–3 in. thick	1,595	1,100	95	625	1,700	1,900,000	
No. 1 and Better	10 in. wide	1,265	852	95	625	1,500	1,800,000	
No. 1		1,100	742	95	625	1,450	1,700,000	
No. 2		962	632	95	625	1,300	1,600,000	
No. 3		550	357	95	625	750	1,400,000	
Select Structural	2–3 in. thick	1,450	1,000	95	625	1,700	1,900,000	
No. 1 and Better	12 in. wide	1,150	775	95	625	1,500	1,800,000	
No. 1		1,000	675	95	625	1,450	1,700,000	
No. 2		875	575	95	625	1,300	1,600,000	
No. 3		500	325	95	625	750	1,400,000	
Select Structural	2–3 in. thick	1,305	900	95	625	1,530	1,900,000	
No. 1 and Better	14 in. and wider	1,035	697	95	625	1,350	1,800,000	
No. 1		900	607	95	625	1,305	1,700,000	
No. 2		787	517	95	625	1,170	1,600,000	
No. 3		450	292	95	625	675	1,400,000	

[a] Notes *a* and *b* from Table A.2 apply to this table.
[b] Adjustment factors: Tabulated values are for normal load duration and dry service conditions, unless otherwise specified. The adjustments for size factor, C_F, are already incorporated in this table. All other applicable factors must be applied. See Chapter 1 for a comprehensive description of design value adjustment factors.
Source: Values in this table are derived from NDS Supplement, Table 4A. Courtesy of American Forest and Paper Association, Washington, D.C.

TABLE A.9 DESIGN VALUES FOR VISUALLY GRADED DIMENSION LUMBER: SOUTHERN PINE[a,b,c]

Species and Commercial Grade	Size Classification	Design Values in Pounds per Square Inch (psi)						Grading Rules Agency
		Bending F_b	Tension Parallel to Grain F_t	Shear Parallel to Grain F_v	Compression Perpendicular to Grain $F_{c\perp}$	Compression Parallel to Grain F_c	Modulus of Elasticity E	
Southern Pine								
Dense Select Structural	2–4 in. thick	3,050	1,650	100	660	2,250	1,900,000	SPIB
Select Structural	2–4 in. wide	2,850	1,600	100	565	2,100	1,800,000	
No. 1		1,850	1,050	100	565	1,850	1,700,000	
No. 2		1,500	825	90	565	1,650	1,600,000	
No. 3		850	475	90	565	975	1,400,000	
Stud		875	500	90	565	975	1,400,000	
Construction	2–4 in. thick	1,100	625	100	565	1,800	1,500,000	
Standard	4 in. wide	625	350	90	565	1,500	1,300,000	
Utility		300	175	90	565	975	1,300,000	
Dense Select Structural	2–4 in. thick	2,700	1,500	90	660	2,150	1,900,000	
Select Structural	5–6 in. wide	2,550	1,400	90	565	2,100	1,800,000	
No. 1		1,650	900	90	565	1,750	1,700,000	
No. 2		1,250	725	90	565	1,600	1,600,000	
No. 3		750	425	90	565	925	1,400,000	
Stud		775	425	90	565	925	1,400,000	
Dense Select Structural	2–4 in. thick	2,450	1,350	90	660	2,050	1,900,000	
Select Structural	8 in. wide	2,300	1,300	90	565	1,900	1,800,000	
No. 1		1,500	825	90	565	1,650	1,700,000	
No. 2		1,200	650	90	565	1,550	1,600,000	
No. 3		700	400	90	565	875	1,400,000	
Dense Select Structural	2–4 in. thick	2,150	1,200	90	660	2,000	1,900,000	
Select Structural	10 in. wide	2,050	1,100	90	565	1,850	1,800,000	
No. 1		1,300	725	90	565	1,600	1,700,000	
No. 2		1,050	575	90	565	1,500	1,600,000	
No. 3		600	325	90	565	850	1,400,000	
Dense Select Structural	2–4 in. thick	2,050	1,100	90	660	1,950	1,900,000	
Select Structural	12 in. wide	1,900	1,050	90	565	1,800	1,800,000	
No. 1		1,250	675	90	565	1,600	1,700,000	
No. 2		975	550	90	565	1,450	1,600,000	
No. 3		575	325	90	565	825	1,400,000	
Dense Select Structural	2–4 in. thick	2,050	1,100	90	660	1,950	1,900,000	
Select Structural	14 in. and wider	1,900	1,050	90	565	1,800	1,800,000	
No. 1		1,250	675	90	565	1,600	1,700,000	
No. 2		975	550	90	565	1,450	1,600,000	
No. 3		575	325	90	565	825	1,400,000	

[a] Lumber dimensions: Tabulated design values are applicable to lumber that will be used under dry conditions, such as in most covered structures. For 2–4 in. thick lumber, the *dry* dressed sizes shall be used (see Table A.1a), regardless of the moisture content at the time of manufacture or use. In calculating design values, the natural gain in strength and stiffness that occurs as lumber dries has been taken into consideration, as well as the reduction in size that occurs when unseasoned lumber shrinks. The gain in load-carrying capacity due to increased strength and stiffness resulting from drying more than offsets the design effect of size reduction due to shrinkage.
[b] Stress-rated boards: Information for various grades of Southern Pine stress-rated boards of nominal 1, $1\frac{1}{4}$, and $1\frac{1}{2}$ in. thickness, 2 in. and wider, is available from the Southern Pine Inspection Bureau (SPIB) in the "Standard Grading Rules for Southern Pine."
[c] Adjustment factors: Tabulated values are for normal load duration and dry service conditions, unless otherwise specified. The adjustments for size factor, C_F, are already incorporated in this table. All other applicable factors must be applied. See Chapter 1 for a comprehensive description of design value adjustment factors.

Source: Values in this table are extracted from NDS Supplement, Table 4B, which contains strength values for additional grades not listed here. Courtesy of American Forest and Paper Association, Washington, D.C.

TABLE A.10 DESIGN VALUES FOR VISUALLY GRADED DECKING: DFL AND SOUTHERN PINE[a,b]

Species and Commercial Grade	Size Classification	Bending Single Member F_b	Bending Repetitive Member $(F_b)(C_r)$	Compression Perpendicular to Grain $F_{c\perp}$	Modulus of Elasticity E	Grading Rules Agency
Douglas Fir-Larch						
Selected	2 in. thick		2,200		1,800,000	WWPA
Commercial	4–12 in. wide		1,815		1,700,000	
Selected	3 in. thick		2,080		1,800,000	
Commercial	4–12 in. wide		1,716		1,700,000	
Selected	4 in. thick		2,000		1,800,000	
Commercial	4–12 in. wide		1,650		1,700,000	
Southern Pine						
Dense standard	2 in. thick	2,200	2,530	660	1,800,000	SPIB
Commercial	4 in. and wider	1,540	1,815	565	1,600,000	
Dense standard	3 in. thick	2,080	2,392	660	1,800,000	
Commercial	4 in. and wider	1,456	1,716	565	1,600,000	
Dense standard	4 in. thick	2,000	2,300	660	1,800,000	
Commercial	4 in. and wider	1,400	1,650	565	1,600,000	

[a] Lumber dimensions: Tabulated design values are applicable to lumber that will be used under dry conditions, such as in most covered structures. For 2–4 in. thick lumber, the *dry* dressed sizes shall be used (see Table A.1a), regardless of the moisture content at the time of manufacture or use. In calculating design values, the natural gain in strength and stiffness that occurs as lumber dries has been taken into consideration, as well as the reduction in size that occurs when unseasoned lumber shrinks. The gain in load-carrying capacity due to increased strength and stiffness resulting from drying more than offsets the design effect of size reduction due to shrinkage.

[b] Adjustment factors: Tabulated values are for normal load duration and dry service conditions, unless otherwise specified. The adjustments for size factor, C_F, repetitive member factor, C_r, and flat use factor, C_{fu}, are already incorporated in this table. All other applicable factors must be applied. See Chapter 1 for a comprehensive description of design value adjustment factors.

Source: Values in this table are extracted from NDS Supplement, Table 4E, which contains strength values for additional species and grades not listed here. Courtesy of American Forest and Paper Association, Washington, D.C.

TABLE A.11 DESIGN VALUES FOR VISUALLY GRADED TIMBERS (5 × 5 IN. AND LARGER): DOUGLAS FIR-LARCH AND SOUTHERN PINE[a,b]

Species and Commercial Grade	Size Classification	Bending F_b	Tension Parallel to Grain[c] F_t	Shear Parallel to Grain[c] F_v	Compression Perpendicular to Grain $F_{c\perp}$	Compression Parallel to Grain F_c	Modulus of Elasticity E	Grading Rules Agency
Douglas Fir-Larch								
Dense Select Structural	Beams and Stringers	1,850	1,100	85	730	1,300	1,700,000	WWPA
Select Structural		1,600	950	85	625	1,100	1,600,000	
Dense No. 1		1,550	775	85	730	1,100	1,700,000	
No. 1		1,350	675	85	625	925	1,600,000	
Dense No. 2		1,000	500	85	730	700	1,400,000	
No. 2		875	425	85	625	600	1,300,000	
Dense Select Structural	Posts and Timbers	1,750	1,150	85	730	1,350	1,700,000	
Select Structural		1,500	1,000	85	625	1,150	1,600,000	
Dense No. 1		1,400	950	85	730	1,200	1,700,000	
No. 1		1,200	825	85	625	1,000	1,600,000	
Dense No. 2		800	550	85	730	550	1,400,000	
No. 2		700	475	85	625	475	1,300,000	
Southern Pine								
Dense Select Structural SR	5 × 5 in. and larger	1,750	1,200	110	440	1,100	1,600,000	SPIB
Select Structural SR		1,500	1,000	110	375	950	1,500,000	
No. 1 Dense SR		1,550	1,050	110	440	975	1,600,000	
No. 1 SR		1,350	900	110	375	825	1,500,000	
No. 2 Dense SR		975	650	100	440	625	1,300,000	
No. 2 SR		850	550	100	375	525	1,200,000	
Dense Structural 86		2,100	1,400	145	440	1,300	1,600,000	
Dense Structural 72		1,750	1,200	120	440	1,100	1,600,000	
Dense Structural 65		1,600	1,050	110	440	1,000	1,600,000	

[a] Lumber dimensions: Tabulated design values are applicable to lumber that will be used under dry conditions, such as in most covered structures. For 5 in. and thicker lumber, the *green* dressed sizes shall be used (see Table A.1a), because design values have been adjusted to compensate for any loss in size by shrinkage that may occur.

[b] Adjustment factors: Tabulated values are for normal load duration and dry service conditions, unless otherwise specified. The adjustments for size factor, C_F, are already incorporated in this table for members 12 in. deep or less. When the depth, d, of a beam, stringer, post, or timber exceeds 12 in., the tabulated bending design value, F_b, shall be multiplied by the size factor, $C_F = (12/d)^{1/9}$. All other applicable factors must be applied. See Chapter 1 for a comprehensive description of design value adjustment factors.

[c] The tabulated shear design values parallel to grain have been reduced to allow for the occurrence of splits, checks, and shakes. Tabulated shear design values parallel to grain, F_v, shall be permitted to be multiplied by the shear factors specified in Table A.12.

Source: Values in this table are extracted from NDS Supplement, Table 4D, which contains strength values for additional grades not listed here. Courtesy of American Forest and Paper Association, Washington, D.C.

TABLE A.12 SHEAR STRESS FACTOR, C_H, FOR TIMBERS[a]

Shear Stress Factor, C_H, for Timbers

Length of split on wide face of 5 in. (nominal) and thicker lumber	C_H	Size of shake[b] in 5 in. (nominal) and thicker lumber	C_H
No split	2.00	No shake	2.00
1/2 × narrow face	1.67	1/6 × narrow face	1.67
3/4 × narrow face	1.50	1/4 × narrow face	1.50
1 × narrow face	1.33	1/3 × narrow face	1.33
1-1/2 × narrow face or more	1.00	1/2 × narrow face or more	1.00

[a]Tabulated shear design values parallel to grain in Table A.11 have been reduced to allow for the occurrence of splits, checks, and shakes. Tabulated shear design values parallel to grain, F_v, shall be permitted to be multiplied by the shear stress factors specified in the table above when length of split or size of check or shake is known and no increase in them is anticipated. When shear stress factors are used for Southern Pine and Mixed Southern Pine, a tabulated design value of $F_v = 90$ psi shall be assigned for all grades of Southern Pine and Mixed Southern Pine dimension lumber. Shear stress factors shall be permitted to be linearly interpolated.

[b]Shake is measured at the end between lines enclosing the shake and perpendicular to the loaded face.

Source: NDS Supplement, Table 4D. Courtesy of American Forest and Paper Association, Washington, D.C.

APPENDIX B

BEAM DESIGN TABLES

TABLE B.1a ALLOWABLE DECK LOADS: NOMINAL 2-INCH-THICK DECK[a]

	Limited by Bending (Total Load, psf) $F_b = 1000$ psi		Limited by Deflection (Live Load Only, psf) $E = 1,000,000$ psi					
Span (ft)	Simple Span; Combination Simple and Two-Span; and Two-Span 1/8	Controlled Random Layup Cantilevered Pieces Intermixed 3/20	Deflection Ratio	Simple Span 5/384	Controlled Random Layup 1/116	Cantilevered Pieces Intermixed 1/105	Combination Simple and Two-Span 1/109	Two-Span 1/185
6	83	69	$L/180$	46	60	63	66	112
			$L/240$	35	45	47	49	84
			$L/360$	23	30	32	33	56
7	61	51	$L/180$	29	38	40	41	70
			$L/240$	22	28	30	31	53
			$L/360$	15	19	20	21	35
8	47	39	$L/180$	20	25	27	28	47
			$L/240$	15	19	20	21	35
			$L/360$	10	13	13	14	24
9	37	31	$L/180$	14	18	19	19	33
			$L/240$	10	13	14	15	25
			$L/360$	7	9	9	10	17
10	30	25	$L/180$	10	13	14	14	24
			$L/240$	8	10	10	11	18
			$L/360$	5	7	7	7	12
11	25	21	$L/180$	8	10	10	11	18
			$L/240$	6	7	8	8	14
			$L/360$	4	5	5	5	9
12	21	17	$L/180$	6	8	8	8	14
			$L/240$	4	6	6	6	10
			$L/360$	3	4	4	4	7

[a] Actual deck size 1.5 in.

Procedure: 1. Select deck layup pattern and trial grade and species. Look up F_b and E in Table A.10.
 2. To find allowable total load controlled by bending, multiply tabulated value (above) by ratio of $F_b/1000$.
 3. To find allowable live load controlled by deflection, multiply tabulated value by ratio of $E/1,000,000$.
 4. Compare allowable loads with the required loads. If inadequate, revise span, grade, or layup.

TABLE B.1b ALLOWABLE DECK LOADS: NOMINAL 3-INCH-THICK DECK[a]

	Limited by Bending (Total Load, psf) $F_b = 1,000$ psi		Limited by Deflection (Live Load Only, psf) $E = 1,000,000$ psi					
Span (ft)	Simple Span; Combination Simple and Two-Span; and Two-Span 1/8	Controlled Random Layup; Cantilevered Pieces Intermixed 3/20	Deflection Ratio	Simple Span 5/384	Controlled Random Layup 1/116	Cantilevered Pieces Intermixed 1/105	Combination Simple and Two-Span 1/109	Two-Span 1/185
8	130	109	L/180	90	137	124	128	218
			L/240	68	102	93	96	163
			L/360	45	68	62	64	109
9	103	86	L/180	64	96	87	90	153
			L/240	48	72	65	68	115
			L/360	32	48	43	45	76
10	83	69	L/180	46	70	63	66	112
			L/240	35	52	47	49	84
			L/360	23	35	32	33	56
11	69	57	L/180	35	53	48	49	84
			L/240	26	39	36	37	63
			L/360	17	26	24	25	42
12	58	48	L/180	27	40	37	38	65
			L/240	20	30	27	29	48
			L/360	13	20	18	19	32
13	49	41	L/180	21	32	29	30	51
			L/240	16	24	22	22	38
			L/360	11	16	14	15	25
14	43	35	L/180	17	25	23	24	41
			L/240	13	19	17	18	30
			L/360	8	13	12	12	20
15	37	31	L/180	14	21	19	19	33
			L/240	10	16	14	15	25
			L/360	7	10	9	10	17
16	33	27	L/180	11	17	15	16	27
			L/240	8	13	12	12	20
			L/360	6	9	8	8	14
17	29	24	L/180	9	14	13	13	23
			L/240	7	11	10	10	17
			L/360	5	7	6	7	11
18	26	21	L/180	8	12	11	11	19
			L/240	6	9	8	8	14
			L/360	4	6	5	6	10
19	23	19	L/180	7	10	9	10	16
			L/240	5	8	7	7	12
			L/360	3	5	5	5	8
20	21	17	L/180	6	9	8	8	14
			L/240	4	7	6	6	10
			L/360	3	4	4	4	7

[a] Actual deck size 2.5 in. See Table B.1a for design procedure.

TABLE B.1c ALLOWABLE DECK LOADS: NOMINAL 4-INCH-THICK DECK[a]

	Limited Bending (Total Load, psf) $F_b = 1000$ psi		Limited by Deflection (Live Load Only, psf) $E = 1{,}000{,}000$ psi					
Span (ft)	Simple Span; Combination Simple and Two-Span; and Two-Span 1/8	Controlled Random Layup; Cantilevered Pieces Intermixed 3/20	Deflection Ratio	Simple Span 5/384	Controlled Random Layup 1/116	Cantilevered Pieces Intermixed 1/105	Combination Simple and Two-Span 1/109	Two-Span 1/185
8	255	213	L/180	248	375	339	352	598
			L/240	186	281	254	264	448
			L/360	124	187	170	176	299
9	202	168	L/180	174	263	238	247	420
			L/240	131	197	179	185	315
			L/360	87	132	119	124	210
10	163	136	L/180	127	192	174	180	306
			L/240	95	144	130	135	230
			L/360	64	96	87	90	153
11	135	112	L/180	95	144	130	135	230
			L/240	72	108	98	102	172
			L/360	48	72	65	68	115
12	113	95	L/180	74	111	101	104	177
			L/240	55	83	75	78	133
			L/360	37	55	50	52	89
13	97	81	L/180	58	87	79	82	139
			L/240	43	65	59	62	104
			L/360	29	44	40	41	70
14	83	69	L/180	46	70	63	66	112
			L/240	35	52	47	49	84
			L/360	23	35	32	33	56
15	73	60	L/180	38	57	51	53	91
			L/240	28	43	39	40	68
			L/360	19	28	26	27	45
16	64	53	L/180	31	47	42	44	75
			L/240	23	35	32	33	56
			L/360	16	23	21	22	37
17	57	47	L/180	26	39	35	37	62
			L/240	19	29	27	28	47
			L/360	13	20	18	18	31
18	50	42	L/180	22	33	30	31	52
			L/240	16	25	22	23	39
			L/360	11	16	15	15	26
19	45	38	L/180	19	28	25	26	45
			L/240	14	21	19	20	33
			L/360	9	14	13	13	22
20	41	34	L/180	16	24	22	23	38
			L/240	12	18	16	17	29
			L/360	8	12	11	11	19

[a] Actual deck size 3.5 in. See Table B.1a for design procedure.

TABLE B.2a   ALLOWABLE SPANS AND REQUIRED BENDING STRENGTH (F'_B) FOR FLOOR JOISTS, 30 PSF

Loads (psf)	Live Load 30	Dead Load 10	Total Load 40	Deflection Limit for Live Load Only $L/360$

			Modulus of Elasticity (psi)											
Member Size	Spacing (in. o.c.)		700,000	800,000	900,000	1,000,000	1,100,000	1,200,000	1,300,000	1,400,000	1,500,000	1,600,000	1,700,000	1,800,000	1,900,000	2,000,000
2 × 6	24		7'-1"	7'-5"	7'-8"	8'-0"	8'-3"	8'-6"	8'-8"	8'-11"	9'-1"	9'-4"	9'-6"	9'-8"	9'-11"	10'-1"
		$F'_b =$	801	875	948	1016	1083	1146	1212	1270	1333	1392	1448	1505	1560	1613
	16		8'-1"	8'-6"	8'-10"	9'-1"	9'-5"	9'-8"	10'-0"	10'-3"	10'-5"	10'-8"	10'-11"	11'-1"	11'-4"	11'-6"
		$F'_b =$	700	764	827	889	946	1003	1058	1111	1164	1215	1265	1314	1361	1409
	12		8'-11"	9'-4"	9'-8"	10'-1"	10'-4"	10'-8"	11'-0"	11'-3"	11'-6"	11'-9"	12'-0"	12'-3"	12'-5"	12'-8"
		$F'_b =$	635	696	753	807	859	912	962	1010	1057	1103	1148	1195	1238	1281
2 × 8	24		9'-4"	9'-9"	10'-2"	10'-6"	10'-10"	11'-2"	11'-6"	11'-9"	12'-1"	12'-4"	12'-7"	12'-10"	13'-0"	13'-3"
		$F'_b =$	801	876	948	1017	1084	1147	1210	1272	1333	1391	1448	1504	1561	1613
	16		10'-8"	11'-2"	11'-7"	12'-1"	12'-5"	12'-10"	13'-2"	13'-6"	13'-9"	14'-1"	14'-4"	14'-8"	14'-11"	15'-2"
		$F'_b =$	699	765	827	889	946	1003	1059	1111	1163	1215	1265	1315	1362	1410
	12		11'-9"	12'-4"	12'-10"	13'-3"	13'-8"	14'-1"	14'-6"	14'-10"	15'-2"	15'-6"	15'-10"	16'-2"	16'-5"	16'-9"
		$F'_b =$	636	696	752	807	859	911	961	1010	1058	1104	1149	1195	1238	1281
2 × 10	24		11'-11"	12'-6"	13'-0"	13'-5"	13'-10"	14'-3"	14'-8"	15'-0"	15'-4"	15'-9"	16'-0"	16'-4"	16'-8"	16'-11"
		$F'_b =$	801	877	948	1016	1083	1148	1212	1273	1333	1392	1449	1504	1560	1613
	16		13'-8"	14'-3"	14'-10"	15'-4"	15'-10"	16'-4"	16'-9"	17'-2"	17'-7"	18'-0"	18'-4"	18'-9"	19'-1"	19'-5"
		$F'_b =$	701	765	828	889	946	1003	1058	1112	1165	1216	1266	1315	1363	1410
	12		15'-0"	15'-9"	16'-4"	16'-11"	17'-6"	18'-0"	18'-6"	18'-11"	19'-5"	19'-10"	20'-3"	20'-7"	21'-0"	21'-4"
		$F'_b =$	636	696	752	807	860	912	961	1010	1057	1105	1150	1194	1238	1282
2 × 12	24		14'-6"	15'-2"	15'-9"	16'-4"	16'-10"	17'-4"	17'-10"	18'-3"	18'-9"	19'-1"	19'-6"	19'-11"	20'-3"	20'-7"
		$F'_b =$	802	876	948	1017	1084	1148	1212	1273	1333	1392	1448	1504	1560	1615
	16		16'-7"	17'-4"	18'-1"	18'-9"	19'-4"	19'-11"	20'-5"	20'-11"	21'-5"	21'-11"	22'-4"	22'-9"	23'-2"	23'-7"
		$F'_b =$	701	765	828	889	947	1003	1058	1112	1164	1216	1266	1315	1363	1410
	12		18'-3"	19'-1"	19'-11"	20'-7"	21'-3"	21'-11"	22'-6"	23'-0"	23'-7"	24'-1"	24'-7"	25'-1"	25'-6"	26'-0"
		$F'_b =$	637	696	752	807	860	912	962	1010	1057	1105	1150	1194	1238	1282
2 × 14	24		17'-1"	17'-10"	18'-7"	19'-3"	19'-10"	20'-6"	21'-0"	21'-6"	22'-0"	22'-6"	23'-0"	23'-5"	23'-10"	24'-3"
		$F'_b =$	802	877	948	1016	1083	1149	1212	1272	1333	1392	1449	1505	1560	1615
	16		19'-7"	20'-6"	21'-3"	22'-0"	22'-9"	23'-5"	24'-1"	24'-8"	25'-3"	25'-9"	26'-4"	26'-10"	27'-4"	27'-9"
		$F'_b =$	701	766	828	889	947	1003	1058	1112	1164	1215	1266	1314	1363	1411
	12		21'-6"	22'-6"	23'-5"	24'-3"	25'-0"	25'-9"	26'-6"	27'-2"	27'-9"	28'-5"	29'-0"	29'-6"	30'-1"	30'-7"
		$F'_b =$	636	696	752	807	860	912	962	1010	1058	1105	1150	1195	1239	1281

Procedure:
1. Select trial size and spacing.
2. Scan horizontally until you find required span.
3. Read vertically to find required modulus of elasticity, E. Directly below the span number, find the required bending strength, F'_b.
4. Refer to Tables A.8 or A.9 to find acceptable species and grade. Apply all appropriate service factors.

TABLE B.S2b   ALLOWABLE SPANS AND REQUIRED BENDING STRENGTH (F_b') FOR FLOOR JOISTS, 40psf

Loads (psf)	Live Load	Dead Load	Total Load	Deflection Limit for Live Load Only
	40	10	50	$L/360$

Member Size	Spacing (in. o.c.)		Modulus of Elasticity (psi)												
			700,000	800,000	900,000	1,000,000	1,100,000	1,200,000	1,300,000	1,400,000	1,500,000	1,600,000	1,700,000	1,800,000	1,900,000	2,000,000
2 × 6	24		6'-5"	6'-9"	7'-0"	7'-3"	7'-6"	7'-8"	7'-11"	8'-1"	8'-3"	8'-6"	8'-8"	8'-10"	9'-0"	9'-1"
		$F_b' =$	827	904	978	1049	1119	1185	1250	1313	1374	1433	1493	1551	1610	1667
	16		7'-4"	7'-8"	8'-0"	8'-3"	8'-7"	8'-10"	9'-1"	9'-3"	9'-6"	9'-8"	9'-11"	10'-1"	10'-3"	10'-5"
		$F_b' =$	721	790	853	916	977	1034	1091	1147	1201	1254	1306	1356	1406	1455
	12		8'-1"	8'-6"	8'-10"	9'-1"	9'-5"	9'-8"	10'-0"	10'-3"	10'-5"	10'-8"	10'-11"	11'-1"	11'-4"	11'-6"
		$F_b' =$	656	717	776	833	887	941	992	1042	1091	1139	1186	1232	1276	1321
2 × 8	24		8'-6"	8'-10"	9'-3"	9'-7"	9'-10"	10'-2"	10'-5"	10'-8"	10'-11"	11'-2"	11'-5"	11'-7"	11'-10"	12'-1"
		$F_b' =$	827	903	977	1048	1118	1185	1249	1311	1373	1434	1493	1551	1610	1667
	16		9'-8"	10'-2"	10'-7"	10'-11"	11'-3"	11'-7"	11'-11"	12'-3"	12'-6"	12'-10"	13'-1"	13'-4"	13'-7"	13'-9"
		$F_b' =$	722	790	854	916	976	1034	1092	1148	1201	1253	1305	1357	1406	1454
	12		10'-8"	11'-2"	11'-7"	12'-1"	12'-5"	12'-10"	13'-2"	13'-6"	13'-9"	14'-1"	14'-4"	14'-8"	14'-11"	15'-2"
		$F_b' =$	656	717	775	833	887	940	993	1042	1091	1139	1186	1233	1277	1322
2 × 10	24		10'-10"	11'-4"	11'-9"	12'-2"	12'-7"	13'-0"	13'-4"	13'-8"	14'-0"	14'-3"	14'-7"	14'-10"	15'-1"	15'-4"
		$F_b' =$	828	904	978	1049	1118	1185	1251	1314	1374	1434	1493	1552	1609	1667
	16		12'-5"	13'-0"	13'-6"	14'-0"	14'-5"	14'-10"	15'-3"	15'-7"	16'-0"	16'-4"	16'-8"	17'-0"	17'-4"	17'-7"
		$F_b' =$	722	790	855	916	977	1034	1092	1147	1201	1254	1305	1356	1406	1456
	12		13'-8"	14'-3"	14'-10"	15'-4"	15'-10"	16'-4"	16'-9"	17'-2"	17'-7"	18'-0"	18'-4"	18'-9"	19'-1"	19'-5"
		$F_b' =$	657	717	776	833	887	940	992	1042	1092	1140	1187	1233	1278	1322
2 × 12	24		13'-2"	13'-9"	14'-4"	14'-10"	15'-4"	15'-9"	16'-2"	16'-7"	17'-0"	17'-4"	17'-9"	18'-1"	18'-5"	18'-9"
		$F_b' =$	827	904	978	1049	1118	1185	1250	1314	1375	1435	1495	1552	1610	1667
	16		15'-1"	15'-9"	16'-5"	17'-0"	17'-6"	18'-1"	18'-6"	19'-0"	19'-6"	19'-11"	20'-3"	20'-8"	21'-1"	21'-5"
		$F_b' =$	723	790	854	917	977	1035	1091	1147	1202	1254	1305	1357	1406	1455
	12		16'-7"	17'-4"	18'-1"	18'-9"	19'-4"	19'-11"	20'-5"	20'-11"	21'-5"	21'-11"	22'-4"	22'-9"	23'-2"	23'-7"
		$F_b' =$	657	717	776	833	888	940	992	1042	1091	1140	1187	1232	1278	1322
2 × 14	24		15'-6"	16'-3"	16'-11"	17'-6"	18'-0"	18'-7"	19'-1"	19'-7"	20'-0"	20'-6"	20'-11"	21'-3"	21'-8"	22'-0"
		$F_b' =$	828	905	978	1049	1118	1186	1250	1313	1376	1436	1495	1552	1611	1667
	16		17'-9"	18'-7"	19'-4"	20'-0"	20'-8"	21'-3"	21'-10"	22'-5"	22'-11"	23'-5"	23'-11"	24'-4"	24'-10"	25'-3"
		$F_b' =$	723	790	854	917	977	1035	1092	1147	1201	1254	1306	1357	1406	1455
	12		19'-7"	20'-6"	21'-3"	22'-0"	22'-9"	23'-5"	24'-1"	24'-8"	25'-3"	25'-9"	26'-4"	26'-10"	27'-4"	27'-9"
		$F_b' =$	657	718	776	833	888	940	992	1042	1091	1139	1187	1232	1278	1323

TABLE B.3 APA GLUED FLOOR SYSTEM: MAXIMUM JOIST SPANS WITH APA-RATED STURD-I-FLOOR PANELS (FT. AND IN.)[a]

Species and Grade of Joist	Joist Size	Joists at 16 in. o.c.		Joists at 19.2 in. o.c.		Joists at 24 in. o.c.
		Sturd-I-Floor 16 or 20 o.c.	Sturd-I-Floor 24 o.c.	Sturd-I-Floor 20 o.c.	Sturd-I-Floor 24 o.c.	Sturd-I-Floor 24 o.c.
Douglas Fir– Larch No. 1	2 × 6	11-0	11-4	10-6	10-6	9-5
	2 × 8	14-3	14-7	13-7	13-10	12-5
	2 × 10	17-11	18-3	17-0	17-4	15-10
	2 × 12	21-7	21-11	20-6	20-10	19-3
Douglas Fir– Larch No. 2	2 × 6	10-6	10-6	9-7	9-7	8-7
	2 × 8	13-10	13-10	12-7	12-7	11-3
	2 × 10	17-7	17-7	16-1	16-1	14-5
	2 × 12	21-3	21-5	19-7	19-7	17-6
Douglas Fir– South No. 1	2 × 6	10-4	10-8	9-11	10-2	9-1
	2 × 8	13-4	13-8	12-8	13-0	12-0
	2 × 10	16-8	17-0	15-11	16-3	15-4
	2 × 12	20-1	20-5	19-1	19-5	18-4
Douglas Fir– South No. 2	2 × 6	10-1	10-1	9-3	9-3	8-3
	2 × 8	13-1	13-4	12-2	12-2	10-11
	2 × 10	16-4	16-8	15-6	15-6	13-11
	2 × 12	19-8	20-0	18-8	18-10	16-11
Hem–Fir No. 1	2 × 6	10-3	10-3	9-5	9-5	8-5
	2 × 8	13-7	13-7	12-5	12-5	11-1
	2 × 10	17-0	17-4	15-10	15-10	14-2
	2 × 12	20-6	20-10	19-3	19-3	17-2
Hem–Fir No. 2	2 × 6	9-4	9-4	8-6	8-6	7-7
	2 × 8	12-4	12-4	11-3	11-3	10-0
	2 × 10	15-8	15-8	14-4	14-4	12-10
	2 × 12	19-1	19-1	17-5	17-5	15-7
Southern Pine KD15 No. 1	2 × 6	11-0	11-4	10-6	10-10	9-8
	2 × 8	14-3	14-7	13-7	13-11	12-9
	2 × 10	17-11	18-3	17-0	17-4	16-3
	2 × 12	21-7	21-11	20-6	20-10	19-7
Southern Pine KD15 No. 2	2 × 6	10-8	10-8	9-9	9-9	8-8
	2 × 8	13-10	14-0	12-10	12-10	11-6
	2 × 10	17-4	17-8	16-4	16-4	14-8
	2 × 12	20-11	21-2	19-10	19-11	17-9
Southern Pine MC19 No. 1	2 × 6	10-10	11-2	10-4	10-4	9-3
	2 × 8	14-0	14-4	13-4	13-8	12-2
	2 × 10	17-8	17-11	16-9	17-1	15-7
	2 × 12	21-3	21-7	20-2	20-6	18-11
Southern Pine MC19 No. 2	2 × 6	10-3	10-3	9-5	9-5	8-5
	2 × 8	13-7	13-7	12-5	12-5	11-1
	2 × 10	17-4	17-4	15-10	15-10	14-2
	2 × 12	20-11	21-1	19-3	19-3	17-2

[a] Based on live load of 40 psf, total load of 50 psf. Deflection limited to 1/360 at 40 psf. Glue to joists and at tongue-and-groove joints. If square-edge panels are used, block panel edges and glue between panels and between panels and blocking. Contact truss manufacturer for floor truss or plywood I-joist span recommendations.

STURD-I-FLOOR and subflooring panels should be installed continuous over two or more spans with the long dimension or strength axis across supports.

Panels recommended for glued floor construction are tongue-and-groove APA RATED STURD-I-FLOOR for single-floor construction, and APA RATED SHEATHING for the subfloor when used with a separate underlayment or with structural finish flooring.

Tongue-and-groove panels are highly recommended for single-floor construction. Before each panel is placed, a line of glue is applied to the joists with a caulking gun. The panel T&G joint should also be glued, although less heavily to avoid squeeze-out. If square-edge panels are used, edges must be supported between joists with 2 × 4 blocking. Glue panels to blocking to minimize squeaks. Blocking is not required under structural finish flooring, such as wood strip flooring, or if a separate underlayment is installed.

Only adhesives conforming with Performance Specification AFG-01 developed by APA are recommended for use with the Glued Floor System.

For a list of qualified adhesives, write to APA. If nonveneer panels with sealed surfaces and edges are to be used, use only solvent-based glues; check with panel manufacturer. Always follow the specific application recommendations of the glue manufacturer.

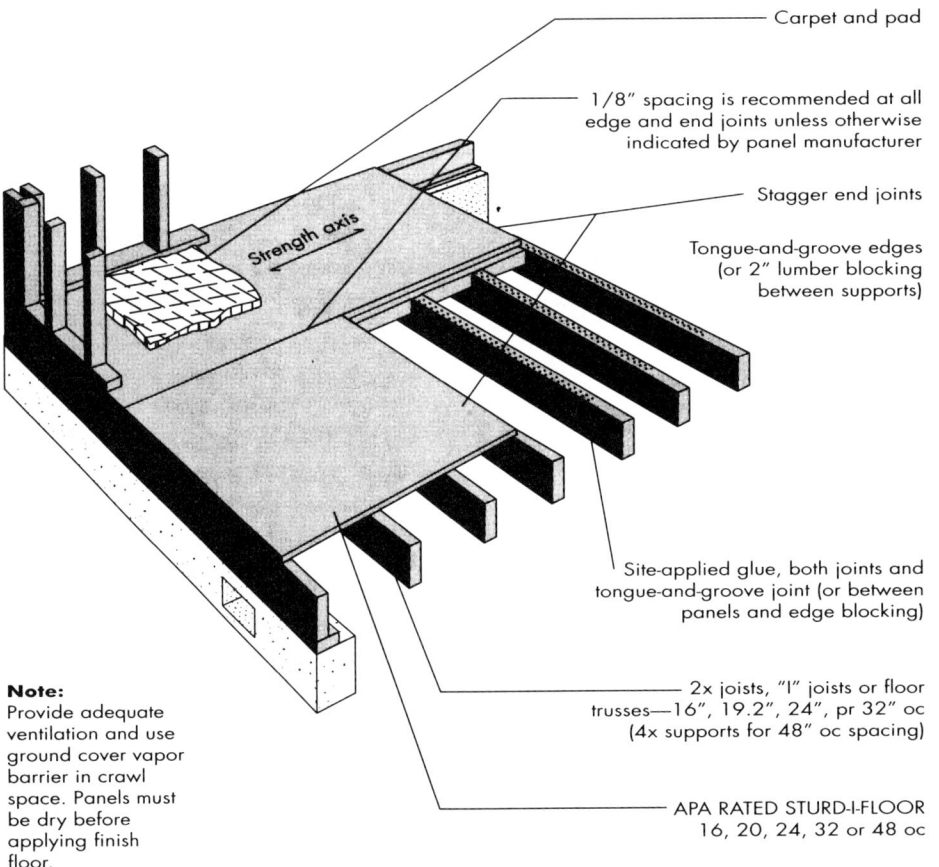

Source: APA Design/Construction Guide. Residential and Commercial. American Plywood Association. April 1993, p.21.

TABLE B.4 MICRO = LAM LUMBER DESIGN DATA

Lumber Design Properties

Size (in.)	Maximum Vertical Shear (lbs)			Maximum Resistive Moment (ft/lb)			Moment of Intertia (in.^{4})	Weight (lb/ft)
	100%	115%	125%	100%	115%	125%		
$1\frac{3}{4} \times 9\frac{1}{2}$	3,160	3,630	3,950	6,620	7,613	8,275	125	4.25
$1\frac{3}{4} \times 11\frac{7}{8}$	3,950	4,540	4,940	10,060	11,570	12,575	245	5.30
$1\frac{3}{4} \times 14$	4,650	5,345	5,810	13,645	15,690	17,055	400	6.25
$1\frac{3}{4} \times 16$	5,320	6,120	6,650	17,435	20,050	21,795	595	7.15
$1\frac{3}{4} \times 18$	5,985	6,880	7,480	21,785	25,050	27,230	850	8.00

Allowable Design Stresses:
Modulus of elasticity $\quad E = 2.0 \times 10^6$ psi *For 12-in. depth; for other depths, multiply by $(12/d)^{1/9}$.

Flexural stress $\quad f_b^* = 2925$ psi
- See NER 126 for additonal design information.

Compression perpendicular to grain parallel to glue line $\quad f_c = 750$ psi
- Assumes lateral support of compression edge at 24. in. o.c. or closer.

Compression parallel to grain $\quad f_{c\text{II}} = 3035$ psi

Horizontal shear perpendicular to glue line $\quad f_v = 285$ psi

Allowable Load (Floor) (lb./linear foot)

Span	One Ply $1\frac{3}{4}'' \times 9\frac{1}{2}''$		One Ply $1\frac{3}{4}'' \times 11\frac{7}{8}''$		One Ply $1\frac{3}{4}'' \times 14''$		One Ply $1\frac{3}{4}'' \times 16''$[1]		One Ply $1\frac{3}{4}'' \times 18''$[1]	
	Deflection $L/360$	Allowable Total Load	Deflection $L/360$	Allowable Total Load	Deflection $L/360$	Allowable Total Load	Deflection $L/360$	Allowable Total Load	Deflection $L/360$	Allowable Total Load
6		1063		1424		1795		2193		2651
8	629	746		979		1207		1443		1701
10	338	502	629	745		909		1074		1251
12	201	297	379	552	599	728		855		988
14	129	188	245	361	390	550	566	706	781	816
16	87	126	167	244	268	394	390	539	542	672
18	62	88	119	172	191	279	280	412	390	529
20	45	63	87	125	141	240	207	303	290	426
22	34	46	66	93	107	153	157	228	221	322
24	26	35	51	71	83	117	122	175	172	249
26						91	97	137	136	196

1. To size a beam for use in a floor, it is necessary to check both deflection and allowable total load.
2. Check local code for deflection criteria.
3. For deflection limits of $L/240$ and $L/480$, multiply loads shown in $L/360$ column by 1.5 and 0.75, respectively.
4. Make sure that the selected beam will work in both columns.

Notes:
- This table is based on uniform loads and simple spans.
- Table is for one beam. When top loaded, double the values for two beams, triple for three, etc., when properly fastened together with a minimum of two rows of 16d nails at 12 in. o.c. For this nailing, the maximum uniform load that can be applied to either outside member is 470 plf. For higher loads, bolted connections are needed.
- Micro = Lam lumber beams are made without camber; therefore, in addition to complying with the deflection limits of the local building code, other deflection considerations should be evaluated, such as ponding (positive drainage is essential) and aesthetics.
- Assumes continuous lateral support of the top edge of beam.
- Bearing area to be calculated for specific application.

[1] 16- and 18-in.-deep beams are to be used in multiple-member units only.

TABLE B.5 GANG-LAM LVL DESIGN DATA

Allowable Working Stresses for Beams (Psi)

Gang-Lam Grade	Bending Stress	MOE ($\times 10^6$)	Tension Stress	Compression Stress	Bearing Stress	Shear Stress
Master Plank equal or better	2800^a	2.0	2300	2400^b	550^b	250

Note: Use National Design Specifications' Group III Lumber Values when designing fasteners.
[a] For 12-in. depth; for other depths, multiply by $(12/d)^{1/9}$.
[b] Where ICBO acceptance is required, compression stress = 2200; bearing stress = 410.

Maximum Uniform Load (plf) at 0% Increase for Floors

Beam Span^a (ft)	1 Ply $1\frac{3}{4}'' \times 9\frac{1}{4}''$			1 Ply $1\frac{3}{4}'' \times 11\frac{1}{4}''$			1 Ply $1\frac{3}{4}'' \times 11\frac{7}{8}''$			1 Ply $1\frac{3}{4}'' \times 14''$			1 Ply $1\frac{3}{4}'' \times 16''$		
	Live-Load Deflection		Total Load	Live-Load Deflection		Total Load	Live-Load Deflection		Total Load	Live-Load Deflection		Total Load	Live-Load Deflection		Total Load
	$L/360$	$L/480$	$L/240$	$L/360$	$L/480$	$L/240$	$L/360$	$L/480$	$L/240$	$L/360$	$L/480$	$L/240$	$L/360$	$L/480$	$L/240$
8	588	441	752	998	748	1074	1151	863	1154	1446	1316	1446	1753	1753	1753
10	315	236	472	545	409	696	632	474	771	985	739	1053	1276	1047	1276
12	187	140	280	327	245	483	381	286	535	602	451	731	864	648	940
14	120	90	179	211	158	316	246	184	369	392	294	537	568	426	690
16	81	61	121	143	107	215	167	126	251	269	201	403	392	294	529
18	57	43	86	102	76	152	119	89	178	192	144	287	281	211	417
20	42	31	63	75	56	112	87	66	131	141	106	212	208	156	311
22	32	24	47	56	42	85	66	50	99	107	80	161	158	118	237
24	24	18	37	44	33	65	51	38	77	83	62	125	123	92	184
26	19	15	29	34	26	52	40	30	61	66	49	98	97	73	146

[a] Beam spans shown are clear spans.

TABLE B.6 TJI/25 SP JOIST DESIGN DATA

Residential Roof Span Chart

o.c. Spacing (in.)	Joist Depth (in.)	Non-Snow (125%)		Snow (115%)			
		20-psf Live Load 10-psf Dead Load	20-psf Live Load 20-psf Dead Load	25-psf Live Load 10-psf Dead Load	30-psf Live Load 10-psf Dead Load	40-psf Live Load 10-psf Dead Load	50-psf Live Load 10-psf Dead Load
16	$9\frac{1}{2}$	22'-7"	20'-3"	21'-5"	20'-6"	18'-9"	17'-4"
	$11\frac{7}{8}$	27'-1"	24'-5"	25'-9"	24'-8"	22'-6"	20'-10"
19.2	$9\frac{1}{2}$	21'-2"	19'-1"	20'-1"	19'-3"	17'-7"	16'-3"
	$11\frac{7}{8}$	25'-6"	22'-11"	24'-2"	23'-2"	21'-2"	19'-7"
24	$9\frac{1}{2}$	19'-7"	17'-7"	18'-7"	17'-10"	16'-3"	15'-0"
	$11\frac{7}{8}$	23'-7"	21'-2"	22'-5"	21'-5"	19'-7"	17'-6"

1. Roof joists to be sloped 1/8 in. in 12 in. minimum. No camber provided.
2. Maximum deflection is limited to $L/180$ at total load, and $L/240$ at live load.
3. The spans above apply to roof slopes 6 in. in 12 in. or less. For slopes greater than 6 in. in 12 in. consideration must be given to the increased dead load and deflection caused by actual slope length.

Note: For loads not shown, refer to allowable uniform load chart.
Source: Specifier's Guide to "Silent Floor" Systems. Trus Joist MacMillan. Sept., 1993. pp. 12, 13.

Allowable Uniform Load (lb/lin. ft)

$9\frac{1}{2}$-in. TJI/25 SP Joist

Span		6	8	10	12	14	16	18	20	22	24
Floor	$L/480$ deflection			164	102	67	46	33	24		
	Allowable Total load	273	206	166	138	119	92	66	49		
Sloped roof	115% Snow	313	236	190	158	136	118	88	65	49	38
	125% Non-snow	341	257	207	172	148	123	88	65	49	38

$11\frac{7}{8}$-in. TJI/25 SP Joist

Span		10	12	14	16	18	20	22	24	26	28
Floor	$L/480$ deflection			111	77	56	41	32	25		
	Allowable total load	166	138	119	104	93	83	63	50		
Sloped roof	115% Snow	190	158	136	119	106	95	83	66	52	42
	125% Non-snow	207	172	148	130	116	103	84	66	52	42

1. Load capacity assumes no composite action provided by sheathing.
2. Roof joists to be sloped 1/4 in. in 12 in. minimum. No camber is provided.
3. Deflection is limited to $L/180$ at total load for roofs.
4. For deflection limits of $L/180$, $L/240$, and $L/360$, multiply $L/480$ deflection loads by 2.67, 2.0, and 1.33, respectively.

Note: Trus Joist MacMillan produces other sizes of TJI Joists including TJI/15 SP, D1/25 SP, TJI/35 SP and TJI/55 SP. For properties and allowable spans, *see Specifiers Guide to "Silent Floor" Systems*. Trus Joist MacMillan. Sept, 1993.

TABLE B.7 GNI JOIST DESIGN DATA

Section Properties for GNI 26 and GNI 30 Joists

Joist Size (In.)	Weight (lb/ft)	$EI \times 10^6$ (in^2-lb)	Maximum Resistive Moment (ft-lb)				Maximum Vertical Shear (lb)			
			0% Inc.	5% Inc.	15% Inc.	25% Inc.	33% Inc.	0% Inc.	5% Inc.	15% Inc.	25% Inc.	33% Inc.
$1\frac{1}{2} \times 9\frac{1}{4}$	1.86	143	2980	3129	3427	3725	3963	960	1008	1104	1200	1276
$1\frac{1}{2} \times 11\frac{1}{4}$	2.04	230	3841	4033	4417	4801	5109	1186	1245	1364	1482	1577
$1\frac{1}{2} \times 11\frac{7}{8}$	2.10	262	4112	4318	4729	5140	5469	1256	1318	1444	1570	1670
$1\frac{3}{4} \times 14\frac{1}{4}$	2.58	470	6087	6391	7000	7609	8096	1533	1609	1763	1916	2039
$1\frac{3}{4} \times 15\frac{3}{4}$	2.65	595	6685	7019	7688	8356	8891	1699	1784	1954	2124	2260

Note: Section properties shown are for GNI Joists spaced at 24 in. on center or less.

Maximum Uniform Load (plf) at 0% Increase for Floors Nailed and Glued Sheathing[a]

Span (ft)	$1\frac{1}{2}'' \times 9\frac{1}{4}''$			$1\frac{1}{2}'' \times 11\frac{1}{4}''$			$1\frac{1}{2}'' \times 11\frac{7}{8}''$			$1\frac{3}{4}'' \times 14\frac{1}{4}''$			$1\frac{3}{4}'' \times 15\frac{3}{4}''$		
	Live-Load Deflection		Total Load	Live-Load Deflection		Total Load	Live-Load Deflection		Total Load	Live-Load Deflection		Total Load	Live-Load Deflection		Total Load
	$L/360$	$L/480$	$L/240$	$L/360$	$L/480$	$L/240$	$L/360$	$L/480$	$L/240$	$L/360$	$L/480$	$L/240$	$L/360$	$L/480$	$L/240$
12	132	99	165	203	154	203	215	173	215	263	263	263	291	291	291
14	85	64	127	133	100	165	150	112	176	224	190	224	249	235	249
16	58	43	86	91	68	125	103	77	134	175	131	196	217	163	217
18	41	31	61	64	48	97	73	55	105	126	94	156	157	118	171
20	—	—	—	47	36	71	54	40	81	93	70	126	117	87	138
22	—	—	—	36	27	54	41	30	61	71	53	104	89	67	114
24	—	—	—	—	—	—	31	24	47	55	41	83	69	52	95
26	—	—	—	—	—	—	—	—	—	44	33	65	55	41	81

[a]Joist spans shown include 2 in. bearing at each end. Assumes plywood (or equal) sheathing nailed and glued to top flange of joist.

TABLE B.8a ALLOWABLE LOADS (plf) FOR BEAMS (LATERALLY UNSUPPORTED)—DOUGLAS FIR-LARCH[a]

	For Built-up Beams					Post and Timbers	Beams and Stringers		
	Grade DF-L No. 2 Fv = 95 psi F_b^* = 962.5 psi E = 1.6 × 1E6 psi					DF-L No. 1 P & T Fv = 85 psi F_b^* = 1,200 psi E = 1.6 × 1E6	DF-L No. 1 B & S Fv = 85 psi F_b^* = 1,350 psi E = 1.6 × 1E6		
	Number of 2 × 10's								
Member	(1)	(2)	(3)	(4)	(5)	6 × 8	6 × 10	6 × 12	8 × 12
b =	1.5	3.0	4.5	6.0	7.5	b = 5.5 in.	5.5	5.5	7.5
d =	9.25	9.25	9.25	9.25	9.25	d = 7.5 in.	9.5	11.5	11.5
S =	21.4	42.8	64.2	85.6	107.0	S = 51.6 in.^{3}	82.7	121.2	165.3
I =	98.9	197.9	296.8	395.7	494.7	I = 193.4 in.^{4}	393.0	697.1	950.5
Beam Length (ft)	Allowable Uniformly Distributed Load (plf)					Allowable Load (plf)	Allowable Load (plf)		
4	439	879	1,318	1,758	2,197	1,169	1,480	1,792	2,444
5	352	703	1,055	1,406	1,758	935	1,184	1,434	1,955
6	293	586	879	1,172	1,465	779	987	1,195	1,629
7	237	502	753	1,004	1,255	668	846	1,024	1,396
8	172	422	639	855	1,070	584	740	896	1,222
9	128	332	505	675	845	507	658	796	1,086
10	97	269	408	547	684	410	592	717	978
11	75	221	337	452	566	339	538	652	889
12	59	185	283	379	475	285	493	597	815
13	48	157	241	323	405	242	436	551	752
14	39	135	208	278	349	209	376	512	698
15	32	117	181	242	304	182	327	478	652
16	27	103	159	213	267	160	287	419	577
18	19	81	125	168	211	118	226	230	455
20	14	65	101	136	171	86	175	267	368
22	11	53	83	112	141	65	131	220	304
24	8	44	70	94	118	50	101	179	244
26	7	37	59	80	100	39	79	141	192

[a] Loads shown are *total loads* (dead plus live). Deflection criterion is L/240. Assume that beams are unbraced against lateral deflections along the compression side (top). All beams are simply supported and uniformly loaded ($L_u = L$ and $L_e = 1.63\ L_u + 3d$.) All loads are shown for normal duration. The duration factor = 1.0.

TABLE B.8b ALLOWABLE LOADS (plf) FOR BEAMS (LATERALLY UNSUPPORTED)—SOUTHERN PINE[a]

	For Built-up Beams					Post and Timbers	Beams and Stringers		
	Grade SP No. 2 $Fv = 90$ psi $F_b^* = 1,050$ psi $E = 1.6 \times 1E6$ psi					SP No. 1 SR $Fv = 110$ psi $F_b^* = 1,350$ psi $E = 1.5 \times 1E6$	SP No. 1 SR $Fv = 110$ psi $F_b^* = 1,350$ psi $E = 1.5 \times 1E6$		
	Number of 2 × 10's								
Member	(1)	(2)	(3)	(4)	(5)	6×8	6×10	6×12	8×12
$b =$	1.5	3.0	4.5	6.0	7.5	$b = 5.5$ in.	5.5	5.5	7.5
$d =$	9.25	9.25	9.25	9.25	9.25	$d = 7.5$ in.	9.5	11.5	11.5
$S =$	21.4	42.8	64.2	85.6	107.0	$S = 51.6$ in.[3]	82.7	121.2	165.3
$I =$	98.9	197.9	296.8	395.7	494.7	$I = 193.4$ in.[4]	393.0	697.1	950.5
Beam Length (ft)	Allowable Uniformly Distributed Load (plf)					Allowable Load (plf)	Allowable Load (plf)		
4	416	833	1,249	1,665	2,081	1,513	1,916	2,319	3,163
5	333	666	999	1,332	1,665	1,210	1,533	1,855	2,530
6	278	555	833	1,110	1,388	1,008	1,277	1,546	2,108
7	238	476	714	951	1,189	864	1,095	1,325	1,807
8	178	416	624	833	1,041	721	958	1,160	1,581
9	131	362	550	736	922	570	851	1,031	1,406
10	99	292	334	596	747	361	738	928	1,265
11	77	241	368	392	617	381	610	843	1,150
12	60	201	309	414	518	320	512	748	1,027
13	48	171	263	352	441	272	436	636	874
14	39	147	226	304	380	235	375	548	754
15	32	127	197	264	331	191	327	477	656
16	27	111	173	232	291	157	287	419	576
18	19	87	136	183	230	111	225	330	455
20	14	70	110	148	186	81	164	266	368
22	11	57	91	122	154	61	123	218	298
24	8	47	76	102	127	47	95	168	229
26	7	39	60	80	100	37	75	132	180

[a]Loads shown are *total loads* (dead plus live). Deflection criterion is L/240. Assume that beams are unbraced against lateral deflections along the compression side (top). All beams are simply supported and uniformly loaded ($L_u = L$ and $L_e = 1.63\ L_u + 3d$). All loads are shown for normal duration. The duration factor = 1.0.

TABLE B.8c ALLOWABLE LOADS FOR BEAMS (CONT. LATERALLY SUPPORTED)—DOUGLAS FIR-LARCH[a]

Member	For Built-up Beams					Post and Timbers	Beams and Stringers		
	Grade SP No. 2 Fv = 95 psi F_b^* = 962.5 psi E = 1.6 × 1E6 psi					SP No. 1 SR Fv = 85 psi F_b^* = 1,200 psi E = 1.6 × 1E6	SP No. 1 SR Fv = 85 psi F_b^* = 1,350 psi E = 1.6 × 1E6		
	Number of 2 × 10's								
	(1)	(2)	(3)	(4)	(5)	6 × 8	6 × 10	6 × 12	8 × 12
b =	1.50	3.00	4.50	6.00	7.50	b = 5.5 in.	5.5	5.5	7.5
d =	9.25	9.25	9.25	9.25	9.25	d = 7.5 in.	9.5	11.5	11.5
S =	21.4	42.8	64.2	85.6	107.0	S = 51.6 in.^{3}	82.7	121.2	165.3
I =	98.9	197.9	296.8	395.7	494.7	I = 193.4 in.^{4}	393.0	697.1	950.5
Beam Length (ft)	Allowable Uniformly Distributed Load (plf)					Allowable Load (plf)	Allowable Load (plf)		
4	439	879	1,318	1,758	2,197	1,169	1,480	1,792	2,444
5	352	703	1,055	1,406	1,758	935	1,184	1,434	1,955
6	293	586	879	1,172	1,465	779	987	1,195	1,629
7	251	502	753	1,004	1,255	668	846	1,024	1,396
8	214	429	643	858	1,072	584	740	896	1,222
9	169	339	508	678	847	509	658	796	1,086
10	137	275	412	549	686	413	592	717	978
11	113	227	340	454	567	341	538	652	889
12	95	191	286	381	477	286	493	597	815
13	81	162	244	325	406	244	441	551	752
14	70	140	210	280	350	210	280	512	698
15	61	122	183	244	305	183	331	478	652
16	54	107	161	214	268	161	291	426	581
18	42	85	127	169	212	118	230	337	459
20	34	69	103	137	172	86	175	273	372
22	28	57	85	113	142	65	131	225	307
24	24	48	71	95	119	50	101	179	244
26	20	40	60	80	100	39	79	141	192

[a] Loads shown are *total loads* (dead plus live). Deflection criterion is L/240. Beams are continuously braced against lateral deflections along the compression side (L_u = 0). All beams are simply supported and uniformly loaded. All loads are shown for normal duration. The duration factor = 1.0.

TABLE B.8d ALLOWABLE LOADS FOR BEAMS (CONT. LATERALLY SUPPORTED)—SOUTHERN PINE[a]

	For Built-up Beams					Post & Timbers	Beams & Stringers		
	Grade SPNo. 2 $Fv = 90$psi $F_b^* = 1050$psi $E = 1.6 \times 1E6$psi					SP No. 1 Sr $Fv = 110$psi $F_b^* = 1350$psi $E = 1.5 \times 1E6$	SP No. 1 SR $Fv = 110$psi $F_b^* = 1350$psi $E = 1.5 \times 1E6$		
	Number of 2×10's								
Member	(1)	(2)	(3)	(4)	(5)	6×8	6×10	6×12	8×12
$b =$	1.50	3.00	4.50	6.00	7.50	$b = 5.5$ in.	5.5	5.5	7.5
$d =$	9.25	9.25	9.25	9.25	9.25	$d = 7.5$ in.	9.5	11.5	11.5
$S =$	21.4	42.8	64.2	85.6	107.0	$S = 51.6$ in.^{3}	82.7	121.2	165.3
$I =$	98.9	197.9	296.8	395.7	494.7	$I = 193.4$ in.^{4}	393.0	697.1	950.5
Beam Length (ft)	Allowable Uniformly Distributed Load (plf)					Allowable Load (plf)	Allowable Load (plf)		
4	416	833	1,249	1,665	2,081	1,513	1,916	2,319	3,163
5	333	666	999	1,332	1,665	1,210	1,533	1,855	2,530
6	278	555	833	1,110	1,388	1,008	1,277	1,546	2,108
7	238	476	714	951	1,189	864	1,095	1,325	1,897
8	208	416	624	833	1,041	725	958	1,160	1,581
9	185	370	555	739	924	573	851	1,031	1,406
10	150	299	449	599	749	464	745	928	1,265
11	124	247	371	495	619	384	615	843	1,150
12	104	208	312	416	520	322	517	758	1,033
13	89	177	266	354	443	275	441	646	880
14	76	153	229	306	382	235	380	557	759
15	67	133	200	266	333	191	331	485	661
16	58	117	175	234	292	157	291	426	581
18	46	92	139	185	231	111	225	337	459
20	37	75	112	150	287	81	164	273	372
22	31	62	93	124	155	61	123	218	298
24	25	51	76	102	127	47	95	168	229
26	20	40	60	80	100	37	75	132	180

[a] Loads shown are *total loads* (dead plus live). Deflection criterion is $L/240$. Beams are continuously braced against lateral deflections along the compression side ($L_u = 0$). All beams are simply supported and uniformly loaded. All loads are shown for normal duration. The duration factor = 1.0.

TABLE B.9a BEAM DESIGN TABLES FOR LATERALLY UNBRACED BEAMS: TRIAL MEMBER SELECTION

	2×4	2×6	2×8	2×10	2×12	(2)2×4	(2)2×6	(2)2×8	(2)2×10	(2)2×12	(3)2×4	(3)2×6	(3)2×8	(3)2×10	(3)2×12
b =	1.50	1.50	1.50	1.50	1.50	3.00	3.00	3.00	3.00	3.00	4.50	4.50	4.50	4.50	4.50
d =	3.50	5.50	7.25	9.25	11.25	3.50	5.50	7.25	9.25	11.25	3.50	5.50	7.25	9.25	11.25
L_e for R_b = 10	5.4	3.4	2.6	2.0	1.7	21.4	13.6	10.3	8.1	6.7	48.2	30.7	23.3	18.2	15.0
L_e for R_b = 30	48.2	30.7	23.3	18.2	15.0	192.9	122.7	93.1	73.0	60.0	433.9	276.1	209.5	164.2	135.0
L_e for R_b = 50	133.9	85.2	64.7	50.7	41.7	535.7	340.9	258.6	202.7	166.7	1205.4	767.0	581.9	456.1	375.0

	4×6	4×8	4×10	4×12	4×14	6×8	6×10	6×12	6×14	8×10	8×12	8×14
b =	3.50	3.50	3.50	3.50	3.50	5.50	5.50	5.50	5.50	7.50	7.50	7.50
d =	5.50	7.25	9.25	11.25	13.25	7.50	9.50	11.50	13.50	9.50	11.50	13.50
L_e for R_b = 10	18.6	14.1	11.0	9.1	7.7	33.6	26.5	21.9	18.7	49.3	40.8	34.7
L_e for R_b = 30	167.0	126.7	99.3	81.7	69.3	302.5	238.8	197.3	168.1	444.1	366.8	312.5
L_e for R_b = 50	464.0	352.0	275.9	226.9	192.6	840.3	663.4	548.0	466.8	1233.6	1019.6	868.1

Procedure:
1. Calculate L_e from $L_e = L \times K$. (L_e units are feet.)
2. Select trial beam size and look up range of L_e. If actual L_e is greater than longest given, R_b will be greater than 50 and beam is not acceptable. If actual L_e is near the shortest given, R_b is near 10 and full bending stress is available. This is usually the most economical design. As actual L_e increases, the beam becomes more slender, the allowable bending strength diminishes, and the design becomes less economical.

TABLE B.9b CALCULATION FOR R_b, SLENDERNESS RATIO

L_e ft	in.	2×4	2×6	2×8	2×10	2×12	(2)2×4	(2)2×6	(2)2×8	(2)2×10	(2)2×12	(3)2×4	(3)2×6	(3)2×8	(3)2×10	(3)2×12
4	48	8.6	10.8	12.4	14.0	15.5	4.3	5.4	6.2	7.0	7.7	2.9	3.6	4.1	4.7	5.2
5	60	9.7	12.1	13.9	15.7	17.3	4.8	6.1	7.0	7.9	8.7	3.2	4.0	4.6	5.2	5.8
6	72	10.6	13.3	15.2	17.2	19.0	5.3	6.6	7.6	8.6	9.5	3.5	4.4	5.1	5.7	6.3
7	84	11.4	14.3	16.5	18.6	20.5	5.7	7.2	8.2	9.3	10.2	3.8	4.8	5.5	6.2	6.8
8	96	12.2	15.3	17.6	19.9	21.9	6.1	7.7	8.8	9.9	11.0	4.1	5.1	5.9	6.6	7.3
9	108	13.0	16.2	18.7	21.1	23.2	6.5	8.1	9.3	10.5	11.6	4.3	5.4	6.2	7.0	7.7
10	120	13.7	17.1	19.7	22.2	24.5	6.8	8.6	9.8	11.1	12.2	4.6	5.7	6.6	7.4	8.2
12	144	15.0	18.8	21.5	24.3	26.8	7.5	9.4	10.8	12.2	13.4	5.0	6.3	7.2	8.1	8.9
14	168	16.2	20.3	23.3	26.3	29.0	8.1	10.1	11.6	13.1	14.5	5.4	6.8	7.8	8.8	9.7
16	192	17.3	21.7	24.9	28.1	31.0	8.6	10.8	12.4	14.0	15.5	5.8	7.2	8.3	9.4	10.3
18	216	18.3	23.0	26.4	29.8	32.9	9.2	11.5	13.2	14.9	16.4	6.1	7.7	8.8	9.9	11.0
20	240	19.3	24.2	27.8	31.4	34.6	9.7	12.1	13.9	15.7	17.3	6.4	8.1	9.3	10.5	11.5
22	264	20.3	25.4	29.2	32.9	36.3	10.1	12.7	14.6	16.5	18.2	6.8	8.5	9.7	11.0	12.1
24	288	21.2	26.5	30.5	34.4	37.9	10.6	13.3	15.2	17.2	19.0	7.1	8.8	10.2	11.5	12.6
26	312	22.0	27.6	31.7	35.8	39.5	11.0	13.8	15.9	17.9	19.7	7.3	9.2	10.6	11.9	13.2
28	336	22.9	28.7	32.9	37.2	41.0	11.4	14.3	16.5	18.6	20.5	7.6	9.6	11.0	12.4	13.7
30	360	23.7	29.7	34.1	38.5	42.4	11.8	14.8	17.0	19.2	21.2	7.9	9.9	11.4	12.8	14.1
32	384	24.4	30.6	35.2	39.7	43.8	12.2	15.3	17.6	19.9	21.9	8.1	10.2	11.7	13.2	14.6
34	408	25.2	31.6	36.3	41.0	45.2	12.6	15.8	18.1	20.5	22.6	8.4	10.5	12.1	13.7	15.1
36	432	25.9	32.5	37.3	42.1	46.5	13.0	16.2	18.7	21.1	23.2	8.6	10.8	12.4	14.0	15.5
38	456	26.6	33.4	38.3	43.3	47.7	13.3	16.7	19.2	21.6	23.9	8.9	11.1	12.8	14.4	15.9
40	480	27.3	34.3	39.3	44.4	49.0	13.7	17.1	19.7	22.2	24.5	9.1	11.4	13.1	14.8	16.3
42	504	28.0	35.1	40.3	45.5	50.2	14.0	17.5	20.1	22.8	25.1	9.3	11.7	13.4	15.2	16.7
47	564	29.6	37.1	42.6	48.2	53.1	14.8	18.6	21.3	24.1	26.6	9.9	12.4	14.2	16.1	17.7
52	624	31.2	39.1	44.8	50.6	55.9	15.6	19.5	22.4	25.3	27.9	10.4	13.0	14.9	16.9	18.6
57	684	32.6	40.9	46.9	53.0	58.5	16.3	20.4	23.5	26.5	29.2	10.9	13.6	15.6	17.7	19.5
62	744	34.0	42.6	49.0	55.3	61.0	17.0	21.3	24.5	27.7	30.5	11.3	14.2	16.3	18.4	20.3
67	804	35.4	44.3	50.9	57.5	63.4	17.7	22.2	25.4	28.7	31.7	11.8	14.8	17.0	19.2	21.1
72	864	36.7	46.0	52.8	59.6	65.7	18.3	23.0	26.4	29.8	32.9	12.2	15.3	17.6	19.9	21.9
77	924	37.9	47.5	54.6	61.6	68.0	19.0	23.8	27.3	30.8	34.0	12.6	15.8	18.2	20.5	22.7

TABLE B.9b Continued

L_e ft	in.	4 × 6	4 × 8	4 × 10	4 × 12	4 × 14	6 × 8	6 × 10	6 × 12	6 × 14	8 × 10	8 × 12	8 × 14
4	48	4.6	5.3	6.0	6.6	7.2	3.4	3.9	4.3	4.6	2.8	3.1	3.4
5	48	5.2	6.0	6.7	7.4	8.1	3.9	4.3	4.8	5.2	3.2	3.5	3.8
6	72	5.7	6.5	7.4	8.1	8.8	4.2	4.8	5.2	5.7	3.5	3.8	4.2
7	96	6.1	7.1	8.0	8.8	9.5	4.6	5.1	5.7	6.1	3.8	4.1	4.5
8	120	6.6	7.5	8.5	9.4	10.2	4.9	5.5	6.0	6.5	4.0	4.4	4.8
9	144	7.0	8.0	9.0	10.0	10.8	5.2	5.8	6.4	6.9	4.3	4.7	5.1
10	168	7.3	8.4	9.5	10.5	11.4	5.5	6.1	6.8	7.3	4.5	5.0	5.4
12	192	8.0	9.2	10.4	11.5	12.5	6.0	6.7	7.4	8.0	4.9	5.4	5.9
14	216	8.7	10.0	11.3	12.4	13.5	6.5	7.3	8.0	8.7	5.3	5.9	6.3
16	240	9.3	10.7	12.0	13.3	14.4	6.9	7.8	8.5	9.3	5.7	6.3	6.8
18	264	9.8	11.3	12.8	14.1	15.3	7.3	8.2	9.1	9.8	6.0	6.6	7.2
20	288	10.4	11.9	13.5	14.8	16.1	7.7	8.7	9.6	10.3	6.4	7.0	7.6
22	312	10.9	12.5	14.1	15.6	16.9	8.1	9.1	10.0	10.9	6.7	7.3	8.0
24	336	11.4	13.1	14.7	16.3	17.6	8.5	9.5	10.5	11.3	7.0	7.7	8.3
26	360	11.8	13.6	15.3	16.9	18.4	8.8	9.9	10.9	11.8	7.3	8.0	8.7
28	384	12.3	14.1	15.9	17.6	19.1	9.1	10.3	11.3	12.2	7.5	8.3	9.0
30	408	12.7	14.6	16.5	18.2	19.7	9.4	10.6	11.7	12.7	7.8	8.6	9.3
32	432	13.1	15.1	17.0	18.8	20.4	9.8	11.0	12.1	13.1	8.1	8.9	9.6
34	456	13.5	15.5	17.6	19.4	21.0	10.1	11.3	12.5	13.5	8.3	9.1	9.9
36	480	13.9	16.0	18.1	19.9	21.6	10.3	11.6	12.8	13.9	8.5	9.4	10.2
38	504	14.3	16.4	18.6	20.5	22.2	10.6	12.0	13.2	14.3	8.8	9.7	10.5
40	528	14.7	16.9	19.0	21.0	22.8	10.9	12.3	13.5	14.6	9.0	9.9	10.7
42	552	15.0	17.3	19.5	21.5	23.3	11.2	12.6	13.8	15.0	9.2	10.2	11.0
47	576	15.9	18.3	20.6	22.8	24.7	11.8	13.3	14.6	15.9	9.8	10.7	11.6
52	600	16.7	19.2	21.7	23.9	26.0	12.4	14.0	15.4	16.7	10.3	11.3	12.2
57	624	17.5	20.1	22.7	25.1	27.2	13.0	14.7	16.1	17.5	10.7	11.8	12.8
62	648	18.3	21.0	23.7	26.1	28.4	13.6	15.3	16.8	18.2	11.2	12.3	13.4
67	672	19.0	21.8	24.6	27.2	29.5	14.1	15.9	17.5	18.9	11.7	12.8	13.9
72	696	19.7	22.6	25.5	28.2	30.6	14.6	16.5	18.1	19.6	12.1	13.3	14.4
77	720	20.4	23.4	26.4	29.1	31.6	15.1	17.0	18.7	20.3	12.5	13.7	14.9

Procedure: 1. Calculate L_e from $L_e = L \times K$.
2. Select trial beam size and find R_b opposite L_e. Units for L_e and R_b are inches.

TABLE B.9c CALCULATION FOR F_{bE} ($K_{bE} = 0.438$)

	Modulus of Elasticity, E (×1E6 psi)																				
R_b	0.5	0.6	0.7	0.8	0.9	1.0	1.1	1.2	1.3	1.4	1.5	1.6	1.7	1.8	1.9	2.0	2.1	2.2	2.3	2.4
2	54,750	65,700	76,650	87,600	98,550	109,500	120,450	131,400	142,350	153,300	164,250	175,200	186,150	197,100	208,050	219,000	229,950	240,900	251,850	262,800
4	13,700	16,450	19,150	21,900	24,650	27,400	30,100	32,850	35,600	38,350	41,050	43,800	46,550	49,300	52,000	54,750	57,500	60,250	62,950	65,700
6	6,100	7,300	8,500	9,750	10,950	12,150	13,400	14,600	15,800	17,050	18,250	19,450	20,700	21,900	23,100	24,350	25,550	26,750	28,000	29,200
8	3,400	4,100	4,800	5,500	6,150	6,850	7,550	8,200	8,900	9,600	10,250	10,950	11,650	12,300	13,000	13,700	14,350	15,050	15,750	16,450
10	2,200	2,650	3,050	3,500	3,950	4,400	4,800	5,250	5,700	6,150	6,550	7,000	7,450	7,900	8,300	8,750	9,200	9,650	10,050	10,500
12	1,500	1,850	2,150	2,450	2,750	3,050	3,350	3,650	3,950	4,250	4,550	4,850	5,150	5,500	5,800	6,100	6,400	6,700	7,000	7,300
14	1,100	1,350	1,550	1,800	2,000	2,250	2,450	2,700	2,900	3,150	3,350	3,600	3,800	4,000	4,250	4,450	4,700	4,900	5,150	5,350
16	850	1,050	1,200	1,375	1,550	1,700	1,875	2,050	2,225	2,400	2,575	2,750	2,900	3,075	3,250	3,425	3,600	3,775	3,925	4,100
18	675	800	950	1,075	1,225	1,350	1,475	1,625	1,750	1,900	2,025	2,175	2,300	2,425	2,575	2,700	2,850	2,975	3,100	3,250
20	550	650	775	875	975	1,100	1,200	1,325	1,425	1,525	1,650	1,750	1,850	1,975	2,075	2,200	2,300	2,400	2,525	2,625
22	450	550	625	725	825	900	1,000	1,075	1,175	1,275	1,350	1,450	1,5450	1,625	1,725	1,800	1,900	2,000	2,075	2,175
24	375	450	525	600	675	750	825	925	1,000	1,075	1,150	1,225	1,300	1,375	1,450	1,525	1,600	1,675	1,750	1,825
26	325	390	455	525	575	650	725	775	850	900	975	1,025	1,100	1,175	1,225	1,300	1,350	1,425	1,500	1,550
28	280	335	390	445	505	560	625	675	725	775	850	900	950	1,000	1,050	1,125	1,175	1,225	1,275	1,350
30	245	290	340	390	440	485	525	575	625	675	725	775	825	875	925	975	1,025	1,075	1,125	1,175
32	215	255	300	340	385	430	470	515	555	600	650	675	725	775	825	850	900	950	975	1,025
34	190	225	265	305	340	380	415	455	495	530	575	600	650	675	725	750	800	825	875	900
36	170	205	235	270	305	340	370	405	440	475	500	550	575	600	650	675	700	750	775	800
38	150	180	210	245	275	305	335	365	395	425	450	475	525	550	575	600	625	675	700	725
40	135	165	190	220	245	275	300	330	355	385	400	450	475	500	525	550	575	600	625	650

Procedure: 1. Select wood species and grade to determine E, modulus of elasticity.
2. Enter table vertically with E. Use R_b from table above to read across horizontally from R_b. Find F_{bE} at the intersection.
3. Note that this table is intended only for visually graded or MEL lumber. ($K_{bE} = 0.438$).

TABLE B.9d   CALCULATION FOR C_L

F_{bE}	\multicolumn{15}{c}{F_b^*}
	100	200	300	400	500	600	700	800	900	1000	1100	1200	1400	1600	1800	2000	2200
100	0.817	0.478	0.325	0.246	0.198	0.165	0.142	0.124	0.110	0.099	0.090	0.083	0.071	0.062	0.055	0.050	0.045
125	0.890	0.584	0.403	0.306	0.246	0.206	0.177	0.155	0.138	0.124	0.113	0.104	0.089	0.078	0.069	0.062	0.057
150	0.925	0.678	0.478	0.365	0.294	0.246	0.211	0.185	0.165	0.149	0.135	0.124	0.107	0.093	0.083	0.075	0.068
175	0.945	0.757	0.550	0.422	0.341	0.286	0.246	0.216	0.192	0.173	0.158	0.145	0.124	0.109	0.097	0.087	0.079
200	0.956	0.817	0.617	0.478	0.388	0.325	0.280	0.246	0.219	0.198	0.180	0.165	0.142	0.124	0.110	0.099	0.090
225	0.964	0.860	0.678	0.532	0.433	0.365	0.314	0.276	0.246	0.222	0.202	0.185	0.159	0.139	0.124	0.112	0.102
250	0.969	0.890	0.733	0.584	0.478	0.403	0.348	0.306	0.273	0.246	0.224	0.206	0.177	0.155	0.138	0.124	0.113
275	0.973	0.911	0.779	0.633	0.522	0.441	0.381	0.335	0.299	0.270	0.246	0.226	0.194	0.170	0.151	0.136	0.124
300	0.976	0.925	0.817	0.678	0.564	0.478	0.414	0.365	0.325	0.294	0.268	0.246	0.211	0.185	0.165	0.149	0.135
325	0.979	0.936	0.848	0.720	0.604	0.514	0.446	0.393	0.352	0.318	0.290	0.266	0.229	0.201	0.179	0.161	0.146
350	0.981	0.945	0.871	0.757	0.642	0.550	0.478	0.422	0.377	0.341	0.311	0.286	0.246	0.216	0.192	0.173	0.158
375	0.983	0.951	0.890	0.789	0.678	0.584	0.509	0.450	0.403	0.365	0.333	0.306	0.263	0.231	0.206	0.185	0.169
400	0.984	0.956	0.905	0.817	0.712	0.617	0.540	0.478	0.428	0.388	0.354	0.325	0.280	0.246	0.219	0.198	0.180
425	0.985	0.960	0.916	0.841	0.743	0.649	0.569	0.505	0.453	0.411	0.375	0.345	0.297	0.261	0.233	0.210	0.191
450	0.986	0.964	0.925	0.860	0.771	0.678	0.598	0.532	0.478	0.433	0.396	0.365	0.314	0.276	0.246	0.222	0.202
475	0.987	0.967	0.933	0.876	0.795	0.707	0.626	0.558	0.502	0.456	0.417	0.384	0.331	0.291	0.259	0.234	0.213
500	0.988	0.969	0.939	0.890	0.817	0.733	0.653	0.584	0.526	0.478	0.438	0.403	0.348	0.306	0.273	0.246	0.224
550	0.989	0.973	0.949	0.911	0.853	0.779	0.703	0.633	0.573	0.522	0.478	0.441	0.381	0.335	0.299	0.270	0.246
600		0.976	0.956	0.925	0.879	0.817	0.747	0.678	0.617	0.564	0.518	0.478	0.414	0.365	0.325	0.294	0.268
650		0.979	0.962	0.936	0.899	0.848	0.785	0.720	0.659	0.604	0.556	0.514	0.446	0.393	0.352	0.318	0.290
700		0.981	0.966	0.945	0.914	0.871	0.817	0.757	0.697	0.642	0.593	0.550	0.478	0.422	0.377	0.341	0.311
750		0.983	0.969	0.951	0.925	0.890	0.844	0.789	0.733	0.678	0.629	0.584	0.509	0.450	0.403	0.365	0.333
800		0.984	0.972	0.956	0.934	0.905	0.865	0.817	0.765	0.712	0.662	0.617	0.540	0.478	0.428	0.388	0.354
900		0.986	0.976	0.964	0.947	0.925	0.897	0.860	0.817	0.771	0.724	0.678	0.598	0.532	0.478	0.433	0.396

TABLE B.9d Continued

												F_b^*									
F_{bE}	100	200	300	400	500	600	700	800	900	1000	1100	1200	1400	1600	1800	2000	2200
1,000		0.988	0.980	0.969	0.956	0.939	0.918	0.890	0.856	0.817	0.775	0.733	0.653	0.584	0.526	0.478	0.438
1,100		0.989	0.982	0.973	0.963	0.949	0.932	0.911	0.884	0.853	0.817	0.779	0.703	0.633	0.573	0.522	0.478
1,200			0.984	0.976	0.967	0.956	0.942	0.925	0.905	0.879	0.850	0.817	0.747	0.678	0.617	0.564	0.518
1,400			0.987	0.981	0.974	0.966	0.956	0.945	0.931	0.914	0.894	0.871	0.817	0.757	0.697	0.642	0.593
1,600			0.989	0.984	0.978	0.972	0.965	0.956	0.946	0.934	0.921	0.905	0.865	0.817	0.765	0.712	0.662
1,800				0.986	0.982	0.976	0.971	0.964	0.956	0.947	0.937	0.925	0.897	0.860	0.817	0.771	0.724
2,000				0.988	0.984	0.980	0.975	0.969	0.963	0.956	0.948	0.939	0.918	0.890	0.856	0.817	0.775
2,200				0.989	0.986	0.982	0.978	0.973	0.968	0.963	0.956	0.949	0.932	0.911	0.884	0.853	0.817
2,400					0.987	0.984	0.980	0.976	0.972	0.967	0.962	0.956	0.942	0.925	0.905	0.879	0.850
2,600					0.988	0.985	0.982	0.979	0.975	0.971	0.967	0.962	0.950	0.936	0.919	0.899	0.875
2,800					0.989	0.987	0.984	0.981	0.978	0.974	0.970	0.966	0.956	0.945	0.931	0.914	0.894
3,000						0.988	0.985	0.983	0.980	0.976	0.973	0.969	0.961	0.951	0.939	0.925	0.909
3,500						0.990	0.988	0.986	0.983	0.981	0.978	0.975	0.969	0.962	0.954	0.945	0.934
4,000							0.990	0.988	0.986	0.984	0.982	0.980	0.975	0.969	0.963	0.956	0.948
4,500								0.989	0.988	0.986	0.984	0.983	0.979	0.974	0.969	0.964	0.958
5,000									0.989	0.988	0.986	0.985	0.981	0.978	0.974	0.969	0.964
6,000											0.989	0.988	0.985	0.983	0.980	0.976	0.973
7,000												0.990	0.988	0.986	0.983	0.981	0.978
8,000													0.990	0.988	0.986	0.984	0.982
9,000														0.989	0.988	0.986	0.984
10,000															0.989	0.988	0.986

Procedure: 1. Use selected wood species and grade to determine F_b; modify by appropriate C factors to find F_b^*. Enter table vertically using F_b^*.
2. Use F_{bE} from table above to read across horizontally. Find C_L at the intersection.
3. Allowable bending strength is $F_b^* \times C_L$. (Values of $C_L > 0.99$ are left blank; use $C_L = 1.0$.)

TABLE B.9e CALCULATION FOR C_L

F_{bE}/F_b^*	C_L	F_{bE}/F_b^*	C_L
0.00	0.000	1.70	0.942
0.10	0.099	1.80	0.947
0.20	0.198	1.90	0.952
0.30	0.294	2.00	0.956
0.40	0.388	2.25	0.964
0.50	0.478	2.50	0.969
0.60	0.564	2.75	0.973
0.70	0.642	3.00	0.976
0.80	0.712	3.25	0.979
0.90	0.771	3.50	0.981
1.00	0.817	3.75	0.983
1.10	0.853	4.00	0.984
1.20	0.879	4.50	0.986
1.30	0.899	5.00	0.988
1.40	0.914	5.50	0.989
1.50	0.925	6.00	0.990
1.60	0.934	6.50	0.991
		7.00	0.992

Procedure: 1. Select wood species and grade to determine F_{bE}, and F_b^*.
           2. Calculate the ratio of F_{bE}/F_b^*.
           3. Use F_{bE}/F_b^* to find C_L.
           4. Allowable bending strength is $F_b^* \times C_L$.

TABLE B.10 PRELIMINARY MAXIMUM MOMENTS AND SHEARS*

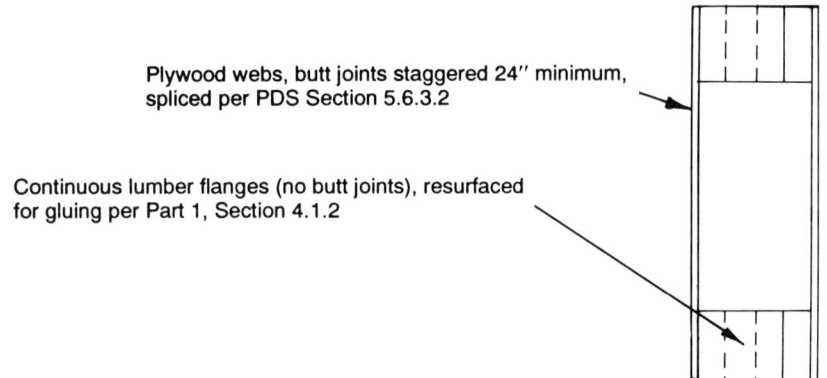

Plywood webs, butt joints staggered 24″ minimum, spliced per PDS Section 5.6.3.2

Continuous lumber flanges (no butt joints), resurfaced for gluing per Part 1, Section 4.1.2

Depth Flange	Max. Moment, $M^{1,2}$ (ft-lb)			Max. Shear, V^1 (lb)	Depth Flange	Max. Moment, $M^{1,2}$ (ft-lb)			Max. Shear, V^1 (lb)
	M_{flange}	$M_{\text{web}}^{3,4}$	M_{total}	$V_{\text{horizontal}}^{5,6}$		M_{flange}	$M_{\text{web}}^{3,4}$	M_{total}	$V_{\text{horizontal}}^{5,6}$
12″ 1-2 × 4	2,375	358	2,733	1,145	30″ 2-2 × 4	17,486	2,309	19,795	3,317
2-2 × 4	4,751	358	5,109	1,159	3-2 × 4	26,229	2,309	28,538	3,381
3-2 × 4	7,126	358	7,484	1,165	4-2 × 4	34,972	2,309	37,281	3,418
16″ 1-2 × 4	3,771	648	4,419	1,612	2-2 × 6	24,358	2,309	26,667	3,189
2-2 × 4	7,543	648	8,191	1,648	3-2 × 6	36,537	2,309	38,846	3,226
3-2 × 4	11,314	648	11,962	1,663	4-2 × 6	48,716	2,309	51,025	3,246
1-2 × 6	4,510	648	5,158	1,486	2-2 × 8	28,417	2,309	30,726	3,051
2-2 × 6	9,019	648	9,667	1,497	3-2 × 8	42,626	2,309	44,935	3,074
3-2 × 6	13,529	648	14,177	1,502	4-2 × 8	56,835	2,309	59,144	3,086
20″ 1-2 × 4	5,217	1,022	6,239	2,073	2-2 × 10	31,270	2,309	33,579	2,898
2-2 × 4	10,434	1,022	11,456	2,135	3-2 × 10	46,904	2,309	49,213	2,910
3-2 × 4	15,652	1,022	16,674	2,162	4-2 × 10	62,539	2,309	64,848	2,917
1-2 × 6	6,646	1,022	7,668	1,950	2-2 × 12	32,693	2,309	35,001	2,768
2-2 × 6	13,291	1,022	14,313	1,978	3-2 × 12	49,039	2,309	51,348	2,773
3-2 × 6	19,937	1,022	20,959	1,990	4-2 × 12	65,385	2,309	67,694	2,776
1-2 × 8	7,193	1,022	8,215	1,845	36″ 2-2 × 4	21,894	3,343	25,237	4,014
2-2 × 8	14,386	1,022	15,408	1,856	3-2 × 4	32,842	3,343	36,185	4,104
3-2 × 8	21,580	1,022	22,602	1,861	4-2 × 44	43,789	3,343	47,132	4,156
24″ 1-2 × 4	6,554	1,465	8,019	2,515	2-2 × 6	31,324	3,343	34,667	3,915
2-2 × 4	13,108	1,465	14,573	2,606	3-2 × 6	46,985	3,343	50,328	3,971
3-2 × 4	19,662	1,465	21,127	2,647	4-2 × 6	62,647	3,343	65,990	4,003
1-2 × 6	8,761	1,465	10,226	2,407	2-2 × 8	37,442	3,343	40,785	3,785
2-2 × 6	17,522	1,465	18,987	2,457	3-2 × 8	56,163	3,343	59,506	3,823
3-2 × 6	26,283	1,465	27,748	2,477	4-2 × 8	74,884	3,343	78,227	3,844
1-2 × 8	9,850	1,465	11,315	2,295	2-2 × 10	42,368	3,343	45,711	3,626
2-2 × 8	19,699	1,465	21,164	2,321	3-2 × 10	63,553	3,343	66,896	3,650
3-2 × 8	29,549	1,465	31,014	2,331	4-2 × 10	84,737	3,343	88,080	3,663
1-2 × 10	10,411	1,465	11,876	2,183	2-2 × 12	45,487	3,343	48,830	3,475
2-2 × 10	20,822	1,465	22,287	2,192	3-2 × 12	68,231	3,343	71,574	3,489
3-2 × 10	31,234	1,465	32,699	2,196	4-2 × 12	90,975	3,343	94,318	3,497

Depth Flange	Max. Moment, $M^{1,2}$ (ft-lb)			Max. Shear, V^1 (lb)	Depth Flange	Max. Moment, $M^{1,2}$ (ft-lb)			Max. Shear, V^1 (lb)
	M_{flange}	$M_{\text{web}}^{3,4}$	M_{total}	$V_{\text{horizontal}}^{5,6}$		M_{flange}	$M_{\text{web}}^{3,4}$	M_{total}	$V_{\text{horizontal}}^{5,6}$
42″ 2-2 × 6	38,362	4,569	42,931	4,632	48″ 2-2 × 6	45,446	5,985	51,431	5,340
3-2 × 6	57,354	4,569	62,112	4,711	3-2 × 6	68,170	5,985	74,155	5,443
4-2 × 6	76,725	4,569	81,294	4,755	4-2 × 6	90,893	5,985	96,878	5,502
2-2 × 8	46,640	4,569	51,209	4,515	2-2 × 8	55,946	5,985	61,931	5,240
3-2 × 8	69,960	4,569	74,529	4,571	3-2 × 8	83,919	5,985	89,904	5,316
4-2 × 8	93,280	4,569	97,849	4,602	4-2 × 8	111,892	5,985	117,877	5,358
2-2 × 10	53,836	4,569	58,405	4,360	2-2 × 10	65,533	5,985	71,518	5,093
3-2 × 10	80,754	4,569	85,323	4,398	3-2 × 10	98,300	5,985	104,285	5,147
4-2 × 10	107,672	4,569	112,241	4,419	4-2 × 10	131,066	5,985	137,051	5,177
2-2 × 12	58,956	4,569	63,525	4,202	2-2 × 12	72,843	5,985	78,828	4,935
3-2 × 12	88,434	4,569	93,003	4,227	3-2 × 12	109,265	5,985	115,250	4,937
4-2 × 12	117,912	4,569	122,481	4,241	4-2 × 12	145,686	5,985	151,471	4,996

1 Basis: Normal duration of load.
   Adjustments. 0.90 for permanent load (over 50 years)
                1.15 for 2 months, as for snow
                1.25 for 7 days
                1.33 for wind or earthquake
                2.00 for impact
2 Basis: F_t of flange = 1000 psi
   Adjustment: $F_t/1000$ for other allowable tension stresses.
                Also see PDS Section 5.7.3 for adjustments
                due to butt joints.
3 Basis: One web effective in bending
   Adjustment: 2.0 for web splices per PDS 5.6.1
4 Basis: $\frac{15}{32}$ in. or $\frac{1}{2}$ in. APA-rated sheathing
         EXP 1 (CDX)
   Adjustments: 0.81 for $\frac{3}{8}$ in.
                0.97 for $\frac{3}{8}$- in. Structural I
                1.19 for $\frac{15}{32}$- or $\frac{1}{2}$-in. Structural I
                1.02 for $\frac{19}{32}$- or $\frac{5}{8}$-in.
                1.51 for $\frac{19}{32}$- or $\frac{5}{8}$-in. Structural I
                1.42 for $\frac{23}{32}$- or $\frac{3}{4}$-in.
                1.84 for $\frac{23}{32}$- or $\frac{3}{4}$-in. Structural I
5 Basis: $\frac{15}{32}$- or $\frac{1}{2}$-in. APA-rated sheathing
         EXP 1 (CDX)
   Note: Adjustments below may in some cases cause rolling
         shear to control final design.
   Adjustments: 0.93 for $\frac{3}{8}$ in.
                1.24 for $\frac{3}{8}$-in. Structural I
                1.80 for $\frac{15}{32}$-in. or $\frac{1}{2}$-in. Structural I
                1.07 for $\frac{19}{32}$ or $\frac{5}{8}$-in.
                2.37 for $\frac{19}{32}$-in. or $\frac{5}{8}$-in. Structural I
                1.49 for $\frac{23}{32}$ or $\frac{3}{4}$-in.
                2.48 for $\frac{23}{32}$-in. or $\frac{3}{4}$-in. Structural I
6 Basis: Plywood edges parallel to face grain glued
         to continuous framing per PDS 3.8.1
   Adjustments: 1.12 for all plywood edges glued to
                framing per PDS 3.8.1
                0.84 for all other conditions.
Source: APA PDS Supplement 2.

APPENDIX C

COLUMN DESIGN TABLES

TABLE C.1 ALLOWABLE COLUMN LOADS (LB): DOUGLAS FIR–LARCH NO. 1[a]

Column Length (ft)	For Studs			For Posts			For built-up Columns							
	$E = 1.7 \times 1E6$			$E = 1.6 \times 1E6$			Nailed $K_f = 0.6$				Bolted $K_f = 0.75$			
Member	(1)2×4	(1)2×6	(1)4×4	(1)6×6	(1)8×8	(2)2×4	(3)2×4	(2)2×6	(3)2×6	(2)2×4	(3)2×4	(2)2×6	(3)2×6
$b =$	1.5	1.5	3.5	5.5	7.5	3.0	4.5	3.0	4.5	3.0	4.5	3.0	4.5
$d =$	3.5	5.5	3.5	5.5	7.5	3.5	3.5	5.5	5.5	3.5	3.5	5.5	5.5
eff.d =	3.5	5.5	3.5	5.5	7.5	3.5	3.5	5.5	5.5	3.5	3.5	5.5	5.5
$F_c^* =$	1668	1595	1523	1000	1000	1668	1668	1595	1595	1668	1668	1595	1595
4	7,250	12,436	15,790	29,194	55,240	8,700	13,049	14,923	22,385	10,875	16,312	18,654	27,981
5	6,168	11,941	13,694	28,511	54,625	7,402	11,103	14,329	21,493	9,253	13,879	17,911	26,867
6	4,997	11,254	11,295	27,586	53,827	5,997	8,995	13,505	20,257	7,496	11,244	16,881	25,321
7	3,983	10,362	9,108	26,373	52,815	4,780	7,170	12,434	18,651	5,975	8,962	15,542	23,314
8	3,192	9,309	7,349	24,841	51,556	3,831	5,746	11,171	16,756	4,789	7,183	13,963	20,945
9	2,594	8,200	5,994	23,015	50,012	3,112	4,669	9,840	14,760	3,890	5,836	12,300	18,450
10	2,139	7,143	4,956	20,989	48,157	2,567	3,851	8,571	12,857	3,209	4,814	10,714	16,071
11	1,790	6,200	4,154	18,906	45,982	2,148	3,223	7,440	11,160	2,686	4,028	9,300	13,950
12	1,518	5,390	3,526	16,902	43,513	1,822	2,733	6,468	9,702	2,277	3,416	8,085	12,128
13	1,302	4,706	3,027	15,061	40,817	1,563	2,344	5,647	8,471	1,953	2,930	7,059	10,588
14	1,129	4,131	2,626	13,421	37,995	1,354	2,032	4,958	7,436	1,693	2,540	6,197	9,296
15		3,648		11,984	35,156			4,378	6,567			5,473	8,209
16		3,241		10,734	32,398			3,889	5,833			4,861	7,292
18		2,600		8,710	27,366			3,120	4,680			3,900	5,850
20		2,128		7,179	23,135			2,553	3,830			3,191	4,787
22		1,771		6,005	19,674			2,125	3,188			2,657	3,985
24					16,866								
26					14,581								

[a] Assume that all 2× studs are braced in the weak direction by the wall sheathing. All loads are shown for normal duration. The duration factor = 1.0. Allowable loads are omitted where L/d exceeds 50.

TABLE C.2 ALLOWABLE COLUMN LOADS (LB): SOUTHERN PINE NO. 1[a]

Column Length (ft)	For Studs			For Posts			For Built-up Columns						
	$E = 1.7 \times 1E6$			$E = 1.5 \times 1E6$			Nailed $K_f = 0.6$				Bolted $K_f = 0.75$		
Member	$(1)2 \times 4$	$(1)2 \times 6$	$(1)4 \times 4$	$(1)6 \times 6$	$(1)8 \times 8$	$(2)2 \times 4$	$(3)2 \times 4$	$(2)2 \times 6$	$(3)2 \times 6$	$(2)2 \times 4$	$(3)2 \times 4$	$(2)2 \times 6$	$(3)2 \times 6$
$b =$	1.5	1.5	3.5	5.5	7.5	3.0	4.5	3.0	4.5	3.0	4.5	3.0	4.5
$d =$	3.5	5.5	3.5	5.5	7.5	3.5	3.5	5.5	5.5	3.5	3.5	5.5	5.5
eff.d =	3.5	5.5	3.5	5.5	7.5	3.5	3.5	5.5	5.5	3.5	3.5	5.5	5.5
$F_c^* =$	1850	1750	1850	825	825	1850	1850	1750	1750	1850	1850	1750	1750
4	7,812	13,556	18,229	24,198	45,677	9,375	14,062	16,267	24,401	11,718	17,578	20,334	30,501
5	6,493	12,943	15,149	23,716	45,238	7,791	11,687	15,531	23,297	9,739	14,608	19,414	29,121
6	5,158	12,091	12,036	23,068	44,671	6,190	9,285	14,509	21,764	7,738	11,606	18,137	27,205
7	4,064	11,000	9,482	22,223	43,958	4,876	7,315	13,199	19,799	6,096	9,143	16,499	24,749
8	3,236	9,752	7,550	21,153	43,076	3,883	5,824	11,702	17,553	4,854	7,280	14,627	21,941
9	2,619	8,489	6,110	19,854	42,000	3,142	4,714	10,187	15,280	3,928	5,892	12,734	19,100
10	2,155	7,238	5,028	18,364	40,708	2,586	3,879	8,794	13,190	3,232	4,848	10,992	16,488
11	1,800	6,320	4,201	16,767	39,186	2,161	3,241	7,584	11,377	2,701	4,051	9,481	14,221
12	1,525	5,470	3,558	15,163	37,433	1,830	2,745	6,565	9,847	2,287	3,431	8,206	12,309
13	1,307	4,761	3,050	13,636	35,477	1,569	2,353	5,714	8,571	1,961	2,941	7,142	10,713
14	1,132	4,171	2,642	12,236	33,368	1,359	2,038	5,005	7,507	1,698	2,548	6,256	9,384
15		3,677		10,983	31,178			4,412	6,619			5,516	8,273
16		3,262		9,876	28,982			3,915	5,872			4,893	7,340
18		2,613		8,057	24,814			3,135	4,703			3,919	5,878
20		2,136		6,663	21,168			2,563	3,844			3,203	4,805
22		1,776		5,586	18,111			2,132	3,197			2,664	3,997
24					15,589								
26					13,516								

[a] Assume that all 2 × studs are braced in the weak direction by the wall sheathing. All loads are shown for normal duration. The duration factor = 1.0. Allowable loads are omitted where L/d exceeds 50.

TABLE C.3 COLUMN CALCULATIONS, L/d^a

Column Effective Length, L_e (in.)		2 × 4	2 × 6	4 × 4	6 × 6	8 × 8
	$b =$	1.50	1.50	3.50	5.50	7.50
	$d =$	3.50	5.50	3.50	5.50	7.50
	eff. $d =$	3.50	5.50	3.50	5.50	7.50
48		13.71	8.73	13.71	8.73	6.40
60		17.14	10.91	17.14	10.91	8.00
72		20.57	13.09	20.57	13.09	9.60
84		24.00	15.27	24.00	15.27	11.20
96		27.43	17.45	27.43	17.45	12.80
108		30.86	19.64	30.86	19.64	14.40
120		34.29	21.82	34.29	21.82	16.00
132		37.71	24.00	37.71	24.00	17.60
144		41.14	26.18	41.14	26.18	19.20
156		44.57	28.36	44.57	28.36	20.80
168		48.00	30.55	48.00	30.55	22.40
180			32.73		32.73	24.00
192			34.91		34.91	25.60
204			37.09		37.09	27.20
216			39.27		39.27	28.80
228			41.45		41.45	30.40
240			43.64		43.64	32.00
252			45.82		45.82	33.60
264			48.00		48.00	35.20
276						36.80
288						38.40
300						40.00
312						41.60
324						43.20
336						44.80

[a]All 2 × studs are braced in the weak direction by the wall sheathing. Values are omitted where L/d greater than 50.

Procedure:
1. Calculation for L/d. Calculate L_e from $L_e = L \times K_e$. L_e is in inches.
2. Select trial member size and enter Table C.3 with member and length, L_e.
3. Find L/d.
4. Go to Table C.4 to find F_{cE}.

TABLE C.4 COLUMN CALCULATIONS, F_{cE}[a]

L/d	\multicolumn{16}{c}{Modulus of Elasticity, $E \times 1{,}000{,}000$ psi}
	0.5	0.6	0.7	0.8	0.9	1.0	1.1	1.2	1.3	1.4	1.5	1.6	1.7	1.8	1.9	2.0
5	6,000	7,200	8,400	9,600	10,800	12,000	13,200	14,400	15,600	16,800	18,000	19,200	20,400	21,600	22,800	24,000
10	1,500	1,800	2,100	2,400	2,700	3,000	3,300	3,600	3,900	4,200	4,500	4,800	5,100	5,400	5,700	6,000
11	1,240	1,488	1,736	1,983	2,231	2,479	2,727	2,975	3,223	3,471	3,719	3,967	4,215	4,463	4,711	4,959
15	667	800	933	1,067	1,200	1,333	1,467	1,600	1,733	1,867	2,000	2,133	2,267	2,400	2,533	2,667
20	375	450	525	600	675	750	825	900	975	1,050	1,125	1,200	1,275	1,350	1,425	1,500
25	240	288	336	384	432	480	528	576	624	672	720	768	816	864	912	960
30	167	200	233	267	300	333	367	400	433	467	500	533	567	600	633	667
35	122	147	171	196	220	245	269	294	318	343	367	392	416	441	465	490
40	94	113	131	150	169	188	206	225	244	263	281	300	319	338	356	375
45	74	89	104	119	133	148	163	178	193	207	222	237	252	267	281	296
50	60	72	84	96	108	120	132	144	156	168	180	192	204	216	228	240

[a]This table is for visually graded and MEL timber only. $K_{cE} = 0.3$.

Procedure:

1. Select the wood species and grade, and find the modulus of elasticity, E ($\times 1{,}000{,}000$ psi).
2. To find F_{cE}, enter table with appropriate E and L/d found in Table C.3.
3. Go to Table C.5 to find C_P.

TABLE C.5 COLUMN CALCULATIONS, $C_P{}^a$

F_{cE}	Allowable Compression Stress, F_c^*													
	100	200	300	400	500	600	700	800	900	1000	1250	1500	1750	2000
50	0.43	0.24	0.16	0.12	0.10	0.08	0.07	0.06	0.05	0.05	0.04	0.03	0.03	0.02
100	0.69	0.43	0.31	0.24	0.19	0.16	0.14	0.12	0.11	0.10	0.08	0.07	0.06	0.05
150	0.81	0.59	0.43	0.34	0.28	0.24	0.20	0.18	0.16	0.15	0.12	0.10	0.08	0.07
200	0.87	0.69	0.54	0.43	0.36	0.31	0.27	0.24	0.21	0.19	0.15	0.13	0.11	0.10
250	0.90	0.76	0.63	0.52	0.43	0.37	0.33	0.29	0.26	0.24	0.19	0.16	0.14	0.12
300	0.92	0.81	0.69	0.59	0.50	0.43	0.38	0.34	0.31	0.28	0.23	0.19	0.16	0.15
350	0.93	0.84	0.74	0.64	0.56	0.49	0.43	0.39	0.35	0.32	0.26	0.22	0.19	0.17
400	0.94	0.87	0.78	0.69	0.61	0.54	0.48	0.43	0.39	0.36	0.30	0.25	0.22	0.19
450	0.95	0.89	0.81	0.73	0.65	0.59	0.53	0.48	0.43	0.40	0.33	0.28	0.24	0.21
500	0.95	0.90	0.83	0.76	0.69	0.63	0.57	0.52	0.47	0.43	0.36	0.31	0.27	0.24
600	0.96	0.92	0.87	0.81	0.75	0.69	0.64	0.59	0.54	0.50	0.42	0.36	0.31	0.28
700	0.97	0.93	0.89	0.84	0.79	0.74	0.69	0.64	0.60	0.56	0.47	0.41	0.36	0.32
800	0.97	0.94	0.91	0.87	0.82	0.78	0.74	0.69	0.65	0.61	0.52	0.46	0.40	0.36
900	0.98	0.95	0.92	0.89	0.85	0.81	0.77	0.73	0.69	0.65	0.57	0.50	0.44	0.40
1,000	0.98	0.95	0.93	0.90	0.87	0.83	0.80	0.76	0.73	0.69	0.61	0.54	0.48	0.43
1,100	0.98	0.96	0.94	0.91	0.88	0.85	0.82	0.79	0.76	0.72	0.65	0.58	0.52	0.47
1,200	0.98	0.96	0.94	0.92	0.89	0.87	0.84	0.81	0.78	0.75	0.68	0.61	0.55	0.50
1,400	0.99	0.97	0.95	0.93	0.91	0.89	0.87	0.84	0.82	0.79	0.73	0.67	0.61	0.56
1,600	0.99	0.97	0.96	0.94	0.92	0.91	0.89	0.87	0.85	0.82	0.77	0.71	0.66	0.61
1,800	0.99	0.98	0.96	0.95	0.93	0.92	0.90	0.89	0.87	0.85	0.80	0.75	0.70	0.65
2,000	0.99	0.98	0.97	0.95	0.94	0.93	0.91	0.90	0.88	0.87	0.82	0.78	0.74	0.69
2,500	0.99	0.98	0.97	0.96	0.95	0.94	0.93	0.92	0.91	0.90	0.87	0.83	0.80	0.76
3,000	0.99	0.99	0.98	0.97	0.96	0.95	0.95	0.94	0.93	0.92	0.89	0.87	0.84	0.81
3,500	0.99	0.99	0.98	0.98	0.97	0.96	0.95	0.95	0.94	0.93	0.91	0.89	0.87	0.84
4,000	0.99	0.99	0.98	0.98	0.97	0.97	0.96	0.95	0.95	0.94	0.92	0.91	0.89	0.87
4,500	1.00	0.99	0.99	0.98	0.98	0.97	0.97	0.96	0.95	0.95	0.93	0.92	0.90	0.89
5,000	1.00	0.99	0.99	0.98	0.98	0.97	0.97	0.96	0.96	0.95	0.94	0.93	0.91	0.90
6,000	1.00	0.99	0.99	0.99	0.98	0.98	0.97	0.97	0.97	0.96	0.95	0.94	0.93	0.92
7,000	1.00	0.99	0.99	0.99	0.99	0.98	0.98	0.98	0.97	0.97	0.96	0.95	0.94	0.93
8,000	1.00	0.99	0.99	0.99	0.99	0.98	0.98	0.98	0.98	0.97	0.97	0.96	0.95	0.94
10,000	1.00	1.00	0.99	0.99	0.99	0.99	0.99	0.98	0.98	0.98	0.97	0.97	0.96	0.95
12,500	1.00	1.00	1.00	0.99	0.99	0.99	0.99	0.99	0.98	0.98	0.98	0.97	0.97	0.96
15,000	1.00	1.00	1.00	0.99	0.99	0.99	0.99	0.99	0.99	0.99	0.98	0.98	0.97	0.97
17,500	1.00	1.00	1.00	1.00	0.99	0.99	0.99	0.99	0.99	0.99	0.99	0.98	0.98	0.98
20,000	1.00	1.00	1.00	1.00	0.99	0.99	0.99	0.99	0.99	0.99	0.99	0.98	0.98	0.98

[a] This table is for visually graded and MEL timber only; $c = 0.8$.

Procedure:
1. For selected material calculate F_c^* by multiplying F_c by C Factors.
2. To find the column stability factor, C_P, enter the table with F_c^* and F_{cE} found in Table C.4.

TABLE C.6 COLUMN CALCULATIONS, C_P

	c Values		
F_{cE}/F_c^*	For Sawn Lumber 0.8	Round Piles 0.85	Glued Laminated 0.9
0.1	0.098	0.098	0.099
0.2	0.191	0.193	0.195
0.3	0.278	0.283	0.288
0.4	0.360	0.368	0.377
0.5	0.434	0.446	0.461
0.6	0.500	0.517	0.538
0.7	0.559	0.580	0.607
0.8	0.610	0.635	0.667
0.9	0.653	0.681	0.718
1.0	0.691	0.721	0.760
1.2	0.750	0.781	0.822
1.4	0.793	0.824	0.862
1.6	0.825	0.854	0.889
1.8	0.849	0.876	0.908
2.0	0.867	0.892	0.921
2.2	0.882	0.905	0.932
2.4	0.894	0.915	0.940
2.8	0.912	0.931	0.951
3.2	0.925	0.941	0.959
3.6	0.934	0.949	0.965

Procedure:
1. Calculate F_{cE} and F_c^*.
2. Calculate F_{cE}/F_c^* and enter table to find C_P.

APPENDIX D

MOMENT CONNECTIONS

Figure D.1 shows the forces on a bolted connection with an applied moment. To determine the capacity of this connection we may replace the applied force with an equivalent force acting through the centroid of the connection and a pure moment equal to the applied force times the eccentricity measured to the centroid of the connection ($M = P \times e$). This allows us to consider the moment independently from the applied lateral load. Each bolt can be analyzed to determine its contribution to the total moment. From that analysis we will find the moment-induced force at each bolt. The total force at each bolt is the force due to the moment plus the force due to the separately considered lateral load. The force due to moment and the lateral force are added vectorially. Each total bolt force is compared with the allowable design value as calculated from the NDS.

The analysis of the moment capacity of a bolted group is based on a few assumptions related to elastic analysis of the connection. These assumptions are all acceptable for bolt sizes shown in the NDS tables.

1. The connection is rigid. Specifically, the side plates form a rigid plate.
2. The rigid plate rotates about a center of rotation and that center is the centroid of the connection group. This implies that each bolt will sweep through a distance that is proportional to the bolts distance (or radius) from the centroid. If the bolts are all tight-fitting, they can only move by deforming the wood around them; therefore, the deformation at each bolt is proportional to its radius from the centroid.
3. The force at each bolt is proportional to the deformation. The bolt strengths are based on the load-slip ratio being linear.
4. The force at each bolt acts perpendicular to the radius from the centroid. Each bolt produces a moment equal to the force times the distance

352 MOMENT CALCULATIONS

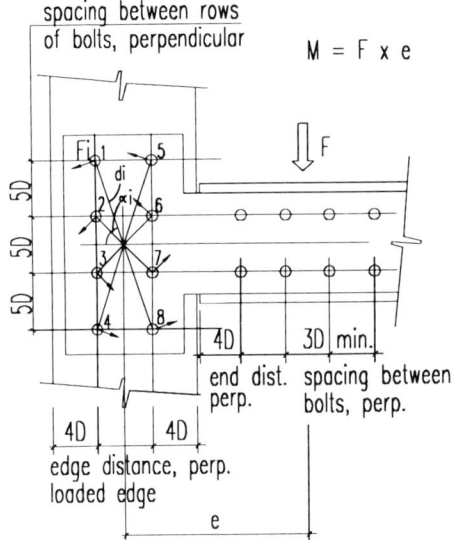

FIGURE D.1 Eccentric bolt connection.

to the centroid, and the total moment at the connection equals the sum of each bolt's moment.

For bolt patterns that are not biaxially symmetrical, the centroid must first be calculated using the standard procedures for calculating the centroid of area. Select a reference point, such as the top bolt, and calculate

$$\bar{y} = \frac{\Sigma y_i}{n}$$

where y_i is the distance from the reference point to bolt i, and n is the number of bolts. Similarly,

$$\bar{x} = \frac{\Sigma x_i}{n}$$

The relationship in assumption 3 above can be written algebraically as

$$\frac{F_1}{d_1} = \frac{F_2}{d_2} = \frac{F_3}{d_3} \cdots \frac{F_i}{d_i} = \cdots = \frac{F_n}{d_n}$$

We can rewrite this so that the force at any bolt is expressed in terms of the force at one selected bolt, for example F_n at bolt n.

$$F_1 = F_n \times \frac{d_1}{d_n}, \quad F_2 = F_n \times \frac{d_2}{d_n},$$

$$F_i = F_n \times \frac{d_i}{d_n}, \quad \text{etc.}$$

The relationships in assumption 4 are written as

$$m_i = F_i \times d_i$$

$$M = \sum_{i=1}^{n} m_i$$

$$= \sum_{i=1}^{n} F_i \times d_i \text{ where } n = \text{number of bolts}$$

Replace F_i with the term containing F_n and gather constants.

$$M = \frac{F_n}{d_n} \times \sum_{i=1}^{n} d_i^2$$

We may rearrange this expression to find the force in bolt n due to the applied moment, M.

$$F_n = \frac{M \times d_n}{\sum_{i=1}^{n} d_i^2}$$

The denominator, $\sum_{i=1}^{n} d_i^2$, is referred to as the polar moment of inertia.

After having discovered the effect of the moment, consider separately the effect of the lateral load. Since we have applied it at the centroid of the connection, it has no rotational effect. The assumption about the rigid plate requires that all the bolts will carry an equal share of the lateral load.

$$F_n = \frac{F_{\text{lateral}}}{n}$$

The total force for any bolt is the sum of the moment force and the lateral force, added vectorially. For convenience the equations above may be written for each component of the force, parallel to grain and perpendicular to grain.

$$F_{n,x} = \frac{M \times y_n}{\sum_{i=1}^{n} \sqrt{x_i^2 + y_i^2}} + \frac{P_{\text{lateral}}}{n}$$

$$F_{n,y} = \frac{M \times x_n}{\sum_{i=1}^{n} \sqrt{x_i^2 + y_i^2}} + \frac{Q_{\text{lateral}}}{n}$$

$$F_n = \sqrt{F_x^2 + F_y^2}$$

Example D.1 demonstrates the use of these equations.

EXAMPLE D.1—ANALYSIS OF BOLT MOMENT CAPACITY

Determine the capacity of the bolt group shown in Figure D.2 to carry the applied load of 3500 lb. Assume that the connection uses $\tfrac{7}{8}$-in.-diameter bolts in $5\tfrac{1}{2}$-in.-thick Douglas Fir and that metal side plates are used. Recommended spacings should be used. Because some bolts in the group will be loaded close to parallel to grain, and others perpendicular to grain, the maximum spacing in either direction is used.

Determine bolt spacing as follows. The ratio L/D is greater than 6 in. (5.5 in. over $\tfrac{7}{8}$ in. equals 6.25 in.). The bolts farthest from the centroid will carry the greatest load. For these bolts the load is basically perpendicular to grain. The distance between rows of bolts for this condition is given as $5D$, or 4.375 in. Let us use 4.5 in. The spacing perpendicular to grain is normally determined by required spacing in the side pieces, which in this case are steel. The maximum spacing between bolts should be used and this value can be found as the width of the main wood piece minus the edge distances at either edge. As the plate attempts to rotate, the left edge above the centroid is the loaded edge, but the right edge is the loaded edge below the centroid. The value for the loaded edge distance, $4D$ or $3\tfrac{1}{2}$ in., is taken at both sides. The remaining spacing between bolt is $11\tfrac{1}{2}$ in. minus 7 in., or $4\tfrac{1}{2}$ in.

Since the bolt group is bilaterally symmetrical, the centroid can be found by inspection to be at the center of the group. The eccentricity of the load is calculated to be 16 in. The moment is 3500 lb times 16 in., or 56,000 in.-lb.

We use the formulas developed above to calculate the values in the chart.

For $\tfrac{7}{8}$-in.-diameter bolts in $5\tfrac{1}{2}$-in.-thick Douglas Fir with steel side plates and double shear, use NDS Table 8.3B. Assume a normal duration of load. The design values parallel and perpendicular to the grain are $Z_{S\perp}$ or $Z_{m\perp}$.

$$Z_\parallel = 4260 \text{ lb}$$

$$Z_\perp = 2310 \text{ lb}$$

Find P_α from Hankinson's formula. Find F_i from the equation

$$F_i = \frac{M \times d_i}{\sum_{i=1}^{n} d_i^2}$$

The values in Table D.1 consider only the moment. Because the F_i values were calculated using the total applied moment, the sum of the moments will always equal the applied moment. The sum is used in the table to check the math. The column showing P_α is included for comparison with the F_i column. Although this is not necessary at this point, it serves as a quick check to determine if any bolt capacity is nearing its maximum. The four outer bolts are loaded to 68% (1640 divided by 2431) of their capacity, based on moments only.

TABLE D.1 ECCENTRIC MOMENT CALCULATIONS

Bolt Number	X_i	Y_i	d_i	Angle α	P_α	d_i^2	F_i	$m_i = F_i \times d_i$
1	2.25	6.75	7.115	71.57°	2431	50.62	1640	11,667
2	2.25	2.25	3.182	45°	3004	10.12	733	2,333
3	2.25	2.25	3.182	45°	3004	10.12	733	2,333
4	2.25	6.75	7.115	71.57°	2431	50.62	1640	11,667
5	2.25	6.75	7.115	71.57°	2431	50.62	1640	11,667
6	2.25	2.25	3.182	45°	3004	10.12	733	2,333
7	2.25	2.25	3.182	45°	3004	10.12	733	2,333
8	2.25	6.75	7.115	71.57°	2431	50.62	1640	11,667
						243		56,000

The next step is to add the horizontal load component at every bolt and to check if all bolts are loaded within their allowable limits. By referring to the diagram of bolt forces, we can see that bolts 4 and 8 will carry the greatest load when the moment and lateral components are added. For either bolt the force due to moment resolves into the components

$$F_x = F_1 \times \sin \alpha = 1640 \times \sin(71.5°) = 1556 \text{ lb}$$

$$F_y = F_1 \times \cos \alpha = 1640 \times \cos(71.5°) = 519 \text{ lb}$$

Each bolt will carry one-eighth of the horizontal load,

$$F_{x,\text{lateral}} = \frac{F}{n} = \frac{3500}{8} = 437.5 \text{ lb}$$

$$F_{x,1} = F_x + F_{x,\text{lateral}} = 1556 + 437.5$$
$$= 1993.5 \text{ lb}$$

$$F_1 = \sqrt{F_x^2 + F_y^2} = \sqrt{1993.5^2 + 519^2}$$
$$= 2056 \text{ lb}$$

The new angle, t, is arctan $(1993.5/519) = 75.4°$. Using this angle to check the allowable bolt load yields

$$P_{75.4°} = \frac{4260 \times 2320}{4260 \times \sin^2(75.4°) + 2320 \times \cos^2(75.4°)}$$
$$= 2389 \text{ lb}$$

The actual load of 2056 lb is less than the allowable load 2389 lb. Bolts 4 and 8 are satisfactory and therefore the connection is acceptable. Table D.2 verifies our assessment that bolts 4 and 8 control. The outer bolts, 1, 4, 5 and 8, carry the same amount of moment and combined account for about 83% of the total moment.

Use eight $\frac{7}{8}$-in.-diameter bolts.

Design of Bolt Group to Carry Moment. A bolt group can be readily analyzed to determine its moment capacity. However, the design of a grouping to meet a required load does not present a convenient closed-form solution. It is often a trial-and-error procedure in which the number of rows, number of bolts, and bolt size and spacing must be revised to arrive at a workable solution. The following discussion is included to provide lower and upper bounds for such a design. This procedure is used in Example D.2.

To understand the forces being developed in the connection, it will be helpful to look at a simplified case consisting of one vertical row of bolts. In this situation all the bolts will produce a load that is perpendicular to the grain. The maximum

MOMENT CALCULATIONS

TABLE D.2 ECCENTRIC BOLT FORCES

Bolt Number	F_i	Angle α	$F_{y,i}$	$F_{x,i}$	$F_{x,i} + F_{x,\text{lat}}$	$F_{i,\text{total}}$	Angle α	P_α	Percent stress
1	1640	71.56°	−519	−1556	−1118	1232	65.1°	2533	49%
2	733	45°	−519	−519	−81	525	9°	4177	13%
3	733	45°	−519	519	956	1088	61.5°	2588	42%
4	1640	71.56°	−519	1556	1993	2059	75.4*	2389	86%
5	1640	71.56°	519	−1556	−1118	1232	65.1°	2533	49%
6	733	45°	519	−519	−81	525	9°	4177	13%
7	733	45°	519	519	956	1088	61.5°	2588	42%
8	1640	71.56°	519	1556	1993	2059	75.4°	2389	86%
Sum			0	0	3500				OK!

bolt capacity at the farthest bolt is the tabulated design value times any modification factors. In the equation above we may replace $F_{\max}$ with $Q_{\max}$ to represent this value.

Replacing F_i in the summation results in

$$M_{\text{total}} = \frac{Q_{\max}}{r_{\max}} \times \sum_{i=1}^{n} (r_i^2)$$

The distance r_i to each bolt is a function of the spacing between rows of bolts, which we may designate as s. Rewriting this equation and simplifying it results in the relationship

$$M_{\text{total}} = \frac{Q_{\max} \times s \times (n)(n+1)}{6}$$

where $Q_{\max}$ is the design value, s the spacing between bolts, and n the number of bolts in the row. Bolts in a single row will only develop forces perpendicular to grain. This arrangement will be the least efficient arrangement. The relationship can be rewritten to solve for n, the upper bound of number of bolts, by solving the quadratic,

$$n_{\text{upper bound}} = \frac{-1 + \sqrt{1 + 24M/Q \times s}}{2}$$

By the same logic a lower bound can be arrived at in which all the bolts will be aligned in one horizontal row so that they develop forces parallel to the grain.

$$n_{\text{lower bound}} = \frac{-1 + \sqrt{1 + 24M/P \times s}}{2}$$

In using these two equations, note that the value s may not be the same in each equation since it refers to the parallel-to-grain distance between bolts in the first equation, and the perpendicular distance in the second. Similarly, the value M, the moment, will not be the same in each case since the moment equals the applied force times the distance from the force to the centroid of the bolt group.

A bolted connection group that uses only one row of bolts will require a longer length of connection than one using a number of rows, so is often not practical. Comparing a single row to multiple rows demonstrates that in the single row, although the moment arm to the farthest bolt will increase, only two bolts will develop the full strength. Also, if the load is perpendicular to the grain, a single row will not make use of the increased strength from parallel-to-grain loading, whereas using more rows of bolts will result in the possibility of some bolts near the centroid

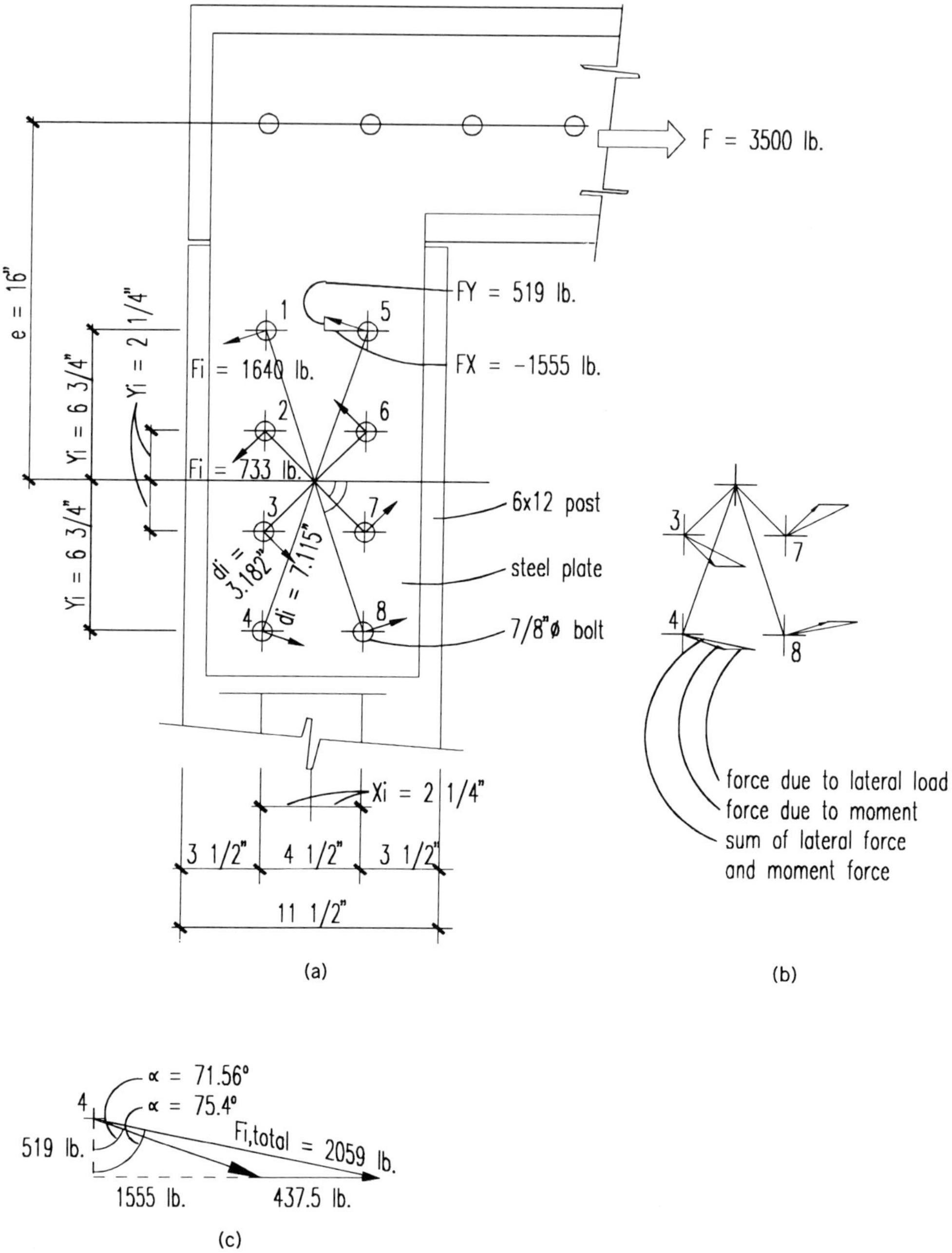

356 MOMENT CALCULATIONS

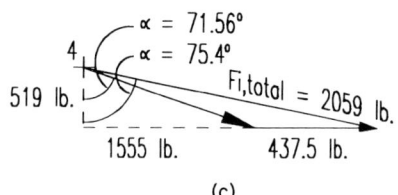

FIGURE D.2 Eccentric post to beam connection.

developing full parallel-to-grain loading (*P*). Example D.2 will explore this trade-off.

EXAMPLE D.2—DESIGN OF BOLT GROUP TO CARRY MOMENT.

Design the beam-to-column connection for the frame in Figure D.3, taking the moment into account. The connection carries a horizontal force, due to wind loads, of 3500 lb. Assume that $\frac{7}{8}$-in.-diameter bolts are acceptable. Assume that steel side plates are used and that the beam is an 8 × 12 and that the column is an 8 × 8 Douglas Fir–Larch.

The moment is dependent on the eccentricity, the distance between the center of the bolt group on the column, and the center of the bolt group on the beam. To establish the eccentricity, we must make some good guesses about the number of bolts in the column and in the beam and about the spacing requirements.

Start with the spacings for the column. Edge distance for the loaded edge of the post is 4*D* or $3\frac{1}{2}$ in. This must be applied to both sides, so only one row of bolts in the post is possible. The bolts will all be loaded perpendicular to grain. The minimum end distance for the post, loaded perpendicular to the grain, is 4*D* or $3\frac{1}{2}$ in., assuming full load on these bolts. The minimum spacing between bolts in a row is 3*D* or $1\frac{3}{8}$ in.; we will use 3 in. Increasing this spacing increases all the moment arms of the bolts and thereby increases the resistance. It also increases the eccentricity and thereby the overall moment, but the eccentricity goes up at a slower rate. It is advantageous to increase these spacings. Let us guess that six bolts will be needed. The distance from the top of the post to the center of the bolt group can be established as $3\frac{1}{2}$ in. plus $2\frac{1}{2}$ spacings of 3 in. each, for a total of 11 in.

Next turn to the beam. A $\frac{7}{8}$-in.-diameter bolt, parallel to the grain, in $7\frac{1}{2}$-in.-thick Douglas Fir–Larch can carry 4260 lb (NDS Table 8.3B). This is increased by the wind duration factor of 1.6, so

$$Z_\parallel = P = 1.6 \times 4260 \text{ lb} = 6816 \text{ lb}$$

The number of bolts required in the beam is

$$N = \frac{3500}{6816} = 0.513$$

If we use one bolt, it will be loaded to 51% of its allowable load. The reduced required end distance on the beam, 3.5*D*, can be used, so the bolt in the beam will line up directly over the bolts in the column. Using one bolt will also induce no moment into the beam. Place the bolt at the midpoint of the beam. The resulting eccentricity is found as 11 in. plus $5\frac{3}{4}$ in. equals $16\frac{3}{4}$ in. The moment on the connection is

$$M = F \times \text{ecc.} = 3500 \text{ lb} \times 16\frac{3}{4} \text{ in.}$$
$$= 58{,}625 \text{ lb-in.}$$

The bolts in the post will be loaded perpendicular to grain. Their capacity, including the wind duration increase, is

$$Z_\perp = 2320 \text{ lb} \times 1.6 = 3712 \text{ lb}$$

We can now solve for the required number of connectors.

$$M_{\text{total}} = \frac{Q_{\max} \times s \times (n)(n+1)}{6}$$

$$n = \frac{-1 \pm \sqrt{1 + 24M/Qs}}{2}$$

$$= \frac{-1 \pm \sqrt{1 + \dfrac{(24 \times 58{,}625 \text{ lb-in.})}{(3712 \text{ lb} \times 3 \text{ in.})}}}{2}$$

$$= \frac{-1 \pm \sqrt{126.3}}{2} = \frac{-1}{2} + \frac{11.24}{2}$$

$$= 5.12$$

If six bolts are used, the top and bottom bolts will be loaded the most. They will carry 5.12/6, or 85% of the allowable load of 3712 lb. Their load from the moment will be 3167 lb perpendicular to the grain.

We must consider the perpendicular-to-grain loading of the 3500 lb directly on the bolts. Each bolt carries an equal share:

$$Q = \frac{3500 \text{ lb}}{6} = 583.3 \text{ lb}$$

This amount is added to the amount due to the moment. The total load on the bottom bolt is the greatest. It must carry 3167 lb plus 583.3 lb, or a total of 3751. It is loaded to about 1% over the total capacity. It should be clear that six bolts is the minimum to suffice. The design is shown in Figure D.2.

Use six $\frac{7}{8}$-in.-diameter bolts with steel side plates.

This example shows the approach of determining the upper bounds to determining the number of bolts needed. The lower bound could be established using the lower-bound formula. This would require a wider column since more than one row of bolts would be needed. Although we can quickly set the limits to the problem, there remains some finesse in choosing an appropriate design that comes with experience.

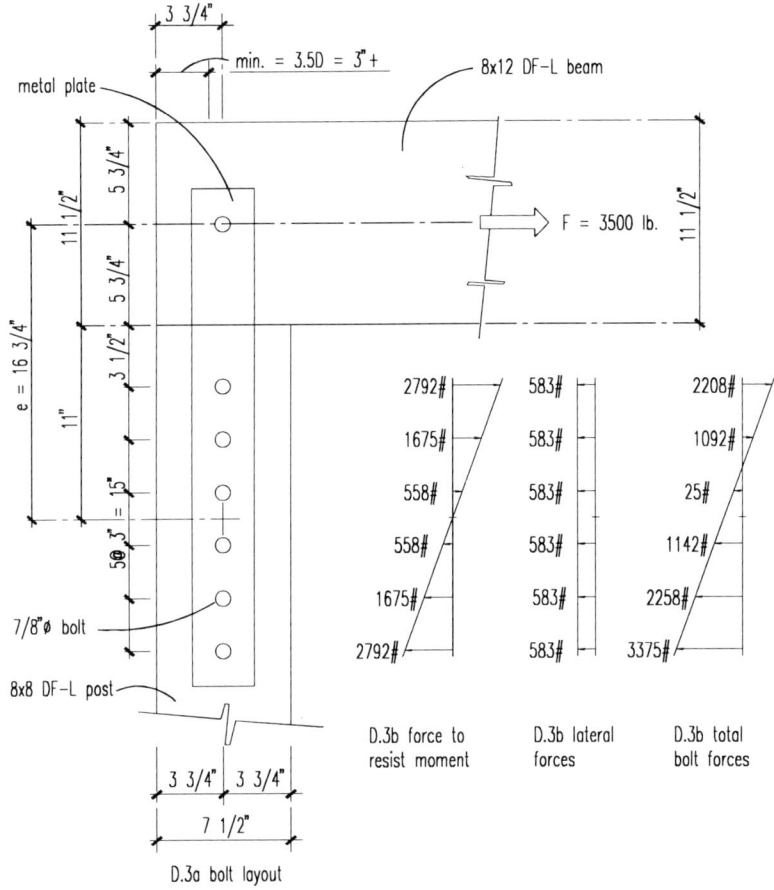

FIGURE D.3 Bolt group to carry moment.

APPENDIX E

WOOD SPECIES AND SPECIFIC GRAVITIES

TABLE E.1 SPECIES AND SPECIFIC GRAVITIES

Species Combination	Specific Gravity G^1	Species Group for Split Rings and Shear Plates (1991)	Species Group for Screws, Nail, and Lag Bolts (Pre-1991)
Aspen	0.39	D	IV
Balsam Fir	0.36	D	IV
Beech–Birch–Hickory	0.71	A	I
Coast Sitka Spruce	0.39	D	IV
Cottonwood	0.41	D	IV
Douglas Fir–Larch	0.50	B	II
Douglas Fir–Larch (Dense)		A	II[5]
Douglas Fir–Larch (North)	0.49	B	II
Douglas Fir–South	0.46	C	III
Eastern Hemlock	0.41	D	III
Eastern Hemlock–Tamarack	0.41	D	III
Eastern Hemlock–Tamarack (North)	0.47	C	III
Eastern Softwoods	0.36	D	III
Eastern Spruce	0.41	D	IV
Eastern White Pine	0.36	D	IV
Engelman Spruce–Lodgepole Pine[2]			
MSR 160f and higher grades	0.46	B[6]	III[3]
MSR 1500f and lower grades	0.38	C[6]	III[3]
Hem–Fir	0.43	C	III
Hem–Fir (North)	0.46	C	III
Mixed Maple	0.55	B	I
Mixed Oak	0.68	A	I
Mixed Southern Pine	0.51	B	II[4]
Mountain Hemlock	0.47	C	III
Northern Pine	0.42	C	III
Northern Red Oak	0.68	A	I
Northern Species	0.35	D	IV
Northern White Cedar	0.31	D	IV
Ponderosa Pine	0.43	C	III
Red Maple	0.58	B	I[4]
Red Oak	0.67	A	I
Red Pine	0.44	C	III
Redwood, close grain	0.44	C	III
Redwood, open grain	0.37	D	III
Sitka Spruce	0.43	C	III
Southern Pine	0.55	B	II[7]
Southern Pine (Dense)		A	II[5]
Spruce–Pine–Fir	0.42	C	III
Spruce–Pine–Fir (South)	0.36	D	III
Western Cedars	0.36	D	IV
Western Cedars (North)	0.35	D	IV
Western Hemlock	0.47	C	III
Western Hemlock (North)	0.46	C	III[4]
Western White Pine	0.40	D	IV
Western Woods	0.36	D	IV
White Oak	0.73	A	I
Yellow Poplar	0.43	C	III

1. Specific gravity based on weight and volume when oven-dry.
2. Applies only to Engelmann Spruce–Lodgepole Pine machine stress rated (MSR) structural lumber.
3. Englemann Spruce is group IV, Lodgepole Pine is group III. The combination is IV.
4. Specie not listed in NDS 1986 Table 8.1A. Placement based on specific gravity.
5. Dense material is defined by rings per inch, not by specific gravity. NDS Table 8.1A does not list dense species. The value for medium-grain material is shown.
6. Englemann Spruce (MSR) not listed in NDS 1992 Table 10A. Placement based on specific gravity.
7. Coarse-grain Southern Pine, as used in some glued-laminated timber combinations, is in group C.

Sources: NDS Tables 8A and 10A (1992) and NDS Table 8.1A (1986).

APPENDIX F

REQUIRED CONNECTORS FOR UPLIFT, OVERTURNING, AND SLIDING

TABLE F.1 REQUIRED CONNECTIONS FOR 24-FT-WIDE BUILDING WITH 6-IN-12 ROOF PITCH (EXTERIOR ONLY, FRAMING MEMBERS AT 16 IN. O.C.)

Wind velocity (mph)	70	80	90	100	110	120
Wind pressure (psf)	16	21	27	33	40	48

UPLIFT/OVERTURNING (pounds per connector)

	70	80	90	100	110	120
Roof to wall plate	19.2	62.2	113.8	165.4	225.6	294.4
Wall plate to stud	19.2	62.2	113.8	165.4	225.6	294.4
Sill plate to 2nd floor	(38.0)	6.0	58.9	111.7	173.3	243.8
2nd floor to wall plate	(96.3)	(48.8)	8.1	65.1	131.5	207.5
Sill plate to 1st floor	(33.8)	78.1	232.7	387.3	567.7	773.8
1st floor to foundation	(83.4)	8.7	119.1	229.6	358.4	505.7

Required Connector

Location	Recom'd	70	80	90	100	110	120
Roof to wall plate	3-16 toenail	1-16d toenail	1-16d toenail	2-16d toenail	2-16d toenail	3-16d toenail	Tydown sr
Wall plate to stud	2-16d toenail	1-16d toenail	1-16d toenail	2-16d toenail	2-16d toenail	3-16d toenail	Tydown sr
Stud to 2nd sill plate	4-8d toenail	None needed	1-8d toenail	2-8d toenail	3-8d toenail	4-8d toenail	24″ strap
Sill plate to 2nd floor	16d @ 16″ o.c.	None needed	16d @ 28′ o.c.	16d @ 35″ o.c.	16d @ 18″ o.c.	16d @ 12″ o.c.	16d @ 8″ o.c.
1st floor stud to wall plate	2-16d toenail	None needed	None needed	1-16d toenail	1-16d toenail	2-16d toenail	3-16d toenail
Stud to 1st sill plate	4-8d toenail	None needed	2-8d toenail	24″ strap	24″ strap	Simpson LFTA	Simpson LFTA
Sill plate to 1st floor	16d @ 16″ o.c.	None needed	16d @ 20′ o.c.	16d @ 17″ o.c.	16d @ 9″ o.c.	Addt'l conn.	Addt'l conn.
1st floor to foundation	1/2″ bolt @ 6′	None needed	1/2″ bolt @ 146′	1/2″ bolt @ 11′	1/2″ bolt @ 6′	1/2″ bolt @ 4′	Addt'l conn.
		Recom'd OK	Recom'd OK				

Wind velocity (mph)		70	80	90	100	110	120
Wind pressure (psf)		16	21	27	33	40	48

Sliding (pounds per connector; friction factor = 0.25)

		70	80	90	100	110	120
Roof to wall plate		25.4	47.2	73.3	99.5	130.0	164.9
Wall plate to stud		25.4	47.2	73.3	99.5	130.0	164.9
Sill plate to 2nd floor		95.7	146.6	207.6	268.7	339.9	421.3
2nd floor to wall plate		81.3	135.8	201.2	266.6	342.9	430.1
Sill plate to 1st floor		151.7	235.2	335.5	435.8	552.8	686.5
1st floor to foundation		32.1	130.7	249.0	367.3	505.3	663.1

Location	Recom'd			Required Connector			
Roof to wall plate	3-16d toenail	1-16d toenail	1-16d toenail	1-16d toenail	1-16d toenail	1-16d toenail	1-16d toenail
Wall plate to stud	2-16d toenail	1-16d toenail	1-16d toenail	1-16d toenail	1-16d toenail	1-16d toenail	1-16d toenail
Stud to 2nd sill plate	4-8d toenail	1-8d toenail	2-8d toenail	2-8d toenail	3-8d toenail	3-8d toenail	Addt'l conn.
Sill plate to 2nd floor	16d @ 16" o.c.	16d @ 43" o.c.	16d @ 28" o.c.	16d @ 20" o.c.	16d @ 15" o.c.	16d @ 12" o.c.	16d @ 10" o.c.
1st floor stud to wall plate	2-16d toenail	1-16d toenail	1-16d toenail	2-16d toenail	2-16d toenail	2-16d toenail	3-16d toenail
Stud to 1st sill plate	4-8d toenail	2-8d toenail	3-8d toenail	3-8d toenail	Addt'l conn.	Addt'l conn.	Addt'l conn.
Sill plate to 1st floor	16d @ 16" o.c.	16d @ 11" o.c.	16d @ 31" o.c.	16d @ 16" o.c.	16d @ 11" o.c.	16d @ 8" o.c.	16d @ 6" o.c.
1st floor to foundation	1/2" bolt @ 6'	1/2" bolt @ 25'	1/2" bolt @ 6'	1/2" bolt @ 3'	Addt'l conn.	Addt'l conn.	Addt'l conn.
		Recom'd OK	Recom'd OK				

Notes:
1. Compare uplift/overturning with sliding. Use the larger connector.
2. Required forces are shown as pounds per connector. Only one metal connector per wood member is assumed. To use two connectors on a member, see manufacturer's requirements on member sizes.
3. Negative values (shown in parentheses) show that no connector is required.
4. At a minimum, *always use the CABO-recommended fastening schedule.*

TABLE F.2 REQUIRED CONNECTIONS FOR 24-FT-WIDE BUILDING WITH 10-IN-12 ROOF PITCH (EXTERIOR ONLY, FRAMING MEMBERS AT 16 IN. O.C.)

Wind velocity (mph)	70	80	90	100	110	120
Wind pressure (psf)	16	21	27	33	40	48

Uplift/Overturning (pounds per connector)

Roof to wall plate	(56.2)	(33.7)	3.5	46.9	97.6	155.5
Wall plate to stud	(56.2)	(33.7)	3.5	46.9	97.6	155.5
Sill plate to 2nd floor	(53.9)	(11.9)	38.6	89.0	147.9	215.1
2nd floor to wall plate	(106.2)	(58.9)	(2.0)	54.8	121.1	196.9
Sill plate to 1st floor	4.5	158.0	339.6	521.1	732.9	974.9
1st floor to foundation	(39.1)	69.8	200.5	331.2	483.6	657.9

Location	Recom'd		Required Connector			
Roof to wall plate	3-16d toenail	None needed	1-16d toenail	1-16d toenail	2-16d toenail	2-16d toenail
Wall plate to stud	2-16d toenail	None needed	1-16d toenail	2-16d toenail	2-16d toenail	2-16d toenail
Stud to 2nd sill plate	4-8d toenail	None needed	1-8d toenail	2-8d toenail	3-8d toenail	24″ strap
Sill plate to 2nd floor	16d @ 16″ o.c.	None needed	16d @ 4′ o.c.	16d @ 23″ o.c.	16d @ 14″ o.c.	16d @ 10″ o.c.
1st floor stud to wall plate	2-16d toenail	None needed	None needed	1-16d toenail	2-16d toenail	3-16d toenail
Stud to 1st sill plate	4-8d toenail	1-8d toenail	24″ strap	Simpson LFTA	Simpson LFTA	Simpson LFTA
Sill plate to 1st floor	16d @ 16″ o.c.	16d @ 29″ o.c.	16d @ 10″ o.c.	16d @ 6″ o.c.	Addt'l conn.	Addt'l conn.
1st floor to foundation	1/2″ bolt @ 6′	1/2″ bolt @ 18′	1/2″ bolt @ 6′	1/2″ bolt @ 4′	Addt'l conn.	Addt'l conn.

Recom'd OK *Recom'd OK*

Wind velocity (mph)		70	80	90	100	110	120
Wind pressure (psf)		16	21	27	33	40	48

Sliding (pounds per connector; friction factor = 0.25)

		70	80	90	100	110	120
Roof to wall plate		86.6	128.6	179.1	229.5	288.4	355.7
Wall plate to stud		86.6	128.6	179.1	229.5	288.4	355.7
Sill plate to 2nd floor		156.9	228.0	313.4	398.7	498.3	612.1
2nd floor to wall plate		142.5	217.3	307.0	396.7	501.3	620.9
Sill plate to 1st floor		212.9	316.7	441.3	565.9	711.2	877.3
1st floor to foundation		82.1	198.6	388.4	478.2	641.3	827.7

Location	Recom'd			Required Connector			
Roof to wall plate	3-16d toenail	1-16d toenail	1-16d toenail	1-16d toenail	2-16d toenail	2-16d toenail	2-16d toenail
Wall plate to stud	2-16d toenail	1-16d toenail	1-16d toenail	1-16d toenail	2-16d toenail	2-16d toenail	2-16d toenail
Stud to 2nd sill plate	4-8d toenail	2-8d toenail	3-8d toenail	3-8d toenail	Addt'l conn.	Addt'l conn.	Addt'l conn.
Sill plate to 2nd floor	16d @ 16″ o.c.	16d @ 26″ o.c.	16d @ 18″ o.c.	16d @ 13″ o.c.	16d @ 10″ o.c.	16d @ 8″ o.c.	16d @ 7″ o.c.
1st floor stud to wall plate	2-16d toenail	1-16d toenail	2-16d toenail	2-16d toenail	3-16d toenail	Addt'l conn.	Addt'l conn.
Stud to 1st sill plate	4-8d toenail	2-8d toenail	3-8d toenail	Addt'l conn.	Addt'l conn.	Addt'l conn.	Addt'l conn.
Sill plate to 1st floor	16d @ 16″ o.c.	16d @ 4′ o.c.	16d @ 21″ o.c.	16d @ 12″ o.c.	16d @ 9″ o.c.	16d @ 6″ o.c.	Addt'l conn.
1st floor to foundation	1/2″ bolt @ 6′	1/2″ bolt @ 10′	1/2″ bolt @ 4′	Recom'd OK	Addt'l conn.	Addt'l conn.	Addt'l conn.

Notes: 1. Compare uplift/overturning with sliding. Use the larger connector.
2. Required forces are shown as pounds per connector. Only one metal connector per wood member is assumed. To use two connectors on a member, see manufacturer's requirements on member sizes.
3. Negative values (shown in parentheses) show that no connector is required.
4. At a minimum, *always use the CABO-recommended fastening schedule.*

TABLE F.3 REQUIRED CONNECTIONS FOR 28-FT-WIDE BUILDING WITH 6-IN-12 ROOF PITCH (EXTERIOR ONLY, FRAMING MEMBERS AT 16 IN. O.C.)

Wind velocity (mph)	70	80	90	100	110	120
Wind pressure (psf)	16	21	27	33	40	48

Uplift/Overturning (pounds per connector)

Roof to wall plate	28.0	79.9	142.2	204.5	277.2	360.3
Wall plate to stud	28.0	79.9	142.2	204.5	277.2	360.3
Sill plate to 2nd floor	(37.3)	13.1	73.6	134.1	204.7	285.3
2nd floor to wall plate	(108.5)	(55.1)	9.0	73.1	147.9	233.4
Sill plate to 1st floor	(61.4)	38.5	195.2	351.8	534.6	743.5
1st floor to foundation	(125.0)	(32.6)	78.3	189.2	318.6	466.5

Location	Recom'd			Required Connector			
Roof to wall plate	3-16d toenail	1-16d toenail	1-16d toenail	2-16d toenail	3-16d toenail	Tydown sr	Tydown sr
Wall plate to stud	2-16d toenail	1-16d toenail	1-16d toenail	2-16d toenail	3-16d toenail	Tydown sr	Tydown sr
Stud to 2nd sill plate	4-8d toenail	None needed	1-8d toenail	2-8d toenail	3-8d toenail	4-8d toenail	24" strap
Sill plate to 2nd floor	16d @ 16" o.c.	None needed	16d @ 13' o.c.	16d @ 28" o.c.	16d @ 15" o.c.	16d @ 10" o.c.	16d @ 7" o.c.
1st floor stud to wall plate	2-16d toenail	None needed	None needed	1-16d toenail	1-16d toenail	2-16d toenail	3-16d toenail
Stud to 1st sill plate	4-8d toenail	None needed	1-8d toenail	4-8d toenail	24" Strap	Simpson LFTA	Simpson LFTA
Sill plate to 1st floor	16d @ 16" o.c.	None needed	None needed	16d @ 26" o.c.	16d @ 11" o.c.	16d @ 6" o.c.	Addt'l conn.
1st floor to foundation	1/2" bolt @ 6'	None needed	None needed	1/2" bolt @ 16'	1/2" bolt @ 7'	1/2" bolt @ 4'	Addt'l conn.
	Recom'd OK	Recom'd OK	Recom'd OK				

Wind velocity (mph)	70	80	90	100	110	120
Wind pressure (psf)	16	21	27	33	40	48

Sliding (pounds per connector; friction factor = 0.25)

Roof to wall plate	26.2	50.6	79.9	109.1	143.3	182.3
Wall plate to stud	26.2	50.6	79.9	109.1	143.3	182.3
Sill plate to 2nd floor	98.3	152.3	217.0	281.8	357.4	443.8
2nd floor to wall plate	79.8	137.5	206.7	275.9	356.7	449.0
Sill plate to 1st floor	151.8	239.1	343.9	448.6	570.8	710.5
1st floor to foundation	19.2	123.9	249.4	375.0	521.5	688.9

Location	Recom'd			Required Connector		
Roof to wall plate	3-16d toenail	1-16d toenail	1-16d toenail	1-16d toenail	1-16d toenail	1-16d toenail
Wall plate to stud	2-16d toenail	1-16d toenail	1-16d toenail	1-16d toenail	1-16d toenail	1-16d toenail
Stud to 2nd sill plate	4-8d toenail	1-8d toenail	2-8d toenail	3-8d toenail	4-8d toenail	Addt'l conn.
Sill plate to 2nd floor	16d @ 16″ o.c.	16d @ 42″ o.c.	16d @ 27″ o.c.	16d @ 19″ o.c.	16d @ 15″ o.c.	16d @ 11″ o.c.	16d @ 9″ o.c.
1st floor stud to wall plate	2-16d toenail	1-16d toenail	1-16d toenail	2-16d toenail	2-16d toenail	3-16d toenail
Stud to 1st sill plate	4-8d toenail	2-8d toenail	3-8d toenail	4-8d toenail	Addt'l conn.	Addt'l conn.
Sill plate to 1st floor	16d @ 16′ o.c.	16d @ 18′ o.c.	16d @ 33″ o.c.	16d @ 16″ o.c.	16d @ 11″ o.c.	16d @ 8″ o.c.	Addt'l conn.
1st floor to foundation	1/2″ bolt @ 6′	1/2″ bolt @ 41′	1/2″ bolt @ 6′	1/2″ bolt @ 3′	Addt'l conn.	Addt'l conn.

Recom'd OK   Recom'd OK

Notes: 1. Compare uplift/overturning with sliding. Use the larger connector.
2. Required forces are shown as pounds per connector. Only one metal connector per wood member is assumed. To use two connectors on a member, see manufacturer's requirements on member sizes.
3. Negative values (shown in parentheses) show that no connector is required.
4. At a minimum, *always use the CABO-recommended fastening schedule.*

TABLE F.4 REQUIRED CONNECTIONS FOR 28-FT-WIDE BUILDING WITH 10-IN-12 ROOF PITCH (EXTERIOR ONLY, FRAMING MEMBERS AT 16 IN. O.C.)

Wind velocity (mph)	70	80	90	100	110	120
Wind pressure (psf)	16	21	27	33	40	48

Uplift/Overturning (pounds per connector)

Roof to wall plate	(65.5)	(39.3)	4.1	54.7	113.9	181.4
Wall plate to stud	(65.5)	(39.3)	4.1	54.7	113.9	181.4
Sill plate to 2nd floor	(73.7)	(31.1)	19.9	70.9	130.5	198.5
2nd floor to wall plate	(138.6)	(91.0)	(34.0)	23.0	89.6	165.6
Sill plate to 1st floor	(40.7)	84.5	259.1	433.8	637.5	870.3
1st floor to foundation	(98.0)	6.4	131.7	257.0	403.1	570.1

Location	Recom'd			Required Connector			
Roof to wall plate	3-16d toenail	none needed	none needed	1-16d toenail	1-16d toenail	2-16d toenail	2-16d toenail
Wall plate to stud	2-16d toenail	none needed	none needed	1-16d toenail	1-16d toenail	2-16d toenail	2-16d toenail
Stud to 2nd sill plate	4-8d toenail	none needed	none needed	1-8d toenail	2-8d toenail	3-8d toenail	4-8d toenail
Sill plate to 2nd floor	16d @ 16" o.c.	none needed	none needed	16d @ 9' o.c.	16d @ 29' o.c.	16d @ 16" o.c.	16d @ 10" o.c.
1st floor stud to wall plate	2-16d toenail	none needed	none needed	none needed	1-16d toenail	1-16d toenail	2-16d toenail
Stud to 1st sill plate	4-8d toenail	none needed	none needed	24" strap	24" strap	Simpson LFTA	Simpson LFTA
Sill plate to 1st floor	16d @ 16" o.c.	none needed	2-8d toenail	16d @ 16" o.c.	16d @ 8" o.c.	Addt'l conn.	Addt'l conn.
1st floor to foundation	1/2" bolt @ 6'	none needed	16d @ 27' o.c.	1/2" bolt @ 19.7'	1/2" bolt @ 10'	1/2" bolt @ 5'	1/2" bolt @ 3'
		Recom'd OK	Recom'd OK				

Wind velocity (mph)	70	80	90	100	110	120
Wind pressure (psf)	16	21	27	33	40	48

Sliding (pounds per connector; friction factor = 0.25)

Roof to wall plate	101.0	150.0	208.9	267.8	336.5	415.0
Wall plate to stud	101.0	150.0	208.9	267.8	336.5	415.0
Sill plate to 2nd floor	173.0	251.7	346.1	440.5	550.6	676.5
2nd floor to wall plate	154.5	236.9	335.7	434.6	549.9	681.6
Sill plate to 1st floor	226.6	338.6	472.9	607.3	764.0	943.1
1st floor to foundation	79.9	206.1	357.6	509.0	685.7	887.7

Location	Recom'd			Required Connector		
Roof to wall plate	3-16d toenail	1-16d toenail	2-16d toenail	2-16d toenail	2-16d toenail	3-16d toenail
Wall plate to stud	2-16d toenail	1-16d toenail	2-16d toenail	2-16d toenail	2-16d toenail	3-16d toenail
Stud to 2nd sill plate	4-8d toenail	3-8d toenail	4-8d toenail	Addt'l conn.	Addt'l conn.	Addt'l conn.
Sill plate to 2nd floor	16d @ 16" o.c.	16d @ 16" o.c.	16d @ 12" o.c.	16d @ 9" o.c.	16d @ 7" o.c.	16d @ 6" o.c.
1st floor stud to wall plate	2-16d toenail	1-16d toenail	2-16d toenail	3-16d toenail	Addt'l conn.	Addt'l conn.
Stud to 1st sill plate	4-8d toenail	3-8d toenail	Addt'l conn.	Addt'l conn.	Addt'l conn.	Addt'l conn.
Sill plate to 1st floor	16d @ 16" o.c.	16d @ 4" o.c.	16d @ 20" o.c.	16d @ 8" o.c.	Addt'l conn.	Addt'l conn.
1st floor to foundation	1/2" bolt @ 6'	1/2" bolt @ 10'	1/2" bolt @ 4'	Addt'l conn.	Addt'l conn.	Addt'l conn.

Recom'd OK

Notes: 1. Compare uplift/overturning with sliding. Use the larger connector.
2. Required forces are shown as pounds per connector. Only one metal connector per wood member is assumed. To use two connectors on a member, see manufacturer's requirements on member sizes.
3. Negative values (shown in parentheses) show that no connector is required.
4. At a minimum, *always use the CABO-recommended fastening schedule*.

TABLE F.5 REQUIRED CONNECTIONS FOR 32-FT-WIDE BUILDING WITH 6-IN-12 ROOF PITCH (EXTERIOR ONLY, FRAMING MEMBERS AT 16 IN. O.C.)

Wind velocity (mph)		70	80	90	100	110	120
Wind pressure (psf)		16	21	27	33	40	48

Uplift/Overturning (pounds per connector)

Roof to wall plate		38.4	99.7	173.3	246.9	332.8	430.9
Wall plate to stud		38.4	99.7	173.3	246.9	332.8	430.9
Sill plate to 2nd floor		(31.4)	27.0	97.0	167.1	248.9	342.3
2nd floor to wall plate		(115.2)	(54.0)	19.3	92.6	178.2	276.0
Sill plate to 1st floor		(78.1)	20.2	185.1	349.9	542.3	762.1
1st floor to foundation		(154.9)	(58.4)	57.4	173.1	308.2	462.6

Location	Recom'd			Required Connector			
Roof to wall plate	3-16d toenail	1-16d toenail	2-16d toenail	2-16d toenail	3-16d toenail	Tydown sr	Tydown sr
Wall plate to stud	2-16d toenail	1-16d toenail	2-16d toenail	2-16d toenail	3-16d toenail	Tydown sr	Tydown sr
Stud to 2nd sill plate	4-8d toenail	none needed	1-8d toenail	2-8d toenail	4-8d toenail	24″ strap	24″ strap
Sill plate to 2nd floor	16d @ 16″ o.c.	none needed	16d @ 6′ o.c.	16d @ 21″ o.c.	16d @ 12″ o.c.	16d @ 8″ o.c.	Addt'l conn.
1st floor stud to wall plate	2-16d toenail	none needed	none needed	1-16d toenail	2-16d toenail	2-16d toenail	3-16d toenail
Stud to 1st sill plate	4-8d toenail	none needed	1-8d toenail	4-8d toenail	24″ Strap	Simpson LFTA	Simpson LFTA
Sill plate to 1st floor	16d @ 16″ o.c.	none needed	none needed	16d @ 36″ o.c.	16d @ 12″ o.c.	16d @ 7″ o.c.	Addt'l conn.
1st floor to foundation	1/2″ bolt @ 6′	none needed	none needed	1/2″ bolt @ 22′	1/2″ bolt @ 7′	1/2″ bolt @ 4′	Addt'l conn.
		Recom'd OK	*Recom'd OK*	*Recom'd OK*			

Wind velocity (mph)	70	80	90	100	110	120
Wind pressure (psf)	16	21	27	33	40	48

Sliding (pounds per connector; friction factor = 0.25)

	70	80	90	100	110	120
Roof to wall plate	26.1	52.8	84.8	116.8	154.1	196.8
Wall plate to stud	26.1	52.8	84.8	116.8	154.1	196.8
Sill plate to 2nd floor	102.5	160.1	229.2	298.3	378.9	471.1
2nd floor to wall plate	80.2	141.6	215.4	289.1	375.2	473.5
Sill plate to 1st floor	156.5	248.9	359.8	470.6	600.0	747.8
1st floor to foundation	11.5	123.8	258.5	393.2	550.4	730.0

Location	Recom'd			Required Connector		
Roof to wall plate	3-16d toenail	1-16d toenail	1-16d toenail	1-16d toenail	1-16d toenail	2-16d toenail
Wall plate to stud	2-16d toenail	1-16d toenail	1-16d toenail	1-16d toenail	1-16d toenail	2-16d toenail
Stud to 2nd sill plate	4-8d toenail	1-8d toenail	2-8d toenail	3-8d toenail	Addt'l conn.	Addt'l conn.
Sill plate to 2nd floor	16d @ 16″ o.c.	16d @ 40″ o.c.	16d @ 26″ o.c.	16d @ 18″ o.c.	16d @ 14″ o.c.	16d @ 11″ o.c.	16d @ 9″ o.c.
1st floor stud to wall plate	2-16d toenail	1-16d toenail	1-16d toenail	2-16d toenail	2-16d toenail	3-16d toenail
Stud to 1st sill plate	4-8d toenail	2-8d toenail	3-8d toenail	4-8d toenail	Addt'l conn.	Addt'l conn.
Sill plate to 1st floor	16d @ 16′ o.c.	16d @ 30′ o.c.	16d @ 33′ o.c.	16d @ 12″ o.c.	16d @ 10″ o.c.	16d @ 7″ o.c.	Addt'l conn.
1st floor to foundation	1/2″ bolt @ 6′	1/2″ bolt @ 69′	1/2″ bolt @ 6′	1/2″ bolt @ 3′	Addt'l conn.	Addt'l conn.	Addt'l conn.

	Recom'd OK	*Recom'd OK*

Notes: 1. Compare uplift/overturning with sliding. Use the larger connector.
2. Required forces are shown as pounds per connector. Only one metal connector per wood member is assumed. To use two connectors on a member, see manufacturer's requirements on member sizes.
3. Negative values (shown in parentheses) show that no connector is required.
4. At a minimum, *always use the CABO-Recommended fastening schedule.*

TABLE F.6 REQUIRED CONNECTIONS FOR 32-FT-WIDE BUILDING WITH 10-IN-12 ROOF PITCH (EXTERIOR ONLY, FRAMING MEMBERS AT 16 IN. O.C.)

Wind velocity (mph)	70	80	90	100	110	120
Wind pressure (psf)	16	21	27	33	40	48

Uplift/overturning (pounds per connector)

Roof to wall plate	(74.9)	(44.9)	4.6	62.6	130.1	207.4
Wall plate to stud	(74.9)	(44.9)	4.6	62.6	130.1	207.4
Sill plate to 2nd floor	(91.8)	(48.3)	4.0	56.2	117.2	186.8
2nd floor to wall plate	(168.9)	(120.5)	(62.5)	(4.5)	63.1	140.5
Sill plate to 1st floor	(78.5)	25.5	197.3	369.2	569.7	798.8
1st floor to foundation	(148.6)	(46.2)	76.8	199.7	343.2	507.1

Location	Recom'd			*Required Connector*			
Roof to wall plate	3-16d toenail	none needed	none needed	1-16d toenail	1-16d toenail	2-16d toenail	3-16d toenail
Wall plate to stud	2-16d toenail	none needed	none needed	1-16d toenail	1-16d toenail	2-16d toenail	3-16d toenail
Stud to 2nd sill plate	4-8d toenail	none needed	none needed	1-8d toenail	2-8d toenail	3-8d toenail	4-8d toenail
Sill plate to 2nd floor	16d @ 16" o.c.	none needed	none needed	16d @ 34" o.c.	16d @ 35" o.c.	16d @ 17" o.c.	16d @ 11" o.c.
1st floor stud to wall plate	2-16d toenail	none needed	none needed	none needed	none needed	1-16d toenail	2-16d toenail
Stud to 1st sill plate	4-8d toenail	none needed	1-8d toenail	4-8d toenail	24" Strap	Simpson LFTA	Simpson LFTA
Sill plate to 1st floor	16d @ 16" o.c.	none needed	none needed	16d @ 27" o.c.	16d @ 10" o.c.	Addt'l conn.	Addt'l conn.
1st floor to foundation	1/2" bolt @ 6'	none needed	none needed	1/2" bolt @ 16'	1/2" bolt @ 6'	1/2" bolt @ 4'	Addt'l conn.
		Recom'd OK	Recom'd OK	Recom'd OK			

Wind velocity (mph)	70	80	90	100	110	120
Wind pressure (psf)	16	21	27	33	40	48

Sliding (pounds per connector; friction factor = 0.25)

Roof to wall plate	115.4	171.5	238.8	306.0	384.5	474.2
Wall plate to stud	115.4	171.5	238.8	306.0	384.5	474.2
Sill plate to 2nd floor	191.7	278.7	383.1	487.5	609.3	748.5
2nd floor to wall plate	169.4	260.3	369.3	478.4	605.6	751.0
Sill plate to 1st floor	245.8	367.6	513.7	659.9	830.4	1025.3
1st floor to foundation	83.4	221.1	386.4	551.7	744.5	964.4

Location	Recom'd			Required Connector			
Roof to wall plate	3-16d toenail	1-16d toenail	2-16d toenail	2-16d toenail	3-16d toenail	3-16d toenail	Addt'l conn.
Wall plate to stud	2-16d toenail	1-16d toenail	2-16d toenail	2-16d toenail	3-16d toenail	Addt'l conn.	Addt'l conn.
Stud to 2nd sill plate	4-8d toenail	2-8d toenail	3-8d toenail	Addt'l conn.	Addt'l conn.	Addt'l conn.	Addt'l conn.
Sill plate to 2nd floor	16d @ 16″ o.c.	16d @ 21″ o.c.	16d @ 15″ o.c.	16d @ 11″ o.c.	16d @ 8″ o.c.	16d @ 7″ o.c.	Addt'l conn.
1st floor stud to wall plate	2-16d toenail	1-16d toenail	2-16d toenail	2-16d toenail	Addt'l conn.	Addt'l conn.	Addt'l conn.
Stud to 1st sill plate	4-8d toenail	3-8d toenail	4-8d toenail	Addt'l conn.	Addt'l conn.	Addt'l conn.	Addt'l conn.
Sill plate to 1st floor	16d @ 16″ o.c.	16d @ 4″ o.c.	16d @ 12″ o.c.	16d @ 11″ o.c.	16d @ 7″ o.c.	Addt'l conn.	Addt'l conn.
1st floor to foundation	1/2″ bolt @ 6′	1/2″ bolt @ 4′	1/2″ bolt @ 4′	Addt'l conn.	Addt'l conn.	Addt'l conn.	Addt'l conn.

Notes: 1. Compare uplift/overturning with sliding. Use the larger connector.
2. Required forces are shown as pounds per connector. Only one metal connector per wood member is assumed. To use two connectors on a member, see manufacturer's requirements on member sizes.
3. Negative values (shown in parentheses) show that connector is required.
4. At a minimum, *always use the CABO-recommended fastening schedule.*

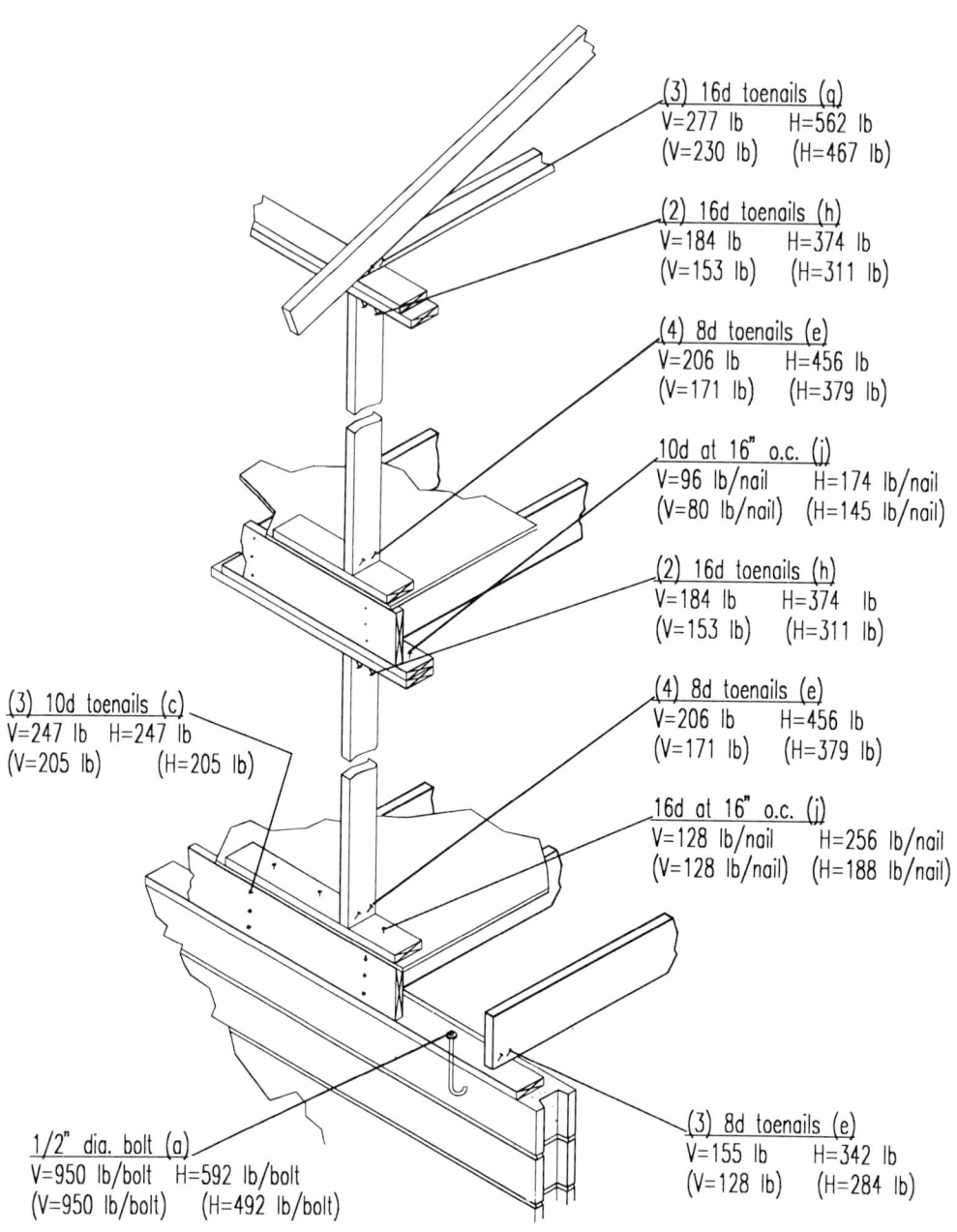

note: values in parentheses use a duration factor of 1.33 as per UBC. Values without parentheses use a duration factor of 1.60 as per SBC and BBC

APPENDIX G

HORIZONTAL DIAPHRAGM AND VERTICAL SHEARWALL TABLES

TABLE G.1 ALLOWABLE SHEAR FOR HORIZONTAL NAILED PLYWOOD DIAPHRAGMS

shear in pounds per foot (plf)

Plywood Grade	Common Nail Size	Min. Nominal Penetration in Framing (inches)	Min. Nominal Plywood Thickness (inches)	Min. Nominal Width of Framing Member (inches)	Blocked Diaphragm				Unblocked Dia'm	
					Nail spacing at diaphragm boundaries (all cases), at continuous panel edges parallel to load (Cases 3 and 4) and at all panel edges (Cases 5 and 6). (inches)				Nail spacing at 6″ max. at supported end.	
					6	4	$2\frac{1}{2}$	2[2]	Load perp. to unblocked edges & continuous panel joints (Case 1)	Other config's (Cases 2, 3, & 4)
					Nail spacing at other plywood panel edges. (inches)					
					6	6	4	3		
STRUCTURAL I	6d	1-1/4	5/16	2 3	185 210	250 280	375 420	420 475	165 185	125 140
	8d	1-1/2	3/8	2 3	270 300	360 400	530 600	600 675	240 265	180 200
	10d	1-5/8	15/32	2 3	320 360	425 480	640 720	730[2] 820	285 320	215 240

6d	1-1/4	5/16	2 3	170 190	225 250	335 380	380 430	150 170	110 125
		3/8	2 3	185 210	250 280	375 420	420 475	165 185	125 140
8d	1-1/2	3/8	2 3	240 270	320 360	480 540	545 610	215 240	160 180
C-D, C-C STRUCTURAL II & other grades covered in UBC Standard 25-9 or PS-1		15/32	2 3	270 300	360 400	530 600	600 675	240 265	180 200
10d	1-5/8	15/32	2 3	290 325	385 430	575 650	655[2] 735	255 290	190 215
		19/32	2 3	320 360	425 480	640 720	730[2] 820	285 320	215 240

1. Values are for short-term loads due to wind or seismic and must be reduced by 25% for normal loads. Space nails in field 10″ o.c. for floors and 12″ o.c. for roofs. Allowable shear values for framing of Douglas Fir-Larch or Southern Pine. For other species, for either grade of plywood, multiply values for STRUCT I by 0.82 for Group III species and by 0.65 for Group IV species.
2. Framing shall be 3-inch nominal or wider and nails shall be staggered where nails are spaced 2 inches or 2-1/2 inches on center, and where 10d nails having penetration into the framing of more than 1-5/8 inches are spaced 3 inches on center.

Source: UBC Table 25-J-1

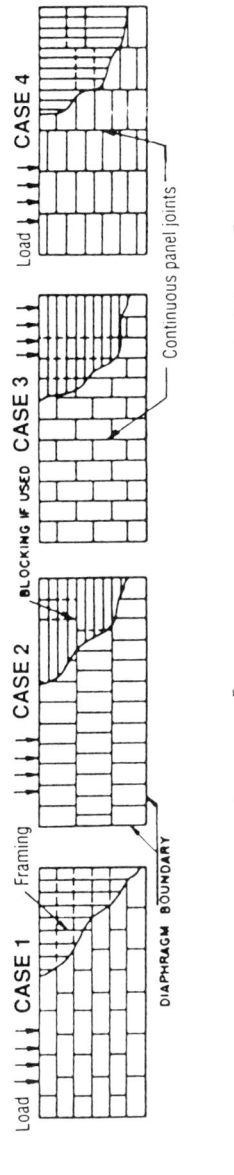
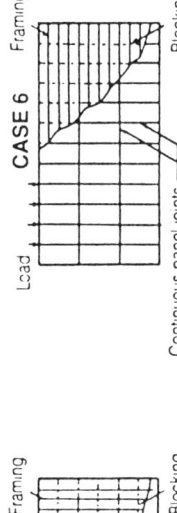
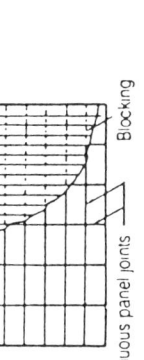

TABLE G.2 ALLOWABLE SHEAR FOR HORIZONTAL PLYWOOD DIAPHRAGMS (plf) STAPLED CONNECTIONS—STRUCTURAL I PLYWOOD

Min. Nominal Plywood Thickness (inches)	Staple Size	Min. Length of Fastener Req'd (inches)	Min. Nominal Width of Framing Member (inches)	Blocked Diaphragm				Unblocked Diaphragm	
				Staple spacing at diaphragm boundaries (all cases), at continuous panel edges parallel to load (Cases 3 and 4) and at all panel edges (Cases 5 and 6).				Staple spacing at 6" max. at supported end.	
				6	4	$2\frac{1}{2}$	2^2	Load perp. to unblocked edges and continuous panel joints (Case 1)	Other config's (Cases 2, 3, & 4)
				Staple spacing at other plywood panel edges.					
				6	6	4	3		
5/16	14 ga or 15 ga	1-3/8	2 3	185 210	250 280	375 420	420 475	165 185	125 140
	16 ga	1-3/8	2 3	155 175	205 230	310 345	345 390	135 155	105 115

3/8	13 ga	1-3/8	2	270	360	530	600	240	180
			3	300	400	600	675	265	200
	14 ga	1-3/8	2	265	355	520	590	235	180
			3	295	395	590	655	260	200
	15 ga	1-3/8	2	220	290	430	485	195	145
			3	240	325	485	545	215	160
	16 ga	1-3/8	2	180	240	350	400	160	120
			3	200	265	400	450	175	135
1/2	12 ga	1-5/8	2	320	425	640	730[2]	285	215
			3	360	480	720	820	320	240
	13 ga	1-1/2	2	315	415	625	715	280	210
			3	350	470	705	800	315	235
	14 ga	1-1/2	2	260	340	510	585	230	170
			3	290	380	575	655	255	190
	15 ga	1-1/2	2	215	285	430	490	190	140
			3	240	320	480	550	215	160
	16 ga	1-1/2	2	175	235	355	405	155	120
			3	200	265	400	455	175	135

1. Values are for short-term loads due to wind or seismic and must be reduced by 25% for normal loads. Space staples in the field at 10″ o.c. for floors and 12″ o.c. for roofs. Allowable shear values for framing of Douglas Fir-Larch or Southern Pine. For other species, for either grade of plywood, multiply values for STRUCTURAL I by 0.82 for Group III species and by 0.65 for Group IV species.
2. Framing shall be 3-inch nominal or wider and staples shall be staggered where staples are spaced 2 inches or 2-1/2 inches on center.
3. Staples shall have a minimum crown width of 7/16 inches. Lengths and penetrations for Group II species. Multiply penetration by 1.2 for Group III and by 1.3 for Group IV.

Source: NER Report No. NER-272

TABLE G.3 ALLOWABLE SHEAR FOR HORIZONTAL PLYWOOD DIAPHRAGMS (plf) STAPLED CONNECTIONS — STRUCTURAL II C-C, C-D PLYWOOD

Min. Nominal Plywood Thickness (inches)	Staple Size	Min. Length of Fastener Req'd (inches)	Min. Nominal Width of Framing Member (inches)	Blocked Diaphragm				Unblocked Diaphragm	
				Staple spacing at diaphragm boundaries (all cases), at continuous panel edges parallel to load (Cases 3 and 4) and at all panel edges (Case 5 and 6).				Staple spacing at 6" max. at supported end.	
				6	4	$2\frac{1}{2}$	2^2	Load perp. to unblocked edges and continuous panel joints (Case 1)	Other config's (Cases 2, 3, & 4)
				Staple spacing at other plywood panel edges.					
				6	6	4	3		
3/8	13 ga	1-3/8	2	240	320	480	345	215	160
			3	270	360	540	610	240	180
	14 ga	1-3/8	2	235	315	470	530	210	155
			3	265	355	530	600	235	175
	15 ga	1-3/8	2	185	250	375	420	165	125
			3	210	280	420	475	185	140
	16 ga	1-3/8	2	160	215	315	360	145	110
			3	180	240	360	405	155	120

1/2	12 ga	1-5/8	2 3	290 325	385 430	575² 650	655² 735	255 290	190 215
	13 ga	1-1/2	2 3	270 300	360 400	530 600	600 675	240 265	180 200
	14 ga	1-1/2	2 3	235 265	315 355	470 530	530 600	210 235	155 175
	16 ga	1-1/2	2 3	160 180	215 240	315 360	360 405	145 155	110 190
5/8	12 ga	1-3/4	2 3	320 360	425 480	640² 720	730² 820	285 320	215 240
	13 ga	1-5/8	2 3	315 350	415 470	625² 705	715² 800	280 315	210 235
	14 ga	1-5/8	2 3	260 290	340 380	510 575	585 655	230 255	170 190
	16 ga	1-5/8	2 3	160 180	215 240	315 360	360 405	145 155	110 120

1. Values are for short-term loads due to wind or seismic and must be reduced by 25% for normal loads. Space staples in the field at 10″ o.c. for floors and 12″ o.c. for roofs. Allowable shear values for framing of Douglas Fir-Larch or Southern Pine. For other species, for either grade of plywood, multiply values for STRUCTURAL I by 0.82 for Group III species and by 0.65 for Group IV species.
2. Framing shall be 3-inch nominal or wider and staples shall be staggered where staples are spaced 2 inches or 2-1/2 inches on center.
3. Staples shall have a minimum crown width of 7/16 inches. Lengths and penetrations for Group II species. Multiply penetration by 1.2 for Group III and by 1.3 for Group IV

Source: NER Report No. NER-272

TABLE G.4 ALLOWABLE SHEAR FOR HIGH-LOAD HORIZONTAL PLYWOOD DIAPHRAGMS FOR LOAD CASE 1 AND CASE 2[1] shear in pounds per foot (plf)

Plywood Grade	Fastener	Min. Nominal Penetration in Framing (inches)	Min. Nominal Plywood Thickness (inches)	Min. Nominal Width of Framing Member (inches)	Lines of Fastener	Fastener spacing per line at diaphragm boundaries (inches) (all cases)			Fastener spac'g per line at other panel edges (in)[3]			
						4	$2\frac{1}{2}$	2	4	3	2	
						6	4	4	3			
STRUCTURAL I	10d common nail	$1\frac{5}{8}$	$\frac{1}{2}$	3	2	650	870	940	1230	—	—	
				4	2	755	980	1080	1305	—	—	
				4	3	940	1305	1340	1415	—	—	
			$\frac{5}{8}$	3	2	650	870	940	1230	—	—	
				4	2	755	980	1080	1410	—	—	
				4	3	940	1305	1375	1810	—	—	
	14 ga. staples	2	$\frac{1}{2}$	3	2	600	600	840	900	1040	1200	
				4	3	840	900	1140	1310	1310	1415	
			$\frac{5}{8}$	3	2	600	600	840	900	1040	1200	
				4	3	840	900	1140	1350	1440	1800	
C-D, C-C, STRUCTURAL II	10d common nails	$1\frac{5}{8}$	$\frac{1}{2}$	3	2	585	785	845	1110	—	—	
				4	2	680	880	975	1555	—	—	
				4	3	845	1175	1240	1360	—	—	
			$\frac{5}{8}$	3	2	645	870	935	1225	—	—	
				4	2	750	980	1075	1340[4]	—	—	
				4	3	935	1305	1370	1445[4]	—	—	
	14 ga. staples	2	$\frac{1}{2}$	3	2	540	540	770	810	950	1080	
				4	3	770	810	1040	1215	1255	1365	
			$\frac{5}{8}$	3	2	600	600	820	900	1020	1200	
				4	3	820	900	1120	1340	1340[4]	1450[4]	

Notes: 1. For blocked diaphragms only, Case 1 and Case 2. See Table G.1 for case illustrations.
2. Allowable shear values for framing of Douglas Fir-Larch or Southern Pine. For other species, multiply values for STRUCTURAL I by 0.82 for Group III species and by 0.65 for Group IV species.
3. Values are for short-term loads due to wind or seismic and must be reduced by 25% for normal loads. Space nails 10" o.c. in the field for floors and 12" o.c. in the field for roofs, except 6" o.c. for spans greater than 32", along intermediate framing members. Space staples 6" o.c. along intermediate framing members.
4. Allowable shear value may be increased 60 plf when 3/4 inch plywood is used.

Source: APA Research Report No. 138. See also ICBO Research Report No. 1952

TABLE G.5 ALLOWABLE SHEAR FOR HORIZONTAL *PARTICLEBOARD* DIAPHRAGMS
shear in pounds per foot (plf)

Panel Grade	Common Nail Size	Min. Nominal Penetration in Framing (inches)	Min. Nominal Panel Thickness (inches)	Min. Nominal Width of Framing Member (inches)	Blocked Diaphragm				Unblocked Dia'm	
					Nail spacing at diaphragm boundaries (all cases), at continuous panel edges parallel to load (Cases 3 and 4) and at all panel edges (Cases 5 and 6).				Nail spacing at 6" max. at supported end.	
					6	4	$2\frac{1}{2}^2$	2^2		
					Nail spacing at other panel edges. (Cases 1, 2, 3, & 4)				Load perpendicular to unblocked edges and continuous panel joints (Case 1)	All other config's (Cases 2, 3, 4, 5, & 6)
					6	6	4	3		
2-M-W	6d	$1\frac{1}{4}$	$\frac{5}{16}$	2	170	225	335	380	150	110
				3	190	250	380	430	170	125
			$\frac{3}{8}$	2	185	250	375	420	165	125
				3	210	280	420	475	185	140
	8d	$1\frac{1}{2}$	$\frac{3}{8}$	2	240	320	480	545	215	160
				3	270	360	540	610	240	180
			$\frac{7}{16}$	2	255	340	505	575	230	170
				3	285	380	570	645	255	190
			$\frac{1}{2}$	2	270	360	530	600	240	180
				3	300	400	600	675	265	200
	10d[3]	$1\frac{5}{8}$	$\frac{1}{2}$	2	290	385	575	655	255	190
				3	325	430	650	735	290	215
			$\frac{5}{8}$	2	320	425	640	730	285	215
				3	360	480	720	820	320	240
2-M-3	10d[3]	$1\frac{5}{8}$	$\frac{3}{4}$	2	320	425	640	730	285	215
				3	360	480	720	820	320	240

Notes: 1. Values are for short-term loads due to wind or seismic and must be reduced by 25% for normal loads. Space nails 10" o.c. in the field for floors and 12" o.c. in the field for roofs. Allowable shear values for framing of Douglas Fir-Larch or Southern Pine. For other species, for either grade of particleboard, multiply values by 0.82 for Group III species and by 0.65 for Group IV species.
       2. Framing at adjoining panel edges shall be 3-inch nominal or wider and nails shall be staggered where nails are spaced 2 inches or 2-1/2 inches on center.
       3. Framing at adjoining panel edges shall be 3-inch nominal or wider and nails shall be staggered where 10d nails having penetration into the framing of more than 1-5/8 inches are spaced 3 inches or less on center.

Source: UBC Table 25-J-2

TABLE G.6 ALLOWABLE SHEAR FOR HORIZONTAL *LUMBER* DIAPHRAGMS (plf)

Nominal Width of Sheathing Lumber, (in.)	Number of 8d Common Nails per Board per Crossing of Framing Member and at Butted Ends	Allowabled Shear, (plf)		
A. Transversely Sheathed Diaphragms		Framing Member Spacing		
		12 in.	16 in.	24 in.
6	2	85	65	45
8	2	100	75	50
10	2	105	80	55

Nominal Width of Sheathing Lumber (in.)	Number of 8d Common Nails per Board per Crossing		Allowable Shear (plf)
B. Diagonally Sheathed Diaphragms			
	Perimeter Members and Butted Ends of Boards	Stud or Joist Crossing	
6	2	2	290
8	2	2	220
8	3	2	320

Source: Western Wood Product Assoc. *Western Woods Use Book*, 1973. These sources contain additional tables for Special Perimeter Framed Diagonal Sheathing and for Double Diagonal sheathing.

TABLE G.7 ALLOWABLE SHEAR FOR NAILED PLYWOOD SHEAR WALLS[1,4]

shear in pounds per foot (plf)

Plywood Grade	Min. Nominal Plywood Thickness (inches)	Min. Nail Penetration in Framing (inches)	Nail Size (Common or Galvanized Box)	Plywood Applied Directly to Framing				Nail Size (Common or Galvanized Box)	Plywood Applied Over 1/2 Inch Gypsum Sheathing			
				Nail spacing at Plywood Panel Edges					Nail spacing at Plywood Panel Edges			
				6	4	3	2^2		6	4	3	2^2
STRUCTURAL I	5/16	1-1/4	6d	200	300	390	510	8d	200	300	390	510
	3/8	1-1/2	8d	230^3	360^3	460^3	610^3	$10d^5$	280	430	550	730
	15/32	1-1/2	8d	280	430	550	730	$10d^5$	280	430	550	730
	15/32	1-5/8	$10d^5$	340	510	665	870	—	—	—	—	—
C-D, C-C STRUCT II- & other grades covered in UBC Standard 25-9 or PS-1	5/16	1-1/4	6d	180	270	350	450	8d	180	270	350	450
	3/8	1-1/4	6d	200	300	390	510	8d	200	300	390	510
	3/8	1-1/2	8d	220^3	320^3	410^3	530^3	$10d^5$	260	380	490	640
	15/32	1-1/2	8d	260	380	490	640	$10d^5$	260	380	490	640
	15/32	1-5/8	$10d^5$	310	460	600	770	—	—	—	—	—
	19/32	1-5/8	$10d^5$	340	510	665	870	—	—	—	—	—

385

		Nail Size Galvanized Casing						Nail Size Galvanized Casing				
		6d				8d		8d				10d[5]
Plywood panel siding in grades covered in UBC Standard 25-9 or PS-1	5/16	1-1/4	140	210	275	360		140	210	275	360	
	3/8	1-1/2					130[3]	200[3]	260[3]	340[3]	160	240	310	410

1. All panel edges backed with 2-inch nominal or wider framing. Plywood installed either horizontally or vertically. Space nails at 6 inches on center along intermediate framing members for 3/8-inch plywood installed with the face grain parallel to studs spaced 24 inches on center and 12 inches on center for other conditions and plywood thicknesses. These values are for short-term loads due to wind or seismic and must be reduced by 25% for normal loads. Allowable shear values for framing of Douglas Fir-Larch or Southern Pine. For other species, values shall be calculated for all grades by multiplying values for common and galvanized box nails in STRUCTURAL I and galvanized casing nails in other grades by the following factors: Group III, 0.82; Group IV, 0.65.
2. Framing at adjoining panel edges shall be 3-inch nominal or wider and nails shall be staggered where nails are spaced 2 inches on center.
3. The values for 3/8-inch plywood applied directly to framing may be increased 20 percent, provided studs are spaced a maximum of 16 inches on center or plywood is applied with face grain across studs.
4. Where plywood is applied on both faces of a wall and nail spacing is less than 6 inches on center on either side, panel joints shall be offset to fall on different framing members or framing shall be 3-inch nominal or thicker and nails on each side shall be staggered.
5. Framing at adjoining panel edges shall be 3-inch nominal or wider and nails shall be staggered where 10d nails having penetration into framing of more than 1 5/8-inches are spaced 3 inches or less on center.

Source: UBC Table 25-K-1

TABLE G.8 ALLOWABLE SHEAR FOR STAPLED PLYWOOD SHEAR WALLS[1,6,7]

shear in pounds per foot (plf)

Plywood Grade	Min. Nominal Plywood Thickness (inches)	Staple Size (Gauge)[8]	Plywood Applied Directly to Framing					Plywood Applied Over 1/2 Inch Gypsum Sheathing				
			Min. Req'd. Length of Fastener (inches)[2,9]	Nail spacing at Plywood Panel Edges (inches)				Min. Req'd. Length of Fastener (inches)[2,9]	Nail spacing at Plywood Panel Edges (inches)			
				6	4	3	2[3]		6	4	3	2[3]
STRUCTURAL I	5/16	13 ga	1-3/8	200	300	385	510	1-7/8	200	300	385	510
		14 ga		—	—	390	510		190	290	370	490
		15 ga		165	250	315	420		165	245	315	420
		16 ga		—	—	—	—		135	200	255	340
	3/8	13 ga	1-3/8	—	—	—	—	1-7/8	270	420	535	715
		14 ga		230	360	460	610		225	345	440	580
		15 ga		190	295	375	500		190	285	365	490
		16 ga		160	240	300	400		155	240	305	405
	1/2	13 ga	1-1/2	340	510	665[3]	870	—	—	—	—	—
		14 ga		270	405	525	690		—	—	—	—
		15 ga		230	345	445	590		—	—	—	—
		16 ga		190	280	365	480		—	—	—	—

Plywood thickness (inch)		Gauge	Penetration					Penetration				
5/16		13 ga	1-3/8	180	270	350	450	1-7/8	180	270	345	450
		14 ga		145	220	285	370		170	260	330	430
		15 ga		—	—	—	—		150	220	285	370
		16 ga		—	—	—	—		120	180	235	300
3/8	C-D, C-C STRUCT II- & other grades covered in UBC Standard 25-9 or PS-1	13 ga	1-3/8	—	—	—	—	1-7/8	260	380	485	640
		14 ga		220	320	410	530		205	300	390	510
		15 ga		180	260	335	440		145	220	280	370
		16 ga		140	210	270	350		145	210	270	355
1/2		13 ga	1-1/2	310	460	600[3]	770		—	—	—	—
		14 ga		250	370	475	615		—	—	—	—
		15 ga		210	310	405	525		—	—	—	—
		16 ga		170	255	330	425		—	—	—	—

1. All panel edges backed with 2-inch nominal or wider framing. Plywood installed either horizontally or vertically. Space nails at 6 inches on center along intermediate framing members for 3/8-inch plywood installed with the face grain parallel to studs spaced 24 inches on center and 12 inches on center for other conditions and plywood thicknesses. These values are for short-term loads due to wind or seismic and must be reduced by 25% for normal loads.
2. Where fastener lengths are not tabulated, the allowable wall shear values do not apply to that wall construction.
3. Framing shall be 3-inch nominal or wider and fasteners shall be staggered where staples are spaced 2 inches on center.
4. Staples exposed to weather shall be zinc coated by hot-dip galvanized zinc, mechanically deposited zinc or electrodeposited zinc.
5. The values for 3/8-inch plywood applied directly to framing may be increased 20 percent, provided studs are spaced a maximum of 16 inches on center or plywood is applied with face grain across studs, or if the plywood thickness is increased to 1/2-inch or more.
6. The tabulated values are for staples installed in Douglas Fir-Larch or Southern Pine (Group II species). For other species, values shall be calculated for all grades by multiplying values for Group II species by the following factors: Group I, 1.00; Group III, 0.82; Group IV, 0.65.
7. Plywood not exceeding 1-1/8 inch in thickness may be connected, provided penetration into framing is at least 1 inch for all sizes of staples.
8. Staples shall have a minimum crown width of 7/16-inch O.D.
9. The tabulated values are for staples installed in Group I or Group II species. Penetration for Group III shall be 13 diameters and for Group IV, 14 diameters.

Source: Taken from National Evaluation Report, NER-272

TABLE G.9 ALLOWABLE SHEAR FOR NAILED PARTICLEBOARD SHEAR WALLS[1,4]

shear in pounds per foot (plf)

Panel Grade	Min. Nominal Panel Thickness (inches)	Min. Nail Penetration in Framing (inches)	Nail Size (Common or Galvanized Box)	Panels Applied Directly to Framing				Nail Size (Common or Galvanized Box)	Panels Applied Over 1/2 Inch Gypsum Sheathing			
				Nail spacing at Panel Edges					Nail spacing at Panel Edges			
				6	4	3	2[2]		6	4	3	2[2]
2-M-W	5/16	1-1/4	6d	180	270	350	450	8d	180	270	350	450
	3/8			200	300	390	510		200	300	390	510
	3/8	1-1/2	8d	220[3]	320[3]	410[3]	530[3]	10d[5]				
	7/16			240[3]	350[3]	450[3]	585[3]		260	380	490	640
	1/2			260	380	490	640					
	1/2	1-5/8	10d[5]	310	460	600	770	—	—	—	—	—
	5/8			340	510	665	870					

1. All panel edges backed with 2-inch nominal or wider framing. Panels installed either horizontally or vertically. Space nails at 6 inches on center along intermediate framing members for 3/8-inch panel installed with the long dimension parallel to studs spaced 24 inches on center and 12 inches on center for other conditions and panel thicknesses. These values are for short-term loads due to wind or seismic and must be reduced by 25% for normal loads. Allowable shear values for framing of Douglas Fir-Larch or Southern Pine. For other species, values shall be calculated for all grades by multiplying values for common and galvanized box nails by the following factors: Group III, 0.82; Group IV, 0.65.
2. Framing at adjoining panel edges shall be 3-inch nominal or wider and nails shall be staggered where nails are spaced 2 inches on center.
3. The allowable shear values may be increased to the values for 1/2-inch-thick sheathing with the same nailing, provided (a) studs are spaced a maximum of 16 inches on center, or (b) the panels are applied with the long dimension perpendicular to the studs.
4. Where panels are applied on both faces of a wall and nail spacing is less than 6 inches on center on either side, panel joints shall be offset to fall on different framing members or framing shall be 3-inch nominal or thicker and nails on each side shall be staggered.
5. Framing at adjoining panel edges shall be 3-inch nominal or wider and nails shall be staggered where 10d nails having penetration into framing of more than 1 5/8-inches are spaced 3 inches or less on center.

Source: UBC Table 25-K-2

TABLE G.10 SHEARWALLS WITH OTHER MATERIALS[1,2]

Type of Material	Thickness (inches)	Wall Construction	Maximum Nail Spacing[3]	Shear Value (plf)	Minimum Nail Size[3]	Source
Stucco or Portland Cement on Expanded metal or woven wire	7/8"	Unblocked	6"	180	No. 11 gauge nail, 1-1/2" long, 7/16" head No. 16 gauge staple 7/8" legs	UBC Table 47-I notes 1, 6
Gypsum lath	3/8" lath & 1/2" plaster	Unblocked	5"	100	No. 13 gauge nail, 1-1/8" long, 19/64" head, plasterboard blued nail	
Gypsum sheathing board	1/2" × 2' × 8'	Unblocked	4"	75	No. 11 gauge nail, 1-3/4" long, 7/16" head diamond point galvanized	
	1/2" × 4"	Blocked	4"	175		
		Unblocked	7"	100		
Gypsum wallboard or veneer base	1/2"	Unblocked	7"	100	5d cooler or wallboard nail	
			4"	125		
		Blocked	7"	125		
			4"	150		
	5/8"	Unblocked	7"	115	6d cooler or wallboard nail	
			4"	145		
		Blocked	7"	145		
			4	175		
		Blocked Two ply	Base ply 9" Face ply 7"	250	Base ply-6d cooler or wallboard Face ply-8d cooler or wallboard	
Let-in Corner Bracing	1 × 4 (wood)	Unblocked	2 per stud 2 per plate	(60)[5] Tension	8d nails	FPL 301 note 7

Metal Strap Corner Bracing	16 ga / 18 ga / 22 ga		1-8d per stud / 2-16d per plate (typ)	(0)[5]	8d nails or 16d nails	Silver Simpson TECO[8] etc.
Diagonal Wood	3/4" × 1" × 6"	Unblocked	3 at boundary / 2 at each stud	300	8d common	Goers ATC note 9
Sheathing	3/4" × 1" × 8"	Unblocked	4 at boundary / 3 at each stud	300	8d common	
Fiberboard	7/16" × 4' × 8'	Applied Vertically	3" at all edges	125	No. 11 ga. galv roofing nail, 1-1/2" long, 7/16" head	
	25/32" × 4' × 8'			175	No. 11 ga. galv roofing nail, 1-3/4" long, 7/16" head	
	1/2" × 4' × 8'	Blocked	6" at all field studs	175	No. 11 ga. galv roofing nail, 1-1/2" long, 7/16" head	
Hardboard	7/16"	Applied Vertically	4" at all edges	230	6d box galvanized nails shiplap vert. joints	
	(notched to 1/4")	Blocked	8" at field studs	300	butted vert. joints	

1. Vertical shearwalls with gypsum materials shall not be used to resist loads imposed by masonry or concrete construction. See UBC Section 4714(b).
2. These values are for short-term loads due to wind or seismic and must be reduced by 25% for normal loads. Allowable shear values for framing of Douglas Fir-Larch or Southern Pine. For other species, values shall be calculated for all grades by multiplying values by the following factors: Group III, 0.82; Group IV, 0.65.
3. Applies to nailing at all studs, top and bottom plats and blocking, unless otherwise noted.
4. Alternate nails may be used if their dimensions are not less than the specified dimensions. If the diameter is larger, check minimum spacing requirements.
5. Both wood and metal corner bracing is intended to prevent walls from racking during construction. These components are not to be used as shearwall components during the occupancy of the building. Other sheathing must be applied. Values (where shown) are for comparison only.

Sources:
6. Uniform Building Code Table 47-I
7. USDA Forest Service Research Paper, FPL 301, 1977. Racking Strength of Walls, Roger L. Tuomi and David S. Gromala
8. Silver Metal Products, Inc; Simpson Strong-Tie Connectors; and TECO Products Catalogs
9. A Method for Seismic Design and Construction of Single-Family Dwellings, Ralph W. Goers & Associates, Applied Technology Council. Published by Department of Housing and Urban Development, September, 1976

TABLE G.11 ALLOWABLE TRANSVERSE LOADS FOR SANDWICH PANELS (PSF)

Panels are made of two equal layers of APA rated sheathing, either OSB or 5-ply plywood. The core is 0.9 pcf density EPS (expanded polystyrene foam) adhered to the sheathing with glue and set under pressure. The gluing is assumed to develop greater shear strength than the foam itself.

SANDWICH PANEL DIMENSIONS					
Skin thickness Core thickness Panel depth	3/8" 3-5/8" 4-3/8"	3/8" 5-5/8" 6-3/8"	15/32" 7-3/8" 8-5/16"	15/32" 9-3/8" 10-5/16"	15/32" 11-3/8" 12-5/16"
SPAN (ft)	ALLOWABLE TRANSVERSE LIVE LOAD (psf)				
4	138	212	283	358	432
5	106	165	222	282	342
6	84	133	181	231	280
7	68	109	151	194	236
8	56	92	129	166	203
9	47	78	111	144	176
10	39	66	97	126	155
11	33	57	85	111	137
12	28	50	75	98	122
13	24	43	66	88	110
14	21	38	59	79	99
15	18	33	53	71	89
16	16	30	47	64	81
17	14	26	43	58	74
18	12	23	38	53	67
19		21	35	48	62
20		19	32	44	56
21		17	29	40	52
22		15	26	37	48
23		14	24	34	44
24		12	22	31	41
25			20	28	38
26			18	26	35
27			17	24	32
28			16	22	30

Allowable loads are based on a deflection criterion of L/180 for roof loads.

For deflection criteria of L/240, multiply values by 0.75.

For floor loads, use deflection criteria of L/360, and multiply values by 0.50.

All loads are based on deflection criteria. No duration factors should be applied for any type of load.

TABLE G.12 ALLOWABLE TRANSVERSE LOADS FOR STRUCTURAL INSULATED PANELS (PSF)

Panels are made of two equal layers of APA rated sheathing, either OSB or 5-ply plywood. The core is nominal 1.0 pcf density (min. 0.9 pcf) EPS (expanded polystyrene foam adhered to the sheathing with glue and set under pressure. Each panel has splines that are nailed to the skin as described below.

SPLINE PARAMETERS				
Spline Configuration Single Spline		Spline spacing 48" o/c	Spline material SYP #2	Spline nailing 6d @ 6" o/c
STRUCTURAL INSULATED PANEL DIMENSIONS				
Skin thickness	3/8"	3/8"	15/32"	15/32"	15/32"
Core thickness	3-5/8"	5-5/8"	7-3/8"	9-3/8"	11-3/8"
Panel depth	4-3/8"	6-3/8"	8-5/16"	10-5/16"	12-5/16"
Spline size	2 × 4	2 × 6	2 × 8	2 × 10	2 × 12
SPAN (ft)	ALLOWABLE TRANSVERSE LOAD (psf)				
4	142	221	290	375	463
5	114	177	232	300	370
6	95	147	194	250	309
7	81	126	166	214	265
8	71	110	145	187	232
9	63	98	129	166	206
10	57	88	116	150	185
11	52	80	106	136	168
12	47	74	97	125	154
13	44	68	89	115	142
14	41	63	83	107	132
15	34	59	77	100	123
16	29	52	73	94	116
17	24	46	68	88	109
18	20	41	63	82	103
19	18	37	57	74	92
20	15	33	51	67	83
21	13	30	46	60	75
22		26	42	55	69
23		23	39	50	63
24		20	35	46	58
25		18	33	43	53
26		16	30	39	49
27		14	28	37	46
28		13	26	34	42

Deflection criterion of L/180 was used. Not all loads are based on deflections.

Multipliers for other deflection criteria are not allowed.

All values are for normal duration loads. No increases for other durations are allowed.

INDEX

Adjustment factors, *see* Modification factors
AFPA (American Forest and Paper Association), 3, 102. *See also* NDS (National Design Specifications)
AISC (American Institute of Steel Construction), 158
AITC (American Institute of Timber Construction), 187, 203, 223, 224
Allen, Edward, 167
ALSA (American Lumber Standards Association), 7
ANSI (American National Standards Institute), 237, 289
APA/Engineered Wood Systems (American Plywood Association), 25, 50
    PDS (Plywood Design Specifications), 246, 303–304
    rated sheathing, 239
    Sturd-I-Floor, 25, 239, 325
    Technical Note N375A, 239, 245
    trademarks, 236
Applied Technology Council, 300, 383
Arches, 195–234
    curvature factor, C_c, 207, 218–219
    deflection, 228–234
    equilibrium polygon, 204–206, 221. *See also* Arches, thrust line
    graphic analysis, 204, 210
    moment, 204
    preliminary design, 202–203
    radial tension, 212, 223–227
    radius of curvature, R, 203
    shapes, 196
    shear, 227
    three-hinged, 196–234
    thrust line, 195, 204. *See also* Equilibrium polygon
Area, 2, 17, 307–310
Asymmetric snow load, 198, 207
Axial loads, 55–77
    with bending, 69. *See also* Interaction formula
Axial thrust, 17

Beams:
    beam stability factor, C_L, 20, 339–341
    built-up beam, 46–51
    box beams, *see* Built-up beam
    continuously supported, 333–334
    deep beam, 29–31
    effective length, 34
    flitch beam, 44–46
    glued-laminated, 17, 37, 51–53
    laterally unsupported, 331–332, 335–341
    plywood, preliminary design, 46–51, 342–343
    size factor, C_F, 29–31
Beams and stringers, 3, 317
Bearing, 18, 25–27
Bearing area factor, C_b, 27
Bending moment, 18–22
Bending strength, F_b, 7, 13
Binders, 237
Board feet, 22

Board measure, 2
BOCA (Building Officials Code Administrators),
  National Building Code, 15, 82-89, 266,
  289-291
Bracing, columns, 64
Buckling, 56
Built-up beams, 46-51
Built-up columns, 65-68

CABO, Council of American Building Officials:
  NER Report 272, 114, 376-377, 381
  One and Two Family Dwelling Code, 15, 79,
    88, 98, 160, 362-373
Cambium, 5
Cellulose, 5
Centroid of area, 352
Checks, 8, 10, 41
Clear span, 19
Clinched nails, 107
Coefficient of variation, 5
Columns:
  allowable loads, 345-346
  with bending, 69
  bracing, 64
  buckling, 56
  built-up, 65
  design tables, 62-63, 345-346
  eccentric load, 70, 73
  intermediate, 58-62
  long, 58-62
  short, 58-62
  slender, 57
  slenderness ratio, L/d, 57, 347
  spaced, 66-69, 188-189
  stability factor, C_P, 58, 349-350
Compression, 6, 13, 55-75
Compression wood, 10
Connections:
  design guidelines, 101
  hurricane clips, 100, 286
  lag screws, 120-136
  metal connectors, 83, 176
  moment, 135, 136, 145, 156, 351-358
  nails, 103-114
  recommended schedule, 79-81, 89,
    92-98
  shear plates and split rings, 151-159
  staples, 114-116, 376-377, 381
  through bolts, 136-151
  timber connectors, *see* Shear plates and split
    rings
  toenails, 96, 113
  wood screws, 116-120
Cross-grain stress, 6-7, 255, 256
CSI (combined stress index), 132. *See also*
  Interaction formula
Culmann, Karl, 165

Danube River Bridge, 164

Deception Bay storage facility, 195
Deck construction, 128
Decking, 27-29, 197
  allowable loads, 320-322
  grades, 316
Deemed-to-comply, 159
Deep beams, 29-31
Defects, 7
Deflection:
  arch, 228-234
  beam, 18-25
Density, 2, 103
Design span, 19
Diaphragms, 108, 238, 245-262
  blocked, 241, 251
  boundary nailing, 251-252
  chord, 248, 252
  design, 249-262
  edge nailing, 251-252
  field nailing, 251-252
  flexible, 258-261, 300-303
  irregularities, 258
  load case, 251, 254
  lumber, 261-262, 306, 380
  openings, 256-258
  rigid, 260-261
  size ratio, 248-249
  struts, 248, 267-269
  torsion, 300-303
  transversely sheathed, 261-262, 306, 380
  unblocked, 241, 251, 252
  vertical, *see* Shearwalls
Dimensional lumber, 3
Double shear, 106, 111
Dowel bearing strength, 141
Duration factor, *see* Load duration factor

E, modulus of elasticity, 12-13
Earthquakes, historic, 288-289
Eccentric load, 70, 73
Equilibrium (moisture content), 12
Equilibrium polygon, *see* Arches
Euler, Leonard, 56
Euler buckling formula, 56-59
European yield theory, 102, 141, 144, 147,
  161-163

Factor of safety, 55, 57, 79
Fiberboard, 237
Fiber saturation point, 12
5% exclusion principle, 5
Flakeboard, 236
Flitch beam, 44-46
Forest Products Laboratory, 1, 383
Foundations, 79-96, 140
Friction, 79

Gang-Lam, 25, 328, 330

Garage door opening, 285
Geometry factor, C_Δ, 143, 151
Glued-laminated arches, 195–234
Glued-laminated beams, 17, 37, 51–53
Goers, Ralph W., 300, 383
Grades, 3–13, 311–318
Grading Agencies, 7, 311–318
Green lumber, 11
Gromala, David S., 383
Group-action factor, C_g, 125, 136, 157
Grubenmann, Jean-Ulrich, 165

Hankinson's formula, 123, 133, 141, 150, 159, 191, 353
Hankinson's nomograph, 126, 150
Hardboard, 237
Hardwood, 1–2
Header, 40–43, 80–81
Hold-downs, 270–271
Horizontal diaphragms, *see* Diaphragms
Horizontal shear, 7, 13, 17–24, 37–43, 44–50, 51–52
HUD (Department of Housing and Urban Design), 300, 383
Hurricane Andrew, 114
Hurricane clips, 100, 286

I (moment of inertia), 2, 17, 307–310, 352
Iano, Joseph, 167
ICBO (International Conference of Building Officials):
  Research Report 1952, 378
  Uniform Building Code, 15, 82, 107, 109, 140, 265, 266, 272, 277, 282–284, 375, 379, 380, 382, 383
In-grade testing, 18
Interaction formula, 70–72, 170–171, 188, 212, 222
Intermediate beam, 34
Irregularities, 281–286

Joists, 18–25, 307–308, 312–315
Joists and Planks, 3, 8, 10
Joist tables, 24, 323–324
Jones, Inigo, 164

Kiln-dried, 2
Knots, 8–9

Lag bolts, *see* Lag screws
Lag screws, 120–136
  combined lateral and withdrawal loads, 125
  end and edge distances, 127
  lateral loads, 123
  modification factors, 127–128
  penetration, 123
  withdrawal, 121
Lateral brace, *see* Columns, bracing

Laterally unsupported beams, 35
Lateral resistance, connectors, *see* Connectors by type
Lateral stability factor (beams), C_L, 20, 32, 51, 339–341
Lateral support requirements:
  lumber beams, 32
  plywood beams, 50
Lignin, 5
Lintel, 44
Load duration factor, C_D, 13–16, 20, 31, 54, 289–291
Long beam, 34
LVL (laminated veneer lumber), 25, 328

Marcin, Thomas C., 1
MEL (machine evaluated lumber), 5
Metal side plates, 106, 110, 148
Micro-Lam, 25, 327
Modification factors:
  beam stability factor, C_L, 20, 32, 51, 339–341
  bearing area factor, C_b (Table 2.1), 27
  buckling stiffness factor, C_T, 14, 170–171, 174
  column stability factor, C_P, 58, 349–350
  curvature factor, C_c, 207, 218–219
  flat-use factor, C_{fu}, 312, 316
  form factor, C_f, 14
  geometry factor, C_Δ, 143, 151, 153
  group-action factor, C_g, 125, 136, 155, 157
  load duration factor, C_D, 13–16, 20, 31, 54, 289–291
  penetration factor, C_d, 105–106, 109, 111, 118–120, 127–128, 132, 154
  repetitive member factor, C_r, 13, 20–21, 312
  shear stress factor, C_H, 20, 40–43, 313, 318
  size factor, C_F, 18, 20, 29–31, 52, 207, 312
  steel side plate factor, C_{st}, 152, 154
  temperature, C_t, 14
  volume factor, C_V, 14, 51, 218
  wet service factor, C_m, 14, 53–54, 137, 313
Modulus of elasticity (E), 12–13
Moisture content, 11
Moment connections, *see* Connections, moment
Moment of inertia (I), 2, 17, 307–310, 352
Moment magnification, 70
Moment resisting frame, 294–300
MSR (machine stress rate), 5

Nail base, 237
Nails, 103–114
  clinched, 107, 112
  double shear, 106, 111
  metal side plates, 106, 110
  modification factors, 105
  plywood gusset plate, 112
  slip, 107
  spacing, 107
  threaded, 104
  toenails, 96, 113

NDS (National Design Specifications), 3, 18, 29, 38, 43, 57, 66, 73, 102, 107, 113, 116–118, 123, 128, 133, 136–137, 140–152, 191, 202, 203, 223, 267, 289, 351, 353
NELMA (Northeastern Lumber Manufacturers Association), 2
Net section, 75
NGR (National Grading Rule), 7–9
Nominal dimensions, 43, 307–310
Non-veneer boards, 236
Notches, 37–39

One and Two Family Dwelling Code, *see* CABO
OSB (Oriented Strand Board), 25, 236
Overturning, 82–100

Palladio, Andrea, 164
Parallel to grain, 6
Particleboard, 236
    diaphragms, 379
    shearwalls, 382
PDS (Plywood Design Specifications), *see* APA
Penetration, 103, 106, 118, 123
Perpendicular to grain, 6
Phenol-formaldehyde binder, 237
Plywood:
    allowable strengths, 245–246
    appearance grades, 236, 244
    beam, preliminary design, 46–51, 342–343
    diaphragms, 245–261, 375–378
    engineering grades, 236, 244
    exposure grades, 244
    exterior grades, 244
    gusset plates, 112, 119
    IMG (intermediate glue), 244
    interior grades, 244
    PII (Panel Identification Index), 238
    plies, 235
    shearwalls, 265–271, 380–381
    span rating system, 238
    species, 242–243
    specifications, 238
    tongue-and-groove, 239
    veneer, 235
    veneer grades, 242–244
Polar moment of inertia, 352
Posts and timbers, 3, 317

Racking, 238
Radius of gyration, 57
Recommended fastening schedule, 79–81, 89, 92–98
Repetitive member factor, C_r, 13, 20–21, 312
Residential construction, 1, 271–303
R_w, seismic factor, 278–282

S4S, 2
Sandwich panels, 303–306
    allowable loads, 385

SBCCI (Southern Building Code Congress International), Standard Building Code, 15, 83, 85, 266, 289–291
Scofield, W. Fleming, 168, 194
Section modulus, S, 2, 17, 307–310
Seismic loads, 277–282
    calculations, 278–282
    compared with wind load, 286–289
Shake, 8, 10, 41
Shear, special design, 39–43. *See also* Horizontal shear
Shear plates and split rings, 151–156
    double shear, 154
    geometry factor, C_Δ, 153
    group action factor, C_g, 155
    modification factors, 152, 154–155
    penetration factor, C_d, 152, 154
    steel side plate factor, C_{st}, 152, 154
    tension, 156
Shear stress factor, C_H, 20, 40–43, 313, 318
Shearwalls, 238, 245, 265–271, 291–295, 380–383
    allowable loads, 380–383
    collector strut, 262, 293–295
    deflections, 271
    design, 265–271, 291–295
    discontinuous, 285–286
    dissimilar materials, 293
    double-sided, 263–265
    drag strut, *see* Collector strut
    foundation connection, 140–141, 266, 269–271
    gypsum wallboard, 293
    overturning, 269–271

Short beam, 34
Shrinkage, 12, 137
Silver TECO, *see* TECO connectors
Simpson Strong-Tie, 91, 160, 270–271, 362–372, 383
Single shear, 106, 108
SIPA (Structural Insulated Panel Association), 304
Size classification, 3, 13, 311–318
Size factor, C_F, 18, 20, 29–31, 52, 207, 312
Slender column, 57
Slenderness ratio:
    beams, 32, 52, 72, 79, 218
    columns, 56, 70, 72, 170, 181, 217
Sliding, 82–100
Slope of grain, 7, 9
Softwood, 1–2
Species:
    plywood, 242–243, 375–378, 380–382
    wood, 1–2, 360–361
Specific gravity, 103, 360–361
SPIB (Southern Pine Inspection Bureau), 13
Spiral grain, 9
Split rings, *see* Shear plates and split rings
Splits, 8, 9, 10, 41
Spreadsheet calculations, 210, 212

Staples, 114–116, 376–377, 381
Stone, Robert N., 1
Stressed-skin panels, 303–306
   allowable loads, 384
Structural light framing, 3
Stud, 3

TECO connectors, 90, 98, 160, 177, 362, 366, 383
Tension, 6–7, 13, 75–77
   analysis, 76
   design, 77
Through bolts, 136–151
   concrete connection, 140
   double shear, 139
   edge and end distances, 142–143
   geometry factor, C_Δ, 138, 143
   masonry connection, 140
   metal side plates, 148
   modification factors, 138
   multiple shear planes, 141
   optimal design, 144–145
   spacing requirement, 142–143
   stiffness, L/D, 144
   tension, 145–148
Timber connectors, *see* Shear plates and split rings
Torsion of diaphragms, 300–303
Transformed beam, 45
Trus Joist Corporation, 25, 329
   TJI/25, 329
Trusses, 164–194
   approximate weights, 168, 186
   bracing, 181–184
   buckling stiffness factor, C_T, 14, 170–171, 174
   compression chord design, 170–175
   connection reduction factor, R_F, 178
   force diagram, 168–169, 185
   heavyweight, 165, 184–194
   lightweight, 165–181
   metal plate connectors, 176–181
   proportions, 186
   residential, 167–181
   split-ring connectors, 190–194
   tension chord design, 175
   types, 166–167
   web design, 176, 190
Truss Plate Institute, 160–178, 181–184
   bracing provisions, 181–184
   design procedures, 171–175
   TPI-85, 172, 177–178
Toenails, 96, 113
Tuomi, Roger L., 383

UBC (Uniform Building Code), *see* ICBO
Unbraced length:
   beams, 34
   columns, 57
Uplift, 82–100
Urea-formaldehyde resin, 237
US Product Standard PS-1, 236

Virtual work, 228–234
Visual grading, 4, 7
Volume factor, C_V, 14, 51, 218

Waferboard, 236
Wane, 8, 10
Warp, 8, 10
Wet service factor, C_m, 14, 53–54, 137, 313
Withdrawal, 96, 104, 119, 123, 125, 127
Wood screws, 116–120
   metal side plates, 120
   modification factors, 118
   penetrations, 118
   plywood gusset plates, 119
Wren, Christopher, 164
WWPA (Western Wood Products Association), 13, 163, 380
   Wood Engineering Handbook, 163, 380

Yield theory, *see* European yield theory
Yoemans, David T., 165